The Age of the Earth

G. BRENT DALRYMPLE

The Age of the Earth

Stanford University Press

STANFORD, CALIFORNIA

Stanford University Press, Stanford, California
© 1991 by the Board of Trustees of the
Leland Stanford Junior University

Printed in the United States of America

CIP data appear at the end of the book

Original printing 1991
Last figure below indicates year of this printing:
01   00   99   98   97   96   95   94

Stanford University Press publications are distributed
exclusively by Stanford University Press within the
United States, Canada, and Mexico; they are distributed exclusively by Cambridge University Press
throughout the rest of the world.

To Stacie, Robynne, and Melinda, of whom I am very proud

Preface

A few years ago I mentioned to a colleague that I was considering writing a book on the age of the Earth. He gave me a strange look, then remarked that it would be a short book, because the answer could be expressed by a single, well-known number. But he had missed the point. The purpose of this volume is not to reveal Earth's age, for as my friend noted, the age of the Earth has been widely known for nearly three decades. What I had in mind was a book that explained in a simple and straightforward way the evidence and logic that have led scientists to conclude that the Earth and the other parts of the solar system are four and one-half billion years old.

I was led to write this book by a rather unique series of events. In December 1980 I received a phone call from Mr. Robert Tyler, Deputy Attorney General for the State of California. He had been given my name by Professor Garniss H. Curtis, a well-known geochronologist and a friend and colleague who was my mentor and thesis advisor while I was engaged in doctoral studies at the University of California, Berkeley, in the early 1960's. Mr. Tyler needed an expert witness to testify for the defense in a civil suit brought against the State by Kelly Segraves et al. on behalf of his sons Kasey, Jason, and Kevin. Mr. Segraves, Director of the Creation-Science Research Center, was alleging that the State was violating his sons' constitutional rights by teaching evolution as fact in the science classrooms of the public schools. Among other things, he seemed to be asking that the state require schools to teach that evolution was "only a theory" and that creationism was an equally viable theory of origins. Tyler asked if I would be willing to testify on radiometric dating and the age of the Earth. I almost said no.

My first encounter with the creationists' peculiar arguments about nature was in 1975, when Henry M. Morris and Duane T. Gish, the Director and Associate Director of the Institute for Creation Research, journeyed from their headquarters in Santee, in Southern California, to Menlo Park to present their case for creationism to several hundred U.S. Geological Survey scientists at an evening seminar.

It was a lively session, but Morris and Gish were speaking to people whose life's work was studying the history of the Earth, and they persuaded no one that evening. The day following the seminar, Morris and Gish visited the Survey's geochronology laboratories for a tour, where a colleague, Marvin Lanphere, and I engaged them in private debate on the efficacy of isotopic dating methods and on their peculiar and unquestionably incorrect interpretation of the Second Law of Thermodynamics as applied to evolution. Morris left several articles and his most recent book, *Scientific Creationism*. I read the book and articles, found them rather bizarre, and didn't give creationism much more than a passing thought until Tyler's call.

Thus, when Tyler asked if I would appear in court, I was already generally familiar with the creationists' arguments for a very young Earth and the nature of their attack on the scientific evidence for Earth's antiquity. But I also knew that the preparations for the trial would divert my energies from productive research for several months. In addition, I was not overly enthusiastic about appearing in court. The goal of both the courts and science is to discover the truth, but the methods of the two are so different that it is difficult for most scientists to enter the legal arena with any degree of confidence. Finally, there was a natural reluctance, common to most scientists, to spend any time dealing with nonsense; the tenets of "scientific" creationism are so absurd that it seems most appropriate simply to ignore them. The counterargument, however, was persuasive. The central issue in the Segraves trial was what children in California would be taught in the science classroom—genuine science or the thinly disguised religious beliefs of "scientific" creationism. Faced with that, I saw that there was no alternative, and I agreed to assist Tyler in whatever way I could. In doing so I joined, on the State's witness list, an illustrious group of scientists that included William Mayer of the Biological Sciences Curriculum Study, Francisco Ayala of the University of California at Davis, Stephen Jay Gould of Harvard University, Arthur Kornberg of Stanford University, Richard Dickerson and Norman Horowitz of the California Institute of Technology, Thomas Jukes, Richard Lemmon, John Huesman, Watson Laetsch, and David Wells of the University of California at Berkeley, Russell Doolittle of the University of California at San Diego, Everett Olson and Rainer Berger of the University of California at Los Angeles, and Carl Sagan of Cornell University. I have never regretted that decision.

I spent the next several months reading creationist literature and assembling notes that would form the basis of my testimony. In March 1981, I drove to Sacramento expecting to appear in what the local news media were by then calling the Scopes II trial. The trial fizzled.

Before the State could begin the scientific part of its defense, Segraves et al. changed their complaint to a minor issue involving a few sentences in the California Science Curriculum Framework, and none of the expert witnesses was permitted to testify on the scientific issues. The Segraveses lost anyway.

Needless to say, we scientists, who had devoted a considerable amount of time preparing for the trial, were disappointed that we were denied the opportunity to testify. What we would have said, we thought, would have had a salutary impact well beyond the local issues. At dinner one evening, after an hour or so of commiseration, Dick Dickerson suggested, and the rest of us agreed, that our individual notes might be expanded into chapters of a book. My contribution was a manuscript titled "Radiometric Dating, Geologic Time, and the Age of the Earth: A Reply to Scientific Creationism."

It took me only a little over a month to complete the first draft of that 77-page manuscript. It contained explanations of several of the most commonly used isotopic-dating techniques, a summary of the evidence for the 4.5-billion-year age of the Earth, and refutations of some common creationist criticisms of dating methods and their arguments for a very young Earth. Somewhere along the way the book project faltered, and that manuscript was never published, although it is now available as a U.S. Geological Survey Open-File Report (Dalrymple, 1986). The manuscript was, however, widely circulated to scientists for review, to other people who requested copies, and to still others I knew were interested in the creationism issue.

As is so often the case, one thing led to another. The widespread circulation of that first manuscript and the inclusion of my name on the witness list for the Segraves trial established me as one of a handful of scientists who were (a) knowledgeable about creation "science," (b) able to refute creationist arguments, and (c) willing to devote some time to the issue, including testifying in court.

Within a few months of the California trial I was asked by the American Civil Liberties Union to appear as one of four expert scientific witnesses (the others were Ayala, Gould, and Harold Morowitz of Princeton University) in their suit to overturn the recently enacted Arkansas Act 590, which required equal treatment of "evolution science" and "creation science" in the science curriculum of the Arkansas public schools. That trial was held in Little Rock in December 1981, my colleagues and I appeared and testified, and U.S. District Court Judge William Overton declared the law unconstitutional—but that is another story (see Gilkey, 1985, for an excellent and informative account of that trial). I was also deposed for the Louisiana creationism trial, but the judge of the U.S. District Court in New Orleans declared that state's "equal-time" law unconstitutional, in 1985, in a summary judgment that was upheld by the U.S. Supreme Court in 1987, and the case never came to trial.

In addition to being involved in the legal proceedings, I received numerous invitations to speak and to write articles on the subject of "scientific" creationism and the age of the Earth. I accepted as many of these invitations as was practical, and over the course of the next year or so authored three more articles dealing with creationism and the age of the Earth (Dalrymple, 1983a,b, 1984).

In the course of doing the research required for the articles, talks, and legal proceedings, I discovered something that I suppose I knew all along—that there was no current book that explained, in a simple, straightforward, and thorough way, the current evidence for the age of the Earth. At least there was no book that did it in the way in which I thought it should be done. I decided that I would write one.

Although the impetus for this book arose from the controversy over creation "science," that subject is not discussed in the body of the text. The creationists' "scientific" arguments for a young Earth are absurd, I and other authors have dealt with them at length elsewhere, and they do not merit further attention here.

The purpose of this book is to explain how scientists have deduced the age of the Earth. It is a fascinating story, but not so simple as a single measurement. Our universe is a large, old, and complicated place. Earth and the other bodies in the Solar System have endured a long and sometimes violent history, the events of which have frequently obscured the record that we seek to decipher. Although in detail the journey into Earth's past requires considerable scientific skill, knowledge, and imagination, the story is not so complicated that it cannot be explained to someone who wants to know and understand the basic evidence.

I have written the book, then, for people with some modest background in science, but I have tried to maintain a level that will allow the material to be useful and informative to those without a deep knowledge of geology or physics. Science has become so specialized that even many scientists have difficulty understanding much of the literature outside their own specialties. The literature of geophysics, for example, is, to a geochemist, difficult and sometimes barely understandable. At the same time, that geochemist probably would find the literature of biochemistry all but incomprehensible. And most biologists find papers in geology, astronomy, or physics difficult if not impossible to understand. Thus, even skilled scientists need a simpler level of communication than is usually to be found in the traditional scientific literature. For the average scientist who is not an expert in what is known about the age of the Earth, the voluminous literature dealing with the topic is difficult both to find and to digest. I hope that I have succeeded in providing a simple, satisfying, and enjoyable account for anyone who is curious about how the age of the Earth has been determined.

Most of the book is organized into chapters that discuss the several types of evidence for Earth's age. Chapter 1 contains a brief history of the universe as science knows it, to provide the reader with a framework sufficient to allow the details of succeeding chapters to be kept in perspective.

Chapter 2 is an account of some of the historical attempts, made before modern radiometric methods were available, to determine the age of the Earth. Chapter 3 provides the background necessary for understanding the remainder of the book. The quantitative evidence for Earth's age is based on measurements involving long-lived, naturally occurring radioactive isotopes, and the chapter explains how these radiometric methods work and how they can be used to date events so far into the distant past.

Chapter 4 is a discussion of the Earth's oldest known rocks. It may surprise some readers to learn that the first 700 million years or so of Earth's history have been effectively erased, and that the oldest rocks found on the Earth are not nearly so old as the Earth itself. I have included in this chapter considerable discussion of the geology of the oldest rocks, because the radiometric results cannot be understood without knowing how these important rocks formed and how they are related geometrically and temporally to other rocks.

Some of the key evidence for the age of the Earth comes from bodies in the Solar System that are not so highly evolved as the Earth. These bodies, the Moon and meteorites, provide a record of the time of some of the earliest events in the Solar System. Because the Earth formed as an integral part of the Solar System, the ages of the oldest meteorites and lunar samples, which are the subjects of chapters 5 and 6, provide estimates of the age of the Earth. As in Chapter 4, these two chapters contain sufficient information about the geology of the Moon and of meteorites to accord the age data their proper meaning.

Chapter 7 explains the *pièce de resistance*—the simple and elegant method that provides us with a reasonably accurate figure for the age of the Earth. This method involves a model for the evolution of lead isotopes in the Solar System, and employs measurements on meteorites, lunar samples, and Earth rocks.

Chapter 8 concerns the evidence for the ages of the Milky Way

Galaxy and the universe, as well as some indirect evidence from radioactive nuclides that indicates that the Solar System is billions, rather than thousands or millions, of years old. The evidence discussed in this chapter does not lead to any specific figure for Earth's age, but it does reinforce the inescapable conclusion that our Earth and the universe of which it is a tiny part are incredibly old compared to the span of the human lifetime.

Chapter 9 offers a synopsis of the information in the preceding chapters as well as some suggestions about the most fruitful directions that research on the age of the Earth might take in the future. It is highly unlikely that future findings will change the current estimate of Earth's age but, like most scientific investigations, it would be nice to tie up what few loose ends remain.

Throughout the book, I have tried to minimize the use of mathematics when it was possible to do so. Mathematics, however, is the language of science, and we must use it when we seek to provide a concise and accurate description of the methods and the evidence. Wherever I have found it necessary to inject formulae or calculations into the text, I have done so in such a way that anyone with a basic, even rusty, understanding of algebra can follow them. I have also tried to use formulae primarily as illustrations, and to structure the text so that the reader who is either uninterested in or allergic to math can skip over the formulae and calculations and still follow the discussion profitably.

Vocabulary can also be a problem for the reader who is unfamiliar with the language of a particular science, but a specialized vocabulary is necessary for concise and precise communication. Some of the specialized terms of physics, chemistry, astronomy, and geology I have been able to explain in the text; the meanings of other terms are made relatively clear by the context in which they are used. I have also included a glossary, and the words explained there are italicized in the text, generally where they are first used.

A word about units. In general, I have used SI units throughout the book except in the historical discussions in Chapter 2, where the units used in the original publications are sometimes more appropriate. Another exception is time. For geologic time, I have followed the recommendations of the North American Stratigraphic Code (*AAPG Bulletin*, vol. 67, pp. 841–75, 1983) and used *Ma* (*mega-annum*) for 10^6 years, *Ga* (*giga-annum*) for 10^9 years, and *ka* (*kilo-annum*) for 10^3 years. "Ago" or "old" is implied where appropriate.

Like any book, this one is a compromise. I have probably included too much information for some readers and too little for others. For the former, I hope that I have made the book sufficiently readable to speed you past those little hurdles of unwanted detail and into the next section. And for those who are still unsatiated after turning the last page, there is an extensive list of references, and I encourage you to explore the subject further on your own. I do hope that after reading this book you will be convinced that we know the age of the Earth to within a few percent or better. Above all, I hope that you have as much fun reading this book as I have had writing it.

Even though my name alone appears on the title page in that space traditionally reserved for the author, no book is the product of a single person. I am grateful to a number of friends and colleagues who, through contributions of their time and energy, helped me to make this book better than it would have been without their efforts. In particular, I would like to thank Stephen Moorbath, Ronald Greeley, Vicki Horner, Jeff Moore, Randy Van Schmus, George Tilton, and Steve Dutch, who reviewed parts of the manuscript and whose excellent and thoughtful suggestions immeasurably improved the work. Konrad Krauskopf read the entire manuscript and suggested a number of valuable improvements. A number of people and institutions generously provided or helped me to locate illustrative material, and I want Roy Clarke, Twyla Thomas, Bill Hartmann, Allen Nutman, Ron Greeley, Vic McGregor, Stephen Moorbath, Bill Brice, the University of California's Lick Observatory, the Geological Survey of Greenland, the Smithsonian Institution, the GEOSAT Corporation, and NASA to know that I am very grateful for their help. I also wish to thank my dear friends Stan and Maggie Skinner for storing and protecting numerous and bulky copies of the manuscript against disaster. Much of the library research was done while I was a Consulting Professor at Stanford University in 1983–85. My long-time friend and colleague Allan Cox was responsible for making that possible—and I wish that he were still here to read the final product. Finally, my thanks to my wife and best friend Sharon, who not only puts up with my time-consuming projects but has provided love, support, and understanding for three decades.

G.B.D.

Contents

The Age of the Earth

The chessboard is the world, the pieces are the phenomena of the universe, the rules of the game are what we call the laws of nature. The player on the other side is hidden from us. We know that his play is always fair, just, and patient. But also we know, to our cost, that he never overlooks a mistake, or makes the smallest allowance for ignorance.   T. H. HUXLEY, 1868
A Liberal Education

CHAPTER ONE

Introduction

*There are few problems more fascinating than those that are bound up
with the bold question: How old is the Earth? With insatiable curiosity
men have been trying for thousands of years to penetrate the carefully
guarded secret.* ARTHUR HOLMES (1927: 5)

Four and one-half billion years. That figure, which represents
the current estimate of the age of the Earth, is so large, so far outside
of our normal, everyday experience, that it is difficult to comprehend
its true scope and meaning. Even scientists who deal with numbers of
that magnitude on a daily basis often find it difficult to grasp the full
significance of that span of time. Analogies help some, but only a
little. If a piece of string 2.5 cm long (about an inch) represents one
year, for example, then a 183-cm length (about 6 feet) is equivalent to
the average lifetime of a person living in the United States. A string
representing all of recorded human history would be fully a kilometer
long, but a piece representing 4.5 billion years would be 114,280 km
long! Four and one-half billion quarters would form a stack nearly
8,000 km high. Can anyone fully visualize a string that would wrap
around the Earth nearly three times, or a stack of quarters that would
reach from here through the center of the Earth and halfway to the
other side?

If a period of time measured in billions of years is difficult to
grasp, then the implications of such a temporal abyss are even more
elusive. For example, a photon of light, traveling at a speed of 300,000
km per second, would be 4.25×10^{22} km from its source, which is
equivalent to 283 trillion times the average distance from Earth to
Sun, in 4.5 billion years. A continent drifting at a constant velocity of
1 cm per year, the rate at which North America currently is moving
away from the mid-Atlantic ridge, would travel 112 times around the
Earth in 4.5 Ga (billions of years; millions of years are indicated by
Ma, thousands of years by ka).

As staggering as these numbers may seem, the evidence clearly shows that the Earth's age is, indeed, 4.5 Ga, and the universe is probably three to four times older. Yet humans are relatively recent inhabitants of our planet and have witnessed only an infinitesimally small percentage of Earth's history. No man, no creature, no plant was present when Earth, her sister planets, and the Sun condensed from a shapeless cloud of primordial matter. How then can we peer back into these seemingly infinite reaches of time and calculate an age for the Earth that requires ten digits?

The answer to that question is, of course, the subject of this book. Today, we can do these things. But it has been only within the last 30 years that the age of the Earth has been known with any reasonable degree of certainty. Even then, the answer emerged only at great expense and after centuries of thought, observation, and experimentation.

Doubtless, we have wondered about the age and origin of our surroundings since we first developed the capacity for abstract curiosity. How and when was the world created? These are natural questions, but the answers are not easy to extract from the meager clues that nature provides. Theologians, philosophers, and scientists have been searching for satisfying answers to these questions for thousands of years.

One might well ask, why have we bothered? Two reasons come immediately to mind. The first is philosophical. History makes it abundantly clear that humankind has struggled with the question of its place in the universe since the beginning of organized thought. Was the cosmos created specifically for us, or are we a minor result of natural processes shaping a universe of unimaginable dimensions over infinite time? Clearly, there is a certain comfort and security in the former explanation, and through most of recorded history Western thought, at least, has focused on the idea that we are central to a grand and purposeful plan. Until Copernicus, Galileo, Kepler, and their successors showed otherwise, it was universally held that the Earth, and by implication humankind, was at the very center of the universe. Timing is critical to this very important philosophical question: even after we discovered that our place in the universe was not geometrically central, we still clung to the belief that our timing was. Indeed, some people still prefer to believe that the universe was created specifically for us, and that it predates our debut by only a few days.

We now know, of course, that the Earth revolves about a rather ordinary star located well out in a spiral arm of a rather ordinary galaxy that occupies an ordinary position in the universe. We also know that we are newcomers on the scene, having appeared so recently that our tenure is but a fleeting moment in the vast span of time since the present universe began. (There may have been other universes; in time there may be new ones.) Thus, knowledge of the age of the Earth (and the universe) puts our lives in temporal perspective and gives us all a better idea of our physical place in the cosmos. Let us be clear about one thing, however: this information provides us no answers to the larger question of whether we are ultimately the result of a grand and purposeful design, or merely an accident of past and current physical processes. Science can attempt to determine how and when the Earth and its surroundings were created, but the question of why it all exists is not one that science can speak to.

The second reason to bother is scientific. We have always sought information about our physical surroundings, simply to satisfy our curiosity and to add to the pool of scientific knowledge from which we draw both intellectual and, frequently, material satisfaction. Thus, the age of the Earth is simply one more interesting fact in the array of information that scientists have gathered in the quest for an ever more precise description of our physical universe. Curiosity! What further justification is needed?

It is one thing, of course, to claim that we know the age of the Earth, and another to demonstrate the formulations for the claim. So where do we begin? Three difficulties arise immediately. First, we must decide exactly what we mean by its "age." The concept of a unique age for the Earth is, in reality, somewhat arbitrary, since the Earth did not appear instantaneously as a planet in space. Like all physical objects, it was "born" through a complex sequence of events over a finite (but quite long) period of time, so we must decide which particular event or stage of evolution in the Earth's early history will reasonably represent its "birth."

The second difficulty is that the physical evidence of the event we choose may not be accessible to us—it may no longer exist, it may be undiscovered, it may even be forever out of reach.

Third, the dating methods available to us may not be capable of directly dating the event we choose. These last two limitations may force us either to choose another event to represent the Earth's birth or to accept upper and lower limits, rather than an explicit value, for its age.

These difficulties might be very serious if the interval of time over which the Earth formed was long relative to the Earth's overall "age." As it turns out, however, the Earth's birth process occurred *very* long ago over a relatively *brief* interval of time. Thus, the errors introduced by the second and third difficulties are very small, if not negligible, at least for the purpose of determining a good approximation of the Earth's age.

The physical processes that gave birth to our galaxy, the Solar

System, and the Earth[1] were set in motion many billions of years ago by an event commonly known as the "Big Bang," or in some quarters as the "Primal Bang." This singularly important event was a sudden decompression of all matter from a single point of infinite density. What followed, at a blinding pace, was a dispersal of phenomenal speed and distance; that it occurred at all is confirmed by back-calculation from the observed trajectories of expansion and by the existence of radiation generated by the process; science therefore marks it, with some confidence, as the beginning of the present universe.

What existed before the Big Bang is not known and may never be discovered, for there is no physical evidence from previous times. Indeed, we cannot even be sure that there *was* a previous time! Whether the present universe is the first universe or just the most recent in a long or infinite sequence of universes is an interesting question, one that science, at least at present, has no way of answering.[2]

The future of the universe is also unknown, but may someday be determined. What we know about these things says that if there is sufficient mass in the universe, then the universe is bounded, the present expansion will eventually stop, and the universe will then begin to contract, owing to gravitation; the decompression process begun by the Big Bang will be reversed, and all matter will once again collapse to a single point. If there is insufficient mass to reverse the expansion, then the universe is unbounded, and, depending on the mass, the universe will either expand forever or reach a state of equilibrium in which it neither expands nor contracts. On the basis of its visible mass, we judge that the universe is unbounded, but we confront the question with considerable uncertainty, as we do the question whether space itself is limitless.

The physical processes and events during and shortly after the Big Bang are subjects of much current interest and research, but the early history of the universe was roughly as follows (Fig. 1.1). At the very beginning, the universe was of infinite temperature and density, and it was filled with light of tremendous energy. The Big Bang yielded a decrease in density and a cooling, both of which are still

1. There are still many uncertainties in these events, but the primary purpose of this book is to discuss the Earth's age, not the origins of the Earth, Solar System, Galaxy, and universe. These are subjects of lively research, are much debated, and are discussed in detail elsewhere. There are many excellent articles and books on these topics and on the probable chronology of events pre-dating the Earth. This abbreviated and simplified summary is primarily from Berlage (1968), Bok (1982), A. G. W. Cameron (1975, 1982), Dermott (1978), Ringwood (1979), Schramm (1974a,b), Scoville and Young (1984), Shu (1982), ter Haar (1967), Weisskopf (1983), and Wetherill (1975, 1980), to whom the reader is referred for additional and more authoritative discussions.

2. But the problem is being considered. See, for example, Hawking (1988).

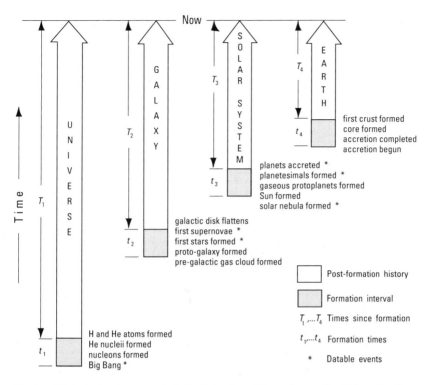

Fig. 1.1. History of various events in the formation of the universe, the Galaxy, the Solar System, and the Earth. Of the "formational" events listed, ages can be either measured or calculated for those indicated with an asterisk. Because the formation intervals $(t_1 - t_4)$ are much shorter than the ages $(T_1 - T_4)$, the age determined for any event can be taken to represent the age of formation. For all practical purposes, the age of the Earth, T_4, is very nearly equal to the age of the Solar System, T_3.

continuing.[3] Within the first 10 seconds or so after the Big Bang, some complex and poorly understood sequence of events created *electrons* and nucleons (*protons*—hydrogen nuclei—and *neutrons*), eliminated excess subatomic particles, and reduced the temperature of all matter to less than a billion degrees Kelvin. By the time 100 seconds had passed, the temperature of the universe had decreased enough to allow neutrons and protons to combine to form helium nuclei. It was not until about 300,000 years later, however, that the temperature had decreased to 10,000 K and the first complete atoms—most of them

3. Weisskopf (1983) makes the point that the Big Bang is often misrepresented in the popular literature as expansion away from a point in space. In fact, the universe was infinite in the beginning as it is now. How can it expand? Infinity times any number you choose is still infinity. Thus, the expansion of the universe is more properly viewed as a decrease in density or an inflation rather than an expansion into space.

hydrogen, some of them helium—formed. The subsequent history of the universe involved continued cooling, continued expansion, a steady decrease in density, and the assembly of matter into galaxies, stars, and planets.

Of all of these events in the birth of the universe, only the time of initiation of the Big Bang itself can be estimated from physical measurements. The ages of the other events can only be calculated theoretically, relative to the time of the Big Bang. Fortunately, the Big Bang is the one event that best represents the time of origin of the universe. Unfortunately, the calculated range of error in determining this age is quite large, and it is known only to within a factor of about two (Fig. 1.2; Chapter 8). For the universe, then, we can date the event we choose, but not with the accuracy we would like.

As the universe expanded, irregularities in the distribution of matter led some of the gas of the early universe to collect into clouds. The cause of these irregularities is uncertain, but it may have been turbulence or pressure waves generated by the Big Bang. In most of these clouds, internal pressure overcame gravitation and prevented collapse. In a few, however, external pressure was sufficient to allow gravity to prevail and thus to cause the gas to contract, forming a proto-galaxy. One such proto-galaxy was the ancestral *Milky Way Galaxy* (often referred to simply as the *Galaxy*), of which our Solar System is a minor part. As contraction proceeded, turbulence caused the proto-Galaxy to begin to rotate and the gas cloud to fragment. Some of the cloud fragments near the center of the Galaxy contracted to form the first stars. The more massive stars evolved quickly, and in time exhausted their nuclear fuel; each ended its existence in a colossal explosion known as a *supernova*. It is probably in supernovae that the bulk of all but the lightest of the elements have been created (Chapter 8). The matter from these supernovae was returned to the galactic pool, where it became available for incorporation into new stars. Most of this matter was gaseous, but some of the new, heavier elements condensed to form dust. As the Galaxy continued to evolve, rotation caused a flattening of the galactic disk and the eventual development of spiral arms. All available evidence indicates that the Galaxy continues to evolve. Stars are still forming in the core and spiral arms, where the density of matter is highest, ultimately to run their course and die. Those stars that perish as supernovae will contribute their remanufactured substance toward the formation of new stars.

In contrast to the beginning of the universe, there is no single, clear-cut event that marks the beginning of the Galaxy (Fig. 1.1). Of the several events discussed above, which should we choose to repre-

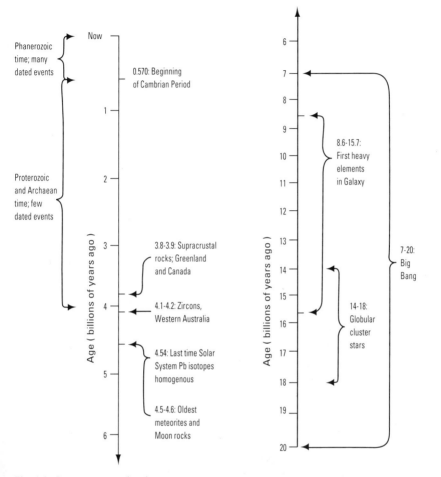

Fig. 1.2. Linear time scale of some important and datable events in the history of the Earth, the Solar System, the Galaxy, and the universe. The bases for the ages are discussed in chapters 4 through 8.

sent the Galaxy's birthday—the formation of the proto-Galaxy, the appearance of the first stars, the flattening of the galactic disk? Any of these would be an arbitrary choice, for none is a discrete event; they are but parts of a continuous process.

As a practical matter, our choice is limited by the fact that the ages of only two of the "events" can be determined. The age of the stars in globular clusters, which occur within a halo that surrounds the galactic disk and contain some of the oldest stars formed in the Galaxy, can be calculated from their stage of development and theoretical rates of star evolution (Chapter 8). We can also estimate the approximate age of the first supernovae from observations of the abundances of the isotopes of certain heavy elements (Chapter 8). Neither of these age estimates is as precise as we might wish, but the ages are consistent with the calculated age of the Big Bang and indicate that our Galaxy is of the order of 14 to 18 Ga.

The Galactic disk is 10^{17} km or so in diameter (about 30,000 light-years across), completes one revolution every 200 Ma, and contains something like 10^8 stars. About 2.6×10^{16} km from the center of the Galaxy, near the inner edge of the spiral arm named after the Orion Nebula, lies the unremarkable star we call the Sun. Orbiting the Sun are nine planets, seven of which have orbiting satellites (moons) of their own, and various debris, including *asteroids, meteoroids, comets,* and dust. The Sun and its various orbiting companions constitute the Solar System.

In many ways, the early history of a star is analogous to that of the Galaxy. Changes in pressure within the spiral arms of the Galaxy cause the interstellar gas and dust to fragment into smaller clouds. Most of these cloud fragments do not form stars because their internal pressure opposes gravity and prevents collapse. Occasionally, however, turbulence and other phenomena, such as nearby supernovae or strong magnetic fields, cause a local increase in pressure within a cloud. If the compression and consequent increase in density are sufficient, then gravitational attraction does cause the cloud to collapse, and a primitive star *nebula* is born. Turbulence and *conservation of angular momentum* cause the nebula to begin to rotate and thus inevitably to flatten into a disk. If the compression of matter at the center of the nebula is sufficient, then the nuclear fires of a new star will ignite. The Sun had such a beginning.

The Sun is the only star known with certainty to have planets, but that is probably because planets are small and dark, and because even the nearest stars are too far away for us to observe directly whether planets are present or not. Most probably, planets are common, perhaps even necessary, features of many star systems. Indeed, there is now some evidence of planets, proto-planets, pre-planetary dust rings, and proto-planetary disks orbiting the nearby stars called Barnard's Star, Van Biesbroeck 8, Vega, T Tauri, Fomalhaut, HL Tauri, Beta Pictoris, and R Monocerotis (Walgate, 1983, Reichardt, 1984, Anonymous, 1984a,b).

How do planetary systems form? Historically, there have been two classes of explanations. Dualistic theories, the first of which was advanced by Comte de Buffon in 1745 (Chapter 2), account for the planets by interaction between the Sun and another star. According to such theories, the planets condensed from matter torn from the Sun by the gravitational attraction of a passing star, or torn *from* the passing star *by* the Sun. Dualistic theories, however, now seem unable to explain the existence of planets. Calculations show that material drawn from the Sun by the gravity of another star would either fall back into the Sun as the other star departs, or be dispersed into space before it could coagulate into planets. Some sort of monistic theory, in which the Sun and planets formed together as an isolated system, as first proposed by René Descartes in 1644, seems much more likely.

Within the primitive *Solar Nebula,* then, dust and the ices of water, ammonia, and methane collided to form aggregate particles perhaps as much as a centimeter or so in diameter. These tiny clumps of material settled through the gas of nebula to the midplane of the flattening disk, where additional collisions and the increasing gravity of the growing clumps caused the particles to adhere to one another, forming larger bodies. Some of these bodies, called *planetesimals,* reached diameters of several hundred kilometers. The accumulation of heat due to gravitational energy and radioactivity caused the interiors of the larger planetesimals to reach temperatures sufficient for them to *differentiate* into *core, mantle,* and *crust,* for *volcanism,* and for *metamorphism.*

Temperature gradients within the Solar Nebula very likely caused considerable segregation of matter, with the more refractory or "rocky" materials being concentrated nearer the Sun and the bulk of the gaseous material farther away. This process would account for the clustering of the higher-density terrestrial planets (Mercury, Venus, Earth, and Mars) close to the Sun, and the lower-density gaseous and icy planets (Jupiter, Saturn, Uranus, Neptune, and Pluto) in more distant orbits.

The final stages of Solar System formation involved the aggregation of the planetesimals into the nine planets and their 34 satellites. As the planets grew they became increasingly efficient at attracting and holding stray bodies, with the result that early in their history they swept their orbits free of most extraneous matter. Evidence of this vigorous cleaning process is evident in the heavy cratering still visible on the Moon, Mars, and Mercury.

Most of the gas and small dust particles not captured by the Sun and larger planets were blown out of the Solar System by solar radiation. Although the "cleaning" of the Solar System was rather efficient, it was not totally so. The meteoroids, the asteroids, which orbit the Sun chiefly between Mars and Jupiter, and the billions of comets, which reside mostly in orbits beyond Pluto, represent the uncollected remainder of the Solar Nebula.

Several events in the formation of the Solar System can be dated with considerable precision (Fig. 1.2). The present composition of isotopes of lead in various bodies in the Solar System provides an index to the last time that these isotopes were uniformly mixed (these matters are explained in chapters 3 and 7). This age, 4.54 Ga, is thought to represent the time that the material of the Solar Nebula was isolated into individual bodies. The ages of the most primitive *meteorites*, which are between 4.5 and 4.6 Ga, provide us the time when planetesimals formed (Chapter 6). The most ancient rocks from the Moon, also 4.5 to 4.6 Ga, represent a time shortly after the satellite (and perhaps all or most of the planets) had grown to nearly its present size, by gathering debris from its path (Chapter 5).

The Earth, however, is another matter. Most people are surprised to learn that the age of the Earth cannot be determined directly by measurements taken from Earth materials. The Earth and Moon probably accreted in a similar way, except that the Moon grew in orbit around the Earth, whereas the Earth grew in orbit about the Sun. Thus both were subjected to intense bombardment as they collected material from their respective orbits. The Earth, however, grew to sufficient size that the combined heat from gravitational energy, meteoritic bombardment, and internal radioactivity caused *partial melting*, chemical segregation, and convection. These processes, which continue still, resulted in a heavy core of iron and nickel, a mantle, and a thin crust of lighter rocks. Furthermore, convection, the mechanism that drives the movement of continents and other *crustal plates*, is an ongoing process by which new crust is continually created and then recycled, continents collide and are later torn apart, and rocks are folded, faulted, buried, changed, or uplifted. As a result of this violent process—not to mention the weathering and eroding that never stops—none of the Earth's earliest-formed rocks have survived—at least none has yet been found. The oldest *rocks* known are 3.7 to 3.9 Ga; the oldest *minerals*, 4.1 to 4.2 Ga (Chapter 4). Although these measurements offer valuable information about the Earth's early history, they provide only minimum indications of the Earth's age.

Thus, the best evidence for the age of the Earth does not come primarily from the Earth at all, but from the Moon and meteorites (meteoroids that did not completely incinerate in the atmosphere and are found on the Earth's surface). Fortunately, the formation of the Solar System, of which the Earth, Moon, and meteoroids are integral parts, occurred over a relatively short period of time, probably less than 100 Ma, and the availability of only indirect evidence from the Earth itself therefore does not introduce a numerically significant ambiguity into the question of the age of the Earth.

The evidence for the age of meteorites, the Moon, and the Earth will be discussed in more detail in succeeding chapters. First, however, it will be instructive to examine some of the many historical attempts to determine the Earth's age.

CHAPTER TWO

Early Attempts:
A Variety of Approaches

*Mathematics may be compared to a mill of exquisite workmanship,
which grinds you stuff of any degree of fineness; but, nevertheless, what
you get out depends on what you put in; and as the grandest mill in the
world will not extract wheat-flour from peascods, so pages of formulae
will not get a definite result out of loose data.*

T. H. HUXLEY (1869: xxxviii)[1]

When and by whom the first attempt to assign an age to the
Earth was made will never be known, but it is clear that man has been
seeking an answer to this interesting question for more than two mil-
lennia (Table 2.1). For most of this time, the question was in the
hands of theologians, who drew their answers from theological theory and sacred writings. These theologically based ages for the Earth
ranged from a few thousand years to infinity, but were commonly in
the range of five to ten thousand years.

Many ancient philosophies, including the Greeks', rejected the
notion of finite time entirely, in favor of cyclic time in which the universe is continually regenerated in an indefinite succession of cycles.
In these systems, there was typically no beginning, no end, no net
gain, no net loss. Time was eternal and closed upon itself like a circle.
The ancient Hindus, for example, believed that a single cosmic cycle
of destruction followed by renewal lasted 4,320,000 years, and that
there were 100,000 such cycles in the life of Brahma (Haber, 1959: 13).
This concept was the basis for their calculation that the world is
1,972,949,094 years old (corrected to 1990) (Holmes, 1947b). The Chal-
deans, who ruled the Neo-Babylonian Empire from 612 to 538 B.C.,
taught that the most recent cycle began when Earth last emerged from
chaos more than 2 million years ago.

1. Commenting on Sir William Thomson's (1871) arguments for the age of
the Earth.

Where religious beliefs dictated a finite time for the cosmos, that time tended to be short compared with the infinity of cycles of the Hindus or the ancient Greeks. The Persian sage Zoroaster, who founded Zoroastrianism in the seventeenth century B.C., estimated that the Earth was 12,000 years old, and Hebrew and Christian calculations invariably resulted in values of less than 10,000 for the age of the Earth.[2]

In the Western world, the dominance of Christian calculations persisted, and it was not until the eighteenth century that the first pioneering naturalists began to formulate methods of estimating the Earth's age from observations, measurements, and scientifically grounded theory. These methods developed very gradually until, during the latter half of the nineteenth century, the age of the Earth had become one of the most hotly debated subjects of science. On the one hand were those scientists, largely physicists, who sought precise answers from rigorous calculations based on presumed initial conditions. Among these methods, the best known were based on the cooling of the Earth and the accumulation of salt in the ocean. The results of such calculations varied by several orders of magnitude, but commonly were of the order of a few tens of millions to a few hundreds of millions of years. On the other hand were those naturalists, primarily geologists and biologists, who were just then beginning to look closely at the rocks and living things of the Earth and think objectively about the ways in which they might have formed. In nature's thick accumulation of *sedimentary rocks* and the fossils contained therein, they saw the products of the processes operating over vast periods of time. From the observed rates of these processes in the present world they could estimate, though not then prove, that billions of years were required for the Earth's history. The geologists were correct, and it was the physicists, ironically, who would later provide the tools to prove them so.

The discovery of radioactivity, isotopes, and the nature of the atom around the turn of the century not only revolutionized physics but provided the theoretical basis for radiometric dating and had a profound effect upon concepts of geological time as well. Although attempts to date rocks and the Earth by using radioactivity followed shortly on these discoveries, they failed because still more tools were needed. Several decades were to pass before radioactivity would be understood, most of the naturally occurring isotopes would be identified and their abundances determined, the half-lives of the long-lived radioactive nuclides would be reasonably well known, and instru-

2. Haber (1959), Dean (1981), and Young (1982) discuss many of the chronologies based on religious tenets.

TABLE 2.1

Some Early (pre-1950) Estimates of the Age of the Earth

Basis	Author	Time of estimate	Age of Earth[a] (years)	Source
RELIGION				
Hindu chronology	priesthood	120–150 B.C.	1,972,949,091	Holmes, 1947b
Biblical chronology	Theophilus of Antioch	169	7,519	Haber, 1959
Biblical chronology	Eusebius of Caesarea	4th century	7,167	Faul, 1966
Biblical chronology	St. Basil the Great	4th century	5,994	Reese, Everett, and Craun, 1981
Biblical chronology	St. Augustine	5th century	6,321	Faul, 1966
Biblical chronology	Alphonso X	13th century	8,952	Faul, 1966
Biblical chronology	Pico della Mirandola	ca. 1490	5,471	Faul, 1966
Biblical chronology	John Lightfoot	1644	5,918	Reese, Everett, and Craun, 1981; Brice, 1982
Biblical chronology	James Ussher	1650	5,994	Brice, 1982
Movement of solar apogee	Johannes Kepler	ca. 1620	5,983	Reese, Everett, and Craun, 1981
SEA LEVEL				
Decline of sea level	Benoit de Maillet	1748	$>2,000 \times 10^6$	Haber, 1959; Albritton, 1980
TEMPERATURE				
Cooling of Earth	Comte de Buffon	1774	0.075×10^6	Haber, 1959; Albritton, 1980
Cooling of Earth	Lord Kelvin	1862	$20–400\ [98] \times 10^6$	W. Thomson, 1862b
Cooling of Earth	S. Haughton	1865	$>1,280 \times 10^6$	Haughton, 1865
Cooling of Earth	Lord Kelvin	1871	$<100 \times 10^6$	W. Thomson, 1871
Cooling of Earth	P. G. Tait	1869	$10–15 \times 10^6$	Albritton, 1980
Cooling of Earth	C. King	1893	24×10^6	King, 1893
Cooling of Earth	Lord Kelvin	1897	$20–40 \times 10^6$	Kelvin, 1897
Cooling of Earth	G. F. Becker	1908	60×10^6	G. F. Becker, 1908
Cooling of Earth	G. F. Becker	1910	$55–70 \times 10^6$	G. F. Becker, 1910c
Cooling of Earth	?. Suzuki	1912	$20–60 \times 10^6$	Holmes, 1913
Cooling of Earth	A. Holmes	1917	$>1,314 \times 10^6$	Barrell, 1917

Cooling of Sun	H. L. F. von Helmholtz	1856	22×10^6	Barrell, 1917; Knopf, 1931b
Cooling of Sun	Lord Kelvin	1862	$10–500 \times 10^6$	W. Thomson, 1862a
Cooling of Sun	P. G. Tait	1876	$<20 \times 10^6$	Hallam, 1983
Cooling of Sun	S. Newcomb	1892	18×10^6	Upham, 1893
Cooling of Sun	A. Ritter	1899	$4.4–5.8 \times 10^6$	Chamberlain, 1899a
ORBITAL PHYSICS				
Earth–Moon tidal retardation	G. Darwin	1879	$>54 \times 10^6$	Darwin, 1879, 1880
Earth–Moon tidal retardation	G. Darwin	1898	$>56 \times 10^6$	Darwin, 1898
Earth tidal effects	Lord Kelvin	1871	$<1,000 \times 10^6$	W. Thomson, 1871
Earth tidal effects	P. G. Tait	1876	$<10 \times 10^6$	Holmes, 1913; Hallam, 1893
Earth tidal effects	Lord Kelvin	1897	$<1,000 \times 10^6$	Kelvin, 1897
Change in eccentricity of orbit of Mercury	H. Jeffreys	1918	$3,000 \times 10^6$	Jeffreys, 1918
Eccentricity of orbit of Mercury	C. Popovici	1940	$1,800,000–3,700,000 \times 10^6$	Marble, 1940
OCEAN CHEMISTRY				
Sulfate accumulation	T. M. Reade	1876	25×10^6	Reade, 1876
Chloride accumulation	T. M. Reade	1876	200×10^6	Reade, 1876
Sodium accumulation	J. Joly	1899	$80–90 [89] \times 10^6$	Joly, 1899
Sodium accumulation	J. Joly	1900	$90–100 \times 10^6$	Joly, 1900
Sodium accumulation	?. Mackie	1902	25×10^6	G. F. Becker, 1908
Sodium accumulation	J. Joly	1909	$<150 \times 10^6$	Barrell, 1917
Sodium accumulation	W. J. Sollas	1909	$80–150 \times 10^6$	Sollas, 1909
Sodium accumulation	G. F. Becker	1910	$50–[70] \times 10^6$	G. F. Becker, 1910c
Sodium accumulation	F. W. Clarke	1910	90×10^6	Barrell, 1917
Sodium accumulation	A. Holmes	1913	$210–340 \times 10^6$	Holmes, 1913
Sodium accumulation	F. W. Clarke	1924	$<100 \times 10^6$	Clarke, 1924
Sodium accumulation	A. Knopf	1931	$>100 \times 10^6$	Knopf, 1931a
Chemical evolution of ocean	E. J. Conway	1943	$700–2,350 \times 10^6$	Marble, 1950
EROSION AND SEDIMENTATION				
Limestone accumulation	E. Dubois	?	$>1,000 \times 10^6$	Geikie, 1903
Limestone accumulation	T. M. Reade	1879	600×10^6	Reade, 1879

(continued on following page)

TABLE 2.1
(*continued*)

Basis	Author	Time of estimate	Age of Earth[a] (years)	Source
Limestone accumulation	A. Holmes	1913	320×10^6	Holmes, 1913
Sediment accumulation	J. Phillips	1860	$38–96 \times 10^6$	G. F. Becker, 1910c; Holmes, 1913
Sediment accumulation	A. Geikie	1868	100×10^6	Geikie, 1899
Sediment accumulation	T. H. Huxley	1869	100×10^6	Holmes, 1913
Sediment accumulation	S. Haughton	1871	$1{,}526 \times 10^6$	Holmes, 1913
Sediment accumulation	T. M. Reade	1876	$53–526 \times 10^6$	Reade, 1876
Sediment accumulation	S. Haughton	1878	$>200 \times 10^6$	Haughton, 1878
Sediment accumulation	A. Winchell	1883	3×10^6	Walcott, 1893; Holmes, 1913
Sediment accumulation	J. Croll	1889	72×10^6	Walcott, 1893; Holmes, 1913
Sediment accumulation	M. A. deLapparent	1890	$67–90 \times 10^6$	G. F. Becker, 1910c; Holmes, 1913
Sediment accumulation	H. H. Hutchinson	1892	600×10^6	Dana, 1895; Walcott, 1893
Sediment accumulation	A. Geikie	1892	$73–680 \times 10^6$	Geikie, 1892
Sediment accumulation	W. J. McGee	1892	$15{,}000 \times 10^6$	Walcott, 1893
Sediment accumulation	A. R. Wallace	1892	28×10^6	Walcott, 1893; Upham, 1893; Holmes, 1913
Sediment accumulation	C. D. Walcott	1893	$35–80\ [55] \times 10^6$	Walcott, 1893
Sediment accumulation	T. M. Reade	1893	95×10^6	Holmes, 1913
Sediment accumulation	T. M. Reade	1893	$100–600 \times 10^6$	Reade, 1893
Sediment accumulation	W. J. McGee	1893	$10–5{,}000{,}000\ [6{,}000] \times 10^6$	McGee, 1893
Sediment accumulation	W. J. McGee	1893	$1{,}584 \times 10^6$	Holmes, 1913
Sediment accumulation	W. Upham	1893	$<100 \times 10^6$	Upham, 1893
Sediment accumulation	W. J. Sollas	1895	17×10^6	Holmes, 1913
Sediment accumulation	J. J. Sederholm	1897	$35–40 \times 10^6$	Holmes, 1913
Sediment accumulation	J. G. Goodchild	1897	$1{,}408 \times 10^6$	Goodchild, 1897
Sediment accumulation	A. Geikie	1899	100×10^6	Geikie, 1899
Sediment accumulation	W. J. Sollas	1900	26.5×10^6	Sollas, 1900

Method	Author	Year	Age	Reference
Sediment accumulation	J. Joly	1908	80×10^6	Holmes, 1913
Sediment accumulation	W. J. Sollas	1909	80×10^6	G. F. Becker, 1910c; Holmes, 1913
Sediment accumulation	A. Holmes	1911	$>325 \times 10^6$	Holmes, 1911b
Sediment accumulation	A. Holmes	1913	$250–350 \times 10^6$	Holmes, 1913
Sediment accumulation	J. Barrell	1917	$1,250–1,700 \times 10^6$	Barrell, 1917
RADIOACTIVITY				
Decay of U to Pb in crust	H. N. Russell	1921	$2,000–8,000 [4,000] \times 10^6$	H. N. Russell, 1921
Decay of U to Pb in crust	A. Holmes	1927	$1,600–3,000 \times 10^6$	Holmes, 1927
Decay of U to Pb in crust	E. Rutherford	1929	$3,400 \times 10^6$	Schuchert, 1931
Decay of U to Pb in crust	S. Meyer	1937	$4,600 \times 10^6$	Koczy, 1943
Decay of U, Th to Pb in crust	F. F. Koczy	1943	$<5,330 \times 10^6$	Koczy, 1943
Decay of U to Pb in crust	I. Starik	1937	$3,000–4,000 \times 10^6$	Starik, 1937
Decay of U to Pb in crust and minerals	A. Holmes	1931	$1,600–3,000 \times 10^6$	Holmes, 1931
Decay of U to Pb in minerals	A. Knopf	1931	$>2,000 \times 10^6$	Knopf, 1949
Pb-isotope ratios in minerals	A. Knopf	1949	$>2,000 \times 10^6$	Knopf, 1949
Pb isotopes in Earth	E. K. Gerling	1942	$3,940 \times 10^6$	Gerling, 1942
Pb isotopes in Earth	A. Holmes	1946	$3,000 \times 10^6$	Holmes, 1946
Pb isotopes in Earth	A. Holmes	1947	$3,350 \times 10^6$	Holmes, 1947a
Pb isotopes in Earth	H. Jeffreys	1948	$1,340 \times 10^6$	Jeffreys, 1948
Decay of K isotope to Ca isotope	A. K. Brewer	1937	$<3,000 \times 10^6$	A. K. Brewer, 1937
Decay of K isotope to Ca isotope	A. K. Brewer	1938	$<10,600 \times 10^6$	A. K. Brewer, 1938
Decay of Rb isotope to Sr isotope	A. K. Brewer	1938	$<15,000 \times 10^6$	A. K. Brewer, 1938
Abundances of radioactive isotopes	H. E. Suess	1949	$4,000–5,000 \times 10^6$	Marble, 1950

NOTE: Not all of the values are ages for the Earth. Some are for very early events in Earth's history, such as the biblical creation of man, the age of the oceans, etc., whereas others are for the age of the Solar System or the age of matter. Ages of less than 10,000 years have been corrected from the year of publication to the year 1990 where appropriate. None of the methods gives the correct age of the Earth.

[a]Original authors' preferred values in brackets.

mentation of the necessary sensitivity would be available. By the 1950's, all of the pieces were in place and the answer to the question of the age of the Earth would soon follow.

A glance at Table 2.1 shows that the values assigned by scientists to the age of the Earth have increased several orders of magnitude over the past century or so. Why such an enormous change? There are two related reasons. First, the change represents an increasing understanding of the Earth and of the nature and rates of geological processes. It represents the search for, and acquisition of, tools, both instrumental and conceptual, with which our planet and its physical place in the cosmos can be investigated and understood. In short, it is the result of scientific progress—the birth and maturation of the science of geology.

Second, the increase in the values for the age of the Earth is a consequence of the slow and sometimes painful separation of scientific and theological thinking on matters related to nature. It is now widely understood that religion and science are separate and complementary fields of intellectual endeavor, each equipped to ask and to seek answers to quite different types of questions about humanity, about nature, and about the cosmos.

This understanding, however, is relatively new. Prior to the Enlightenment, in the mid-eighteenth century, the church dominated Western thinking, not only in affairs of the soul but also in matters of nature. The book of Genesis, as translated and interpreted by church scholars, was the principal and final authority for understanding both humanity's relationship to God and the physical history of the universe. Taken literally, the biblical measures of time were the day and the generation—time reckoned in millions and billions of years was unthinkable, because Scripture, it was then thought, proved it so. Not until the Enlightenment did observation and secular reasoning become widely accepted as an alternative to the Christian interpretation of nature, and modern science was born. But the habits of centuries were difficult to overcome. The process was slow and required difficult and fundamental shifts in both scientific and theological thought. Thus, the increase in the estimates of the age of the Earth represents the gradual liberation of concepts of geological time from the artificial limits imposed by the length of the human lifetime.

Estimates of the age of the Earth made prior to about 1950, biblical or otherwise, are all wrong, because they were based on methods now known to be invalid. Nevertheless, they are important, not only for historical reasons, but because they provide an interesting and valuable perspective on our current understanding of the age of the

Earth. It is, therefore, well worth a few pages to explore a few of the more interesting and authoritative of these estimates and the methodology that supported them.

Biblical Chronology

Bishop Ussher and 4004 B.C. are the name and date most frequently associated with estimates of the Earth's age based on biblical chronology. But although best known today, Ussher was far from the first to make such a determination. The founder of Christian chronological methods based on scholarship was probably Theophilus of Antioch (ca. 115–180), Syrian saint and Christian apologist. Theophilus was converted to Christianity as an adult and was elected sixth Bishop of Antioch in 170. His sole surviving work is contained in an eleventh-century Greek manuscript. Titled *To Autolycus*, it was written in 169 as a defense of Christianity in response to the derision of the faith by a pagan friend. In it, Theophilus attacked non-Christian views of history and, on the basis of a careful investigation of Scripture, calculated that the world had been created 5,698 years ago (5529 B.C.). He did not pretend that his calculations were exact, but did claim accuracy sufficient to refute previous "heathen" estimates of significantly longer duration. "I think I have now, according to my ability, accurately discoursed both on the godlessness of your practices, and on the whole number of the epochs of history. For if even a chronological error has been committed by us, of, for example, 50 or 100, or even 200 years, yet [it is] not of thousands or tens of thousands, as Plato and Apollonius and other mendacious authors have hitherto written."[3]

Christian literature contains numerous chronologies for man, the Earth, and the cosmos based on biblical and supplementary sources. In the preface of his *Chronologie de l'histoire sainte*, written in 1738, Alphonse de Vignolles claimed to have collected more than 200 such computations for the date of Creation, ranging from 3483 B.C. to 6984 B.C. (Reese, Everett, and Craun, 1981). These chronologies used various religious and historical works as sources of data, among which the texts of the Greek Septuagint, the Samaritan Pentateuch, the Hebrew text, and the Vulgate were of primary importance. The common method was to determine the time elapsed between certain key "historical" milestones (e.g. the Flood, the birth of Abraham, etc.) by

---

3. Quoted in Haber (1959: 17). Theophilus is referring to Apollonius of Rhodes (305–235 B.C.), Greek poet and scholar, who was born in Alexandria, Egypt, and eventually became head of the Alexandria library.

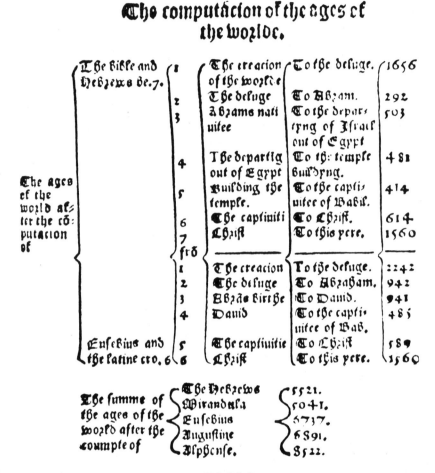

Fig. 2.1. An early compilation of time scales based on biblical chronology. (From *Cooper's Chronicle*, London, 1560, reproduced from Faul, 1966, p. 52.) Count Pico della Mirandola (1463–1494) was an Italian philosopher and scholar whose most ambitious work was an elaborate refutation of astrology. From 1488 until his death, possibly by poisoning, he lived in Florence on request of Lorenzo de' Medici. Eusebius of Caesarea (ca. 260–341) was a Christian theologian who was made Bishop of Caesarea Palestinea ca. 313. He was a pioneer of Christian history, and his scholarly writings are the sole documentation of much of the history of the early church. St. Augustine (354–430) was one of the foremost of the Latin Church Fathers. He was Bishop of Hippo from 395 until his death and, through his writings, has remained an influential figure in Western Christianity to the present. Alphonso is probably Alphonso X the Wise (1221–1284), Spanish monarch and scholar whose main interests were astronomy and civil government. He sought the unification of all human knowledge, a worthy but difficult goal.

summing generations and the reigns of various rulers (Fig. 2.1). Variations in the calculated values were due to the relative importance assigned to the numerous sources of data available as well as to the lengths assigned to the average generation (typically three per century) and the average reign of a monarch (typically 20 to 25 years).

Many early chronologies were based on a mixture of biblical and astronomical evidence. For example, Johannes Kepler (1571–1630), the German astronomer who is known for his discoveries of the laws of planetary motion, thought that Creation occurred at the summer solstice, at a time when the solar apogee was at the head of Aries. Using the known rate at which the apogee moved and its (then) present position, Kepler calculated that creation occurred in 3993 B.C. (Reese, Everett, and Craun, 1981).

Many biblical chronologists were not only prominent men of their day but were also true scholars, especially when viewed in the context of their times and the contemporary import of the church in all matters intellectual. Among the well-known authorities of the seventeenth century was John Lightfoot (1602–1675) (Brice, 1982). A graduate of Christ's College, Cambridge University, Lightfoot was a distinguished biblical and Greek scholar who eventually became Vice-Chancellor of Cambridge. In 1647, Lightfoot published a detailed account of his calculations of the date of creation of the world, which he put at 3928 B.C. (Fig. 2.2). Lightfoot's chronology was based on a careful analysis of the ages of individuals mentioned in the Old Testament, and he even assigned the exact time of Creation as in September, at the autumnal equinox. In an earlier (1642) and brief book with the interesting title *A few, and New Observations upon the Book of Genesis. The Most of them Certain, the rest Probable, all Harmless, Strange and rarely heard of before*, Lightfoot had announced that the creation of man occurred at about nine o'clock in the morning. It is Lightfoot's time of nine o'clock for the creation of man that is often but erroneously cited as the time of creation of the world and attributed to Bishop Ussher (Brice, 1982).

James Ussher was born in Dublin on January 4, 1581. He was educated at Trinity College, Dublin, where he received a B.A. in 1597, an M.A. in 1601, and later became Vice-Chancellor. Ussher was ordained a minister in the Irish Anglican Church in 1601 and became Archbishop of Armagh and Primate of Ireland in 1625. His property was confiscated in 1641 during the Irish rebellion while he was on a trip to England, and he never returned to his homeland. He died in Reigate on March 21, 1656, and was buried in Westminster Abbey.

Ussher was a well-known biblical scholar with a keen interest in the history of Christianity, but he is now remembered primarily for

The first Age of the World: From the Creation to the Flood: This space is called, *Early in the morning, Mat. 3.*

Hilar. in loc.

Ten Fathers before the Flood

Adam hath *Cain* and *Abel*, and Joseth them both, *Gen. 4.* unhappy in his children, the greatest earthly happiness, that he may think of Heaven the more.

Seth born in original sin, *Gen. 5. 2, 3.* a holy man: and father of all men after the Flood, *Numb. 24. 17.* to shew all men born in that estate, *Sorrowful.*

Enosh born: corruption in Religion by Idolatry begun, *Gen. 4. 25.* *Enosh* therefore so named, *Sorrowful.*

Kainan born: *A mourner for the corruption of the times.*

Mahalaleel born: *A praiser of the Lord.*

Jared born: when there is still *a descending from evil to worse.*

Enoch born: and *Dedicated* to God: the seventh from *Adam, Jude 14.*

Methushelah born: his very name foretold the Flood. The lease of the world is only for his life.

Lamech born. A man *smitten* with grief for the present corruption and future punishment.

Adam dieth: having lived 1000 years within 70. Now 70 years a whole age, *Psal. 90. 10.*

Enoch translated: next after *Adam's* death: mortality taught in that, immortality in this.

Seth dieth.

Enosh dieth.

Noah born a comforter.

Kainan dieth.

Mahalaleel dieth.

Jared dieth.

The CXX. years begin, *Gen. 6. 3.*

Japhet born.

Sem born.

Lamech dieth.

Methushelah dieth, and the Flood cometh.

130											55	57	113	
235	105									65	112	168	355	
325	195	90								162	126	210	369	
395	265	160	70					227	252	308	453	596	
460	330	225	135	65				292	414	453	680		
622	492	387	297	227	162			362	479	527	775		
687	557	452	362	292	227	65		549	535	592	830		
874	744	639	549	479	414	252	187	695	662	647	895		
930	800	695	535	539	470	308	243	752	717		962		
987	857	752	662	592	527	365	300	807	731				
1042	912	807	717	647	582		355	821	745				
1056							369	905	815				
1140							453	910	830				
1235							548		895				
1290							603						
1422							735						
1536							849						
1556							869						
1558							871						
1651							964						
1656							969						

Enoch the seventh from *Adam* in the holy line of *Seth* prophecied against the wickedness that *Lamech* the seventh from *Adam* in the cursed line of *Cain* had brought in.

Enoch lived as many years as there be days in a year, *viz.* 365. and finished his course like a Sun on earth.

With the story of this fifth Chapter, read 1 *Chron. 1. 1, 2, 3, 4.* which are an abridgment of it.

Fig. 2.2. Old Testament chronology published by John Lightfoot (1647). (Photo courtesy of William R. Brice.)

his date of 4004 B.C. for the creation of the world (Fig. 2.3). In contrast to Lightfoot, Ussher used a combination of astronomical cycles, historical accounts, and several sources of biblical chronology as data for his calculations, which he published in two volumes in 1650 and 1654 (Brice, 1982). Ussher argued that the beginning of the Julian Period occurred when the Roman indiction of 15 years (a cycle used as a chronological unit in several ancient cultures), the lunar cycle of 19 years, and the solar cycle of 28 years were all at "zero." Using primarily events recorded in the Hebrew text of the Old Testament, but drawing also from the Pentateuch, the Septuagint, and an Ethiopic text, he then calculated that Creation occurred in the year 710 of the Julian calendar. Ussher left unexplained the apparent paradox of how these 710 years could even exist before the Sun was created. Like Lightfoot and many other biblical scholars, Ussher placed the exact time of Creation near the beginning of autumn, and settled on midnight beginning the nearest Sunday as the beginning of Christian chronology.

In the Beginning, God created Heaven and Earth, Gen. I.V.I. Which beginning of time, according to our chronology, fell upon the entrance of the night preceding the twenty third day of October in the year of the Julian calendar, 710 (Ussher, 1658).

Thus, Ussher arrived at the beginning of night (evening), October 22, 4004 B.C., for the date of Creation, giving a universe 5,994 years old in 1990.[4]

The prominence of Ussher's date of 4004 B.C. among biblical chronologies is probably due to William Lloyd, Bishop of Winchester. Lloyd entered the date as a marginal note in the Great (1701) Edition of the English Bible, where it remained without explanation until 1900, when Cambridge University Press, followed a decade later by Oxford University Press, removed it from further editions (Kirkaldy, 1971: 5).

Biblical chronologies are historically important, but their credibility began to erode in the eighteenth and nineteenth centuries when it became apparent to some that it would be more profitable to seek a realistic age for the Earth through observation of nature than through a literal interpretation of parables. Today, scientists and biblical scholars alike agree that science is the proper arena in which to seek the numerical age of the Earth. As we shall see in succeeding sections, however, the answer did not come for several centuries after the search began in earnest.

4. According to Reese, Everett, and Craun (1981) the date of 4004 B.C. is not unique to Ussher but dates at least to St. Basil the Great (330–379).

nished his work which he intended, he then rested from all labour, and blessing the seventh day, he ordained and consecrated the Sabbath, [*Gen.*2. *verf.* 2,3.] because therein he took breath, as himself is pleased to speak of himself, [*Exod.* 31. *verf.* 17.] and, as it were, refreshed himself. Nor as yet (for ought appeareth) had sin set footing into the world. Nor was there any punishment laid by God, either upon man-kinde, or upon Angels. Whence it was, that this afterward was set forth for a signe, as well of our sanctification in this world [*Exod.* 31. *verf.* 13.] as of that eternall Sabbath, to be enjoyed hereafter; wherein we expect a full deliverance and discharge from sin, and the dregs thereof, and all punishments belonging thereunto, [*Heb.* 4. *verf.* 4,9,10.]

After the first week of the world ended, as it seemeth, God brought the new married couple into the garden of Eden, and charged them not to eat of the tree of knowledge of good and evil; but left them free use of all the rest.

But the Devil, envying Gods honour and mans felicity, tempted the woman to sin by the Serpent; whence himself got the name and title of *the old Serpent*. [*Apoc.* 12. *verf.* 9. and 20. *verf.*2.] The woman then beguiled by the Serpent, and the man seduced by the woman, brake the ordinance of God concerning the forbidden fruit; and accordingly being called, and convicted of the crime, had their severall punishments inflicted on them : yet with this promise added, that the Seed of the woman should, one day, break the Serpents head, (i.) That Christ in the fulnesse of time should undo the works of the Devil, [1 *Ioh.* 3. *verf.* 8, *Rom.* 16. *verf.* 20.] From whence it was, that *Adam* then first called his wife *Evah*; because she was then ordained to be the mother, not onely of all that should live this naturall life, but, of those also who should live by faith in her seed; which was the promised Messias: as *Sara* also afterward was counted the mother of the faithfull, [1 *Pet.*3.*verfe* 6. *Gal.*4. *verfe* 31.]

Upon this occasion our first Parents, clad by God with raiment of skinnes, were turned out of *Eden*, and a fierie flaming sword set to keep the way leading to the tree of Life, to the end they should never after eat of that fruit, which hitherto they had not touched *Gen.*3.*verfe* 21.22.&c. whence it is very probable, that *Adam* was turned out of Paradise the self same day that he was brought into it, which seemeth to have been upon the tenth day of the world (answering to our first day of *November*, according to supposition of the Julian Period) upon which day also, in remembrance of so remarkable a thing, as in all reason it should seem, was appointed the solemnity of *Expiation*, or attonement, and the yearly fast, spoken of by Saint *Paul*, *Acts* 27. *verse* 9. termed more especially by the name of *nesias*, wherein all, as well strangers as home born people, were commanded to afflict their souls with a most severe intermination, that *every soul which should not afflict it self upon that day should be destroyed from among his people,* [Lev.16.*v.*29.and23.*verfe* 29.]

After the fall of *Adam*, *Cain* was the first of all mortall men that was born of a woman, [*Gen.*4. *verfe* 1.]

Abel being murthered by his brother Cain, the first born of all man-kind, God gave *Eve* another son in his stead; whence his name was called *Seth*, *c.*4. *v.*25. when *Adam* had now lived 130 years, *c.*5.*v.*3. From whence it is gathered, that between the death of *Abel*, and the birth of *Seth*, there was no other son born to *Eve*; for then he should have been recorded to have been given her instead of him : so that whereas now the race of man-kind had been continued to the terme of 128 years, it is probable, that the number of men was so encreased in the world, that Cain might justly fear, through the conscience of his crime, that every man that met him would also slay him.[*c.*4.*v.*14,15.]

Seth now being 105 years old, begat a son, whom he named *Enoch*; which signifies, the lamentable condition of all mankind. For even then was the worship of God wretchedly corrupted by the race of Cain : whence it came, that men were even then so distinguished, that they who persisted in the true worship of God, were known by the name of the children of God; and they which forsook him, were termed the children of men, *Gen.*4. *v.*26. and 6. 1,2.

Cainan the son of *Enoch* was born when his father was 90 years old, [*c.*5.*v.*10.]

Mahaleel was born when Cainan his father had lived 70 years, [*c.*5.*v.*12.]

Jared was born when his father *Mahaleel* had lived 65 years, [*c.*5.*v.*15.]

Enoch was borne when his father *Jared* had lived 162 years, [*c.*5.*v.*18.]

Mathusalah was born when *Enoch* his father had lived 65 years [*c.*5.*v.*25.]

Lamech was born when his father *Mathusalah* had lived 187 years, [*c.*5. *v.*25.]

Now *Adam* the first father of all man kind, died when he had lived 930 years.

As for *Enoch*, the seventh from *Adam*, God translated him in an instant, whiles he was walking

Year of World	Julian Period	Year before Christ
130. d.		
235. d.	945.	3769.
325. d.	1035.	3679.
395. d.	1015.	3069.
460. d.	1017.	3544.
622. d.	1332.	3382.
687. d.	1397.	3317.
874. d.	1584.	3130.
930. d.	1640.	3074.
987. d.	1697.	3017.

Fig. 2.3. Old Testament chronology published by Bishop James Ussher (1658).

De Maillet and the Decline of the Sea

One of the earliest refutations of the Christian tradition of a very young Earth through reasoning based on observation and scientific theory was by Benoit de Maillet, French diplomat, savant, and amateur naturalist. De Maillet was born in Lorraine in 1656. From age 35 until his retirement in 1720, he held various diplomatic posts around the Mediterranean, which not only led him to travel extensively but also afforded him the opportunity to observe and study the geology and geography of the area. Upon his retirement, he moved to Paris, then to Marseilles, where he died in 1738.

The result of de Maillet's extensive studies and observations was a history of the Earth that required a span of time vastly greater than the few thousand years calculated by the biblical chronologists. De Maillet was well aware of the power and influence of the Church, and his theory was written as a fictitious account of a series of conversations extending over six days between a French missionary and an Indian philosopher named Telliamed (de Maillet spelled backwards). Even with this precautionary device, however, de Maillet did not publish his work. It was circulated among his contemporaries as a handwritten manuscript and did not appear in print until 1748, ten years after his death.

Speaking through Telliamed, de Maillet revealed a history of the Earth that attempted to account for its origin, the formation of strata and fossils, the building of mountains, and the origin of all forms of life, including humans (de Maillet, 1748; for summaries see Geikie, 1905: 84–88, Albritton, 1980: 68–77, and Haber, 1959: 108–12). As his starting point, de Maillet adopted Cartesianism, a fanciful cosmogony described by the French philosopher René Descartes (1596–1650) in his *Philosophiae Principia,* published in 1644.

According to Descartes, all matter in the universe is concentrated in vortices. Each vortex consists of a central sun about which swirl planets. As the sun burns, it expels ashes, which, along with water and dust from the vortex, accumulate on the planets. After the central sun has burned out, it becomes a planet and either another planet takes its place and flares into brightness at the center of the vortex or the entire vortex becomes a comet, perhaps to be incorporated into another vortex. Thus, planets are extinguished suns that accumulate or lose water depending on their position in the vortex. The Earth, according to Descartes, consists of three distinct regions: an incandescent nucleus composed of matter like that of the Sun, an opaque and solid middle region that cooled from a once-liquid state, and an outer region consisting of metals, rocks, sediment, water, and atmosphere gathered from the vortex and stratified according to density.

De Maillet accepted the idea that the Earth was once covered entirely with water. This was not simply an assumption but was based, at least partly, on observations. Telliamed reveals to the missionary that his grandfather had found that the strata of the mountains far inland were not only extensive and varied in type and composition but contained sea shells, certain proof of formation over a long period of time in an ocean more extensive than the present one. He ridicules the idea that these universal marine formations are the result of the flood of Noah, which, he asserts, was a local and transient phenomenon. In contrast, Telliamed observes, the water that once covered the globe is still being lost into the vortex, causing sea level to fall continuously. It is this decline of sea level that is the basis for de Maillet's estimate of the age of the Earth.

Telliamed's grandfather had observed that a part of the shoreline near his village had been awash when he was a boy and had emerged some years later. He reasoned that he could calculate the approximate time when the tallest mountains first emerged from the sea if he knew the rate at which sea level was declining. To this end, Grandfather constructed a hydrographic station on the shore near his village and over 75 years determined that the decline amounted to three inches per century. To confirm these observations, Telliamed cites structures at Carthage, Alexandria, and Acre that were built originally at sea level but now stand several feet above. The age and elevations of the structures above the sea at these localities yield a rate of three to four inches per century, or about three feet in a thousand years. Telliamed asserts that certain ancient settlements, some now standing as much as 6,000 feet above sea level, originated as seaports. From the rate of three inches per century, the age of these oldest marine settlements must be 2,400,000 years or so. Telliamed concludes that on any reasonable assumption two billion years must have passed since the decline of sea level began.

It should be no surprise that de Maillet's ideas were widely criticized, for they represented a radical departure from the commonly accepted notions that the Earth was both young and unchanging, and they defied the intellectual authority of the Church. Voltaire, one of the most prominent figures of the Enlightenment and a firm believer in the immutability of the Earth and all living things, went to considerable lengths to refute such heretical notions: "There is then, no system which can give the least support to this idea so widely prevalent that our globe has changed its face, that the ocean has occupied the earth for a very long time, and that men have formerly lived where porpoises and whales are today."[5] De Maillet's case was not aided by his occasional departures from careful scientific reasoning based on sound evidence. He firmly believed, and with good reason, that the sea was the ultimate birthplace of all life on Earth, an idea now considered correct. But in support of this proposition he cited accounts of mermaids, mermen, giants, and men with tails as evidence of the progression of man from the sea to the land. Elephant seals became the ancestors of elephants; petrified logs found inland, the remains of ships' masts.

De Maillet's vortices and mermaids aside, most of his ideas were not fanciful products of the imagination but an honest and pioneering attempt to interpret Earth history on the basis of observations and the viable scientific theories of his time. There is, after all, abundant evidence that most of the continents have, at one time or another, been inundated by the sea, and many mountains, now thousands of feet above sea level, are composed of undisputed marine sediments.

We now know that the Earth was never covered entirely by water in the way proposed by de Maillet. Most of the changes in sea level cited by him as evidence that the sea was declining are due to uplift of the land, and he failed to recognize that there are also areas around the Mediterranean that are sinking. Thus, de Maillet's basis for calculating ages based on the diminution of the sea is invalid. Nevertheless, he was one of the first to recognize the importance of slow operation of natural processes over vast periods of time in forming Earth's rocks and shaping its features. He also recognized that the temporal setting of events in the history of the Earth could be estimated by observing natural processes, measuring their rates, and making reasonable extrapolations based on logic. In one forceful intellectual blow, he felled the notion that geologic time must be reckoned in terms of the human lifetime and introduced the concept of an Earth that is billions, instead of mere thousands, of years old: "Let us be here content not to fix a beginning to that which perhaps never had one. Let us not measure the past duration of the world by that of our own years."[6]

Cooling of the Earth and Sun

In 1862 Lord Kelvin published his first calculation of the age of the Earth based on the time required for the planet to cool to its pres-

---

5. Voltaire, 1758, quoted in Haber (1959: 108). Voltaire was critical of many popular theories of the Earth, including those of Leibniz and Buffon. For an account, see Carozzi, 1983.
6. De Maillet, 1797, quoted in Haber (1959: 111).

ent state from a white-hot molten globe. For more than half a century, his ideas dominated scientific thinking about the antiquity of the Earth. Kelvin's results, which we shall turn to subsequently, are among the best known of the early attempts to determine a rational age for the Earth based on physical evidence. But the concept that a date for Earth's beginning could be found from calculations of cooling was introduced nearly two centuries before.

Isaac Newton (1642–1727), English physicist and mathematician, was unquestionably the culminating figure of the Enlightenment. Among his many important discoveries and inventions were the composition of light, the laws of motion and gravitation, the principles of the calculus, and the reflecting telescope. Newton was also interested in heat. From experiments on heated bodies cooling in air, he concluded that the rate of cooling was proportional to the difference in temperature between a hot body and the surrounding medium. Newton did not use this information to calculate an age for the Earth, but he came close. In his *Philosophiae Naturalis Principia Mathematica*, published in 1687, he surmised that a comet passing near the Sun would absorb an immense amount of heat, which it would retain for an exceedingly long time. To illustrate his point, he speculated on the length of time required for a globe of hot iron the size of the Earth to cool:

A globe of iron of an inch in diameter, exposed red hot to the open air, will scarcely lose all its heat in an hour's time; but a greater globe would retain its heat longer in proportion of its diameter, because the surface (in proportion to which it is cooled by the contact of the ambient air) is in that proportion less in respect of the quantity of the included hot matter; and therefore a globe of red hot iron equal to our earth, that is, about 40,000,000 feet in diameter, would scarcely cool in an equal number of days, or in above 50,000 years. But I suspect that the duration of heat may, on account of some latent causes, increase in a yet less proportion than that of the diameter; and I should be glad that the true proportion was investigated by experiments (Newton, 1687).

It is apparent that Newton knew, or at least suspected, that the relation between cooling time and size for bodies of planetary dimensions was more complicated than it appeared at first glance. As we shall see elsewhere in this section, it is even more complicated than Newton could possibly have imagined.

The use of molten iron as an analog for the cooling Earth was a natural choice because it was a material of everyday experience. Baron Gottfried Wilhelm von Leibniz (1646–1716), German philosopher and mathematician, codiscoverer (with Newton) of the calculus and one of the universal minds of the Western world, was an early subscriber to the concept of an initially molten Earth. He proposed that the Earth solidified in stages similar to those he had observed in the cooling of large masses of metal. The cooling Earth, according to Leibniz, was sculpted by large bubbles, some of which hardened into mountains while others collapsed to form valleys (Haber, 1959: 84–88). Like Newton, Leibniz did not venture to determine an age for the Earth from cooling. That bold step was left to another prominent figure of the Enlightenment.

Georges-Louis Leclerc, Comte de Buffon, was born September 7, 1707, at Montbard in Burgundy. He was educated in the law at the College of Godrans in Dijon and went on to study medicine, botany, and mathematics at Angers. Buffon was a man of enormous talent and energy. In addition to his scientific interests, he managed his family land holdings at Buffon and Montbard, engaged in the business of harvesting timber, established a commercial tree nursery, and built and operated an iron foundry, which played an important role in his research into the age of the Earth. His business activities, however, did not detract from his interest in science.

Buffon was one of the most productive and well-known scientists of the eighteenth century, and during his career he was elected to both the Royal Society and the Académie Française. He made fundamental contributions to the calculus of probability, plant physiology, and the scientific method, and laid the foundations for what would become the field of paleontology. He is best known, however, for an encyclopedic work in which he attempted, with considerable success, to synthesize all knowledge of nature and natural history into an intelligible whole. *Histoire Naturelle, Générale et Particulière* was originally intended to include an ambitious fifty volumes, of which Buffon actually completed 35 before his death in 1788. Among them were 12 volumes on mammals, nine on birds, and five on minerals as well as three introductory volumes and six lengthy supplements. The fifth supplement and the twentieth volume is *Epochs of Nature*, which was published in 1778 and is probably the best known. In it Buffon divided the history of the Earth into seven epochs. In the first epoch the Earth was a molten globe and the final epoch included the advent of man and the world as it is today; the intervening epochs included the formation of Earth's surface, the appearance of oceans and the beginnings of life, the formation of continents, the development of mammals, and the separation of the American and Eurasian continents.[7]

According to Buffon, the first epoch began when a comet collided with the Sun, causing the ejection of hot gases and liquid to

7. Both Albritton (1980: 84) and Haber (1959: 124–25) suggested that Buffon's use of seven epochs was a device to defuse church criticism of his concept of lengthy geological time, as the epochs are arbitrary and not an essential part of his synthesis of Earth history.

form the planets of the Solar System and their satellites. Buffon's proposition that the Earth began as a molten globe was consistent with accepted cosmogonies of the time. Leibniz had argued for a liquid primitive Earth, and in *Principia* Newton had furnished proof that the Earth was once fluid and had cooled in the shape of an oblate spheroid, thus explaining the equatorial bulge. Buffon also leaned heavily on the work of Jean-Jacques Dortous de Mairan, a French pioneer in atmospheric physics known for his work on the aurora. De Mairan had compiled more than a half century of measurements on temperatures of the atmosphere, of hot springs, in mines, and on the formation of ice in surface waters. These measurements, published in 1749, convinced Mairan and Buffon, with whom Mairan corresponded, that the Earth contained residual heat and was cooling. The next logical step was to calculate the time required for the Earth to cool from its initial to its present state.

Rather than speculate, Buffon had his foundry fabricate ten iron spheres whose diameters varied in half-inch increments up to 5 inches. These he heated to white heat and then observed the time required for them to cool, first to red heat, then to absence of glow, then to a point where they were cool enough to hold in his hand, and finally to room temperature. To ensure uniform conditions and minimize the daily temperature fluctuations caused by the Sun, Buffon performed his experiments in a cellar laboratory. He found an approximately linear relation between diameter and cooling time, which he then logically but naively extrapolated to a sphere the size of the Earth. On this basis he calculated that a mass of molten iron the size of the Earth would require 42,964 years to cool below incandescence and 96,670 years to cool to the present temperature of the Earth.

Buffon then performed similar experiments on a second set of graduated spheres composed of materials nearer the actual composition of the Earth. He corrected his calculations for the delaying effect of the Sun's heat and combined these data with some major events in Earth's history as reconstructed in *Epochs* to deduce the following scale of times, each in years from the beginning (Haber, 1959: 118):

Surface of Earth consolidated	1
Earth consolidated to center	2,936
Earth cool enough to be touched	34,270
Beginning of life	35,983
Temperature of present reached	74,832
End of life	168,123

Although the calculations were detailed and carried an air of exactitude, Buffon was suspicious of his results because he felt that the events of Earth history required a time much longer than the 75,000 years his calculations indicated, perhaps as much as double. He was impressed, for example, by the tremendous thickness of sedimentary rocks exposed in the Alps and by the exceedingly slow rate at which similar modern sediments are formed in the ocean. In unpublished manuscripts not publicized until the century after his death, Buffon detailed several longer chronologies, including one that estimated the age of the Earth at nearly 3 Ga (Albritton, 1980: 85).

As we shall see later in this section, there are numerous complicating factors, most of which could not have been anticipated by Buffon, that preclude calculating an age for the Earth from refrigeration. Buffon's major contribution to science, however, was much more important and lasting than his several ages for the Earth. Buffon believed and demonstrated that nature was rational and could be understood in terms of ordinary physical processes—force, motion, chemical reaction, heat, and other forms of energy—operating over geologic time. Invoking unique, supernatural, or extraordinary causes to explain natural history was to Buffon both unnecessary and unproductive. Thus, Buffon was the first to construct a history of the Earth based on the application of observable processes to explain the effects produced by like processes in the past. He also was the first to apply laboratory experimentation to the problem of the age of the Earth, and in doing so became one of the founders of geophysics. Finally, Buffon clearly separated the appropriate roles of theology and science in explaining nature—science was perfectly capable of answering the questions of how and when; the question of why was reserved for theology. Although Buffon's age for the Earth and much of his detailed history of the planet as set forth in *Epochs* are now known to be incorrect, the techniques he developed for deciphering Earth history through application of the law of cause and effect are a keystone of the modern scientific method.

Buffon had shown that calculations based on the cooling of the Earth held some promise as a method of determining the age of the planet. Nearly a century was to pass, however, before Buffon's approach was explored in detail. In 1862 there appeared two papers, written by a professor at the University of Glasgow, that were to dominate scientific thinking about the age of the Earth for more than five decades. The papers concerned the cooling of the Sun and Earth; the professor would later become Lord Kelvin.

William Thomson was born June 26, 1824, in Belfast. His father, James Thomson, professor of mathematics at the University of Glasgow, provided him with an early education in mathematics, a subject in which young William was especially adept. Thomson entered the

University of Glasgow at the age of ten, where he studied mathematics and became particularly interested in the work of Joseph Fourier, a French physicist who had devised mathematical methods of describing the physics of heat. In 1841 Thomson entered Cambridge University, from which he graduated in 1845 with high honors. Following graduation, he spent several months at the University of Paris, where he gained experience in laboratory methods by working with Henry Regnault, the French physicist. In 1846, largely through the efforts of his father, Thomson was appointed Professor of Natural Philosophy at Glasgow, a chair he held throughout his long and distinguished career.

Thomson was an unusually productive scientist, and at the time of his retirement in 1899 he had authored more than 600 scientific papers and books on electricity, magnetism, *thermodynamics*, hydrodynamics, atmospheric electricity, geomagnetism, *geodesy*, the thermal state and rotation of the Earth, tidal theory, and the age of the Earth. In 1848 he devised and described the absolute temperature scale, which still carries his name (the *Kelvin scale*). One of his most significant scientific achievements, a statement of the second law of thermodynamics, was published in 1851 under the title "On the Dynamical Theory of Heat." The cosmic implications of this fundamental principle later formed the basis for his calculations of the age of the Earth.

Thomson's work on electricity led him into the field of applied physics and to a role in the laying of the first trans-Atlantic telegraph cable. He became a director of the Atlantic Telegraph Company in 1856, served as the electrician aboard the *Agamemnon* in the first unsuccessful attempt to lay an Atlantic cable in 1858, and supervised the successful laying of the first cable by the *Great Eastern* in 1866, for which he was knighted that year by Queen Victoria. Thomson was also a prolific inventor, being responsible for the mirror galvanometer and siphon recorder used to receive telegraph signals, the stranded electrical conductor, the tide gauge, and an improved mariner's compass that allowed compensation for the magnetism of a steel ship. By the end of his career he held some 70 patents.

Thomson was probably the most honored British scientist in history, with countless awards, medals, and degrees. He was first elected president of the Royal Society in 1890 and was reelected to that office for five consecutive terms. In 1882 he was raised to the peerage and became Baron Kelvin of Largs, Ayrshire, and in 1896, on the fiftieth anniversary of his professorship at Glasgow, he was awarded the Grand Cross of the Royal Victorian Order. Because of his renown and the great respect in which he was held throughout the Western world, Kelvin's writings and opinions were accorded special significance. It is against this background of fame and influence that Kelvin's work on the age of the Earth took on a decided air of authority.[8]

Kelvin's interest in the age of the Earth arose from his research in thermodynamics. In 1852, he wrote a brief paper showing that in every transformation of energy from one form to another a small amount of the energy is converted to heat and dissipated (W. Thomson, 1852). This law of dissipation of energy is the basis for the second law of thermodynamics and the reason that perpetual-motion machines are impossible. It also means that natural engines, such as the Earth, Sun, and Solar System, must eventually run down and cease to function; even an injection of new energy can be only temporary. To Kelvin, the application of this principle to nature meant that both the Earth and the Sun must be cooling; from this it follows that they were hotter in the past and must become cooler in the future. The consequences were clear:

Within a finite period of time the earth must have been, and within a finite period of time to come the earth must again be, unfit for the habitation of man as at present constituted, unless operations have been, or are to be performed, which are impossible under the laws to which the known operations going on at present in the material world are subject (W. Thomson, 1852: 306).

The next obvious step was to calculate the time.

Kelvin's first paper on the subject, which appeared in *Macmillan's Magazine* in March of 1862, was about the Sun and the age of its fires (W. Thomson, 1862a). Kelvin began by discussing the notion that the Sun might be receiving new energy through the infall of meteoritic material. He calculated that an amount of new material equivalent to about $\frac{1}{5000}$ of the Sun's mass would be required to supply the Sun with enough gravitational energy to keep it burning for 3,000 years. If such material existed, he argued, it would have to reside between the Earth and the Sun, or the accumulation of material by the Earth would measurably shorten the year, a consequence that direct observations denied. But a mass of meteoritic material within Earth's orbit should have either disturbed the orbits of the inner planets or, if too close to the Sun to do that, altered the paths of comets passing close to the Sun. Kelvin noted that detailed observations of the planet

8. Thomson's involvement in the debate about the Earth's age, his sharp criticisms of the principle of uniformity, and his influence on geological thinking is described in detail by J. D. Burchfield (1975). See also Albritton (1980: 175–90). Thomson's half-century-long feud with some of the most prominent geologists of the time involved primarily the validity of uniformity, a concept that is still widely misunderstood, of little value to modern science, and not covered here. Readers interested in discussions of uniformity should consult Gould (1967), Hubbert (1967), Austin (1979), Shea (1982), and Hallam (1983).

Mercury and of comets passing within one-eighth of the Sun's radius showed that the amount of material in that region of the Solar System was very small. He thus concluded that meteoritic material was not a significant source of energy and that the Sun was cooling.

How much was the Sun cooling? That depended on its *specific heat*. Kelvin reasoned that if the specific heat of the Sun is similar to that of water, then at the present rate of radiation, the Sun would cool about 1.4°C per year. This would result in a contraction of the Sun's diameter of about 1% in 860 years. A change of this magnitude would be easily detectable, and as no such change had been observed, Kelvin concluded that the specific heat of the Sun must be higher than that of water or of any other terrestrial material. A second argument against such a large contraction was that the gravitational energy provided by a contraction of only 0.05% would provide enough energy to equal the heat radiated from the Sun in about 20,000 years. By a series of limiting arguments, Kelvin concluded that the specific heat of the Sun's material must be greater than ten times but less than 10,000 times that of liquid water, which would result in a temperature decrease of 100°C in from 700 to 700,000 years.

Kelvin's next step was to estimate the temperature of the Sun. He thought that there was little reason to suspect that the temperature at the surface was incomparably higher than that attainable in terrestrial laboratories. He noted that the amount of heat radiated from the Sun's surface is equivalent to that radiated from coal burning at a rate of a little less than one pound every two seconds. In locomotives, a pound of coal burns in 30 to 90 seconds. Hence, Kelvin concluded, heat is radiated from the Sun's surface at a rate of about 15 to 45 times that of coal burning on the grate of a locomotive furnace. Kelvin recognized that the Sun's interior was probably hotter than its surface, but because there must be prodigious *convection*, owing to the cooling of the surface, he saw little reason to think that the Sun was not in approximate convective thermal equilibrium. He concluded that the mean temperature of the Sun is about 14,000°C.

Kelvin's final tasks were to determine the origin of the Sun's heat, estimate its total amount, and calculate limits for the age of the Earth. As to origin, he noted that either the Sun's heat was created "by over-ruling decree" or by some natural process. He thought the former improbable provided that the latter could be shown not to contradict known physical laws. Nineteenth-century science knew only two possibilities for a natural origin—either meteoric aggregation or chemical action. Chemical action was ruled out because the most energetic chemical action then known could generate only about 3,000 years' heat before the Sun's mass was exhausted. If the Sun had formed by aggregation of meteoritic material through mutual gravitation, however, the heat generated could easily account for the Sun's continued radiation.

Kelvin calculated that the heat released by smaller bodies coalescing to form the mass of the Sun would be equivalent to about 20 million times that radiated from the Sun in one year. Some heat, no more than half, probably was dissipated by resistance and minor impacts before the final conglomeration, so Kelvin set the lower limit of the Sun's age at 10 million years. On the other hand, noted Kelvin, the Sun's density must increase toward the center, so that the total mass of the Sun must be greater, perhaps by a factor of 10, than that of a sphere of uniform density. Kelvin concluded

It seems, therefore, on the whole most probable that the sun has not illuminated the earth for 100,000,000 years, and almost certain that he has not done so for 500,000,000 years. As for the future, we may say, with equal certainty, that inhabitants of the earth cannot continue to enjoy the light and heat essential to their life, for many million years longer, unless sources now unknown to us are prepared in the great storehouse of creation (W. Thomson, 1862a: 494).

This last statement was prophetic. There were indeed powerful and unknown sources of energy fueling the Sun's fires, but their discovery was still four decades in the future.

Shortly after Kelvin's paper on the Sun appeared in print, he published in the Transactions of the Royal Society of Edinburgh another paper titled "On the Secular Cooling of the Earth" (W. Thomson, 1862b). In it, Kelvin applied to the Earth the same principles of thermodynamics that he had used to place limits on the Sun's age. He began by arguing that the energy of the entire Solar System must suffer "irrevocable loss" by dissipation and that, with sufficient data, the age of the Earth could be determined from the present distributions of temperatures within the Earth's crust. This was not a new idea for Kelvin. In a paper read before the faculty of the University of Glasgow in 1846 (and written in Latin!), he suggested that a perfectly complete geothermal survey would provide data for determining "an initial epoch in the problem of terrestrial conduction," and in another paper read before the British Association in 1855 he urged that such a survey be made (W. Thomson, 1862b: 470).

Although the existing temperature data were far from being as complete as Kelvin desired, measurements from mines and wells were enough to show that heat was indeed flowing outward from the Earth's interior. To Kelvin, these data confirmed that the Earth was cooling. Kelvin easily dismissed alternative hypotheses. He admitted that some heat might be generated by the dissipation of tidal forces or by chemical action, but on the whole he thought that these sources were inadequate to account for any but a small fraction of the heat flowing from the Earth. In addition, he argued that chemically generated heat would result in localized temperature gradients, whereas the measurements showed that there was a temperature gradient everywhere on the Earth. He also rejected Poisson's hypothesis[9] that the heat of the Earth is due to the former passage of the planet through a warm region of space by showing that the existence of life precluded the high surface temperatures required. He calculated, for example, that if such a transit occurred between 1,250 and 5,000 years ago, then the temperature of the proposed hot region would have to have been 14 to 28°C hotter than the present mean temperature of the Earth. Human history precluded that possibility. If the transit occurred more than 20,000 years ago, then the region would have to have been more than 56°C hotter. A temperature that high would clearly have destroyed life on Earth, and the paleontological record showing the continuity of life from the distant past to the present was evidence to the contrary. All things considered, Kelvin preferred the hypothesis that the Earth is a warm, chemically inert planet that is cooling.

The mathematical framework adopted by Kelvin for the cooling Earth was quite simple. He assumed that all of the heat was transferred from the interior to the surface by *conduction* and used Fourier's elementary solution for linear conduction of heat in a cooling solid whose surface is an infinite plane and whose distribution of mass beneath the plane is infinite in all directions (Fig. 2.4). Even though the Earth is a spheroid, Kelvin argued, the errors introduced by using this form of solution were negligible because the cooling effects would be confined to a region very near the surface (within a few hundred kilometers), and thus the finite size of the Earth and the curvature of its surface could be neglected.

Kelvin needed three essential quantities for his calculations: the initial temperature of the Earth, the *thermal conductivity* of the rocks of which the Earth is composed, and the present-day temperature gradient in the outer part of the Earth. The data available to him were not good. As the initial temperature, Kelvin used 3,870°C, which was then thought to be a reasonable estimate for the temperature at which rock melts. This was not much more than a guess, for at the time not a single experimental measurement of this quantity had been made. For the thermal conductivity, Kelvin used a weighted average of three

9. Simeon-Denis Poisson (1781–1840) was a French mathematician known for his work on definite integrals, probability, and electromagnetism. He also wrote a treatise on the mathematical theory of heat.

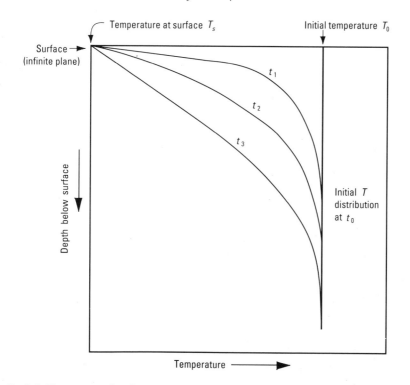

Fig. 2.4. Temperature distribution at various times near the surface of a body of infinite thickness bounded by an infinite plane. If the initial temperature of the body is T_0 at time t_0, the surface is held at constant temperature T_s, the heat is lost by conduction, and no heat is added to the body, then the temperature gradient near the surface of the body will be as shown at successively later times t_1, t_2, and t_3. Theoretically, the elapsed times $t_1 - t_0$, $t_2 - t_0$, and $t_3 - t_0$ can be calculated from the temperature gradients if the initial temperature T_0, the surface temperature T_s, and the thermal conductivity of the body are known. It is from such a model that Lord Kelvin first calculated limiting ages for the Earth.

values measured on sand, *sandstone*, and *"trap rock"* (*diabase*) that he had determined in the laboratory and reported to the Royal Society of Edinburgh two years earlier. A number of measurements of temperature gradients were available, but they varied widely, from $1/60.3°$ C per meter to $1/8.2°$ C per meter, and were insufficient to provide a basis for a meaningful worldwide average. Kelvin, therefore, adopted the commonly accepted value of $1/27.8°$ C per meter. Using these data,[10] Kelvin

10. The data used by Kelvin to obtain this age are initial temperature of Earth, 7,000°F or 3,871°C; temperature of surface, 0°F or −18°C; diffusivity, 400 ft²/yr or 0.01178 cm²/s; thermal gradient, 1°F/50.6 ft or 1°C/27.76 m. W. Thomson (1862b: 473–76).

calculated that the consolidation of the Earth from a molten state occurred 98 million years ago.

Kelvin was well aware of the inadequacies of the data then available. In addition to the uncertainties in the melting temperature of
rocks and the *geothermal gradient*, virtually nothing was known then
about the effect of temperature on the conductivity and specific heat
of rocks, nor were there any data on the amount of heat released by
rocks when they crystallize, i.e., their *latent heat of crystallization*. Accordingly, Kelvin allowed for the errors in the data by broadening the
limits for the time of solidification:

> But we are very ignorant as to the effects of high temperatures in altering the
> conductivities and specific heats of rocks, and as to their latent heat of fusion.
> We must, therefore, allow very wide limits in such an estimate as I have at-
> tempted to make; but I think we may with much probability say that the con-
> solidation cannot have taken place less than 20,000,000 years ago, or we
> should have more underground heat than we actually have, nor more than
> 400,000,000 years ago, or we should not have so much as the least observed
> underground increment of temperature. That is to say, I conclude that Leib-
> nitz's [sic] epoch of emergence of the *consistentior status* was probably between
> those dates (W. Thomson, 1862b: 474).

In spite of the questionable assumptions and the high degree of
uncertainty in the data, Kelvin's calculations of the ages of the Sun
and Earth were, at the time, considered highly authoritative. For
three decades they stood as the best that physics could offer on the
subject. Nineteenth-century geologists, following the innovative concepts of Hutton, Playfair, and Lyell, had finally broken with the
church-inspired doctrine of a recent creation and were formulating
new concepts for the history of the Earth and its biota based on the
availability of seemingly limitless time. Among these geologists was
Charles Darwin, whose remarkable new theory required that geologic
time be long enough to account for the evolution of the myriad species of plants and animals, both living and fossil, through natural selection. Now Kelvin, one of the most imposing scientific authorities
of the day, had imposed on the time available to geologists and biologists narrow limits that were derived from data, physical principles,
and elegant mathematics. The limits on the age of the Earth from thermal considerations, however, were to get narrower still, and a noted
geologist would provide the data, the rationale, and the calculations.

Clarence King was born on January 6, 1842, in Newport, Rhode
Island, to a family engaged in the East India trade. At the age of 17, he
withdrew from school without a diploma and took a position as a
clerk. The following year he enrolled in the new Yale Scientific School,

from which he graduated with a Ph.B. after completing the two-year course with honors. While at Yale, King had become fascinated by the fledgling science of geology and with the pioneering work of the California Geological Survey, headed by J. D. Whitney. Upon graduation from Yale, he headed west, where he joined the California Survey on a volunteer basis, working until 1867 on a field team under the direction of William H. Brewer. In that year he was appointed United States Geologist in Charge of the Geological Survey along the Fortieth Parallel, an expedition he had proposed to the Secretary of War, Edwin M. Stanton, and which lasted until 1869. During the 1870's, there were repeated efforts to persuade Congress and the President to consolidate into a permanent bureau the many successful but *ad hoc* and short-lived surveys engaged in exploring the West, and in 1879 these efforts culminated in the creation of the United States Geological Survey, with King as its first Director.[11] King did not find administration to his liking, however, and resigned from the Survey in 1881 to turn his attentions to mining ventures and consulting, activities that occupied him until his death in 1901. King was an energetic, ambitious, and talented geologist who was highly regarded by both the scientific community and the public. He was elected to membership in numerous national and international scientific societies, and in 1876 was elected to the National Academy of Sciences, being the youngest scientist so honored during his lifetime (Wilkins, 1958).

One of King's accomplishments as Director of the Geological Survey was to recognize the potential importance of the application of physics to the study of the Earth, and he established a program of geophysics within the Survey. To this end he engaged Carl Barus, a Würzburg Ph.D., first to make geophysical measurements over ore bodies, and later to establish a laboratory to measure the effects of temperature and pressure on rocks. The working relationship between King and Barus was quite simple; Barus would make the laboratory measurements and King would interpret them.

At King's behest, and partly supported by King's personal funds, Barus set about determining the melting and solidification temperatures as a function of pressure for diabase, a shallow intrusive rock similar in composition to basalt, by using samples from a railroad cut along the Pennsylvania Railroad in New Jersey. King selected diabase because its density was reasonably close to the average density at the surface and in the outermost part of the Earth. An apparatus to measure temperatures at high pressures was not then available, so Barus

---

11. The Geological Survey, one of the nation's oldest bureaus, still functions as a highly respected scientific research agency. The current Director, Dr. Dallas Peck, is only the Survey's eleventh.

did the next best thing: he melted diabase and measured the changes in volume and the latent heat of fusion as it cooled and solidified. He then used these two quantities in the *Clapeyron–Clausius equation* to calculate the change in melting temperature as a function of the change in pressure. On the basis of his measurements and calculations, he concluded that the change was +0.025°C per atmosphere and was linear (Barus 1891, 1892).[12] Barus then handed the ball to King: "The immediate bearing of all of this on Mr. Clarence King's geological hypothesis is now ripe for enunciation" (Barus, 1892: 57). King took the ball and ran with it.

The purpose of King's (1893) paper, titled "The Age of the Earth," was to advance Lord Kelvin's method of determining the Earth's age from the geothermal gradient and cooling considerations. King approached the problem from a very simple but elegant point of view. Kelvin, G. H. Darwin, and S. Newcomb all had previously argued that the Earth was effectively rigid because it appeared to be tidally stable. King reasoned, therefore, that the criterion of solidity could be used to test the validity of models of temperature distribution within the Earth. Although an initially molten Earth was acceptable, any temperature distribution that implied a present-day layer of fluid rock in the outer one-fourth of the Earth's radius could be excluded from further consideration because tides in such a liquid layer would violently disrupt the overlying solid crust.

King's first step was to determine the melting temperature of diabase as a function of depth. He used a formula developed nearly a century before by the French astronomer Marquis de Laplace (1749–1827) to describe the radial distribution of pressure within the Earth, then extrapolated Barus' value for the change in melting temperature as a function of pressure from the surface to the center of the Earth. This was a bold extrapolation on King's part, for Barus' measurements were all made at a pressure of one atmosphere and at temperatures less than 1,500°C.

One consequence of Barus' new data and King's calculations was the revelation that the Earth was probably never entirely molten. Because the melting temperature of diabase increased with depth, any reasonable initial temperature lay on the solidus side of the melting line except in the outermost regions of the Earth. For example, if King's extrapolation was correct, then an initial temperature of 76,000°C was required to create liquid diabase at Earth's center:[13] "The empire of

12. The Clapeyron–Clausius equation applies to single-component chemical systems. Rocks, of course, are multi-component systems, a fact that was unappreciated then. The application of this formula to diabase is invalid, but neither Barus nor King had any way of knowing that at the time.

13. In general, King was right about the dominance of pressure over temperature in the interior of the Earth and the improbability that the Earth was initially mol-

heat over pressure is thus seen to be purely superficial, while that of pressure over heat begins not far below the surface and extends more and more powerfully to the center" (King, 1893: 4).

Having established the melting conditions for an Earth made of diabase, King then calculated temperature distributions for various initial Earth temperatures and cooling times by using the same assumptions, mathematics, and diffusivity (0.01178 cm²/sec) that Kelvin had used in his 1862 paper. Finally, he tested the validity of these distributions by examining their relationship to his diabase melting line (Fig. 2.5).

The parameters used by Kelvin, i.e. an initial temperature of 3,870°C and a cooling time of 100 million years, were untenable. Kelvin's temperature curve (b in Fig. 2.5) intersected the diabase melting line at 364 km and again at 42 km, resulting in a molten layer 322 km thick overlain by a solid crust only 42 km thick. Increasing the cooling time, i.e. the age of the Earth, to 600 Ma avoided the liquid layer but resulted in a thermal gradient near the surface of only 1°C/68 m (c in Fig. 2.5), which was only half of the average (1°C/35 m) of all measured values in the recent compilation by a committee of the British Association. An initial temperature of less than 2,000°C was clearly necessary to avoid a layer of liquid rock. Any number of curves could be constructed that fell entirely on the solidus side of the melting line (for example, i and f in Fig. 2.5), but King thought that a combination of temperature and cooling time lower than that just required to do the job was unjustified. Any curve that resulted in a near-surface temperature gradient significantly less than the British Association average (for example, i in Fig. 2.5) was also precluded. Curve d, while indicating a thin liquid layer between 53 km and 106 km, could be brought into conformity with the requirements if the cooling time was increased by only 7 to 9 Ma.

From the point of view of age no greater time of cooling is allowable than enough to bring the gradient for any initial excess [= initial temperature] to the mean surface rate [= gradient]. Thus the condition for excess and age exclude a line of over 2,000°C and 24 × 10⁶ years. Conductivity remaining of the value used, any higher excess involves fluidity, and any greater age an inadmissible surface rate [= gradient] (King, 1893: 12).

Thus, King settled on 24 Ma as the age of the Earth.

Lord Kelvin's last word on the age of the Earth from cooling calculations was in 1897, but he added little new to his previous methods

---

ten. We now know, however, that diabase occurs only in the Earth's crust and that King's diabase melting line is incorrect. We also know that the outer core is liquid and so must be composed of material that melts in that region of the Earth, probably a mixture of nickel and iron.

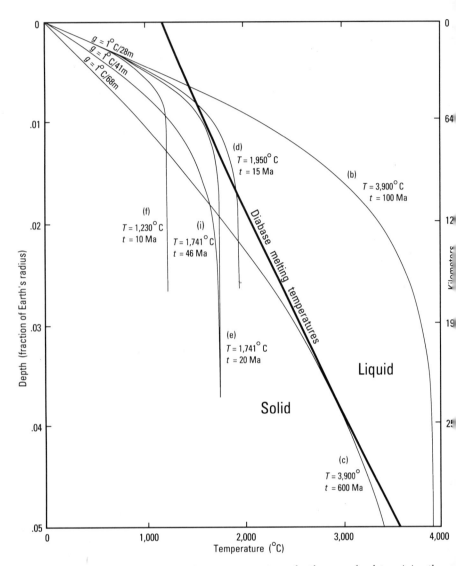

Fig. 2.5. Some of Clarence King's temperature versus depth curves for determining the age of the Earth. The curves are based on loss of heat by conduction for various gradients, *g*, uniform initial temperatures, *T*, and cooling times, *t*. The melting temperatures of diabase as a function of depths are also shown. The letters identifying individual curves correspond to those in King's original figure. (After King, 1893.)

or results. In an address to the Victoria Institute titled "The Age of the Earth as an Abode Fitted for Life" (Kelvin, 1897), he noted that critical data lacking in 1862 had become available and that these new data greatly modified the results of his previous calculations.

During the thirty-five years which have passed since I gave this wide-ranged estimate [of 20–400 million years] experimental investigation has supplied much of the knowledge then wanting regarding the thermal properties of rocks to form a closer estimate of the time which has passed since the consolidation of the earth, and we have now good reason for judging that it was more than 20 and less than 40 million years ago, and probably much nearer 20 than 40 (Kelvin, 1897: 345).

He observed that if he had used 1,200° C for the initial temperature of the Earth, his original estimate of 100 Ma for the Earth's probable age would have been reduced to a little less than 10 Ma.

Kelvin revised his previous calculations using improved methods (although he did not elaborate on exactly what he had done) and the new result, he said, supported that of King.

By an elaborate piece of mathematical bookkeeping I have worked out the problem of the conduction of heat outward from the earth, with specific heat increasing up to the melting point as found by Rucker and Roberts-Austen and by Barus, but with the conductivity assumed constant; and, by taking into account the augmentation of melting temperature with pressure in a somewhat more complete manner than that adopted by Mr. Clarence King, I am not led to differ much from his estimate of 24,000,000 years. But until we know something more than we know at present as to the probable diminution of thermal conductivity with increasing temperature, which would shorten the time since consolidation, it would be quite inadvisable to publish any closer estimate (Kelvin, 1897: 346).

The conclusions of King and Kelvin regarding the probability of a youthful Earth were widely accepted, but not all scientists were convinced. In 1869, for example, T. H. Huxley attacked the assumptions and data used by Kelvin in his first calculations, pointing out the high degree of uncertainty in both (Huxley, 1869).[14] John Perry, a noted physicist and former assistant of Kelvin's, not only questioned the assumptions used by both Kelvin and King, he showed that cooling calculations using different but equally likely assumptions and data resulted in ages for the Earth of as much as 29 Ga (Perry 1895a,b).[15]

14. See also quotation at beginning of Chapter 2.
15. In a letter to Perry reprinted in Perry (1895b), Kelvin admits that he might just as well have set his upper limit for the age of the Earth at 4,000 Ma instead of 400 Ma. This was as close as Kelvin ever came to recanting, although after radioactivity had proved to be a significant source of the Earth's internal heat he privately admitted to

Perry did not lend much credence to his own calculations but used them merely to show that Kelvin's and King's were of little value.

One of the harshest and most carefully measured criticisms of Kelvin's results came from T. C. Chamberlain, a highly respected professor of geology at the University of Chicago. In a paper specifically directed at Kelvin's 1897 address to the Victoria Institute, Chamberlain (1899b) challenged the noted physicist on his own ground. He noted that Kelvin had greatly overstated the certainty of his assumptions and results, and that Kelvin's use of phrases like "no other possible alternative" and "certain truth" were unjustified. Echoing Huxley's earlier criticism, he pointed out that Kelvin's calculations were only as good as the assumptions on which they were based.

The fascinating impressiveness of rigorous mathematical analyses, with its atmosphere of precision and elegance, should not blind us to the defects of the premises that condition the whole process. There is perhaps no beguilement more insidious and dangerous than an elaborate and elegant mathematical process built upon unfortified premises (Chamberlain, 1899b: 224).

In particular, Chamberlain (1899b: 225) challenged Kelvin's "very sure assumption" that the Earth was once a white-hot liquid: "I beg leave to challenge the certitude of this assumption of a white-hot liquid earth, current as it is among geologists, alike with astronomers and physicists. Though but an understudent of physics, I venture to challenge it on the basis of physical laws and physical antecedents." Chamberlain's challenge was on the most fundamental of scientific grounds—Kelvin had not considered other, perhaps even more likely, hypotheses.

Chamberlain noted that the physics of meteoritic aggregation made slow accumulation far more probable than the rapid aggregation required by Kelvin. Slow formation of the Earth would provide the internal heat necessary to explain the measured geothermal gradient but would not result in a liquid planet. Chamberlain also showed that the rapid accumulation assumed by Kelvin would require an equally rapid segregation of metals into the core, which would result in an initial Earth temperature exceeding 3,000°C, a condition that was precluded by the new data on the temperature and pressure of melting rock. This fact alone cast serious doubt on the viability of Kelvin's hypothesis. In contrast, slow accumulation would result in slow segregation of the materials in the Earth's interior, providing a continued release of heat through both conduction and convection over a long period of time.

J. J. Thomson that his cooling calculations were wrong (J. J. Thomson, 1936: 420, J. D. Burchfield, 1975).

Chamberlain did not claim that his hypothesis was proved. On the contrary, he readily acknowledged that it required rigorous testing before acceptance and that it might well be wrong. But, he argued, on the basis of current knowledge of physics and geology, his was a more logical sequence of events than Kelvin's and must be given, at the minimum, equal consideration.

Chamberlain also attacked Kelvin's age for the Sun, pointing out that the existing knowledge of the history and evolution of the Sun was so meager that there was no valid way to determine the amount of energy stored in the Sun or the timing of its release. He also argued that the behavior of matter in the interior of the Sun was completely unknown and that there might be sources of energy in addition to those considered by Kelvin. In particular, he argued that atoms might provide energy in quantities sufficient to prolong the Sun's fires far beyond the limits set by Kelvin.

Is present knowledge relative to the behavior of matter under such extraordinary conditions as obtain in the interior of the sun sufficiently exhaustive to warrant the assertion that no unrecognized sources of heat reside there? What the internal constitution of the atoms may be is yet an open question. It is not improbable that they are complex organizations and the seats of enormous energies (Chamberlain, 1899b: 225).

This last statement was especially prophetic, for although the phenomenon of radioactivity had been discovered by Henri Becquerel three years earlier, a great deal else—the nature of radioactivity, the isolation of radium, the measurements of the amount of energy released in nuclear transmutations, the discovery of isotopes, and a realistic atomic theory—was all in the future.

The question of whether or not calculations based on simple cooling could provide valid estimates for the ages of the Sun and Earth was partly resolved in 1903, when Ernest Rutherford and Frederick Soddy first determined the amount of heat generated by radioactive decay.[16] Rutherford and Soddy (1903: 591) readily appreciated the significance of their discovery for cosmological hypotheses: "It [the energy from radioactive decay] must be taken into account in cosmical physics. The maintenance of solar energy, for example, no longer presents any fundamental difficulty if the internal energy of the component elements is considered to be available, i.e., if processes of sub-atomic change are going on."

We now know, of course, that the Sun's fires are due to nuclear

16. Rutherford, a British physicist, and Soddy, a British chemist, were awarded Nobel Prizes in chemistry for their pioneering work on radioactivity, isotopes, and nuclear transformations. Rutherford's award was in 1908, Soddy's in 1921.

reactions, and that its fuel will last for billions of years. Thus Kelvin's calculations, which were based primarily on the available supply of gravitational energy, are invalid.

Although Kelvin, King, and their followers claimed a high degree of exactitude for their cooling calculations, their results were subject to large errors. By carefully choosing from a wide range of reasonable values for the geothermal gradient, initial temperature, and thermal diffusivity, it is possible to calculate ages for the Earth from simple conductive cooling models that exceed 3 Ga (for example, see Spicer, 1937).

But the most serious flaw in Kelvin's method is that its basic assumptions are invalid. The thermal budget of the Earth is far more complex than Kelvin, King, Chamberlain, or any nineteenth-century scientist could possibly have imagined.[17] One complication is that there are far more sources of heat within the Earth than were known in Kelvin's time. One important source, recognized by Kelvin, is primordial, i.e. heat left over from the formation of the Earth. Radioactivity, gravitational energy from compaction, chemical energy from the segregation of the iron–nickel core, and mechanical energy from meteoritic impacts during the period that the Earth was still sweeping up large quantities of material from its orbital path all contributed to this primordial heat. These sources probably generated enough heat to raise the temperature of the outer part of the Earth to near the melting point within 100 to 200 Ma of its formation. Much of this primordial heat has not yet escaped from the Earth.

A second and probably the most important source of continuing heat production is radioactivity. This heat is generated by the radioactive decay of primarily uranium, thorium, and potassium contained in the rocks of the Earth. Although the exact distribution of these radioactive elements within the Earth is not well known, there is no problem in constructing reasonable models that attribute most or even all of the heat now flowing outward from the Earth to radioactive decay.

In addition to primordial heat and heat from radioactivity, contraction of the Earth due to cooling and release of gravitational energy as the core grows may also be important contributors to the Earth's thermal budget. Thus, even though the relative importance of the various sources of heat is poorly known, there is little doubt that they are sufficient to permit an Earth many billions of years old.

Another factor that invalidates Kelvin's approach is that convec-

17. For discussions of the Earth's thermal budget, see, for example, F. D. Stacey (1977a,b, 1981), Sleep and Langan (1981), and Lubimova and Parphenuk (1981).

tion is a more important mechanism for the loss of heat from within the Earth than conduction. About three-fourths of the heat lost from the Earth leaves through the ocean basins. Virtually all of this heat comes from the mantle and is brought close to the ocean floors by convection. Conduction brings the heat to the surface, where it is lost into space. Approximately two-thirds of the heat lost from the continents is generated within the continental crust by radioactivity. The remaining third is brought to the base of the crust by convection, whereupon it is conducted upward through the crust to the surface. But even though the balance of mechanisms by which heat is transferred from the interior of the Earth outward is known in a semiquantitative way, current knowledge is insufficient to permit an exact description of heat loss from the Earth.

Finally, there are other poorly known factors, including the exact composition and structure of the Earth as well as the relevant physical properties of the rocks at various depths, such as conductivity, specific heat, and *viscosity*. Thus, the thermal history and budget of the Earth are very complicated, and our knowledge of the relevant details remains far too inadequate to permit any valid estimates of the age of the Earth from thermal calculations.

Darwin's Origin of the Moon

George H. Darwin was the second son of Charles R. Darwin, the renowned British naturalist and author. A mathematical astronomer, he was educated at Trinity College, Cambridge, where in 1883 he became Plumian Professor of Astronomy and Experimental Philosophy, a position he occupied until his death in 1912. Although now overshadowed by his famous father, George Darwin pioneered in tidal theory and cosmogony, and he is regarded by many as the father of modern geophysics. Darwin's interests were diverse, and his lengthy bibliography addresses topics ranging from the evolution of the Sun–Earth–Moon system to a statistical analysis of the marriage of first cousins. His methods of tidal prediction still form the basis for modern techniques. Darwin's accomplishments were not unnoticed in his day, and he enjoyed high standing in the scientific community. In 1899 he was elected president of the Royal Astronomical Society, and in 1905 he was elected president of the British Association and made Knight Commander of the Bath by King Edward VII.

George Darwin is probably best remembered as the originator of the hypothesis that the Moon was torn from the Earth by rapid rotation. This idea was far from being one of his most significant accomplishments, and he himself gave the thesis little credibility. Nonetheless, his calculations provided an early and widely cited quantitative estimate for the age of the Earth.

The basis of Darwin's idea was the braking effect of the tides on the rotation of the Earth. As the tides caused by the gravitational pull of the Moon move around the Earth, their energy is lost in the form of heat, which is radiated into space. This loss of energy results in a slowing of the Earth's rotation. Although this phenomenon was discovered by Immanuel Kant in 1754 (Kelvin, 1897: 342), it was not until 1868 that it was used to estimate a limit for the age of the Earth. In that year, Lord Kelvin read a paper to the Geological Society of Glasgow in which he observed that tidal friction was retarding the rate of rotation of the Earth by an average of 22 arc-seconds per century,[18] resulting in the lengthening of the day by about one second each 50,000 years (W. Thomson, 1871: 5–16, Kelvin, 1897: 342–43). If this were really so, then 7.2 Ga ago the Earth would have been rotating twice as fast as now and the additional tidal bulging would have forced all the oceans to the polar regions. According to Kelvin, this meant that the Earth must have solidified when the rate of rotation was not greatly different from today's, and he concluded that the age of the Earth was probably less than 1 Ga.

Darwin carried the concept that tidal friction might provide an estimate of the Earth's age to a much higher degree of refinement than had Kelvin, taking into account the effects of the tides on both the Earth and the Moon. He developed his hypothesis in a series of papers in the Philosophical Transactions of the Royal Society between 1879 and 1882 (Darwin, 1879, 1880, 1898). The calculations supporting his arguments are difficult and lengthy, but his reasoning is easy to follow.

Darwin began by observing the effects of tidal forces acting on a liquid planet with an orbiting satellite. In the absence of friction the tidal bulges in the planet, represented by the points P and P' in Figure 2.6, will be directly opposite the position of the satellite, which is represented by the point M'. But because friction is not absent in the real world, the tidal bulge is delayed in its passage around the planet and is carried beyond its place directly beneath the satellite by the planet's rotation, which is more rapid than the satellite orbits. In the figure this is most conveniently and simply represented by moving the position of the satellite from M' to M. The bulge does not rotate with the planet but with the orbit of the satellite, so the positions of the two bulges, P and P', will remain constant relative to the satellite,

18. The modern value is -23.8 ± 4 arc-seconds per century (Stephenson, 1981).

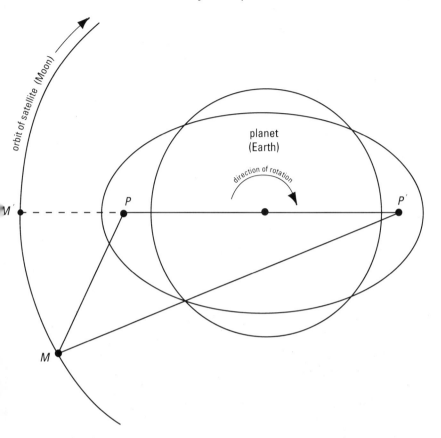

Fig. 2.6. Tidal effect on a rotating liquid planet (Earth) with an orbiting satellite (Moon). The circle represents the shape of the planet without a satellite, while the ellipse shows the tidal bulges caused by the gravitational attraction of the satellite. P and P' represent the centers of mass of the tidal bulges. M and M' represent the relative positions of the satellite in the presence and absence of tidal friction, respectively. (From G. Darwin, *The Tides and Kindred Phenomena in the Solar System*, 1898, W. H. Freeman and Co., copyright © 1962.)

M. Thus, an observer on the planet, which completes one revolution in less time than it takes for the satellite to make a complete orbit, will note that the rise of the tide occurs after the satellite is directly overhead.

In his next step Darwin considered the consequences of this relationship between the satellite and the tidal bulges. The attraction at P by M acts to retard the rotation of the planet, while the attraction at P' by M acts to accelerate rotation. P is nearer M than is P', however, so the net effect is to slow the planet's rotation. Because for every action there must be an equal but opposite reaction, it follows (and can easily be seen from the figure) that M must also accelerate because of the attraction of P. This acceleration of M causes the orbital distance to increase and the *angular velocity* to decrease. Darwin used the analogy of a rock twirled at the end of an elastic string. The more vigorously it is twirled (accelerated), the more the string stretches and the longer it takes to make a 360° circuit (orbit).

To summarize, the effects of tidal friction on this hypothetical liquid planet–satellite system are

1. The tidal bulges are thrown forward of the position directly beneath the satellite;

2. The rotation of the planet is retarded;

3. The orbital period of the satellite is decreased (though at a rate slower than the rotation of the planet is slowed); and

   4. The orbital distance of the satellite is increased.

Darwin then applied these conclusions to the Earth and Moon. He fully realized that the Earth is not liquid, nor does it contain oceans sufficiently deep for the analogy represented by Figure 2.6 to apply strictly. For the case of shallow oceans, as on the Earth, the tidal bulges are thrown backward rather than forward, but the net effects are nevertheless a retardation of both the rotation of the Earth (lengthening of the day) and the orbital period of the Moon (lengthening of the month) and an increase in the Moon's orbital distance from the Earth.

Extrapolating forward in time, Darwin showed that when the lengths of both the day and month equal 55 of our present days, then the Earth–Moon system will be stable and the Earth and Moon will forever present the same side toward each other, as if connected by a rigid bar. What happens when this line of reasoning is followed back in time? Both the day and month shorten until once again the configuration reaches stability. Darwin calculated that this occurs when both the day and the month are about three to five of our present hours in length, at which time the two bodies again revolve as if connected by a rigid bar, and the Moon is only about three Earth radii from the Earth.

At this point, Darwin was forced to engage in speculation, for his mathematical analysis is incapable of seeing beyond this period of initial stability. He suggested that an original molten Earth rotating rapidly might have spun off large pieces that aggregated to form the Moon. A rotation every three to five hours, however, would not be quite sufficient for the task, but if the period were one and one-half to two hours, then the tidal oscillations of the Earth caused by the gravitational pull of the Sun might be amplified by free oscillations of the

Sun–Earth system. Once the Moon has been ejected from the Earth, then the path to the present is not overly difficult, for the conditions of initial stability with the three- to five-hour period of rotation are metastable and easily disturbed. If the Moon orbited only slightly faster than the Earth, the equilibrium would be disturbed and the Moon would be pulled inward to destruction. The very existence of the Moon, Darwin argued, negates this possibility. The Moon, therefore, must originally have circled the Earth slightly slower than the Earth rotated on its axis. Tidal friction would then cause both the rotation of the Earth and the orbital period of the Moon to lengthen, and the orbital distance to increase irreversibly.

Almost as an afterthought, Darwin calculated the length of time required for the Earth and Moon to progress from the initial metastable conditions to the present with a day of 24 hours and a month of $27\frac{1}{2}$ days. Assuming that tidal conditions that would produce the most rapid change remained constant, he calculated that 56 Ma would be the minimum length of time required.

Darwin's value of 56 Ma is often quoted in the literature as his age for the Earth. Usually forgotten, however, is that he considered this to be a minimum. In his classic book, *The Tides and Kindred Phenomena in the Solar System*, Darwin (1898) set the range at 50 to 60 Ma and stated "The actual period, of course, must have been much greater."

It is also clear that he was not strongly attached to the validity of his hypothesis: "There is nothing to tell us whether this theory affords the true explanation of the birth of the moon, and I say that it is only a wild speculation, incapable of verification" (Darwin, 1898: 284).

We now know that Darwin's minimum value for the age of the Earth is much too low and his origin of the Moon is incorrect. As discussed in Chapter 5, it is more likely that the Moon accreted from the debris of a collision between the Earth and an asteroid-sized body, and that neither was ever entirely molten. There are also serious theoretical difficulties with Darwin's fission hypothesis. For example, it cannot account for the loss of nearly one-half of the initial angular momentum of the Earth–Moon system or the present 5° inclination of the lunar orbit to the plane of the ecliptic. Darwin himself calculated that fission under high rotation should have resulted in a body at least three times the mass of the Moon. Recent calculations show that the rate of tidal dissipation of energy can be changed materially by even moderate changes in the shapes of the oceans, and also that the present rate of this dissipation is anomalously high because of resonance effects in the shallow oceans. Thus extrapolation of present rates into the past is likely to result in serious errors. It also now seems unlikely that the Moon was ever closer to the Earth than about 38 Earth radii (compared to the present 60).

Within the framework of late-nineteenth-century understanding of the Earth, Darwin's hypothesis was reasonable and appropriate, but we now know that tidal considerations are incapable of providing a basis for determining the Earth's age. One has the feeling that Sir George wouldn't have minded.

The Salt Clock

Physics was not the only discipline to provide early estimates of Earth's age; chemistry was also set to the task, and in a rather clever way.

Imagine a tub of water to which a chemical is continuously added. If you knew the amounts of the chemical in the water both now and when the tub was first filled, and the rate at which the chemical was added to the water, then you could calculate the time of origin, i.e. the age, of the tub of water. It was the possibility of just such a calculation that Edmund Halley (1656–1742), the Astronomer Royal who predicted the return of the comet that bears his name, had in mind when, in 1715, he proposed that the age of the Earth might be calculated from the salt content of the ocean and of certain kinds of lakes (Halley, 1715).[19]

Halley observed that all lakes that receive runoff from rivers but lack outflow contain salt in varying amounts. The concentration of salt in the waters of these lakes must increase, he said, because salt, picked up by the rivers in their passage over the Earth, is continuously added but not removed.[20] Water is removed by evaporation, but water vapor is fresh so the salt is left behind: "But the vapours thus exhaled are perfectly fresh; so that the saline particles brought in by the rivers remain behind, while the fresh evaporates; and hence it is evident that the salt in the lakes will be continually augmented, and the water grow salter and salter. . . ." (Halley, 1715, in G. F. Becker, 1910a: 460). If this was truly the cause of the saltiness of lakes, Halley reasoned, then it is probable that the same mechanism was responsible for the saltiness of the ocean.

Analytical methods were then incapable of measuring the minute quantities of salt in the rivers that supplied the ocean, but Halley

19. Lengthy excerpts from Halley's paper appear in G. F. Becker (1910a).
20. The idea that the ocean was originally fresh and that its salt was dissolved out of the Earth's crust did not originate with Halley. Leibniz, among others, had incorporated the idea into his model of the Earth.

had another method in mind. If the concentration of salt in the ocean was measured at different times, then the rate of addition and the age of the ocean, which he equated with the age of the Earth, could be determined. A repeat of the experiment at some later date, he noted, would not only check the constancy of the rate at which salt is added but would verify the hypothesis.

Halley observed, however, that the age of the Earth could not yet be calculated. The experiment would require a very long time, and he lamented that the ancient Greek and Latin authors had not provided information on the saltiness of the ocean 2,000 years ago. The only thing that could be done, advised Halley, was for the Royal Society to ensure that future generations would be provided with the necessary data: "I recommend it therefore to the society, as opportunity shall offer, to procure the experiments to be made of the present degree of saltness of the Ocean, and as many of these lakes as can be come at, that they may stand upon record for the benefit of future ages" (Halley, 1715, in G. F. Becker, 1910a: 461). Halley also observed that the method would provide only a maximum age for the Earth, because the ocean and some lakes might well have contained some salt when they first formed. The experiment, he said, was still worthwhile because "[it] is chiefly intended to refute the ancient notion, some have of late entertained, of the eternity of all things; though perhaps by it the world may be found much older than many have hitherto imagined" (Halley, 1715, in G. F. Becker, 1910a: 461).

Halley's idea was never pursued and seems to have been largely forgotten until 1876, when T. Mellard Reade rediscovered the method he called "chemical denudation." Reade proposed that the age of the ocean could be found from the concentrations of *chlorides* and *sulfates*. Instead of determining the rates of addition of these compounds by measuring their concentrations at different times, Reade proposed to find the values by estimating the annual amounts carried into the ocean by the rivers of the world. At the present annual rates of addition, he calculated that it would require 25 Ma for the sulfates of calcium and magnesium to reach their present concentrations in ocean water; for chlorides (principally of sodium) the comparable time was 200 Ma (Reade, 1876, 1879).

Reade's basic idea of dating the Earth from the progressive change in the chemistry of the ocean was carried to a high degree of refinement by John Joly (1857–1933), a professor of geology and mineralogy at the University of Dublin. The son of a clergyman, Joly was born in Holywood, Kings County, Ireland, and educated at Trinity College, Dublin, where he taught throughout his career. Joly was trained as a physicist but developed a keen interest in geology, an evolution that is reflected in his sequence of positions at Trinity College, where he held appointments as demonstrator in civil engineering (1883), in physics (1893), and finally as professor of geology and mineralogy (1897).

Joly's accomplishments were numerous. He developed a method of extracting radium and pioneered its use in the treatment of cancer. He invented a type of thermometer, a steam calorimeter to measure heat energy, and a photometer to measure light frequencies. He was the first to propose that convection, driven by the heat from radioactive decay in the Earth's interior, might play a major role in the energetics and evolution of the Earth's crust. During his distinguished career he was accorded many honors, including election as a Fellow of the Royal Society in 1892.

Joly's (1899) classic paper "An Estimate of the Geological Age of the Earth" was read to the Royal Dublin Society, of which Joly was then Secretary, on May 17, 1899. In it (p. 249) Joly proposed to measure the age of the Earth from the accumulation, not of a salt, but of an element in the waters of the ocean:

Now, if any of the elements entering the ocean is not again withdrawn, but is, in a word, "trapped" therein, reappears as no extensive marine deposit, and is not laid down sensibly upon its floor, and if the amount of uniformity already defined is accepted, evidently in the rate of annual accretion by the ocean, from the rivers, of this substance and the amount of it now in the ocean, the whole period since the beginning of its supply can be estimated.

Such an element, he said, was sodium, and by using the pure element he avoided the questions and uncertainties of ionization and chemical form.

The result of Joly's calculations was an age for the Earth that differed little from Kelvin's:

The quantity of sodium now in the sea, and the annual rate of its supply by the rivers, lead, it will be seen, to the deduction that the age of the earth is 99×10^6 years. Certain deductions from this are, it will be shown, warranted, so that the final result of this paper will be to show that the probable age is about 89×10^6 years. Also, that this is probably a major limit, and that considerable departure from uniformity of activities could hardly amend it to less than 80×10^6 years (Joly, 1899: 249).

How did Joly arrive at these values? The basic equation, lacking certain necessary corrections, is the soul of simplicity:

$$\text{age of Earth} = \frac{\text{total sodium in ocean}}{\text{annual sodium influx from rivers}}$$

Joly began by arguing that the extrapolation of the rates of present processes into the past was warranted unless it could be shown that the rates had been interrupted by catastrophe or change. He pointed out that the approximate constancy of erosion of the land surface throughout geological time was a tenet whose validity had not seriously been questioned. Even so, his calculation required acceptance of this tenet only in part:

. . . that part of it which refers to the removal of the land surface by solution. It has to be accepted as a preliminary step that this, on the whole, has been constant. Herein are involved a constancy, within certain fairly wide limits, of rainfall over the land areas; a constancy, within fairly wide limits (which can be roughly defined), of the exposed land area, and a constancy in the nature and rate of solvent actions going on over the land surfaces (Joly, 1899: 248).

The other tenet that must be accepted was that the primeval ocean did not contain sodium in the quantities now observed.

The determination of a value for the sodium in the modern ocean was not difficult. Sir John Murray, a British oceanographer and founder of the study of submarine geology, had made estimates of the mass and mean depth of the ocean, the total volume of river discharge, and the quantity of dissolved matter in a number of the world's rivers.[21] Joly used Murray's mean ocean depth of 3,797 m combined with Hermann Wagner's ocean area of 1.0372×10^8 km^2 and the density of sea water to calculate that the mass of water in the world's ocean was 1.3245×10^8 *metric tons*.[22] The salinity of the ocean was known to be 3.5%, of which sodium chloride, NaCl, constitutes 77.758% of the total salts. Since 39.32% by weight of NaCl is Na, Joly arrived at a value of 1.4177×10^{16} metric tons for the total mass of Na in the ocean.

Murray's estimate for the total annual discharge of rivers into the ocean was 2.7176×10^4 km^3. His estimate for the dissolved matter in river water was based on analyses of waters from 19 rivers of the world. These analyses showed that of the total salts in river water, three contain sodium, giving a combined mass of 5,250 metric tons of Na per cubic kilometer of river water. The annual discharge multiplied by the sodium content provided Joly with a value of 1.4268×10^8 metric tons of Na supplied per year by rivers to the ocean. Joly's uncorrected age for the Earth thus was

$$\text{age of Earth} = \frac{1.4177 \times 10^{16} \text{ metric tons Na}}{1.4268 \times 10^8 \text{ metric tons Na/yr}} = 99.4 \times 10^6 \text{ yr}$$

21. John Murray (1841–1914) was one of the organizers of the famed *Challenger* expedition of 1872–76 and served on board as a naturalist. He completed the report of the expedition after the death of the leader, Sir Wyville Thomson.

22. Joly's 1899 data were expressed in English units of measurement. I have converted them to metric units.

But this was not the true age of the Earth; some corrections were necessary.

Joly applied a correction to the numerator for the amount of sodium in the original ocean. This correction was based on his estimate of the amount of sodium that would have been dissolved out of the primeval rocks as the crust and ocean were forming. Like Kelvin, Joly presumed that the Earth began as a molten globe. At a temperature of 1,500°C, the crust would have been molten and the material of the future ocean would have consisted primarily of free hydrogen, free oxygen, and HCl gas; NaCl would not exist at that temperature. As the crust and primitive atmosphere cooled, first water vapor then liquid water would form and the hot acidic rains (from the HCl dissolved in the rain) would react with the hot crustal rocks to form salts of Na, K, Mg, Ca, and Fe. From the abundances of these elements in what Joly took to be the average crustal rock, and presuming that all of the HCl would be neutralized by reactions with these elements in the crust, he concluded that only 14% of the Cl in the acidic rains, i.e. in the original ocean, would combine with Na to form NaCl.

The next quantity Joly had to find was the total amount of Cl available in the primeval ocean. To do this Joly assumed that the amount of Cl now in the ocean is the sum of the amount in the primeval ocean and the amount brought in by rivers since the ocean first formed. The amount of Cl in the present ocean was relatively easy to estimate; it was the sum of the amounts in the ocean's NaCl (21.913×10^{15} tons) and $MgCl_2$ (3.775×10^{15} tons).

The amount brought in by rivers over geologic time was a less certain calculation. Joly summed the amounts brought in each year as NaCl, LiCl, and NH_4Cl, but reduced the contribution from NaCl by 10% to account for recycling from the sea by evaporation. This sum, less the 10% correction, was about 69×10^6 tons of Cl per year. But for how many years? Joly chose 86 Ma as a reasonable value, which yielded 5.929×10^{15} tons of Cl added to the ocean by rivers since the world began. To summarize, Joly's estimate of the Cl in the primeval ocean was

now as NaCl	21.913×10^{15} tons
now as $MgCl_2$	3.775×10^{15} tons
from rivers in 86×10^6 years	-5.929×10^{15} tons
Cl in primeval ocean	19.759×10^{15} tons

Fourteen percent of this Cl, or 2.766×10^{15} tons, would combine with 1.789×10^{15} tons of Na, which was Joly's value for the original Na in the primeval ocean and was the amount to be subtracted from the numerator in his age equation.

The correction to the denominator consisted of subtracting the amount of Na in the NaCl that was recycled by evaporation from the ocean and returned to the rivers in rain. As above, Joly used a value of 10% of NaCl, or 0.381×10^8 tons of Na per year. Thus, Joly's age for the Earth was

$$\frac{(1.4177 - 0.1789) \times 10^{16} \text{ tons Na}}{(1.4268 - 0.0381) \times 10^8 \text{ tons Na/yr}} = 89.2 \times 10^6 \text{ years}$$

To this age he added 0.1 Ma to account for the time necessary to neutralize the free hydrochloric acid by reaction with the hot crust, giving a best estimate of 89.3 Ma.

Joly considered other possible sources of error, including the Na permanently removed from the system as salt deposits and the possible violations of his assumption of a constant rate of Na influx, but concluded that, taken as a whole, they probably were insignificant. He concluded

We think that it is at least justifiable to claim that our present knowledge of solvent denudation of the earth's surface points to a period of between eighty and ninety millions of years having elapsed since water condensed upon the earth, and rain and rivers and the actions continually progressing in the soils began to supply ocean with materials dissolved from the rocks (Joly 1899: 287).

The following year Joly (1900) revised his estimate upward. His method was basically the same as in his original calculations, but he used slightly different values for the mass of the ocean and the amount of Na brought in by rivers. In addition, he reduced by nearly half the amount of Na in the primeval ocean and deducted a correction of 4% from the numerator to account for the amount of Na introduced to the ocean by direct marine erosion. Joly's revised data yielded an age of 90.8 Ma, but he also noted that this age would be increased by nearly 6% if he used Professor DeLapparent's new value for the mass of the ocean. He concluded "We sum up the results of our inquiry, then, in the statement that the probable age of the earth, estimated from solvent denudation, is between ninety and one hundred millions of years" (Joly, 1900: 378). This was not, however, Joly's final word on the subject, and in 1909 he calculated an upper limit of 150 Ma (Barrell, 1917: 835).

Joly was not the only one to play the game. For example, William J. Sollas (1909: cxii), using Joly's method and various estimates of the annual influx of sodium from rivers, obtained values that ranged from 47 Ma to 175 Ma, finally concluding "On a review of all the facts, the most probable estimate of the age of the Ocean would appear to lie between 80 and 150 millions of years."

George F. Becker (1910b: 509–10) added a new dimension to the problem when he proposed that the accumulation of Na in the ocean was decreasing with time because the area of exposed crystalline rocks, i.e. the source of the Na, was decreasing:

Thus in the distant past there must have been a time when a far greater mass of massive rock was decomposed each year than now decays in the same period . . . the rate of chemical denudation per unit area may not have changed considerably, but the most rigid uniformitarian would not maintain that the total area of exposed massive rocks has been constant. The inference seems unavoidable that sodium accumulation is an asymptotic process.

He assumed that the rate of accumulation decreased by a constant proportion per unit time (an exponential function), and concluded that the age of the Earth was between 70 and 50 Ma, "probably nearer to the upper limit than to the lower one" (G. F. Becker, 1910c: 17).

What is wrong with the calculations of Joly and those who adopted his method? As might be expected, much better estimates of most of the numerical quantities are available today than were available to Joly. In addition, his assumption of constant rates for Na influx over geologic time is likely invalid. Even Becker's adoption of a systematic change in the rate of Na accumulation was a marginal improvement at best. The rates of erosion and solution, and the values of rainfall, runoff, continental area, and average exposed rock composition over geologic time are so poorly known that age calculations based on these quantities are mostly wishful thinking. But the most serious flaw in the method is in the basic assumption that Na, unlike all other elements, accumulates continuously in the ocean and is not withdrawn. We now know that the composition of sea water is not changing significantly because it is in approximate chemical equilibrium (Goldberg, 1965). All elements are removed from the ocean at approximately the same rate at which they are introduced. In the case of Na, slightly over half (far more than Joly's estimate of 10%) is recycled via the atmosphere by evaporation and rain. The rest is removed as a constituent in sediments, which eventually are subjected to erosion after uplift or reconstitution as crystalline rocks. The result of Joly's calculation was not an age for the ocean or the Earth but the contemporary *residence time* for Na, i.e. the average time that Na remains in sea water before it is removed. The current estimate for the residence time of sodium is 68 Ma (P. G. Brewer, 1975).

"Chemical denudation" was an important method for estimating the age of the ocean around the turn of the century, but it is now known to be incapable of providing even a crude estimate of the age of the Earth.

Sediment Accumulation

Many nineteenth-century geologists were uncomfortable with the application of physics and chemistry to the problems of the origin of the Earth and the length of geologic time. A few, like Chamberlain and King, were not intimidated by the esoteric calculations of the physicists, but they were exceptions. Most preferred the conclusions drawn from the evidence of their own science. Their reaction to the physicists' intrusion into what they considered to be their domain ranged from the defiance of Sir Archibald Geikie (1903: 77): "Until it can be shown that geologists and paleontologists have misinterpreted the records contained in the earth's crust, they may not unreasonably claim as much time for the history revealed in these records as the vast body of accumulated evidence appears to them to demand" to the apologia of T. Mellard Reade (1893: 97): "Geologists can hardly be blamed if they attach greater weight to their own observations and data and to reasoning that is more familiar and appears more certain and satisfactory to their minds."

The use of sediment accumulation as an hourglass was a method based solely in geology and involved estimating the time required for the sedimentary rocks of the Earth to accumulate. It was a game that any geologist could play and led to more estimates of the age of the Earth than any method save biblical chronology. As expressed by Warren Upham (1893: 211), it was considered by many to be the best weapon that geology had to offer in the heated debate with the physicists over the age of the Earth: "Among all the means afforded by geology for direct estimates of the earth's duration, doubtless the most reliable is through comparing the present measured rate of denudation of continental areas with the aggregate of the greatest determined thickness of the strata referable to the successive time divisions."

The basic method, first applied in 1860 by John Phillips (G. F. Becker, 1910c: 2), was deceptively simple. It involved summing the thicknesses of sediment deposited during the various divisions of geologic time (Table 2.2) and determining a rate for sediment deposition; the former divided by the latter was an estimate of the age of the Earth, or at least the time when sediments first began to form in the world's oceans and seas. As we shall see, however, there were many ways to do this and numerous uncertainties in the methods.

Perhaps the most thorough application of the method was by Charles D. Walcott, a noted American geologist renowned for his expertise in the *stratigraphy* and fossils of the Cambrian Period. Born in 1850 in the town of New York Mills, New York, Walcott ended his formal education at the age of 18 when circumstances required that he

TABLE 2.2

The Geologic Time Scale

Eon	Era/subera		Period/subperiod		Epoch	Estimated age (Millions of years ago; Ma)
Phanerozoic	Cenozoic	Tertiary	Quaternary		Holocene	.01
					Pleistocene	1.6
			Neogene		Pliocene	5.1
					Miocene	24
			Paleogene		Oligocene	38
					Eocene	55
					Paleocene	65
	Mesozoic		Cretaceous			144
			Jurassic			213
			Triassic			248
	Paleozoic		Permian			286
			Carboniferous	Pennsylvanian		320
				Mississippian		360
			Devonian			408
			Silurian			438
			Ordovician			505
			Cambrian			570
Proterozoic	Late					900
	Middle					1,600
	Early					2,500
Archean	Late					3,000
	Middle					3,500
	Early					4,000
Priscoan						4,550

SOURCES: Adapted from Harland et al., 1982; Palmer, 1983.

work to help support his family. He had been attracted to geology at an early age, however, and he continued to study the fledgling science on his own, eventually securing employment as an assistant with the New York Geological Survey in 1876. Three years later he joined the newly formed U.S. Geological Survey as an assistant geologist and rose through the ranks to serve as its Director from 1894 to

1907. From 1907 until his death 20 years later, he was Secretary of the Smithsonian Institution. Walcott was an influential figure in American science and instrumental in founding and organizing the Carnegie Institution of Washington, the National Research Council, and the National Advisory Council for Aeronautics. In spite of his heavy administrative duties, he was also a productive and respected researcher. Walcott was the recipient of numerous medals and honors, including election to the National Academy of Sciences. His lack of a traditional college degree was more than offset by ten honorary doctorates from such prestigious institutions as Cambridge, Harvard, and Yale.

Walcott's paper (1893), titled "Geologic Time, as Indicated by the Sedimentary Rocks of North America," provides a detailed description of his methods, assumptions, and uncertainties. He was a cautious scientist, and rather than venture too far afield he used, as the basis of his inquiry, the sedimentary rocks deposited in the *Cordilleran Sea* during the Paleozoic, a rock sequence that was the focus of much of his own research. The Cordilleran Sea was a shallow inland sea that stretched from Mexico to Canada (Fig. 2.7), and Walcott thought that the Paleozoic sedimentary rocks there were deposited nearly without interruption.

Walcott divided the problem into two parts: the time required for deposition of the "mechanical" sedimentary rocks like sandstones and shales, which are composed of particulate debris, and the time required to form the carbonate rocks, or *limestones,* which are formed primarily by chemical and organic precipitation from sea water. The mechanical sedimentary rocks he divided into those formed during the early and middle Cambrian Period and those formed during the rest of the Paleozoic. He did so because the evidence indicated that the land area to the east of the Cordilleran Sea (Fig. 2.7) was depressed below sea level at the end of middle Cambrian time, thereby reducing the supply of sediment for the remainder of the Paleozoic.

The thickness of early and middle Cambrian sedimentary rocks in the Cordilleran area averaged from 3,048 to 4,572 m.[23] For his calculation, Walcott adopted the smaller value. He estimated the area of the Cordilleran Sea to have been 1,036,000 km^2, and the land area supplying sediment about four times that, or 4,144,000 km^2. Then-current estimates of the rate of erosion varied from between one meter in 2,461 years to one meter in 19,685 years and averaged about one meter in 9,843 years. The overall character of the sediments de-

23. Walcott's data were expressed in English units, which I have converted to metric units.

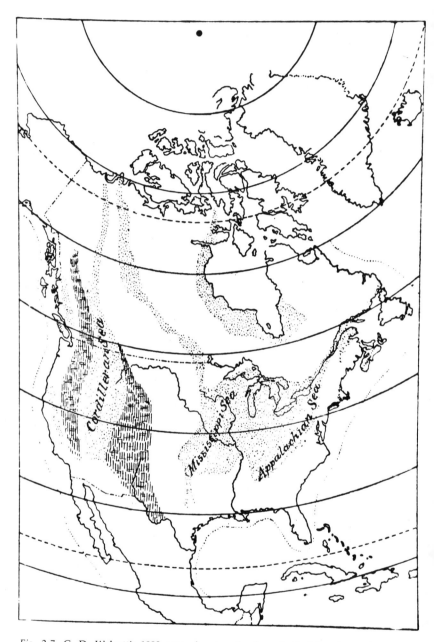

Fig. 2.7. C. D. Walcott's 1893 map showing the hypothetical locations of the Cordilleran, Mississippian, and Appalachian Seas. The thickness of sedimentary rocks deposited in the Cordilleran Sea during the Paleozoic Era formed the basis for his estimate of 55 Ma for the age of the Earth. (From C. D. Walcott, *J. Geol.* 1: 639–76, 1893.)

posited in the Cambrian Cordilleran Sea, however, suggested to Walcott that deposition there had been unusually rapid, so he adopted a rate of one meter in 656 years. The time required to deposit the 3,048 m of sedimentary rocks was

$$3,048 \text{ m} \times \frac{656 \text{ yr}}{\text{m}} \times \frac{1,036,000 \text{ km}^2}{4,144,000 \text{ km}^2} = 500,000 \text{ yr}$$

The thickness of post-middle Cambrian mechanical sedimentary rocks varied from 305 m in Montana to 3,048 m in the Wasatch Mountains of Utah; Walcott adopted 1,524 m as an average. Depression of the continent had reduced the land area supplying the sediment to about 1,554,000 km^2, but the area of deposition and the rate of erosion remained the same. Walcott's calculation of the time required to deposit the post-middle Cambrian mechanical sedimentary rocks, therefore, was

$$1,524 \text{ m} \times \frac{656 \text{ yr}}{\text{m}} \times \frac{1,036,000 \text{ km}^2}{1,554,000 \text{ km}^2} = 660,000 \text{ yr}$$

The calculations involving the carbonate rocks were somewhat more complex than those for the mechanical sedimentary rocks. Walcott estimated that there was an average of about 1,829 m or 4.687×10^9 metric tons of Paleozoic limestones deposited over the 1,036,000 km^2 of the Cordilleran Sea. Limestone, however, is not pure calcium carbonate ($CaCO_3$). Analyses had shown that a typical limestone is only about 75% $CaCO_3$, the rest being largely particulate matter. In addition, all sections of this limestone contain layers of clay and sand; Walcott estimated about 50%. So the mass of Paleozoic $CaCO_3$ deposited in the Cordilleran Sea was

$$\frac{4.687 \times 10^9 \text{ tons}}{\text{km}^2} \times 1.036 \times 10^6 \text{ km}^2 \times 0.5 \times 0.75 = 1.82 \times 10^{15} \text{ tons}$$

T. M. Reade had estimated that 17.5 metric tons of $CaCO_3$ and 7 tons of $CaSO_4$ were eroded from each square kilometer of land per year, all calcium to be deposited as $CaCO_3$. With 1,554,000 km^2 as the drainage area supplying material to the Cordilleran Sea, the total mass eroded was 3.81×10^7 tons per year. If all of this was deposited in the Cordilleran Sea, then the time required to accumulate the Paleozoic $CaCO_3$ would be

$$\frac{1.82 \times 10^{15} \text{ metric tons}}{3.81 \times 10^7 \text{ metric tons/yr}} = 47.8 \times 10^6 \text{ yr}$$

But, argued Walcott, this was a maximum, because the Cordilleran Sea was open to the ocean, and there would have been additional $CaCO_3$ supplied by the free circulation of ocean currents through the sea. Moreover, the water temperature and abundant biota in the Cordilleran Sea doubtless made conditions for deposition of $CaCO_3$ much more favorable, perhaps by a factor of four or so, than in the open ocean. Therefore, Walcott considered the likely contribution from the ocean.

For the total land area during Paleozoic time, Walcott took the world's present land area (1.42×10^8 km^2), less 25% for the area of the continents inundated by inland seas (like the Cordilleran Sea), plus an allowance for possible land areas in the Pacific and Arctic, for a total of 1.30×10^8 km^2. At an erosion rate of 24.5 tons of Ca salts per square kilometer per year, the total material available each year for deposition as limestone was 3.18×10^9 tons. Walcott estimated that conditions were favorable for limestone deposition over an area as follows:

Cordilleran Sea	1,036,000 km^2
other continental seas	1,554,000
subtotal	2,590,000
open ocean	168,350,000
total depositional area	170,940,000 km^2

To determine the rate of deposition, an allowance had to be made for the increased rate in the shallow seas, where conditions for deposition were more favorable by his estimated factor of four, so the rate in the open ocean was

$$\frac{3.18 \times 10^9 \text{ tons/yr}}{(1.6835 \times 10^8 \text{ km}^2) + (4 \times 2.59 \times 10^6 \text{ km}^2)} = 17.8 \text{ tons/km}^2/\text{yr}$$

The rate in the shallow seas was four times that or 71.2 tons/km^2/yr, so for the 1.036×10^6 km^2 of the Cordilleran Sea the rate of limestone deposition from the waters of the open ocean was

$$71.2 \text{ tons/km}^2/\text{yr} \times 1.036 \times 10^6 \text{ km}^2 = 7.38 \times 10^7 \text{ tons/yr}$$

To this Walcott added the 3.81×10^7 tons/yr contributed directly from the drainage area of the Cordilleran Sea for a total of 11.2×10^7 tons/yr. The time required for accumulation of the Paleozoic limestones was therefore

$$\frac{1.82 \times 10^{15} \text{ tons}}{11.2 \times 10^7 \text{ tons/yr}} = 16.3 \times 10^6 \text{ yr}$$

and the duration of the Paleozoic was

Cambrian mechanical sedimentary rocks	0.5×10^6 yr
post-middle Cambrian sedimentary rocks	0.66×10^6 yr
Paleozoic limestone	16.3×10^6 yr
duration of Paleozoic time	17.5×10^6 yr

Walcott considered the 17.5 Ma a minimum value because he had chosen all of the data so that his result would be conservative. For example, he had used rates of sedimentation and erosion much higher than were generally accepted, and the area of the Cordilleran Sea could have been as much as 50% greater than the value he used in the calculations.

So far, so good, but the Paleozoic by no means represented the whole, or even the majority, of geologic time (Table 2.2). Various authors had estimated the relative lengths of the Cenozoic, Mesozoic, and Paleozoic eras from the cumulative thicknesses of sedimentary rocks. Among them were James D. Dana in 1875 (Cenozoic 1, Mesozoic 3, Paleozoic 12) and Henry S. Williams in 1893 (Cenozoic 1, Mesozoic 3, Paleozoic 15) (Walcott, 1893: 673–74).[24] Walcott, however, felt that the estimates of Dana and Williams overestimated the Paleozoic, so he adopted relative lengths of 2, 5, and 12 for the Cenozoic, Mesozoic, and Paleozoic, respectively. For the Precambrian, Walcott estimated that the Algonkian was about equivalent to the Paleozoic, and that 10 Ma was a fair but highly uncertain estimate for the Archean.[25] Walcott then summed these values to determine the whole of geologic time.

Cenozoic	2.9 Ma
Mesozoic	7.24
Paleozoic	17.5
Algonkian	17.5
Archean	10.0(?)
total	55.14 Ma

He thought it unlikely that a post-Archean time could be less than 25 to 30 Ma or more than 60 to 70 Ma, giving limits for the age of the Earth of 35 to 80 Ma. He concluded that "geologic time is of great but not of indefinite duration. I believe that it can be measured by tens of

24. The modern ratios (Table 2.2) are, approximately, Cenozoic 1, Mesozoic 3, Paleozoic 5. Williams also coined the widely used term "geochronology" for the study or measurement of geologic time.

25. The Precambrian was then divided by some geologists into the Algonkian, which was roughly equivalent to the Proterozoic, and the Archean, which represented all of pre-Algonkian time. Compare Table 2.2.

millions, but not by single millions or hundreds of millions of years" (Walcott, 1893: 676).

Walcott had used the average thickness of sedimentary rocks in one depositional basin. Other workers of the time preferred a global approach and followed the maxim of the Rev. Samuel Haughton (1878): "The proper relative measure of geological periods is the maximum thickness of the strata formed during these periods." Among these was William J. Sollas.

In 1895, Sollas estimated the aggregate maximum thickness of post-Archean sedimentary rocks to be 50,000 m (Barrell, 1917: 820). In 1900 he increased his estimate to 80,800 m, and in 1909 to 102,400 m (Sollas, 1909: cxii), distributed as follows:

Cenozoic	19,450 m
Mesozoic	21,050
Paleozoic	36,900
Proterozoic	25,000
total	102,400 m

If these rocks accumulated at a rate of 1 meter in 328 years, a rate Sollas considered not unlikely, especially in those areas where the sedimentary rocks attain their maximum thickness, then the length of time required for their accumulation was about 34 Ma. For Archean time Sollas lacked data, but he noted that many paleontologists had speculated that at least as much time must have elapsed before the Cambrian as after it. If this were so, and Sollas thought it reasonable, then geologic time would be twice the time required for the Phanerozoic sedimentary rocks to accumulate, or

$$\frac{2 \times 77,400 \text{ m}}{1 \text{ m}/328 \text{ yr}} = 51 \times 10^6 \text{ yr}$$

But because this was less than the minimum estimate of 80 Ma he had based on the accumulation of Na in the ocean (Table 2.1), Sollas allowed some time for the major (18–24 Ma) and minor (5 Ma) unconformities that represented the sedimentary rocks missing from the record, and obtained a final age for the Earth of 80 Ma.

The ways of estimating the age of the Earth from sedimentation and erosion were as varied as the scientists who devised them. Warren Upham (1893), for example, sought to determine the duration of the Phanerozoic by multiplying the time that had elapsed since the great ice sheets disappeared from North America by ratios for the remainder of the Phanerozoic. He surveyed the various estimates of postglacial time that had been offered by others. Among these were N. H. Winchell's estimate of 8,000 yr based on the rate of erosion of the gorge and falls of St. Anthony, G. K. Gilbert's 7,000 yr from the retreat of Niagara Falls, E. Andrews' 7,500 yr from wave erosion on the shores of Lake Michigan, B. K. Emmerson's 10,000 yr from a study of the glacial valleys in Massachusetts, and D. Mackintosh's 6,000 yr from the erosion of limestone beneath glacial boulders. Taken together, these estimates indicated that the ice sheets had disappeared 6,000–10,000 years ago. He then applied the average of 8,000 yr to the time ratios of W. M. Davis[26] to obtain

post-glacial Quaternary	8,000 yr
Triassic Period	2.4 Ma
since Carboniferous	4–5 Ma
since middle Cambrian	10–20 Ma
earliest Cambrian fossils	20–25 Ma
beginning of life	40–50 Ma

Trying another tack, Upham then estimated the length of the Quaternary as some 100,000 yr and, on the basis of the change of faunas since the beginning of the Cenozoic, 3 Ma for the length of Cenozoic time. Applying this value to Dana's ratios (mentioned above) he obtained 48 Ma for the time since the beginning of the Cambrian. The diversity of life in the Cambrian suggested a long antecedent time for its development, leading Upham (1893: 218) to conclude that "the time needed for the deposition of the earth's stratified rocks and the unfolding of its plant and animal life must be about a hundred millions of years."

T. Mellard Reade (1893) adopted a different and less detailed approach. He estimated that the land area that supplied the bulk of Phanerozoic sediment was equal to the present land area less the area now occupied by Phanerozoic sedimentary rocks plus the area now occupied by Phanerozoic igneous rocks. He eliminated the Phanerozoic sedimentary rocks in this way because they were simply being recycled and did not contribute fresh material. The appropriate area of denudation, he estimated, was one-third of the present land area.

The thickness of sedimentary rocks accumulated since the beginning of the Cambrian Period he based on the data from the few deep drill holes available. He noted that there were several such holes that had penetrated 805 m and one that had penetrated 1,609 m without reaching the base of the Cambrian section.[27] He adopted 1,609 m as the average thickness. For the area that these sedimentary rocks

26. In an article in *Atlantic Monthly*, July 1891 (cited by Upham, 1893: 214), Davis estimated that if postglacial time was represented by a distance of two inches, then the beginning of Tertiary time was expressed by 10 feet, the Triassic by 50 feet, the Pennsylvanian by 80–100 feet, and Middle Cambrian by 200, 300, or 400 ft.

27. I have converted Reade's English units to metric units.

covered, he took the present land area plus an equivalent amount to allow for the rocks in and under the ocean. These two values combined, the Phanerozoic sedimentary rocks amounted to a mass equivalent to the present land area 3,218 m thick.

Finally, he used the commonly accepted rate of erosion of the land of one meter in 9,843 yr to calculate the time of accumulation, i.e. the time since the beginning of the Cambrian:

$$1,609 \text{ m} \times 2 \times 9,843 \text{ yr/m} \times 3 = 95 \times 10^6 \text{ yr}$$

He concluded that "Checked by the method adopted in this paper it would appear that the earth's age geologically speaking must be, as inferred in the Presidential Address [of A. Geikie in 1892], somewhere between 100 million and 600 million years" (Reade, 1893: 100).

Despite the popularity of the method among geologists and the confidence with which most presented their results, many of the errors in using sediment accumulation to find the age of the Earth or any other period of geologic time were fully appreciated at the time. Reade (1893: 98) enumerated some of the more important ones in justifying his own variation of the method:

If our modulus is the maximum thickness of all the formations we are naturally met with the objection that such a pile of strata nowhere occurs in nature. If we take as our gauge the actual thickness in any one place it may be reasonably objected that 10 feet of one set of strata may chronologically represent 1,000 feet or more of another. In addition to this, how can the hiatus represented by unconformity be estimated? If, again, we start with the average thickness of the sedimentary rocks as our principal measure, it is obvious on the slightest consideration that this will give a result much below the actual time, as the particles of such sedimentary rocks have been used up over and over again.[28]

Probably the most serious flaw in the method, however, was the one about which Reade had no qualms: the assumption of uniform rates of erosion and deposition. These factors are now known to vary so much, because of the high degree of variation in the conditions on which they depend, that it is simply impossible to determine an accurate average rate for virtually any period of geologic time. Calculations based on such rates, therefore, are little more than educated guesses—better than nothing perhaps but subject to a high degree of uncertainty. Moreover, the Precambrian, for which the record is both highly incomplete and nearly intractable to detailed stratigraphic analysis, constitutes the bulk of geologic time. Thus, ages based on sedi-

28. Louderback (1936) also presented an interesting discussion of the inadequacy of sedimentation and erosion for determining the age of the Earth.

mentation are, at the very most, useful in dealing primarily with the Phanerozoic, or only the last 10–15% of the history of the Earth.

Sediment accumulation as a method of determining the age of the Earth was no better than the other contemporary methods. But in the end it didn't matter, for waiting in the wings was a family of methods that would eventually stretch geologic time to its true proportions and yield an answer beyond the imagination of most nineteenth-century scientists.

Early Appeals to Radioactivity

For centuries the dream of the medieval alchemists was to turn base metals into gold. They never succeeded, but at about the turn of the twentieth century a handful of European scientists made the startling discovery that nature performs such elemental transmutations regularly through the phenomenon of radioactivity. It was this discovery that led directly to the methods now used to determine the ages of rocks, the Earth, the planets, and the Solar System.

In 1895 the German physicist Wilhelm K. Roentgen (1845–1923) observed that a mysterious source of energy produced within a cathode-ray tube caused a sheet of specially coated paper to luminesce even when the tube was covered with cardboard and the coated paper had been removed to another room. Roentgen's discovery of the mysterious rays, which he called *X rays*, prompted physicists to investigate the relationship between the phenomenon of luminescence and the X rays. In 1896 the French physicist A. Henri Becquerel (1852–1908) discovered that uranium salts emitted invisible rays similar to X rays even when the salts were not made to luminesce. This new radiation would darken a wrapped photographic plate and induce electrical conductivity in gases. Two years later, in Paris, Marie S. Curie (1867–1934) and her husband Pierre (1859–1906) discovered that thorium also emitted radiation, and they named the new phenomenon "radioactivity." The Curies also determined that radioactivity was a property of atoms, not molecules, being dependent only on the presence of the radioactive element and not on the chemical *compound* in which it occurs. They also observed that some minerals containing U and Th are more radioactive than pure U or Th, deduced that other radioactive substances must, therefore, be present, and discovered two of these additional radioactive elements: polonium and radium.[29]

29. Roentgen was awarded the Nobel Prize for physics in 1901. In 1903, Becquerel and the Curies shared the physics prize for their work on radioactivity. Marie Curie was awarded the Nobel Prize for chemistry in 1911 for the isolation of radium

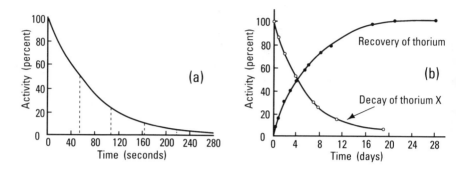

Fig. 2.8. Observations by Rutherford and Soddy that were the basis for their exponential theory of radioactive decay and elemental transmutation. (a) Decay of activity of the thorium "emanation," later identified as ^{220}Rn, a gas with a half-life of 55 seconds. (b) Decay of the activity of thorium X (^{224}Ra, half-life 3.6 days) and the corresponding recovery of the activity of thorium. (After Rutherford, 1906.)

In 1902, Ernest Rutherford and Frederick Soddy of McGill University, in Montreal, published the results of a series of experiments that led them to formulate a general theory governing the rates of radioactive change (Rutherford and Soddy, 1902a–d). They began by isolating the radioactive gas emitted by Th compounds and observing the change in its activity when left to itself. They discovered that after 54.5 seconds the activity was only one-half of its initial value, after 109 seconds only one-fourth of its initial value, after 163.5 seconds only one-eighth of its initial value, and so forth (Fig. 2.8). They had discovered that the decay of the radioactive gas from Th was exponential. A similar experiment showed that Th-X, which they isolated from Th and was later identified as ^{224}Ra, decayed in a similar manner but with a different time constant. Moreover, the activity of the Th from which the Th-X had been removed increased at the same rate at which the Th-X decayed, indicating that Th-X was being created by the decay of Th (Fig. 2.8). These observations were similar to those by Becquerel on U and U-X the previous year, and led Rutherford and Soddy to propose that the atoms of radioactive elements are unstable and decay spontaneously to other elements accompanied by the emission of alpha or beta particles. They also proposed that the activity of a substance is directly proportional to the number of atoms present, a proposition from which the formula for radioactive decay can be derived (as will be shown in the next chapter).

Another important suggestion offered by Rutherford and Soddy was that helium might be the product of the decay of radioactive ele-

and her research on its compounds, thus becoming the first and one of the few scientists to be awarded two Nobel Prizes.

ments. This was confirmed the following year by Ramsay and Soddy (1903), who showed that He is formed by the decay of Ra. In 1905 B. B. Boltwood examined the composition of naturally occurring U minerals. Invariably, he noted, they contain lead. Moreover, there was more Pb and He in the older minerals than in the younger. He concluded that Pb might be a decay product of U (Boltwood, 1905). The stage was now set for the first attempts to apply the newly discovered and poorly understood phenomenon of radioactivity to the problem of geologic time.

In March of 1905 Ernest Rutherford, then the Macdonald Professor of Physics at McGill University, delivered the Silliman Lectures at Yale University. In them, he (1906: 187–92) offered the possibility of using radioactivity as a geological timekeeper:

The helium observed in the radioactive minerals is almost certainly due to its production from the radium and other radioactive substances contained therein. If the rate of production of helium from known weights of the different radioelements were experimentally known, it should thus be possible to determine the interval required for the production of the amount of helium observed in radioactive minerals, or, in other words, to determine the age of the mineral (Rutherford, 1906: 187–88).

Rutherford offered two examples of the proposed radioactive method of calculating ages. The first was a sample of the mineral *fergusonite*, an oxide of the elements yttrium and niobium that contains 7% U and 1.81 cm^3 of He per gram of the mineral. From a production rate of 5.2×10^{-8} cm^3 of He per year for each gram of U and its associated Ra, the age of the mineral was

$$\frac{1.81}{0.07 \times 5.2 \times 10^{-8}} = 497 \text{ Ma}$$

This age, cautioned Rutherford, was a minimum, for some of the He had probably escaped. A calculation for a second mineral, a *uraninite* from Glastonbury, Connecticut, also yielded a minimum U–He age of about 500 Ma.

The same year R. J. Strutt measured the U, Th, Ra, and He content of 22 radioactive minerals and found that when He was abundant, the mineral also contained Th, but the converse did not hold. This strengthened the conclusions that Th, in addition to U and Ra, produced He and that the He quantity was a function of age. The production rate of He from Th was not known, but Strutt used his data to calculate a minimum age of 2.4 Ga for a thorianite from Ceylon on the simplifying assumptions that He was produced only by Ra and that the amount of Ra remained constant. If Th also produced He, and

Strutt noted that the evidence it did was strong, then the age was invalid. This objection did not apply, however, to minerals free of Th, such as a Cornish pitchblende (a U mineral), which gave a Ra–He age of 16.8 Ma (Strutt, 1905).[30]

In 1908 Strutt compared the He/U ratios of 13 samples of *phosphate nodules* and *phosphatized bone* as a function of geologic age. He selected these materials because they occurred in sedimentary rocks, and so their stratigraphic ages were known. He found that the ratios of He to U did not uniformly follow the known relative ages but also that high ratios did not occur in the younger samples. He hypothesized that He was imperfectly retained within the samples. Using assumptions similar to those in his 1905 calculations and Rutherford's latest estimate of the production rate of He from Ra, Strutt calculated minimum ages ranging from 0.225 Ma to 141 Ma for four of the samples.

In his 1905 lectures at Yale, Rutherford (1906: 192) also suggested that age calculations based on Pb might be superior to those based on He: "If the production of lead from radium is well established, the percentage of lead in radioactive minerals should be a far more accurate method of deducing the age of the mineral than the calculation based on the volume of helium, for the lead formed in a compact mineral has no possibility of escape." This possibility was first put to the test by Boltwood.

Boltwood (1907) designed an experiment to test the hypothesis that Pb was the disintegration product of U. He reasoned that if the hypotheses was true, then it followed that

1. Different types of unaltered U-bearing minerals of equal geologic age should have a constant ratio of Pb to U;

2. The Pb/U ratio in unaltered minerals of different age should be a function of age; and

3. Altered minerals should have a Pb/U ratio lower than that of unaltered minerals of the same age.

He compiled analyses of 43 U-bearing minerals from 10 different localities. His predictions were correct. The Pb/U ratios in unaltered minerals from the same locality (and presumably of the same age) were within a few percent of each other, and altered samples had ratios lower than did unaltered samples from the same locality. As for the variation of the Pb/U ratios with age:

It is beyond the writer's province to discuss the data bearing on the geological ages of the different deposits, but he is indebted to Professor Joseph Barrell of Yale University for the statement that, so far as the knowledge of the latter

---

30. Robert J. Strutt was the son of the third Baron Raleigh, John W. Strutt, who was awarded the Nobel Prize for physics in 1904. R. J. Strutt later succeeded his father as the fourth Baron Raleigh.

extends, the relative values of the ratios are not contradictory to the order of the ages attributed by geologists to the formations in which the different minerals occur (Boltwood, 1907: 83–84).

He concluded that the assumption that Pb is the final decay product of U appeared to be justified.

Boltwood was now in a position to calculate the ages of the minerals from their Pb/U ratios, but there was a difficulty: the rate of disintegration of U was not known. The rate of decay of Ra, however, had been estimated by Rutherford. If the decay of U to Ra to Pb was in equilibrium in these minerals, then in any given period of time the number of U atoms that decayed equaled the number of Ra atoms that decayed. Using Rutherford's estimate of the fraction of Ra that decays per year (2.7×10^{-4}) and the Ra/U ratio at equilibrium (3.8×10^{-7} g Ra/g U), and ignoring the small difference in the atomic weights of the two elements, Boltwood calculated that the fraction of uranium decaying per year was

$$2.7 \times 10^{-4} \times 3.8 \times 10^{-7}/\text{yr} = 1 \times 10^{-10}/\text{yr}$$

An estimate of the age of the minerals could be calculated from this approximate decay constant by the method suggested to Boltwood by Rutherford.[31]

$$\text{age in years} = \frac{\text{Pb}/\text{U}}{1 \times 10^{-10}/\text{yr}}$$

Using the average Pb/U ratio Boltwood calculated the ages for each locality (Table 2.3). Boltwood was cautious about the significance of these ages:

The actual values obtained for these ages are, of course, dependent on the value taken for the rate of disintegration of radium. When the latter has been determined with certainty, the ages as calculated in this manner will receive a greater significance, and may perhaps be of considerable value for determining the actual ages of certain geological formations (Boltwood, 1907, Holmes, 1911a).

In 1911 Arthur Holmes, about whom we shall hear more later, reevaluated Boltwood's data, added 17 new analyses of U-bearing minerals from near Oslo, Norway, and calculated U/Pb ages for nine localities (Table 2.3) by using an improved estimate of the decay constant:

$$\text{age in years} = \frac{\text{Pb}/\text{U}}{1.22 \times 10^{-10}/\text{yr}}$$

31. Even though they knew that radioactive decay was exponential, the early pioneers of radiometric dating invariably used simple linear formulae for their calcula-

TABLE 2.3

Comparison of Boltwood's and Holmes' Calculated Pb/U Ages
with the Geological Ages as Determined by Holmes

Locality	Geologic age	Pb/U age (Ma)	
		Boltwood, 1907	Holmes, 1911a
Glastonbury, Conn.	Carboniferous	410	340
Norway	Devonian	—	370
Spruce Pine, N.C.	pre-Carboniferous	510	410
Marietta, S.C.	pre-Carboniferous	460	
Branchville, Conn.	Silurian–Devonian	535	430
Sweden & Norway	Precambrian	1,300	1,025
Sweden & Norway	Precambrian	1,700	1,270
Texas	Precambrian	1,800	1,310
Colorado	Precambrian	1,900	1,435
Ceylon	Precambrian	2,200	1,640

N O T E : Except for that from the Devonian locality, the Pb/U data are from Boltwood (1907: 87). The difference between the two sets of ages is due primarily to the use of different decay constants.

Holmes (1911a: 255–56) concluded that

Wherever the geological evidence is clear, it is in agreement with that derived from lead as an index of age. Where it is obscure, as, for example, in connection with the pre-Cambrian rocks, to correlate which is an almost hopeless task, the evidence does not, at least, contradict the ages put forward. Indeed, it may confidently be hoped that this very method may in turn be applied to help the geologist in his most difficult task, that of unravelling the mystery of the oldest rocks of the earth's crust; and, further, it is to be hoped that by the careful study of igneous complexes, data will be collected from which it will be possible to graduate the geological column with an ever-increasingly accurate time scale.

Holmes devoted most of his career to the pursuit of these goals.

These first radiometric ages by Rutherford, Boltwood, Strutt, and Holmes were what are termed "*chemical ages*." They were calculated before *isotopes* were discovered, before the decay rates and intermediate products of U were known, and before it was discovered that Pb is also produced by the decay of Th. These factors combined to produce ages that were usually too high. For example, modern U–Pb isotopic ages of samples from the Glastonbury, Branchville, and Spruce Pine localities (Table 2.3) show that Boltwood's ages were excessive by 120 to 170 Ma, Holmes' by 20 to 65 Ma (Dalrymple and Lanphere, 1969: 19). Ages based on He were further complicated by He loss, a problem first recognized by Rutherford and one that renders U–He methods largely useless to this day.

tions. In view of the uncertainties about the methods and constants then, this simplification was appropriate.

For all their imperfections, these early and highly experimental mineral ages based on the decay of U were at least as firmly grounded in both theory and empirical evidence as those methods that relied on the cooling of the earth, the accumulation of salt in the ocean, or sedimentation and erosion. Although these first radiometric experiments did not attempt to date the Earth's origin, their importance to scientific thought about the age of the Earth cannot be overestimated. They were the first quantitative indication, based on physical principles rather than scientific intuition, that the Earth might be billions, rather than a few tens or hundreds of millions, of years old.

The discovery of radioactivity had made Kelvin's simple conductive-cooling calculations obsolete. That same phenomenon was now providing quantitative evidence that some geologists may have been right in insisting that the Earth was very, very old.

Not all geologists, however, greeted this new evidence for the vastness of geologic time and the antiquity of the Earth with equal enthusiasm. Many had reconciled their thinking to the shorter time scales and were highly skeptical of evidence to the contrary based on radioactivity. George Becker (1910c), for example, after finding that both cooling and sodium accumulation indicated that the Earth's age was 70 Ma or less, concluded that "this being granted, it follows that radioactive minerals cannot have the great ages which have been attributed to them." Becker was not alone, nor did the skepticism end quickly. F. W. Clarke (1924: 323) of the U.S. Geological Survey commented:

From chemical denudation, from paleontological evidence, and from astronomical data the age [of the Earth] has been fixed with a noteworthy degree of concordance at something between 50 and 150 millions of years. The high values found by radioactive measurements are therefore to be suspected until the discrepancies shall have been explained.

Some workers, however, were quick to see the potential value of radioactivity to the study of geologic time. Barrell used the radioactive age measurements available in 1917 to calibrate a detailed Phanerozoic time scale based otherwise on sediment thicknesses. That scale was remarkably close to the modern time scale. For example, Barrell (1917: 884–85) placed the base of the Cenozoic at 55 to 65 Ma, the base of the Mesozoic at 135 to 180 Ma, and the base of the Cambrian at 360 to 540 Ma (compare with Table 2.2).[32]

32. Arthur Holmes wrote the sections in Barrell's paper dealing with radiometric ages, but the time scale was Barrell's. This method of using semi-quantitative methods to interpolate between radiometric data in the construction of the geological time scale continues to this day, although at least for the Phanerozoic the ages of virtually all of the period and most of the epoch boundaries have been determined by radiometric dating.

In addition to providing methods of determining the ages of rocks, radioactivity also offered the possibility of calculating the age of the Earth from the relative abundances of the radioactive elements and their decay products. The first such calculation appeared in a paper by Henry N. Russell (1921), Professor of Astronomy at Princeton University, titled "A Superior Limit to the Age of the Earth's Crust." Joly had estimated that the proportion by weight of Ra in the crust was about 2.5×10^{-12}. Using an equilibrium ratio of U to Ra of 3.1×10^6, Russell estimated the proportion of U in the crust to be

$$2.5 \times 10^{-12} \times 3.1 \times 10^6 = 7 \times 10^{-6}$$

The proportion of Pb in the crust was, according to Clarke, 22×10^{-6}. By 1921 it was known that Pb was produced by the decay of both U and Th, with "half-periods" of 5×10^9 and 1.3×10^{10} yr, respectively. Russell calculated that the necessary 22×10^{-6} parts of Pb would be formed from the decay of U and Th in 8 Ga. This was an upper limit, however, because Pb may have been present initially in the crust and there was then speculation that U itself might be produced in the crust by the decay of some other element. The lower limit of the age of the crust, Russell noted, must be considerably greater than 1.1 Ga, which was the approximate age of the oldest Precambrian minerals that had been dated by their U/Pb ratio. He concluded

Taking the mean of this and the upper limit found above from the ratio of uranium to lead, we obtain 4×10^9 years as a rough approximation to the age of the Earth's crust. The radio-active data alone indicate that this estimate is very unlikely to be in error in either direction by a factor as great as three. Indeed, it might be safe to say that the age of the crust is probably between two and eight thousand millions of years (H. N. Russell, 1921: 86).

This estimate, Russell observed, was consistent with H. Jeffreys' 3 Ga for the age of the Solar System, which was based on entirely different data relating to the eccentricities of the present orbits of the planets.

In a popular 1927 booklet titled *The Age of the Earth: An Introduction to Geological Ideas*, Holmes revised Russell's calculation by using current estimates of crustal composition (pp. 71–72).[33] One million grams of the average crustal rock, according to Holmes, contained 7.5 gm of Pb, 6 gm of U, and 15 gm of Th. If all of the Pb in the crust is produced by the decay of U and Th, then the age can be calculated from Holmes' version of the age equation:

$$\text{age} = \frac{\text{Pb}}{\text{U} + 0.38\,\text{Th}} \times 7.4 \times 10^9\ \text{yr} \times \left(1 - \frac{X}{2} + \frac{X^2}{3}\right)$$

33. Holmes (1931) contains a nearly identical calculation with similar results.

where

$$X = 1.155 \times \frac{Pb}{U + 0.38\,Th}$$

and the quantities Pb, U, and Th are expressed in grams. The last term of the equation (in parentheses) is a factor to correct for the decrease in the amounts of U and Th over time because of decay. Substituting the quantities of Pb, U, and Th in the average crustal rock, we have

$$\text{age} = \frac{7.5}{6 + 0.38 \times 15} \times 7.4 \times 10^9 \times (0.740) = 3.629 \times 10^9 \, \text{yr}$$

Holmes did not actually produce this number in his paper but instead simply stated that the age of the Earth

is just over 3,000 million years. The age of the earth cannot exceed this figure, because some of the lead in rocks may be ordinary lead of atomic weight 207.2, and because the radio-active elements may have existed in the sun and have there generated lead before the earth was born. We should therefore expect the age of the earth to be nearer 1,600 million years than 3,000 million years (Holmes, 1927: 72).

The lower limit of 1,600 Ma Holmes obtained by adding the approximate length of an era (300 to 400 Ma) to the age of the oldest dated mineral (then 1,260 Ma, from Western Australia).

At the time Holmes published his 1927 booklet, there were only a handful of data on mineral ages in existence, and Holmes was able to summarize them all in a brief table (Holmes, 1927: 73–74) containing entries for only 23 localities. Yet these ages, ranging from 35 to 1,260 Ma, were so consistent with the geologically determined ages of the localities that they were difficult to doubt. In addition, if rocks in the Earth's crust were more than one billion years old, then Holmes' age of 1.6 to 3.0 Ga for the Earth was credible. In his final chapter Holmes tabulated the physical evidence for the age of the Earth (Table 2.4) and concluded that it was consistent with the age of 1.6 to 3.0 Ga based on radioactivity. Results from sodium accumulation and sediment thickness were relegated to insignificance; cooling calculations were not even listed. The methods so important to the pioneers in the search for the Earth's age had been rendered obsolete by the new evidence from the decay of radioactive elements—the short geological time scale was disproved once and for all. Any lingering doubts were put to rest in 1931 with the publication of a treatise on the age of the Earth by a committee of the National Research Council of the National Academy of Sciences (Subsidiary Committee on the Age of the Earth,

TABLE 2.4

Arthur Holmes' Summary of the Evidence for the Age of the Earth

Method	Result (Ma)
Eccentricity of the orbit of Mercury	1,000–5,000
Tidal theory of origin of Moon	<5,000
Journey of Solar System from Milky Way	2,000–3,000
Uranium and lead in average rocks	<3,000
Oldest analyzed radioactive minerals	>1,400
Sodium in the oceans	$n \times 300$ (where n is a multiplier such as 5)
Thickness of geological formations	incalculable
Cycles and revolutions in Earth's history	>1,400

SOURCE: Holmes, 1927: 76–77.

1931). Holmes (1931), Knopf (1931a,b) and others contributed papers. In the face of the data from radioactivity, the old methods were shown to be untenable.

The discovery between 1930 and 1950 of new radioactive elements and the measurement of their rates of decay provided the means for additional estimates of the maximum age of the Earth or, more exactly, of the Earth's matter. The methods using the new radioactive elements all followed more or less the general approach established by Russell, and involved estimating the crustal or earthly abundances of a radioactive isotope and its ultimate decay product, then calculating the time required for all of the product isotope to be generated by the decay of the parent. It was, however, a method based on Russell's original idea—the decay of U to Pb—that finally provided a precise value for the age of the Earth and Solar System. But the telling of that is left until Chapter 7.

CHAPTER THREE

Modern Radiometric Methods: How They Work

Geology needs an independent time clock that runs at a uniform rate, just as we need it in our daily life, and the physicist needs it in his laboratory.   G. D. LOUDERBACK (1936: 245)

The ages of meteorites and of rocks from the Earth and the Moon are measured by *radiometric dating*, a family of techniques based on the spontaneous decay of long-lived naturally occurring radioactive isotopes. These radioactive *parent* isotopes decay to stable *daughter* isotopes at rates that can be measured experimentally and are effectively constant over time regardless of physical or chemical conditions. Each parent–daughter pair constitutes an independent clock in which atoms of the parent isotope are transformed at a constant and predictable rate into atoms of its daughter isotope. In principle, then, the amounts of parent and daughter isotopes in a rock, along with the known rate of decay, provide the information necessary to calculate the time that has elapsed since the rock formed.

There are a number of long-lived radioactive isotopes used in radiometric dating (Table 3.1) and a variety of ways in which they are used. In addition, each of the various dating methods has unique characteristics that make it applicable to particular geologic circumstances and not to others.

In this chapter, the principal radiometric dating methods that provide the evidence for the age of the Earth will be explained briefly. It is important to understand, however, that because most of these methods have been the subjects of entire books, what follows is abbreviated and, in places, oversimplified. Anyone interested in a thor-

TABLE 3.1

*Principal Parent and Daughter Isotopes Used to
Determine the Ages of Rocks and Minerals*

Parent isotope (radioactive)	Daughter isotope (stable)	Half-life (Ma)	Decay constant (yr^{-1})	
^{40}K	$^{40}Ar^{a}$	1,250	5.81	$\times 10^{-11}$
^{87}Rb	^{87}Sr	48,800	1.42	$\times 10^{-11}$
^{147}Sm	^{143}Nd	106,000	6.54	$\times 10^{-12}$
^{176}Lu	^{176}Hf	35,900	1.93	$\times 10^{-11}$
^{187}Re	^{187}Os	43,000	1.612	$\times 10^{-11}$
^{232}Th	^{208}Pb	14,000	4.948	$\times 10^{-11}$
^{235}U	^{207}Pb	704	9.8485	$\times 10^{-10}$
^{238}U	^{206}Pb	4,470	1.55125	$\times 10^{-10}$

^{a} ^{40}K also decays to ^{40}Ca, for which the decay constant is 4.962×10^{-10} yr^{-1}, but that decay is not used for dating. The half-life is for the parent isotope and so includes both decays.

ough understanding of any particular method should consult one or more of the authoritative texts on radiometric dating methods.[1]

Radioactivity, Nature's Timepiece

Of the 339 isotopes of 84 elements found in nature, 269 are stable and 70 are radioactive. Eighteen of the radioactive isotopes have long *half-lives* and have survived since the elements of the Solar System were manufactured. These long-lived radioactive nuclides are the basis for radiometric dating. The remaining 52 radioactive isotopes have short half-lives but are continuously created by nuclear reactions in nature or by decay of the long-lived radioactive isotopes.[2] For example, ^{14}C (carbon) is continually created from ^{14}N (nitrogen) in the upper atmosphere by reaction with cosmic-ray neutrons, ^{234}Th results from the radioactive decay of ^{238}U and decays into ^{234}Pa (protactinium), and so forth.

In addition to the 339 naturally occurring nuclides, more than 1,650 isotopes of the elements have been created in the laboratory

1. In particular, the book by Faure (1986) is current, comprehensive, and covers all of the methods in use as well as the fundamentals. Faul (1966) and York and Farquhar (1972) are lighter fare and include the principal methods. For particular techniques, I recommend Dalrymple and Lanphere (1969) and Jäger and Hunziker (1979) for the K–Ar method, Faure and Powell (1972) for the Rb–Sr method, and Doe (1970) and Jäger and Hunziker (1979) for U–Th–Pb methods. The Sm–Nd method is reviewed by O'Nions et al. (1979) and DePaolo (1981, 1988).

2. There are some radioactive nuclides now found in nature that have been introduced by man-made nuclear reactions, especially those in nuclear weapons, but these nuclides are not included in the 70.

with nuclear reactors and particle accelerators. These are sometimes referred to as "artificial" nuclides, but there is nothing artificial about them. They don't exist in nature because they are radioactive with relatively short half-lives. Enough time has elapsed since the elements of the Solar System were created that these "artificial" nuclides have simply decayed out of existence. As we shall see in Chapter 8, the existence of nuclides with long half-lives and the absence of those with short half-lives provides some semi-quantitative information about the age of the elements in the Solar System.

Only certain combinations of N (number of *neutrons*) and Z (number of *protons*, = atomic number) yield stable nuclides; all other combinations are unstable. *Radioactive decay* is the process whereby an unstable nucleus either ejects or captures particles, transforming the radioactive *nuclide* into an isotope of another element. Sometimes the daughter nuclide is also unstable, in which case decay continues through as many steps as necessary to generate a stable nuclide.

There are a variety of ways in which radioactive decay occurs, but only three are important in radiometric dating (Fig. 3.1). The most common mechanism is *beta* (β^-) *decay*, in which the nucleus ejects a β^- particle, an energetic electron. The β^- particle is created by the breakup of a neutron into a proton, the β^- particle, and an *antineutrino*. The β^- particle and antineutrino are ejected from the atom entirely, while the proton remains in the nucleus. Thus, the number of protons in the nucleus is increased by one, the number of neutrons is decreased by one, and the atomic mass remains unchanged. As may be obvious, β^- decay is the mechanism by which nuclides with an excess of neutrons (or too few protons) tend toward stability.

A second type of decay that is more or less the opposite of β^- decay is *electron capture*, abbreviated e. c. As the name implies, in e. c. decay an electron from the innermost electron shell falls into the nucleus and a proton is converted into a neutron.[3] As with β^- decay, the atomic mass remains the same but the number of neutrons is increased by one at the expense of a proton (Fig. 3.1). Electron capture is not a common mode of decay, but it is one way in which a nuclide deficient in neutrons can increase N. The other way is by β^+ *emission*, in which a *positron* (the antimatter equivalent of an electron) is ejected from the nucleus and a proton is converted into a neutron. Decay by β^+ emission and e. c. occur primarily in the lighter isotopes. None of the isotopes used in radiometric dating decays by β^+ emission.

---

3. Because the electron involved in e. c. comes from the innermost, or k, shell, e. c. is also called k-capture decay. The captured electron is replaced by one from an outer shell.

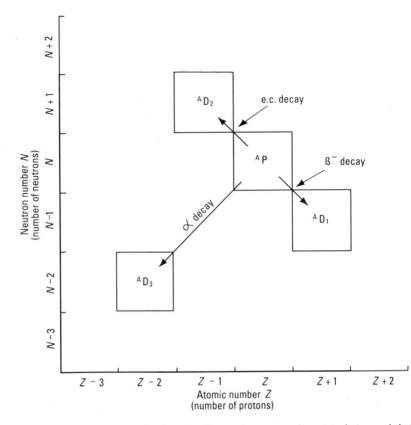

Fig. 3.1. The three types of radioactive decay relevant to radiometric dating and their effect on mass number *A*, atomic number *Z*, and neutron number *N*. P, parent isotope; D, daughter isotope.

The third type of decay important to radiometric dating is *alpha* (*α*) *decay*, which occurs primarily among the heavy elements. In *α* decay the nucleus ejects an *α* particle, a particle consisting of two protons and two neutrons. As a result of *α* decay, both the atomic number and the neutron number are decreased by two and the atomic mass is decreased by four (Fig. 3.1).

When some isotopes decay, the daughter nucleus is left with an excess of energy. This excess is usually shed immediately by emission of a *gamma* (*γ*) *ray*. In some isotopes, however, the energy is released at a later time by *internal conversion* (also called *isomeric transition*), which takes place either by emission of a gamma ray or by transfer of the excess energy to an orbital electron. Internal conversion in any particular isotope takes place with a characteristic half-life and in that respect it is a form of radioactivity, but no elemental transmutation occurs.

There are other types of radioactive decay that are unimportant to radiometric dating but are worth mentioning nonetheless. Some of the heavy elements decay by *spontaneous fission*, in which the nucleus of the parent isotope breaks into nuclei of various other elements plus neutrons. An example is ^{238}U, which may fission into one ^{133}Sn (tin) nucleus, one ^{103}Mo (molybdenum) nucleus, and two neutrons. Other nuclides can also be formed by ^{238}U fission. Other types of decay involve ejection of one or more neutrons, one or more protons, and various combinations of beta particles (positive or negative), neutrons, protons, and alpha particles. None of the isotopes used in radiometric dating (Table 3.1) decays in these ways, however, and so they will not be discussed further.

Even though the physical mechanisms governing radioactivity are complex and not completely understood, the mathematics used to predict how decay proceeds is simple and straightforward. Radioactive decay is most easily understood as a statistical process in which each atom of a given radioactive nuclide has exactly the same probability of decaying in some particular period of time as any other atom of that nuclide. That characteristic probability is known as the *decay constant*, λ, and is expressed as probability per unit time. Decay constants range from zero for a stable isotope to 1.0 for an isotope that decays instantaneously. For example, suppose that in a jar there are 100 atoms of an isotope with a decay constant of 0.1 yr^{-1} (10% per year). This means that *on the average* we can expect 10 of the atoms to have decayed by the end of the first year, 9 (10% of the remainder of 90) at the end of the second year, 8.1 (10% of the remainder of 81) at the end of the third year, and so on.

Because each atom decays independently of all others, the more atoms there are, the more decays occur, and so the number of atoms that decay per unit time is directly proportional to the number of atoms of that species present. For a given quantity of decaying atoms, as the number decreases, the fewer parent atoms remain to decay, so the rate decreases proportionately. Of course, there is a corresponding change in the rate of growth of the number of daughter atoms. Radioactivity is not the only phenomenon in which the rate of growth (or decrease, which is simply negative growth) is directly related to the size of the growing quantity. The growth of the world's population is another example—the more people there are, the faster the population grows, provided that the birth rate remains constant. This type of growth is sometimes called "organic growth" because so many organic processes exhibit it.

The formula that expresses the simple statistical principle that governs radioactivity and is the basis for all radiometric dating is

$$P_t = P_0 e^{-\lambda t} \tag{3.1}$$

where P_0 is the number of parent atoms at some starting time, P_t is the number of parent atoms at some later time t, and λ is the decay constant. This equation was deduced empirically by Rutherford and Soddy in 1902 (Chapter 2) from their experiments on the decay of ^{220}Rn, and by Egon von Schweidler in 1905 purely from statistical considerations (Dalrymple and Lanphere, 1969: 16–17).[4]

Because decay is a statistical process, it is not possible to tell when any particular atom will decay. For a small number of atoms, therefore, it is virtually impossible to determine the exact number of decays in a given time. In the case of the hundred-atoms-in-a-jar discussed above, 13 might decay the first year, 6 the second, 8 the third, and so on. For large numbers of atoms, however, the statistical uncertainty becomes vanishingly small, and Equation 3.1 can be used to predict, with great accuracy, the number of atoms that will decay in a given period of time. Fortunately, the numbers of atoms in even small amounts of matter are very large—as little as 0.00001 gram of potassium contains 150,000 trillion atoms!—so the statistical nature of radioactive decay is of no practical concern to the accuracy of radiometric dating.

Most people are familiar with the term *half-life*, $t_{1/2}$, another constant characteristic of each radioactive nuclide. The half-life is the time required for exactly one-half of any given quantity of parent atoms to decay. As one might expect, the half-life is related to the decay constant in a simple and straightforward way. The relationship can be found by setting P_t equal to one-half of P_0 and solving Equation 3.1 for t. The result is

$$t_{1/2} = \frac{\log_e 2}{\lambda} \tag{3.2}$$

We can now find the half-life of the hypothetical isotope represented by the 100 atoms with a decay constant 0.1 yr^{-1}.

$$t_{1/2} = \frac{0.693}{0.1 \text{ yr}^{-1}} = 6.93 \text{ yr}$$

This means that every 6.93 years we can expect one-half of the remaining atoms of the parent isotope to decay and a corresponding

4. For derivations of Equation 3.1 see Dalrymple and Lanphere (1969: 16) or Faure (1986: 38–39).

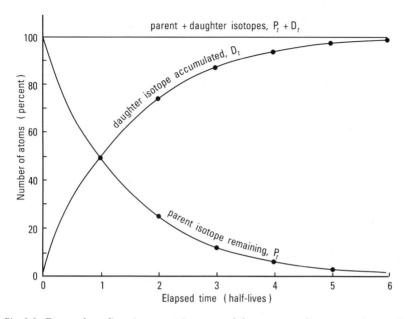

Fig. 3.2. Decay of a radioactive parent isotope and the corresponding accumulation of its daughter isotope. In a closed system, the sum of the parent and daughter isotopes at any time equals the original amount of the parent isotope.

number of daughter atoms to appear. Thus, 50% of the parent will remain at the end of 6.93 years, 25% at the end of 13.86 years, 12.5% at the end of 20.79 years, and so on. When the percentages of the parent and daughter isotopes are graphed as a function of the number of elapsed half-lives (Fig. 3.2) the result is two exponential curves—one showing the decrease of the parent and another the increase of the daughter. Provided that none of the parent or daughter atoms escapes, the sum of the two curves (parent + daughter) will always equal the original amount of the parent isotope. Another interesting feature about exponential decay is that for very large numbers of atoms the parent never completely disappears: it just keeps decreasing by half every half-life.[5] As a practical matter, however, most parent isotopes become extremely difficult to detect after about 10 to 15 half-lives have passed.

The half-lives of the more than 1,700 known radioactive nuclides range from a very small fraction of a second to billions of years. The half-lives and decay constants of the isotopes used for radiometric dating are given in Table 3.1.

5. Eventually, of course, only one atom will remain and that will also decay, but exponential decay is virtually a never-ending process for the numbers of atoms involved in nature.

As it stands, Equation 3.1 is not very useful for radiometric dating, because there is no way to determine the original amount of the parent, P_0. Fortunately, there is an easy solution to this problem because the sum of the parent atoms left, P_t, and the daughter atoms formed, D_t, must always equal the original number of parent atoms, P_0 (Fig. 3.2):

$$P_0 = P_t + D_t \qquad (3.3)$$

If we now substitute $P_t + D_t$ for P_0 in Equation 3.1, then

$$P_t = (P_t + D_t)e^{-\lambda t} \qquad (3.4)$$

and

$$D_t = P_t (e^{\lambda t} - 1) \qquad (3.5)$$

This equation can be solved for t, thus

$$t = \frac{1}{\lambda}\log_e\left(\frac{D_t}{P_t} + 1\right) \qquad (3.6)$$

which is the basic radiometric-age equation and contains only quantities that can be measured today in the laboratory.

If the rock or mineral to be dated contained none of the daughter isotope at the time of formation, then at any later time the age of the rock or mineral could be found from Equation 3.6 after D_t and P_t are measured. But if the rock incorporated some of the daughter isotope when it formed, then this initial amount of the daughter, D_0, would have to be subtracted from the total amount measured for the equation to yield the correct age:

$$t = \frac{1}{\lambda}\log_e\left(\frac{D_t - D_0}{P_t} + 1\right) \qquad (3.7)$$

where D_t is now the total number of daughter atoms present. At first glance it may seem that the requirement of knowing D_0 is a formidable limitation to the accuracy of radiometric dating. As we shall see, however, for the principal methods D_0 is either zero, negligible, or not required.

Is Decay Constant?

One of the principal requirements of a radiometric clock is that decay be constant and predictable from the time the clock was first "set," i.e. from the time the rock or mineral was formed, to the time the measurements are made. This involves the assumption that the rate at which decay in any given nuclide proceeds is either unalterable or that variations are negligible over a wide variety of physical and chemical conditions and for millions or billions of years. Just how reasonable is this assumption? The answer is that unless there has been some undiscovered change in the fundamental nature of matter and energy since the universe formed, the presumption of constancy for radioactive decay is, for all practical purposes, eminently reasonable.

There are two basic reasons why significant changes in rates of decay are not expected. First, the nuclei of atoms are extremely small and well insulated by their cloud of orbiting electrons. These electrons not only separate nuclei sufficiently that they cannot interact, they also provide a "shield" that prevents ordinary chemical or physical factors from affecting the nucleus. Chemical activity in an atom, for example, occurs almost entirely among the outermost electrons and does not involve the nucleus at all. Likewise the "compressibility" of a substance may result in slight changes in the configuration of electrons but has no effect on the nucleus.

Second, the energies involved in nuclear changes are 10^6 times greater than those involved in chemical activity and 10^4 to 10^5 times greater than the energies that bind the electrons to the nucleus. Chemical forces, which bind atoms together into molecules, are on the order of 1 electron-volt (eV), while the forces required to remove an electron from an atom are typically in the range of 10 to 100 eV. In contrast, the forces that hold nuclei together are on the order of 10^6 eV, and those that hold quarks, the elementary constituents of protons and neutrons, together are on the order of 10^9 eV (Weisskopf, 1983: 474). This is the reason why nuclear reactors and powerful particle accelerators are required to penetrate and make changes in atomic nuclei. Except in nuclear reactions, such energies are generally unavailable in natural processes such as those that form, change, and destroy rocks on the Earth and in the Solar System.

It was not long after the discovery of radioactivity that experiments were conducted to determine whether or not the rates of radioactive decay could be changed (Emery, 1972: 166, Hopke, 1974: 517). In 1907, Rutherford and a colleague, J. E. Petavel, placed a sample of radium "emanation" (^{220}Rn, radon) in a steel-encased cordite bomb. They observed no change in the activity of the sample even though the explosion generated an estimated temperature of 2,500°C and pressure of 1,013 bars. Madame Curie and M. Kamerlingh Onnes found in 1913 that lowering the temperature of a radium compound to the boiling point of liquid hydrogen (-252.8°C) did not change the radium activity by more than 0.05%, if at all. Other experiments involved varying gravity by measuring the rates on mountain tops and in the depths of mines or by whirling samples in a centrifuge; in yet others the samples were subjected to magnetic fields of as much as 8.3 teslas.[6] These early experiments and subsequent ones involving extremes of temperature, pressure, chemical state, and electric and magnetic fields have uniformly failed to induce any changes in the decay rates of a wide variety of alpha and beta emitters.

Although variations in the rates of alpha and beta decay have not been observed, theoretical considerations suggest that small changes may be possible. For a β^- particle to leave the atom, it must overcome the shielding effect of the negative electrical field caused by the orbiting electrons. Chemical interactions between atoms affect electron configuration and density and thus may change the strength of shielding, making it either easier or more difficult for a β^- particle to bid farewell. For example, ^{187}Re (rhenium) is normally radioactive and decays by β^- emission to ^{187}Os (osmium) with a half-life of 43 Ga (Table 3.1), but when stripped entirely of electrons the bare nucleus theoretically is stable (Emery, 1972: 195). This is an extreme case, however, that has never been produced in the laboratory. For theoretical reasons the maximum change expected because of chemical effects in a β^- emitter is on the order of 0.01%. Under certain environmental conditions the decay characteristics of ^{14}C, ^{60}Co (cobalt), and ^{137}Ce (cesium), all of which decay by β^- emission, deviate slightly from the ideal random distribution predicted by current theory (Anderson, 1972, Anderson and Spangler, 1973), but actual changes in the decay constants have not been detected.

The theoretical effects of external factors on rates of α decay are even smaller than those for β^- decay. For an α particle to leave the nucleus, it must overcome the Coulomb barrier, an electrical field surrounding the nucleus. Although this field is due primarily to charged particles (protons) within the nucleus, there is a small effect attributable to the orbiting electrons that can be changed by chemical and other environmental interactions. The maximum effects of this phenomenon on rates of α decay can be calculated but are exceedingly small. For example, the changes induced in the decay of ^{226}Ra and ^{147}Sm (samarium) by chemical effects are predicted to be less than 0.000015 and 0.0000017%, respectively (Emery, 1972: 194–95).

Changes in the rates of electron capture decay have been observed. This is not surprising because this type of decay involves an electron from the innermost electron shell. If the density of the electrons surrounding the nucleus could be changed, then we might ex-

6. For comparison, the Earth's magnetic field at the surface is about 5×10^{-5} Teslas.

pect a change in the probability that an electron would be captured by the nucleus. We might further expect that the effect is most noticeable in elements of low Z, i.e. in those elements whose innermost electrons are also the ones that interact with the outside world.

The possibility of such an effect was predicted in 1947 by Emilio Segre, who suggested that changes in the decay rate of ^{7}Be (beryllium) might be most easily induced (Emery, 1972: 166, Hopke, 1974: 517). ^{7}Be is an "artificial" isotope that decays to ^{7}Li (lithium) with a half-life of 54 days. It has only four electrons in the neutral state, and the two usually involved in chemical reactions are quite close to the nucleus. The activity of ^{7}Be has been both increased and decreased by combining it with other elements into different chemical compounds. For example, the decay rate of ^{7}Be in Be metal is 0.013 to 0.015% higher than in BeO. The maximum difference in activity yet found between any two Be compounds is 0.18%.[7] Although chemically induced changes in the electron capture decay rates of ^{89}Zr (zirconium) (0.08%) and ^{85}Sr (0.005%) have been reported (Emery, 1972: 180–81), ^{7}Be is the only isotope for which changes have been observed by more than one investigator (Hopke, 1974: 517).

An effect of pressure on the decay rate of ^{7}Be in BeO has also been reported (Hensley, Bassett, and Huizenga, 1973). A change in pressure of from 0 to 270 kbar resulted in a linear increase of 0.6% in the decay constant. Note that a pressure of 270 kbar is the pressure at a depth of about 750 km below the Earth's surface, a depth considerably greater than that of formation of any rock available to scientists for dating.

Changes in the rates of internal conversion decay have been induced in several isotopes by chemical combination, electric fields, pressure, and transitions to the superconducting state (Emery, 1972, Hopke, 1974). The maximum change reported is 5.7% for an excited state of ^{235}U. As mentioned above, however, internal conversion decay does not involve an elemental transmutation and so these changes are of no importance to radiometric dating.

In summary, both theory and experiment show that changes in α, β^-, and e. c. decay rates are not only rare but exceedingly small. Even the largest observed change of 0.18% in ^{7}Be would have a negligible effect on a calculated radiometric age. Also of importance is the fact that no changes have ever been detected in any of the isotopes used for dating and none of significance are theoretically expected. Of the physical and chemical processes that affect meteorites and rocks from the Earth and Moon, including pressure, temperature, gravity,

---

7. Emery (1972: 176–78) and Hopke (1974: 517) both tabulated the experimental results on ^{7}Be.

and magnetic and electric fields, none should affect radioactive decay to any significant degree. Thus, we can be confident that for all practical purposes the radiometric clocks used for geologic dating "tick" at unchanging rates. As discussed below, the assumption of constant decay is further strengthened by the consistency of dating results and thus is supported by evidence from the past as well as from the present.

Simple Accumulation Clocks

In its simplest form, radiometric dating relies on the accumulation with time of a daughter isotope, created by the radioactive decay of a specific amount of its parent, in a closed system.[8] This relationship is expressed by either Equation 3.6 or 3.7, depending on the absence or presence of an initial quantity of the daughter isotope. In principle, then, the age of a rock or mineral should be determinable by measuring the amount of parent and daughter isotopes present in a single sample for any of the several decay schemes listed in Table 3.1. In practice, however, the nearly ubiquitous presence of an unknown amount of initial daughter generally prevents the use of such simple accumulation methods. The exceptions are one particular parent–daughter pair (^{40}K and ^{40}Ar, potassium and argon) and special but rare cases of the other decay schemes. This "initial daughter" problem is easily solved by the application of *isochron* and similar methods that will be discussed later in the chapter. To understand these somewhat more esoteric variants of radiometric dating methods, however, it is useful to learn the simple accumulation methods, their uses, and their limitations.

The K–Ar Method

The K–Ar method, which is based on the decay of ^{40}K to ^{40}Ar, is probably the most commonly used radiometric dating technique available to geologists. An interesting feature of the radioactivity of ^{40}K is that this parent isotope undergoes decay by two modes: by electron capture to ^{40}Ar and by β^- emission to ^{40}Ca (Fig. 3.3). The ratio of ^{40}K atoms that decay to ^{40}Ar to those that decay to ^{40}Ca is 0.117, which is called the *branching ratio*. Because ^{40}Ca is the most abundant (97%) iso-

8. A *closed system* is one in which matter neither enters nor leaves. An *isolated system* is one in which neither matter nor energy enters or leaves. A system that is not closed or isolated is an *open system*. A "system" may be of any size, from very small, like a mineral grain, to very large, like the entire universe. For radiometric dating the system, usually a rock or some specific mineral grains, need only be closed to the parent and daughter isotopes.

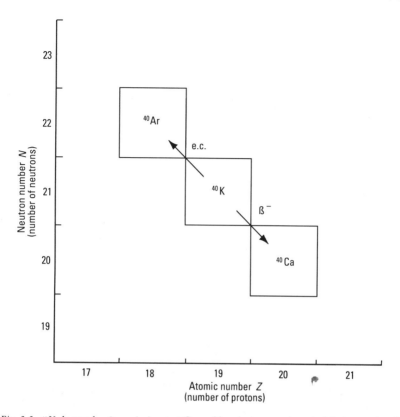

Fig. 3.3. ^{40}K decays by β^- emission to ^{40}Ca and by electron capture to ^{40}Ar in a ratio of 8.5 to 1. Only the decay to Ar is used for radiometric dating.

tope of Ca and because Ca is both abundant and nearly ubiquitous in rocks and minerals, it is usually not possible to determine the amount of ^{40}Ca present initially, and the ^{40}K–^{40}Ca method is therefore rarely used for dating.

The K–Ar method is the only decay scheme that can be used with little or no concern for the initial presence of the daughter isotope. This is because ^{40}Ar is an inert gas that does not combine chemically with any other element and so escapes easily from rocks when they are heated. Thus, while a rock is molten the ^{40}Ar formed by the decay of ^{40}K escapes from the liquid. After the rock has solidified and cooled, the radiogenic ^{40}Ar is trapped within the crystal structures of the minerals like a bird in a cage and accumulates with the passage of time. If the rock is heated or melted at some later time, then some or all of the ^{40}Ar may escape and the K–Ar clock is partially or totally reset. Some cases of initial ^{40}Ar remaining in rocks have been docu-

TABLE 3.2

Natural Abundances of the Isotopes Used in Radiometric Dating

Isotope	Abundance (%)	Isotope	Abundance (%)
^{39}K	93.26	^{36}Ar	0.337
^{40}K	0.0117	^{38}Ar	0.063
^{41}K	6.73	^{40}Ar	99.60
^{85}Rb	72.17	^{84}Sr	0.56
^{87}Rb	27.83	^{86}Sr	9.87
		^{87}Sr	7.04
		^{88}Sr	82.53
^{144}Sm	3.0	^{142}Nd	27.3
^{147}Sm	14.9	^{143}Nd	12.3
^{148}Sm	11.2	^{144}Nd	23.8
^{149}Sm	13.8	^{145}Nd	8.3
^{150}Sm	7.4	^{146}Nd	17.1
^{152}Sm	26.8	^{148}Nd	5.7
^{154}Sm	22.9	^{150}Nd	5.6
^{175}Lu	97.4	^{174}Hf	0.17
^{176}Lu	2.6	^{176}Hf	5.2
		^{177}Hf	18.5
		^{178}Hf	27.2
		^{179}Hf	13.8
		^{180}Hf	35.1
^{185}Re	37.40	^{184}Os	0.02
^{187}Re	62.6	^{186}Os	1.6
		^{187}Os	1.6
		^{188}Os	13.3
		^{189}Os	16.1
		^{190}Os	26.4
		^{192}Os	41.0
^{232}Th	100.0	^{204}Pb	1.4
		^{206}Pb	25.2
^{234}U	0.0057	^{207}Pb	21.7
^{235}U	0.72	^{208}Pb	51.7
^{238}U	99.27		

NOTE: Abundances are for the Earth's crust except for argon, which is for the atmosphere. The isotopic abundances for those elements that include a daughter isotope vary because of decay of the corresponding parent isotope. The isotope pairs used in radiometric dating are indicated by arrows.

SOURCES: Lederer, Holland, and Perlman, 1967; Faure, 1986.

mented but they are uncommon, as noted by Dalrymple and Lanphere (1969: 121–44), who also described studies of historic lava flows showing that "excess" argon is rare in these rocks.

Another reason why the K–Ar method is so useful to geologists is that potassium is common in rocks and there are a number of minerals in which it is a principal element. Thus, even though ^{40}K is the least abundant isotope of potassium (Table 3.2), it occurs in measurable quantities in most rocks. As a result, the K–Ar method enjoys a wide range of applicability.

A third beneficial feature of the K–Ar method is the nearly ideal half-life of ^{40}K, 1.25 Ga. This half-life is long enough that the "clock" is still running in a rock formed 4.5 Ga, i. e. the quantity of parent has not become vanishingly small even in the Solar System's oldest rocks, and short enough so that measurable quantities of ^{40}Ar accumulate in as little as 50 ka or so.[9]

Because newly solidified rocks rarely retain initial ^{40}Ar, Equation 3.6 applies to the K–Ar method, but it must be modified to take into account the branching decay of ^{40}K:

$$t = \frac{1}{\lambda_{ec} + \lambda_{\beta^-}} \log_e \left(\frac{^{40}\text{Ar}}{^{40}\text{K}} \frac{\lambda_{ec} + \lambda_{\beta^-}}{\lambda_{ec}} + 1 \right) \qquad (3.8)$$

where λ_{ec} and λ_{β^-} are the decay constants of the e. c. and β^- decays, respectively. Note that this modification involves only the constants and is not a change in the fundamental nature of the age equation. If we now substitute the values of the decay constants from Table 3.1 into Equation 3.8, the K–Ar age equation becomes

$$t = 1.804 \times 10^9 \log_e \left(9.54 \frac{^{40}\text{Ar}}{^{40}\text{K}} + 1 \right) \qquad (3.9)$$

and t will be in years.

In the process of analysis, a correction must be made for the atmospheric Ar present in most minerals and in the experimental apparatus.[10] This correction is easily made by measuring the amount of ^{36}Ar in the sample to be dated and, using the known isotopic composition of atmospheric argon (Table 3.2), subtracting the appropriate amount of ^{40}Ar due to atmospheric contamination. What is left is the daughter or radiogenic ^{40}Ar. This correction can be made very accurately and has no appreciable effect on the calculated age unless the atmospheric argon is a very large percentage of the total. The uncertainty due to this correction is taken into account when the error to the calculated age is assigned.

Like all radiometric methods, the K–Ar method does not work on all rocks and minerals under all geologic conditions. By many experiments over the past three decades, geologists have learned which rocks and minerals act as closed systems and under what geologic conditions they do so. The K–Ar clock works particularly well on *ig-*

9. The younger limit depends on the potassium content of the sample and other factors. High-potassium minerals as young as 6 ka have been successfully dated (Dalrymple, 1969).

10. Approximately 1% of the Earth's atmosphere is Ar, of which 99.6% is ^{40}Ar. Most of the ^{40}Ar is due to the decay of ^{40}K in the Earth over time and its subsequent release into the atmosphere.

neous rocks that have not been heated significantly since their formation. It does not work on most sedimentary rocks because these rocks are composed of the debris from older rocks. It also does not work well on many *metamorphic* rocks because rocks of this type form from other rocks under heat and pressure but without undergoing complete melting. Many metamorphic rocks have complex histories involving several heatings that were not quite sufficient to melt the rock but were sufficient to release some of the Ar within it. As a result, it is difficult to know when or if the K–Ar clock in a metamorphic rock was last completely reset.

Because of its susceptibility to resetting by later heating, the K–Ar method is of limited use for measuring the ages of meteorites, lunar rocks, and the oldest rocks from the Earth. The $^{40}Ar/^{39}Ar$ variant of the method (described later), however, overcomes many of the problems of postformation heating and provides important data concerning the age of bodies in the Solar System.

The Rb–Sr Method

The Rb–Sr method is based on the radioactivity of ^{87}Rb (rubidium), which undergoes simple β^- decay to ^{87}Sr (strontium) with a half-life of 48.8 Ga (Fig. 3.4). Rb is a major constituent of very few minerals and never forms minerals of its own, but the chemistry of Rb is similar to that of K and Na, both of which do form many common minerals, and so Rb occurs as a trace element in most rocks. Sr, like Rb, occurs in minerals primarily as a trace element. Its chemistry is similar to that of Ca, a major element in rocks for which Sr may substitute.

Under the right circumstances, Equation 3.6 can be used for calculating Rb–Sr ages, and when the ^{87}Rb decay constant (Table 3.1) is substituted for λ, the Rb–Sr age equation is

$$t = 7.042 \times 10^{10} \log_e \left(\frac{^{87}Sr}{^{87}Rb} + 1 \right) \qquad (3.10)$$

Unlike Ar, which escapes easily and entirely from most molten rocks, Sr is present as a trace element in most minerals when they form. For this reason, accurate Rb–Sr accumulation ages can be calculated with Equation 3.10 only for those rare minerals that are high in Rb and contain a negligible amount of initial Sr. For most rocks, however, initial Sr is present in significant amounts. An equation like 3.7, which allows for the initial daughter isotope, is generally of little use in the Rb–Sr method because the amount of initial ^{87}Sr cannot be determined. So Rb–Sr dating is done primarily by the isochron method, which completely eliminates the problem of initial Sr.

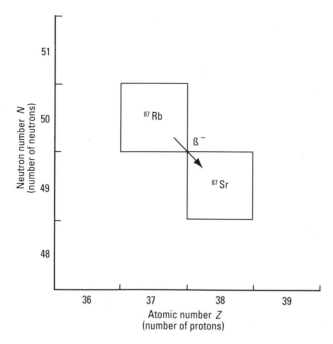

Fig. 3.4. ^{87}Rb decays to ^{87}Sr by β^- emission.

Because of the long half-life of ^{87}Rb, Rb–Sr dating is used mostly on rocks older than about 50 to 100 Ma, for only in these rocks has sufficient time elapsed for measurable quantities of radiogenic ^{87}Sr to accumulate. The method is very useful on rocks with complex histories because the daughter element, Sr, is chemically bound within the crystals and does not escape from minerals nearly as easily as Ar. As a result, a sample can obey the closed-system requirements for Rb–Sr dating over a wider range of geologic conditions than can a sample for K–Ar dating. Because of its relative resistance to postformation events, Rb–Sr dating (mostly by the isochron method) is used extensively to determine the ages of the oldest rocks in the Solar System.

The Sm–Nd, Lu–Hf, and Re–Os Methods

Since about 1980, improvements in the precision and sensitivity of analytical techniques have permitted geochronologists to exploit three decay schemes that were previously of little value. These include the decay of ^{147}Sm to ^{143}Nd (neodymium), of ^{176}Lu (lutetium) to ^{176}Hf (hafnium), and of ^{187}Re to ^{187}Os (Table 3.1). The combination of the long parental half-lives and the very low natural abundances of the parent isotopes means that the daughter isotopes accumulate very slowly. Nevertheless, these three dating methods have some unique

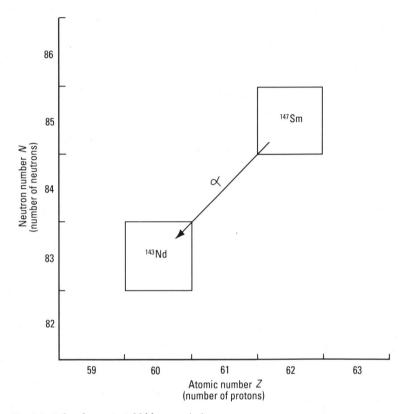

Fig. 3.5. ^{147}Sm decays to ^{143}Nd by α emission.

advantages not found in the more commonly used methods. In recent years, a considerable number of age measurements by the Sm–Nd method have appeared in the literature, and the method has become a common tool for geochronologic studies of old rocks and meteorites. The Lu–Hf and Re–Os methods are still infrequently used but, as we shall see in later chapters, provide some valuable data relevant to the age of meteorites.

Sm and Nd are two of the 14 *rare earth* or lanthanide elements, a group of metals of atomic numbers 57 (lanthanum, La) through 71 (lutetium, Lu) that are chemically very similar. ^{147}Sm decays by α emission to ^{143}Nd with a half-life that exceeds those of the other isotopes used in radiometric dating by a factor of more than two (Fig. 3.5, Table 3.1). Although there are minerals in which the rare earths are principal elements, they are uncommon. On the other hand, the rare earth elements do occur in trace amounts in nearly all rocks and minerals, although their concentrations are usually only a few parts per million or less.

Because both parent and daughter are rare earth elements, natural geochemical processes do not favor their separation, and variations in the concentrations of Sm and Nd of more than a factor of two are uncommon. This means that significant quantities of initial Nd are present in all samples and equations 3.6 or 3.7 must be abandoned in favor of the isochron method. Another consequence of the geochemical similarity between Sm and Nd is that the Sm–Nd method is more resistant to metamorphism than are other dating methods. This is a decided advantage for age measurements of very old rocks. Another advantage of this method is that it can be used to determine the ages of certain rocks that are difficult to date by other methods. In particular, ancient basalt and the achondrite meteorites contain so little K, Rb, and U that precise dating can be done only with the Sm–Nd method.

Lu, like Sm and Nd, is a rare earth element, but its daughter, Hf, is not. The Lu–Hf method is based on the β^- decay of ^{176}Lu to ^{176}Hf with a half-life of 35 Ga (Fig. 3.6, Table 3.1). At one time, ^{176}Lu was thought to undergo branching decay, with about 3% of the decay occurring by e. c. to ^{176}Yb (ytterbium), but subsequent measurements indicate that e. c. decay of ^{176}Lu probably does not occur (Lederer, Hollander, and Perlman, 1967: 103).

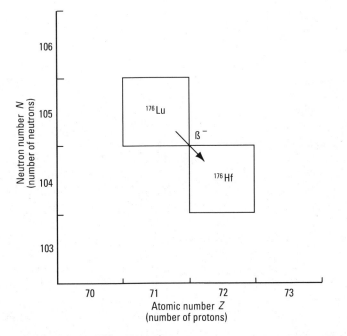

Fig. 3.6. ^{176}Lu decays to ^{176}Hf by β^- emission.

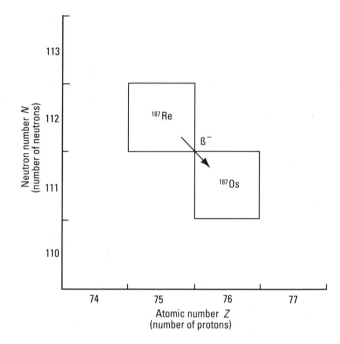

Fig. 3.7. ¹⁸⁷Re decays to ¹⁸⁷Os by β^- emission.

Like the other rare earth elements, Lu occurs as a trace element in most rocks but usually in concentrations of less than one part per million. Exceptions include some lunar rocks in which Lu concentrations may reach several parts per million. Minerals high enough in Lu and low enough in Hf that the Lu–Hf method can be applied as a simple accumulation clock are very rare, so the isochron method is more useful.

The Re–Os method is based on the β^- decay of ^{187}Re to ^{187}Os with a half-life of 43 Ga (Fig. 3.7). Both Re and Os are metals whose average abundances in igneous rocks are on the order of only 0.5 parts per billion. Because of the extremely low abundance of the parent isotope, the method is generally not applicable to common rocks and minerals. Re, however, is chemically similar to Mo, and like that common metal is often concentrated in areas of Cu mineralization. Thus, the method has been used to date some ore deposits that contain molybdenite, which is also low in the platinum metals, including Os. Although of little use with most rocks, the Re–Os method has proven of value in dating the metallic phases of meteorites, where the concentration of Re may be as much as 0.5 parts per million but that of the other parent isotopes listed in Table 3.1 may be negligible.

The age equations for the Sm–Nd, Lu–Hf, and Re–Os methods are of the form of equations 3.6, 3.7, and 3.10.

The U–Th–Pb Methods

These methods are based on the radioactivity of ^{235}U, ^{238}U, and ^{232}Th, all of which decay to different isotopes of Pb (Tables 3.1 and 3.2). These three decays differ from the others because they involve a *decay series* with intermediate radioactive daughter products rather than simple or branching decay. The decay series of ^{238}U, for example, includes 13 principal intermediate daughter isotopes of eight elements between the parent and the stable end-product daughter, ^{206}Pb (Fig. 3.8). Similarly, the decay series of ^{235}U and ^{232}Th include ten and eight principal intermediate daughter isotopes, respectively. In addition to the principal intermediate daughter isotopes, each of the three decay series includes several quantitatively minor isotopes formed by branching decay of some of the intermediate daughters.

In spite of the complexities seemingly introduced by the existence of intermediate daughter products, the decays of ^{238}U to ^{206}Pb, ^{235}U to ^{207}Pb, and ^{232}Th to ^{208}Pb can be treated as if they were a simple one-step decay. The intermediate members of the series can be ignored. This is because of two fortunate circumstances. The first is that each of the three series is entirely independent of the others, i.e. none of the intermediate daughter isotopes occurs in more than one series. Thus, each series ultimately results in a unique isotope of Pb. The second is that the half-lives of the intermediate daughters are very much shorter than those of the three parents, so that *secular equilibrium* is established very shortly after a new rock is formed.[11] Once secular equilibrium is established within a closed system, the number of atoms of each intermediate member of the series is inversely proportional to its decay constant. This can be expressed mathematically as

$$N_1\lambda_1 = N_2\lambda_2 = N_3\lambda_3 = N_4\lambda_4 \ldots \qquad (3.11)$$

where N_x represents the number of atoms of the isotope and λ_x is its decay constant. The ultimate result of equilibrium is that the production rate of the stable daughter (in this case Pb) is exactly equal to the decay rate of the primary parent (U or Th). The length of time necessary for the various intermediate daughters to reach secular equilibrium within the decay series can actually be used as a dating tool for very young rocks, but these *disequilibrium* dating methods will not be discussed because they have no bearing on the age of the Earth.

---

11. In the ^{238}U series, the longest intermediate half-life is 0.25 Ma (^{234}U), the shortest 0.00016 s (^{214}Po). For the ^{235}U series, the corresponding values are 33 ka (^{231}Pa) and 0.0018 s (^{215}Po); for the ^{232}Th series, 5.8 yr (^{228}Ra) and 0.0000003 s (^{212}Po).

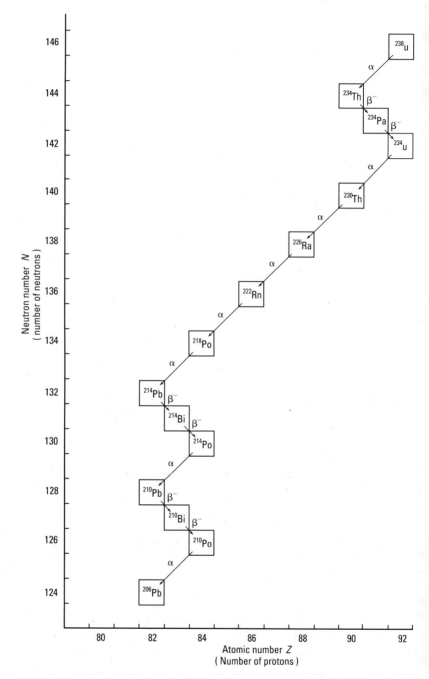

Fig. 3.8. The main decay series of ^{238}U. Not shown are some minor (<1%) branches in the chain beginning at ^{218}Po that also result in stable ^{206}Pb.

In Chapter 2 we learned of some of the early attempts to use the decay of U to the inert gas He as a dating method. The He atoms produced are, in fact, the α particles ejected during the decay of the parent and the intermediate daughters. If you look at Figure 3.8, you will see that the decay series of ^{238}U to ^{206}Pb entails eight α decays and six β^- decays. This means that the decay of one atom of ^{238}U to one atom of ^{206}Pb generates eight atoms of helium and six free electrons. This can be described concisely by

$$^{238}U \rightarrow \, ^{206}Pb + 8^4He + 6\beta^- + energy \qquad (3.12)$$

In a similar manner, the decay of a ^{235}U atom results in seven atoms of ^{4}He and four β^- particles; that of a ^{232}Th atom results in six atoms of ^{4}He and four β^- particles. The production of He cannot be used reliably for dating because a He atom is very tiny and, unlike the larger Ar atom, escapes easily from rocks and minerals. Note, however, that the escape of He has no effect on the accumulation of Pb because the He is a by-product and not part of the decay series.

U and Th are both members of the actinide group of elements, which includes all elements with atomic numbers from 89 to 103. Members of this group are chemically very similar, so U and Th can substitute extensively for each other in minerals. There are a number of minerals in which Th and U are principal elements, but these occur primarily in ore deposits. Many common rock-forming minerals contain these elements but the amounts are usually a few parts per million or less. There are a few common minerals, however, that contain much larger amounts of U and Th. These minerals, of which *zircon* is the most common, do not occur in large volumes but do occur in small amounts in many rocks. Zircon is also exceedingly low in initial Pb, so the U–Pb and Th–Pb methods can be applied to this mineral, and a few others, using Equation 3.6 with little concern for initial Pb.

Three simple independent age calculations can be made from the three U–Pb and Th–Pb decays. If these age calculations agree, then that age represents the age of the mineral. More often than not, however, the three ages do not agree. This is primarily because Pb is a volatile element and is lost easily if the mineral is reheated at some later date. The problem of Pb loss can be minimized by calculating an age based on the ratio of any pair of the three Pb isotopes. Most often, the ratio of ^{207}Pb to ^{206}Pb is used so that the slight differences in the chemical behavior of U and Th are eliminated. The equation for this age calculation is

$$\frac{^{207}Pb}{^{206}Pb} = \frac{1}{137.88} \frac{e^{\lambda_2 t} - 1}{e^{\lambda_1 t} - 1} \qquad (3.13)$$

where 137.88 is the constant ratio of ^{238}U to ^{235}U in rocks of the Moon, the Earth, and meteorites today and λ_1 and λ_2 are the decay constants of ^{238}U and ^{235}U, respectively. This equation cannot be solved algebraically for t, but a unique value for t can be found by calculating various values for the ratio $^{207}Pb/^{206}Pb$ as a function of t until there is agreement between the calculated ratio and the measured ratio. This equation is the starting point for the method that gives the most precise value for the age of the Earth and will be discussed further in Chapter 7.

Because of the problems of initial Pb and Pb loss the U–Th–Pb methods of dating are most often applied with the aid of *isochron* or *concordia* diagrams, both of which circumvent these problems and are discussed below. We will, therefore, leave the U–Th–Pb methods for now but will return to them after we have discussed the isochron method in its simplest form.

Age-Diagnostic Diagrams

As we have seen in the preceding discussion, most of the simple accumulation methods are encumbered with two requirements: (1) that the amount of initial daughter be known, and (2) that the rock has remained a closed system since formation. But how can we find the quantity of initial daughter or determine that the system has been open or closed? These seemingly formidable problems can be solved by the use of simple graphic devices that we will collectively call *age-diagnostic diagrams*. These diagrams and their mathematical equivalents not only provide an age, but some provide an exact measure of the initial daughter, some provide an age for systems that have not remained closed, and all are self-checking. They are especially useful on old rocks with complex histories, and are the basis of most of the radiometric evidence for the age of the Earth. They are important and worth the time needed to understand how they work.

There is a rather wide variety of age-diagnostic diagrams, each used for a particular purpose. In the remainder of this section, only those diagrams that are especially useful in determining the age of the Earth will be discussed.

Simple Isochrons

The *isochron diagram, correlation diagram*, or *isotope evolution diagram* is a device of magnificent power and simplicity. The most common form of isochron diagram was conceived in 1961 by L. O. Nicolaysen of the Bernard Price Institute of Geophysical Research, University of Witwatersrand, South Africa, who applied it to Rb–Sr data and suggested that it could be used for U–Pb data as well (Nicolaysen, 1961). It is now a widely used geochronological tool, applicable in one form or another to all of the decay schemes used for radiometric dating.

The isochron method has two significant advantages over the simple accumulation clock. First, it circumvents the problem of the amount of the initial daughter. That information need not be known—it is one of the answers provided by the method. Second, the method is self-checking, providing the user with information about the degree to which the sample has behaved as a closed system.

As the name implies, an isochron is a line of equal time. It is obtained by analyzing several minerals from the same rock, or several rocks that formed from the same source at the same time but with differing amounts of the parent and daughter elements. On a simple graph, the amount of the parent isotope is plotted on the abscissa (x axis) and the amount of the daughter isotope is plotted on the ordinate (y axis), both values being normalized to (divided by) the amount of a nonradiogenic isotope of the same element as the daughter. If the samples have been closed systems since they formed, the points will fall on a line whose slope is a function of the age of the rock. The intercept of the line on the abscissa gives a measure of the initial daughter. At the moment, this description of the isochron may seem a bit cryptic, but the method is really quite simple, as the next few paragraphs will make clear.

The trick to the isochron diagram is the normalization of both parent and daughter isotopes to a third isotope of the daughter element. To see exactly what normalization does and how the isochron works, let us first consider what happens when the data are not normalized and consist solely of the amounts of the parent and daughter isotopes, using the Rb–Sr decay scheme to illustrate.

Suppose that we separate three minerals, P, Q, and R, from newly formed rock ($t = 0$), determine their contents of ^{87}Rb and ^{87}Sr, and plot this information on a graph. For simplicity, imagine that the results can be expressed in small numbers of atoms. Mineral P is low in both Rb and Sr, mineral Q is low in Rb and high in Sr, and mineral R is high in Rb and low in Sr, so the graph looks like Figure 3.9a. At $t = 0$, the positions of the points on this graph depend only on the amounts of Rb and Sr in the minerals, i.e. they depend on the chemical composition of the minerals. If we reanalyze these minerals after letting them sit for some length of time, until their age is t', the points on our graph will have moved because of the decay of ^{87}Rb to ^{87}Sr. In each sample, the decay of one atom of ^{87}Rb results in an increase of exactly one atom of ^{87}Sr, and so the points will have moved along tra-

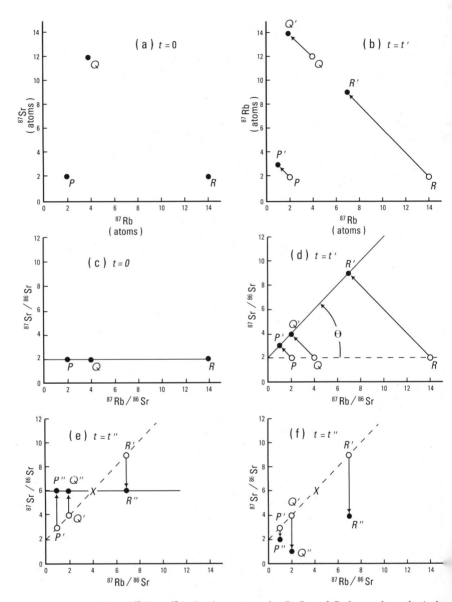

Fig. 3.9. (a) Plot of ^{87}Rb vs ^{87}Sr for three minerals, P, Q, and R, from a hypothetical rock of zero age. (b) Because of the decay of ^{87}Rb, points P, Q, and R move along trajectories of decreasing ^{87}Rb and increasing ^{87}Sr to P', Q', and R' after passage of time t'. The amount of movement is proportional to the ^{87}Rb content of the minerals, but this type of plot gives no information about the age of the rock. (c) The same data at $t = 0$ but normalized to ^{86}Sr. (d) After time t' has passed, the points still fall on a line, an isochron, whose slope is a function of age. (e) Complete resetting of the Rb–Sr clock at time t'' moves the points to a new "zero-age" isochron. The composition of the total rock is indicated by x. (f) Partial resetting, in this example due to loss of ^{87}Sr, results in the points scattering.

jectories of 45° toward decreasing ^{87}Rb and increasing ^{87}Sr (Fig. 3.9b). Since the number of atoms of ^{87}Rb that decay in any given period of time is proportional to the number present, the distance the points move along the trajectories is a direct function of their ^{87}Rb content. Thus, if P decreases by one atom of ^{87}Rb, it will increase by one atom of ^{87}Sr and end up at P'. Point Q will move two units to Q' because it has twice the ^{87}Rb content of P. Point R has seven times as much ^{87}Rb as P so it will move seven times farther along its trajectory.

What information does a diagram like that in Figures 3.9a and 3.9b provide? Not much. If we had all six of the values shown in Figure 3.9b then we could calculate an age for the rock because we would know how far each point had moved along its trajectory, and that distance is a function of time. The fact is, however, that because we cannot determine P, Q, and R, only P', Q', and R', we have no way of calculating the age, t'. In short, we would be stymied by the initial daughter problem.

Let us see what happens when these same hypothetical data are normalized. Normalization converts the isotopic data into ratios by measuring the amounts of the parent and daughter isotopes relative to that of another stable isotope of the daughter element, in this case Sr. Any of the available Sr isotopes could be used (Table 3.2), but there is some advantage in using an isotope as close as possible to the daughter in both mass and natural abundance, and by convention ^{86}Sr is universally used for normalization. For simplicity, let us assume that for every two atoms of ^{87}Sr in sample P there is one atom of ^{86}Sr. From Figure 3.9 we can see that there are two atoms of ^{87}Rb in sample P, so both ratios ^{87}Sr/^{86}Sr and ^{87}Rb/^{86}Sr for sample P are equal to 2. If the ratio ^{87}Sr/^{86}Sr is 2 for sample P, then it must also be 2 for samples Q and R, so the data plot on a straight, horizontal line with ^{87}Sr/^{86}Sr equal to 2 for all values of ^{87}Rb/^{86}Sr (Fig. 3.9c). The value of ^{87}Sr/^{86}Sr when ^{87}Rb/^{86}Sr is zero, i.e. the intersection of the line on the ordinate, is the initial isotopic ratio of ^{87}Sr to ^{86}Sr in the rock—in this case 2.

Why are the Sr isotopic ratios the same even though the Rb and Sr contents of the minerals vary? The reason is that when a rock forms, all of the isotopes of any given element in the rock are homogenized. Consider, for example, the formation of a new rock by melting of an old rock. When the old rock is melted, all of the constituents, isotopes and elements, are thoroughly mixed by convection of the liquid and diffusion of the atoms throughout the melt. As the new rock begins to solidify, the minerals crystallize according to definite chemical rules that are governed by the composition, temperature, and pressure of the melt. Some minerals form early and extract from

the liquid those elements that constitute their crystals. This early crystallization changes the elemental composition of the remaining liquid,
which leads to the formation of other minerals of different composition. Mineral species that form early in the process tend to incorporate more Sr than Rb, whereas those that form late tend to be high in
Rb and low in Sr. Even though the relative amounts of Rb and Sr vary
from mineral to mineral, however, the chemical processes of crystallization do not fractionate isotopes of the same element, so the isotopic
composition of the Sr and Rb is the same in all of the minerals. This
means that for a rock whose age is zero, the ratios of ^{87}Sr to ^{86}Sr in all
of the minerals will be identical, whereas the ratios of ^{87}Rb to ^{86}Sr will
vary from one mineral species to another.

As our hypothetical rock ages, the isotopic compositions of the
samples P, Q, and R will move along their respective trajectories as
before (Fig. 3.9d). Each decay of an atom of ^{87}Rb results in the addition of an atom of ^{87}Sr, so as the ratio ^{87}Rb/^{86}Sr decreases there is a
corresponding increase in ^{87}Sr/^{86}Sr, and the magnitude of change for
any given period of time is a function of the Rb content, or in this case
of the ratio ^{87}Rb/^{86}Sr. With our normalized data, however, the points
P', Q', and R' will always fall on a straight line whose slope is a direct
function of the age of the rock, i.e. the older the rock, the steeper the
slope of the line. Furthermore, this line will always intersect the ordinate at the value of the initial isotopic composition of Sr. This must be
so because ^{87}Rb is zero at this intersection, so there can be no increase
in ^{87}Sr over time. Thus, the isochron method gives both the age and
the initial amount of daughter isotope in a rock solely from its current
isotopic composition.

The equation for the isochron diagram is based on a simple
modification of Equation 3.5. For the Rb–Sr system Equation 3.5 is

$$^{87}\text{Sr} = (^{87}\text{Sr})_0 + (e^{\lambda t} - 1)\,^{87}\text{Rb} \tag{3.14}$$

where ^{87}Sr and ^{87}Rb are the total amounts at time t and $(^{87}\text{Sr})_0$ is the
amount of initial ^{87}Sr at $t = 0$. We can normalize the isotopic values in
this equation by dividing all terms by the constant ^{86}Sr:

$$\frac{^{87}\text{Sr}}{^{86}\text{Sr}} = \left(\frac{^{87}\text{Sr}}{^{86}\text{Sr}}\right)_0 + (e^{\lambda t} - 1)\,\frac{^{87}\text{Rb}}{^{86}\text{Sr}} \tag{3.15}$$

This is the equation of a straight line of the form

$$y = b + mx \tag{3.16}$$

where the initial Sr ratio is the intercept, b, on the y axis at a value, x,
of ^{87}Rb/^{86}Sr $= 0$, and the term $(e^{\lambda t} - 1)$ is the slope, m.

When Equation 3.15 is solved for t and the appropriate value is used for the decay constant, we have

$$t = 7.042 \times 10^{10} \log_e \left[\frac{\frac{^{87}Sr}{^{86}Sr} - \left(\frac{^{87}Sr}{^{86}Sr}\right)_0}{\frac{^{87}Rb}{^{86}Sr}} + 1 \right] \qquad (3.17)$$

First, note the similarity of this equation to Equation 3.10. It is simply the Rb–Sr age equation with the isotope values expressed as ratios and a term for the initial ^{87}Sr added.

Second, note what happens if we rearrange Equation 3.15 slightly.

$$e^{\lambda t} - 1 = \left[\frac{\frac{^{87}Sr}{^{86}Sr} - \left(\frac{^{87}Sr}{^{86}Sr}\right)_0}{\frac{^{87}Rb}{^{86}Sr}} \right] = \text{slope} \qquad (3.18)$$

Now look again at equations 3.10 and 3.17 and Figure 3.9d. It should be obvious that the slope of the isochron is simply the net change in the ratio $^{87}Sr/^{87}Rb$ over time t. It may also be apparent that the tangent of the angle θ between the sloping isochron at $t = t'$ and the horizontal isochron at $t = 0$ is equal to $^{87}Sr/^{87}Rb$, so that another way to express the relationship between age and slope is

$$t = 7.042 \times 10^{10} \log_e(\tan \theta + 1) \qquad (3.19)$$

Next, let's see what happens to the isochron when the Rb–Sr clock is reset. Suppose that the rock is completely melted and allowed to recrystallize at some time t'', just shortly after t' (Fig. 3.9e). Melting will rehomogenize the Sr isotopes, and the new minerals will once again share the same initial ratio of ^{87}Sr to ^{86}Sr, but that ratio will be higher than it was at $t = 0$ because of the decay of ^{87}Rb from time $t = 0$ to $t = t''$. Graphically, it is as if the isochron pivots about the isotopic composition of the total rock (point x in Figure 3.9e) and becomes horizontal again. The remelting and recrystallization have completely reset the Rb–Sr clock. A later age measurement will reflect the most recent time of recrystallization and "initial" Sr isotopic composition— the previous "age" and initial composition have been erased.

If the rock is not heated sufficiently to rehomogenize the Sr isotopes completely but enough so that Sr or Rb is allowed to move about, then the clock may not be completely reset. In such a *disturbed system*, there are a variety of things that can happen to the isotopic composition of and within the rock. Either Sr or Rb isotopes, or both, may move in or out of the system, or the isotopes may simply be redistributed among the different minerals within the system. In such instances the results are unpredictable in detail, but the isotopic ratios for a system that has been disturbed almost invariably do not fall on any sort of isochron.

To see why, consider an oversimplified example that involves only the loss of ^{87}Sr from our hypothetical rock (Fig. 3.9f). Mineral R is high in Rb and low in Sr because a Rb atom is chemically and physically more compatible with the particular chemistry and crystal structure of that mineral than is a Sr atom. This means that a ^{87}Sr atom resulting from the decay of ^{87}Rb may find itself at a location within the crystal at which it is less firmly bound than either the original ^{87}Rb atom or a ^{86}Sr atom incorporated into the crystal when it formed. When the crystal is reheated, therefore, radiogenic ^{87}Sr atoms may be lost more easily than either ^{86}Sr or ^{87}Rb atoms. The same will not necessarily be true for mineral P, in which Sr is a more natural constituent. The ease with which radiogenic ^{87}Sr is lost from minerals P, Q, and R will be a function of many factors, such as chemical composition, crystal structure, and crystal defects, most of which are not directly related to the ratio ^{87}Rb/^{86}Sr. Thus the movement of the compositions from P', Q', and R' to P'', Q'', and R'' will not be an exact function of ^{87}Rb/^{86}Sr and so P'', Q'', and R'' will not fall on a line except by a highly unlikely coincidence.

If we analyze the three minerals in our hypothetical rock and obtain data P'', Q'', and R'', we cannot determine exactly which isotopes have moved, where they have gone, or what amounts were involved. The data do, however, give us some very valuable information—they clearly reveal that the rock has been disturbed since it was formed, that the conditions of a closed system have been violated, that the Rb–Sr clock was partially reset at some unknown time, and that the age of the rock cannot be found from these particular data.

The effects of an incomplete resetting or disturbance are usually much more complicated than the simple example shown in Figure 3.9f. Isotopes may enter the system as well as leave, and there may be exchange of ^{87}Rb, ^{87}Sr, and ^{86}Sr between different minerals. These complications make it even less likely that the ratios will coincidentally fall on a straight line or isochron. For all practical purposes, the only way to move the isotopic compositions of samples from one isochron to another is by either radioactive decay through time or complete isotopic rehomogenization. Points that fall on an isochron, therefore, can confidently be interpreted as indicating the time of last isotopic homogenization, i.e. formation or reformation of the rock.

Thus, the isochron method is self-checking, providing not only the prospect of an age but also a statement on its validity.

An example of a valid Rb–Sr isochron is shown in Figure 3.10a in which data for the chondrite meteorite Tieschitz are plotted. Compare this isochron with the scattered data for the dikes and sills of the Precambrian Pahrump Group (Fig. 3.10b). For Tieschitz we can confidently conclude that its age is 4.52 Ga; for the data to fall on an isochron for any reason other than decay of ^{87}Rb within a closed system over time would be a highly improbable coincidence. For the Pahrump dikes and sills, however, we can only conclude that the samples are of different ages,[12] have not remained closed systems since their formation, or both. There is no way to determine the age of the rocks from these data, but neither do we face the prospect of calculating a Rb–Sr age that is incorrect and misleading. If we had made only a single analysis of any of the samples in Figure 3.10b our conclusions might have been quite different. It would have been necessary to make some estimate of the initial Sr composition and then calculate a simple accumulation age (sometimes called a *model age*) for the sample. The chances are great that the calculated age would have been incorrect, but, more seriously, we would have no way of knowing if it was. Thus, the isochron method, while more work because it requires multiple analyses, is worth the effort because it is self-checking.

The simple isochron can be used for virtually any of the decay schemes in Table 3.1. In practice, it is used extensively for the Sm–Nd, Lu–Hf, and Re–Os systems in exactly the way in which it is used for Rb–Sr dating, the only difference being the ratios expressed on the ordinate and abscissa (Table 3.3). The mathematics (Equations 3.14 through 3.19) are also identical, requiring only the substitutions of the proper isotope ratios (Table 3.2) and constants (Table 3.1). For the U–Pb and K–Ar systems, slightly different diagrams are used for reasons that we will examine.

The ^{40}Ar/^{39}Ar Age Spectrum and Isochron

The ^{40}Ar/^{39}Ar method is a form of K–Ar dating in which the sample is irradiated with *fast neutrons*[13] in an atomic reactor to convert a fraction of the ^{39}K, which is the most common isotope of K (Table 3.2), to ^{39}Ar. The reaction of a fast neutron with a ^{39}K nucleus results in the addition of a neutron and the ejection of a proton, which

---

12. Whenever data are plotted on an isochron diagram, the initial assumption is that the samples (minerals or rocks) are of the same age and thus members of the same system. If this is not true, then the data will not fall on an isochron.

13. "Fast" neutrons are more energetic than "slow" neutrons. The terms are not precise ones, but energies above 0.02 MeV usually qualify neutrons as fast.

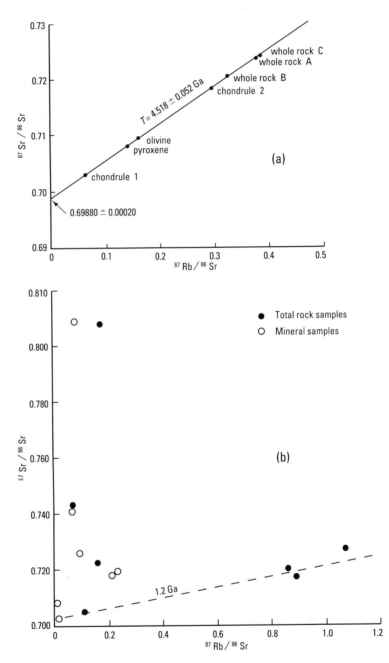

Fig. 3.10. (a) A mineral and whole-rock Rb–Sr isochron for the chondrite meteorite Tieschitz. (After Minster and Allègre, 1979a.) (b) Plot of Rb–Sr data for samples from igneous dikes and sills that intrude the Pahrump Group of the Panamint Mountains, California. The scatter of data shows clearly that these samples have been open systems, did not form at the same time, or both. Regardless, the ages of these rocks cannot be determined from these data. Other evidence indicates that these rocks are all about 1.2 Ga (dashed reference isochron). (After Wasserburg, Albee, and Lanphere, 1964.)

TABLE 3.3
Parameters of the Common Radiometric Age-Diagnostic Diagrams

Diagram name	Ordinate (y axis)	Abscissa (x axis)
Rb–Sr isochron	$^{87}Sr/^{86}Sr$	$^{87}Rb/^{86}Sr$
Sm–Nd isochron	$^{143}Nd/^{144}Nd$	$^{147}Sm/^{144}Nd$
Lu–Hf isochron	$^{176}Hf/^{177}Hf$	$^{176}Lu/^{177}Hf$
Re–Os isochron	$^{187}Os/^{186}Os$	$^{187}Re/^{186}Os$
$^{40}Ar/^{39}Ar$ age spectrum	age or $^{40}Ar/^{39}Ar$	^{39}Ar released
$^{40}Ar/^{39}Ar$ isochron	$^{40}Ar/^{36}Ar$	$^{39}Ar/^{36}Ar$
Pb–Pb isochron	$^{207}Pb/^{204}Pb$	$^{206}Pb/^{204}Pb$
U–Pb concordia	$^{206}Pb/^{238}U$	$^{207}Pb/^{235}U$

changes the ^{39}K to ^{39}Ar. Instead of the amounts of K and Ar being measured in separate experiments by different methods, as is done in the conventional K–Ar method, the exact ratio of daughter to parent in the sample is determined by measuring the ratio of ^{40}Ar to ^{39}Ar in one experiment. Corrections must be made for atmospheric Ar and for certain interfering Ar isotopes produced by unwanted neutron reactions with Ca and other K isotopes, but these corrections can be made quite precisely and for old rocks are usually very small or negligible. This method of measuring all of the Ar in a sample in one experiment generally gives an age comparable to one determined by the conventional K–Ar method (including necessary corrections, described by Brereton, 1970, Dalrymple and Lanphere, 1971, Dalrymple et al., 1981, Faure, 1986, and McDougall and Harrison, 1988).[14]

An age calculation from $^{40}Ar/^{39}Ar$ data is done with an equation similar to Equation 3.9:

$$t = 1.804 \times 10^9 \log_e\left(J \frac{^{40}Ar}{^{39}Ar} + 1 \right) \tag{3.20}$$

where J is a constant that includes a factor for the fraction of ^{39}K converted to ^{39}Ar during the irradiation. J is determined for each irradiation by irradiating a sample of known age, a monitor, alongside the unknown sample and using Equation 3.20 to calculate J for the monitor. The value of J for the monitor applies to the unknown sample as well because both the monitor and the unknown sample receive the same dose of neutrons.

The $^{40}Ar/^{39}Ar$ technique has some advantages over the conventional K–Ar method, including increased precision and applicability to smaller samples. But the primary advantage of this technique is

---

14. Dalrymple and Lanphere (1971) also compared $^{40}Ar/^{39}Ar$ results with those obtained by the conventional K–Ar method on the same samples.

that the sample can be heated to progressively higher temperatures and the Ar released at each temperature can be collected and analyzed separately. An age is then calculated for each gas increment from Equation 3.20. The data from this incremental heating technique have some useful properties, the most important of which are that they are self-checking and can provide a valid age for some samples that have not been closed systems.

The series of ages from an incremental heating experiment are most often plotted as a function of the percent of the ^{39}Ar released. This type of diagram is called an *age spectrum* or an *Ar-release diagram*. For an ideal, undisturbed sample, the calculated ages for the successive gas increments are all the same, and the age spectrum is a horizontal line at the value corresponding to the age of the rock (Fig. 3.11a). These same data can also be plotted on an ^{40}Ar/^{39}Ar isochron or correlation diagram (Fig. 3.11b) and will fall on a straight line whose slope is equal to the ratio ^{40}Ar/^{39}Ar in Equation 3.20 and whose intercept is the ^{40}Ar/^{36}Ar ratio of nonradiogenic, or air, Ar. The only difference between the age spectrum and isochron diagrams is that the isochron treatment does not require any assumption about the composition of nonradiogenic Ar; otherwise, the two diagrams are just two methods of visually displaying the same data.

For a sample that is heated at some later time t' after formation at time t, the ages calculated for the individual gas increments may not all be the same. It is the increments released at the lower temperatures, however, that are usually disturbed. The high-temperature increments may not be affected and may form a "plateau" that still reflects the original formation age (Fig. 3.11c). The plateau increments will still fall on a meaningful isochron, whereas the disturbed increments will not (Fig. 3.11d). Disturbed samples may give patterns more complicated than the one shown in Figure 3.11, but a high-temperature plateau age can usually be safely interpreted as the formation age of the sample.

An age spectrum and an isochron for the meteorite Menow are shown in Figure 3.12. From the shape of the age spectrum, one can calculate that approximately 25% of the Ar was lost, perhaps because of heating induced by collision with another body, about 2.5 Ga (Turner, Enright, and Cadogan, 1978). The high-temperature increments, however, still reflect the age of the meteorite—an age confirmed by Rb–Sr dating of Menow (Minster and Allègre, 1979a: 337).

The theory of ^{40}Ar/^{39}Ar age spectra is poorly understood, but a simple diffusion model seems to explain the data obtained from many samples, especially meteorites and many lunar lava flows. One way

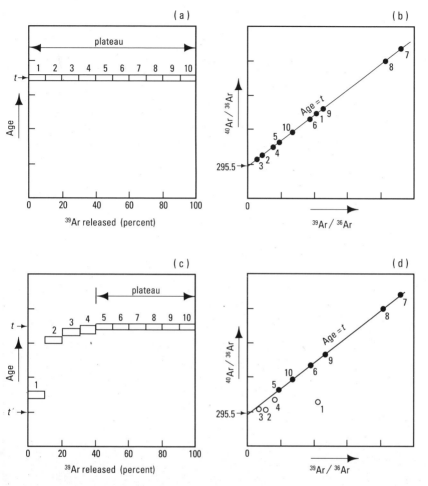

Fig. 3.11. (a) Hypothetical $^{40}Ar/^{39}Ar$ age spectrum for an undisturbed sample. Because the sample has been a closed system since its formation, each of the gas increments (1 through 10) gives the same age, t. Increment 1 is released at the lowest temperature and 10 at the highest. (b) $^{40}Ar/^{39}Ar$ isochron plot for the same undisturbed sample. The isochron intercepts the $^{40}Ar/^{36}Ar$ axis at the composition of atmospheric Ar. (c) Age spectrum for a hypothetical sample of age t that lost 15% of its Ar when it was heated at time t'. The increments (5 through 10) released at high temperatures still give the formation age and form a "plateau." (d) Isochron for sample in (c). The high-temperature increments still fall on an isochron, whereas the low-temperature increments do not, having been released from disturbed portions of the sample.

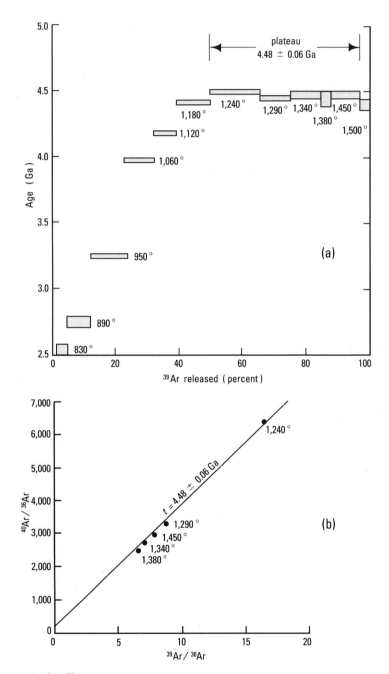

Fig. 3.12. $^{40}Ar/^{39}Ar$ age spectrum (a) and isochron (b) for the meteorite Menow, which lost 25% of its Ar at about 2.5 Ga. The temperature at which each gas increment was released is shown in degrees Celsius. (After Turner, Enright, and Cadogan, 1978.)

to visualize this mechanism is to use the age spectrum itself but consider the abscissa to represent distance from the surface of a mineral grain, where 0 represents the surface and 100 represents the center of the grain. When the grain is heated, the first gas to diffuse out will be from the region of the grain nearest the surface. The outer regions will also be those to lose the most Ar if the sample is disturbed by postformation heating. The incremental heating in the laboratory taps progressively deeper regions of the mineral grain, which are those parts of the crystal most resistant to disturbance simply because the Ar located there has farther to travel to reach freedom. This explanation is greatly oversimplified, but calculations based on the diffusion model give theoretical age spectra virtually identical to that of Menow and many other disturbed samples (Turner, 1968).

The data from an ^{40}Ar/^{39}Ar incremental heating experiment are, like the Rb–Sr isochron method, self-checking. The fact that a sample has been disturbed at some time after formation is nearly always revealed by some departure of the data from a totally horizontal age spectrum and some scatter in the isochron plot. If the sample has not been too badly disturbed, however, the method may still provide a valid age and thus be of use on rocks that have been open systems. A high-temperature plateau usually gives the age of the rock, although it might reflect a complete resetting of the K–Ar clock rather than the very first time the rock formed.

U–Pb Concordia and Discordia

One of the most powerful and reliable dating methods available is the U–Pb *concordia–discordia method*. It was devised in 1956 by G. W. Wetherill, a pioneer in radiometric dating and now Director of the Department of Terrestrial Magnetism of the Carnegie Institution of Washington (Wetherill, 1956). Like the ^{40}Ar/^{39}Ar age spectrum, this method can be used on open, as well as closed, systems—a feature of special value for dating old rocks with complex histories—and is self-checking. The U–Pb concordia–discordia method differs from the simple isochron methods in a fundamental way because it utilizes the simultaneous decay and accumulation of two parent–daughter pairs—^{238}U–^{206}Pb and ^{235}U–^{207}Pb.

To understand how the concordia–discordia method works, we must return to one of the simplest of the radioactive decay equations. Let us rewrite Equation 3.5 in terms of ^{238}U and its final daughter product ^{206}Pb:

$$^{206}Pb = {}^{238}U(e^{\lambda_1 t} - 1) \qquad (3.21)$$

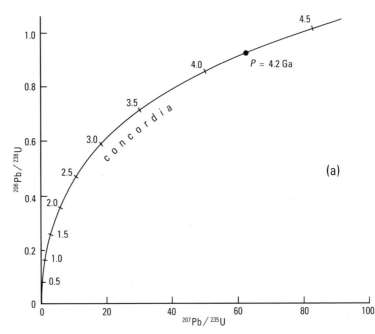

Fig. 3.13. (a) The U–Pb concordia diagram. The concordia is the locus of all points representing equal ^{235}U/^{207}Pb and ^{238}U/^{206}Pb ages. The location of a sample on concordia is a function of age; point *P*, for example, is at 4.2 Ga. (b) The episodic loss of Pb moves a point, *P*, off concordia along a straight line (discordia) connecting *P* and the origin to *P'*. The dashed lines show the position of the origin of the graph as it would have been drawn at 2.0 Ga. At some later time (2.0 Ga later in this example) samples from the same rock (*P'*, *Q*, and *R*) will plot on a line whose upper intercept with concordia gives the age of the rock and whose lower intercept gives the time of episodic Pb loss.

and rearrange it to express the relationship of daughter to parent as a ratio:

$$\frac{^{206}\text{Pb}}{^{238}\text{U}} = e^{\lambda_1 t} - 1 \tag{3.22}$$

where λ_1 is the decay constant of ^{238}U (Table 3.1). In these two equations the quantity of ^{206}Pb represents only the amount of the Pb isotope produced by the decay of its U parent since the rock formed, so initial Pb must either be zero or be eliminated by being subtracted from the total ^{206}Pb. The comparable equation for ^{235}U and ^{207}Pb is

$$\frac{^{207}\text{Pb}}{^{235}\text{U}} = e^{\lambda_2 t} - 1 \tag{3.23}$$

where λ_2 is the decay constant of ^{235}U. We can now substitute various

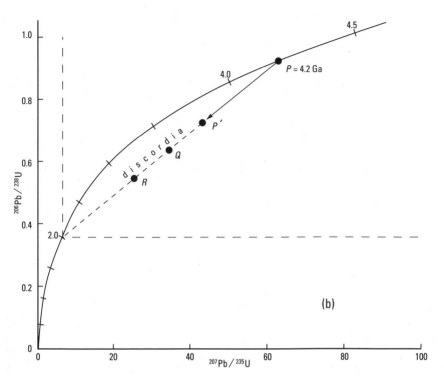

values of t into Equations 3.22 and 3.23 and graph the resulting ratios $^{206}\text{Pb}/^{238}\text{U}$ and $^{207}\text{Pb}/^{235}\text{U}$ (Fig. 3.13a). These ratios for all values of t plot on a single curve called *concordia*, which is the locus of all concordant U–Pb ages. All minerals that formed at 4.2 Ga and have remained closed systems, for example, will plot at point P in the figure and will also have identical $^{206}\text{Pb}/^{238}\text{U}$ and $^{207}\text{Pb}/^{235}\text{U}$ ages of 4.2 Ga. Concordia is curved because ^{238}U and ^{235}U decay at different rates (Table 3.1) and so the relative rates of production of the two isotopes of Pb have changed with the passage of time.

The graphing of concordant U–Pb data on a concordia diagram does not provide any information that is not obvious from the individual U–Pb ages themselves. The principal value of the concordia diagram is its unique ability to yield crystallization ages from open systems. Let us examine how this aspect of the method works.

As mentioned in the section on U–Th–Pb methods, Pb is a volatile element and is rather easily lost from minerals when they are heated. Loss of Pb, however, does not fractionate the isotopes in the remaining Pb because they are chemically identical and their masses are very nearly the same. Thus, Pb loss from a mineral with a composition at point P will result in the ratios $^{206}Pb/^{238}U$ and $^{207}Pb/^{235}U$ changing from P to some point P' along a straight line connecting P and the time of the Pb loss (Fig. 3.13b).

To visualize why this happens, suppose, for example, that a mineral formed at 4.2 Ga was heated at 2 Ga and lost some of its Pb. If we could transport ourselves back in time and make some Pb and U measurements while the event was in progress, we would see P move off the concordia and progress on a straight-line path toward the origin, i.e. toward $^{206}Pb/^{238}U = 0$ and $^{207}Pb/^{235}U = 0$. This happens because the two Pb isotopes are lost in the same proportion as their relative compositions at P. The result of complete Pb loss would be the total absence of Pb in the sample and the composition would then plot exactly on the origin. In this latter case, the U–Pb clock would be completely reset. In Figure 3.13b, the straight line path is a line connecting P with the 2-Ga point on concordia because the origin of the graph at 2 Ga, shown by the dashed lines, was at what is now the 2-Ga point. As the 2 Ga elapsed between then and the present, the origin of the graph moved along the concordia to its present position. The Pb compositions along the ordinate and abscissa, of course, also "follow" the origin and change with time to reflect the change in the composition of Pb in the sample as well as the evolution of concordia. Thus, today point P' lies on a chord connecting P, the original age of the sample, and the age of Pb loss, i.e. 2 Ga. This chord is called a *discordia*.

As may be obvious, a discordia cannot be determined by a single point such as P'. We must find at least two and preferably three or more points along a discordia before we can find either the original age of the rock, P, or the time of Pb loss. As it turns out, however, this is not especially difficult. The amount of Pb lost from a sample is controlled by a variety of factors, including grain size, original crystal imperfections, composition, and radiation damage from the decaying U. Thus, different crystals from the same rock sample will lose differing amounts of Pb, so analyses of different forms (colors, shapes, sizes) of the same mineral species from a sample will usually do the trick and provide sufficient differences in Pb/U composition to define a discordia. Different minerals from the same sample, mineral separates from different samples of the same rock body, or different zones within a single crystal also can be used to determine a discordia. Thus, Pb loss from a sample at P would not only result in the point P' but would generate other points, such as Q and R, as well.

Episodic loss of Pb, discussed above, is only one of several theoretical Pb-loss models that have been proposed to explain discordia. All of the models predict that the upper intercept of a discordia with concordia is the original age of the rock, or, more precisely, the time of the last complete resetting of the U–Pb clock. The interpretation of the lower intercept, however, is not so clear and differs according to the model used, although the current weight of evidence favors the episodic model. We need not be concerned with this point, however, because the evidence for the age of the Earth does not depend on the interpretation of the lower intercept.

Like the simple isochron methods, the U–Pb concordia–discordia method is self-checking. Disturbance of the sample due to partial Pb loss leads to a discordia and permits the determination of the age of the rock. Uranium gain, if such were to occur, would have exactly the same effect as Pb loss and the data could be interpreted in the same way. The loss of U drives the point *P* along an extension of the discordia above the concordia and toward higher Pb–U values; the original age is still recorded by the intersection. The result of the addition of Pb to the sample is unpredictable because it depends on the isotopic composition of the added Pb.

The concordia–discordia method can be used only on minerals that contain either no initial Pb or initial Pb in such small quantities that a correction for its presence can be accurately made. Such minerals are not abundant in rocks, but fortunately there are several that occur frequently, although in only small amounts, in most igneous and metamorphic rocks. The most common is zircon, whose crystal structure accepts U but rejects Pb so that initial Pb is invariably negligible or very small. The presence of initial Pb can be detected and corrected for if necessary by using the nonradiogenic ^{204}Pb. The crystal and chemical properties of zircon and of other minerals used in this method make U loss, or the gain of either Pb or U, very unlikely, so that Pb loss is the dominant effect of heating.

The U–Pb concordia–discordia method is especially resistant to heating and metamorphism and thus is extremely useful in rocks with complex histories. Quite often this method is used in conjunction with the K–Ar and Rb–Sr isochron methods to unravel the history of metamorphic rocks, because each of these methods responds differently to metamorphism and heating. For example, the U–Pb discordia age might give the age of initial formation of the rock, whereas the K–Ar method, which is especially sensitive to heating, might give the age of the latest heating event. There are examples of concordia–discordia diagrams in succeeding chapters.

The Pb–Pb Isochron

Data from U–Pb systems can be treated solely in terms of the product isotopes of Pb normalized to nonradiogenic ^{204}Pb. This method, known as the Pb–Pb isochron method, is based on a simple extension of Equation 3.13. By normalizing the quantities ^{207}Pb and ^{206}Pb in the equation to the quantity of ^{204}Pb and subtracting the initial amount of Pb, Equation 3.13 becomes

$$\left[\dfrac{\dfrac{^{207}Pb}{^{204}Pb} - \left(\dfrac{^{207}Pb}{^{204}Pb}\right)_0}{\dfrac{^{206}Pb}{^{204}Pb} - \left(\dfrac{^{206}Pb}{^{204}Pb}\right)_0}\right] = \dfrac{1}{137.88} \times \dfrac{e^{\lambda_2 t} - 1}{e^{\lambda_1 t} - 1} \qquad (3.24)$$

This equation describes a family of straight lines whose slope is expressed by the right-hand side of the equation and thus is a function of the age of the system, t. When plotted on a ^{207}Pb/^{204}Pb vs ^{206}Pb/^{204}Pb isochron diagram (Fig. 3.14a), this family of lines passes through a common point, P, that represents the composition of the initial Pb in the system. Thus, as a system (rock) grows older, its Pb isochron "rotates" about point P to increasing slopes. Unlike the simple isochron method, the Pb–Pb isochron method does not allow point P to be determined from the isochron data alone because P is not defined by the intersection of the isochron with one of the coordinates of the graph. Nevertheless, the age is determined only by the slope of the isochron, and the method can be used to determine the age of a system when either the composition or the amount of initial Pb remains unknown. Like the other isochron methods, this one is self-checking, and unreliable data are indicated by scatter.

An example of an application of the Pb–Pb isochron to the meteorite St. Severin can be seen in Figure 3.14b. The Pb compositions of plagioclase, whitlockite,[15] and a whole-rock sample indicate that the age of this former traveler of the Solar System is 4.54 Ga.

The ^{207}Pb/^{204}Pb vs. ^{206}Pb/^{204}Pb isochron diagram is the basis for what is thought to be the most precise value for the age of the meteorites and of the Earth, and will be discussed in more detail in Chapter 7.

As will be shown in Chapter 6, St. Severin is one of the few meteorites whose age has been determined by more than one radiometric dating method; the Pb–Pb isochron age shown in Figure 3.14b is consistent with the Rb–Sr and Sm–Nd isochron ages and slightly older (ca 0.13 Ga) than the Ar–Ar ages of St. Severin.

---

15. Whitlockite, a calcium and magnesium phosphate, is a common accessory mineral in meteorites and in certain types of granite. It has also been found in lunar rocks.

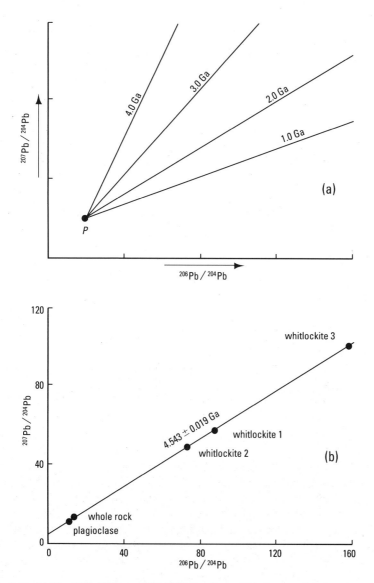

Fig. 3.14. (a) The $^{206}Pb/^{204}Pb$ vs $^{207}Pb/^{204}Pb$ isochron. Data from a closed system will fall on an isochron whose slope is a function of age. The point P represents the composition of the initial Pb for this family of isochrons. (b) Pb data for five samples from the St. Severin meteorite fall on an isochron. (After Manhes, Minster, and Allègre, 1978.)

The Accuracy of Radiometric Dating

We have looked in some depth at radiometric methods and the principles on which they are based, the rationale for concluding that radioactive decay is constant over time, the geological factors that can partially or totally reset radiometric clocks, and the various ways in which the graphic methods are self-checking. But there are some other factors that are worth a few pages of attention.

First, all of the decay constants given in Table 3.1 have been determined by direct laboratory counting experiments and, with the possible exception of ^{187}Re, are known to within an accuracy of about 2% or better. The activities, i.e. the number of decays per unit time, of the parent isotopes with the longest half-lives are very low, and the measurements are difficult to make. Nevertheless, the decay constants of ^{87}Rb, ^{147}Sm, ^{176}Lu, and ^{187}Re are known to within about 2%. The decay constants of ^{40}K, ^{232}Th, ^{235}U, and ^{238}U are known to an accuracy of better than 1%. Thus, although the uncertainties in some of the decay constants are still significant sources of error when we attempt to distinguish between the ages of early events in the Solar System as measured by different methods, these uncertainties do not significantly affect the values for the ages of the Earth, Moon, or meteorites.

Second, the various isotopic compositions used as constants in age calculations, e.g. the ratio of ^{235}U to ^{238}U today or the present composition of atmospheric Ar, have been measured to accuracies of better than 1% and are not significant sources of error in age calculations.

It is also worth noting that virtually all investigators worldwide use the same decay constants and isotopic compositions for their calculations (Steiger and Jäger, 1977). The currently accepted decay constants are those listed in Table 3.1. All ages in this book that were originally calculated with other decay or abundance constants have been recalculated with the values in this table. These conventional values are used so that radiometric ages from different laboratories can be compared directly without tedious recalculations. From time to time, new laboratory measurements improve the constants, but the new constants are not necessarily adopted immediately. The reason for this is that the changes are usually small and the improvement in accuracy is not worth the confusion that would result from the continual adoption of new constants. Thus, the constants in use at any particular time are not necessarily the best ones available but are close enough that they do not introduce significant errors. There is no standardized time when constants are evaluated and new ones are adopted, but in general this updating happens only every decade or so.

Third, modern analytical instruments, especially the *mass spec-*

trometers used for isotopic measurements, have been refined to the point where the precision of the laboratory measurements usually far exceeds the errors introduced by geological factors. Most isotope ratios can now be measured to an accuracy of a few tenths of a percent or better.

Fourth, isochron slopes, concordia–discordia intercepts, and other relevant quantities are not determined graphically but are calculated by appropriate formulae and statistical methods. The age-diagnostic diagrams discussed above are used solely as illustrative devices. It is also much easier and quicker to evaluate a set of data from a graphic plot than from a set of numbers, although the numbers are still the most rigorous way to determine how well a set of data fit or do not fit an isochron. The various mathematical methods used to solve these "graphic" problems will not be described in this book because they are primarily of interest to the specialist in radiometric dating. For more information on this subject, the reader should consult the references listed in note 1 of this chapter.

Fifth, the uncertainty in a radiometric age is usually an estimate of the precision of the age measurement at the two-thirds confidence level. It is assigned by the scientist who generated the data and is his or her best estimate of the range within which two-thirds of the data would fall if the measurements were repeated many times. For a simple accumulation age, the uncertainty is usually calculated from an appropriate error formula into which goes a combination of rigorously determined errors and educated guesses. For an isochron or other such age, the assigned uncertainty is usually a calculated statistical parameter that accurately describes the "goodness" with which the data fit the calculated line. Sometimes, the 95% confidence level will be used instead of the two-thirds confidence level. The main thing to keep in mind is that the errors, although expertly determined, highly useful, and generally realistic, are statistically based approximations and do not define precise limits within which the "real" age must fall.

Finally, it is important to realize that geochronologists do not rely entirely on the error estimates and the self-checking features of age-diagnostic diagrams to evaluate the accuracy of radiometric ages. Whenever possible and practical, the experiment is designed to take advantage of other ways of checking the reliability of the age measurements. The simplest device is to repeat the analytical measurements in order to minimize analytical errors. The use of multiple measurements not only helps to identify spurious results caused by human error and malfunctioning equipment, it also provides data with which analytical precision can be determined.

Another method is to make age measurements on several samples (minerals or rocks) from the same rock unit. This technique helps to identify postformation geological disturbances because different mineral species usually respond differently to heating and chemical changes. The isochron techniques are based partly on this principle.

The use of different decay schemes on the same rock is an excellent way to check the accuracy of age results. If two or more radiometric clocks running at different rates give the same age, this is powerful evidence that the ages are correct. This approach is similar to one we have all used from time to time to check a clock in our house or a watch on our wrist. The suspect clock is checked against other timepieces that may operate on different principles and run at different rates. Thus, a clock with a mainspring may be checked by comparing the time it indicates with that of a pendulum clock, an electric clock running off of house electrical current, a *quartz* wristwatch that operates on batteries, or a time signal on the radio. If all agree within a few minutes, then we are confident that we know the approximate time of day. In the same way, different radiometric methods may be compared. If all give the same age, our confidence in the accuracy of the results is greatly increased.

Geologic relationships also constitute an important way to evaluate radiometric data. For example, a series of age measurements on rock bodies whose relative ages are known because they are stacked one on top of another should fall in the same sequence. If they do not, then it is a clear signal that something is amiss with either the radiometric data or the interpretation of the geologic relationships.

In the next four chapters we will examine a sample of the radiometric data relevant to the age of the Earth, and we will see examples of the various ways in which the age measurements are confirmed.

CHAPTER FOUR

Earth's Oldest Rocks: The Direct Evidence

Here, then, we have the *oldest known* rocks. Are they, then, absolutely the oldest—the *primitive rocks*, as some imagine? By no means. They are *stratified* rocks, and therefore consolidated sediments, and therefore, also, the *debris* of still older rocks, of which we know nothing. Thus, we seek in vain for the absolutely oldest, the primitive crust.   J. LECONTE (1884: 263–64)

The Precambrian represents 88% of geologic time (Table 2.2), and Precambrian rocks underlie more than half of the Earth's present-day land area (Fig. 4.1), yet far less is known about the geologic history of the Precambrian than about that of the Phanerozoic. There are three reasons for this. First, and perhaps foremost, is the general paucity of fossils in Precambrian rocks. Precambrian organisms apparently lacked the hard skeletal parts necessary for preservation as fossils except under extraordinary conditions, as well as the abundance and diversity required for them to be used as distinctive stratigraphic age indicators. Thus biostratigraphy, one of the primary tools used for determining the relative sequence of rock units and geologic events in the Phanerozoic, is largely unavailable for use in Precambrian studies.

Second, most Precambrian rocks have been subjected to one or more episodes of metamorphism and *deformation*. As a result, Precambrian rocks tend to be deformed on all scales, from microscopic to regional, and the relative ages of rock units are commonly obscure or decipherable only with difficulty and uncertainty (Fig. 4.2). Early Precambrian rock formations, especially, are often distorted, thickened, thinned, or removed entirely by intense deformation, and commonly it is not possible to trace individual formations for any distance. Mineralogical and textural changes induced by metamorphism also mask many of the original features of Precambrian rock units, making con-

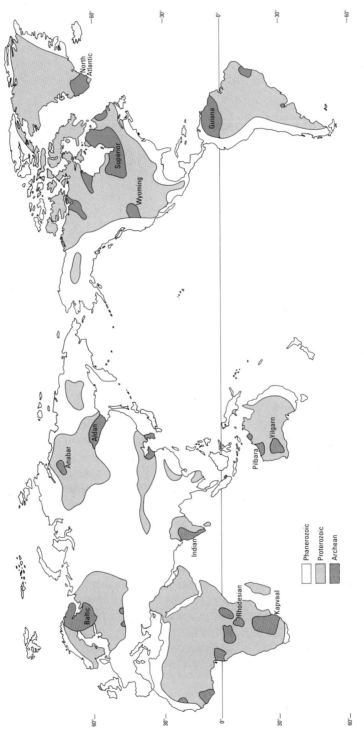

Fig. 4.1. Distribution of Archean and Proterozoic rocks. More than half of Antarctica (not shown) is also underlain by rocks of Precambrian age. (After various sources, including Condie, 1976, 1981, B.C. Burchfield, 1983, and Windley, 1984.)

Fig. 4.2. (above) Strongly deformed Ameralik mafic dike (now amphibolite) cutting Amîtsoq gneiss near the south coast of Ameralik Fjord. (Photo courtesy of V. R. McGregor.) (right) The inclusion of Akilia amphibolite, derived by metamorphism of mafic supracrustal rocks and so named because amphibole is a major constituent, is older than the gneiss that surrounds it. (Photo courtesy of A. Nutman and the Geological Survey of Greenland.)

ditions of formation difficult to determine. Often the metamorphic changes are so severe that the original rock type is indeterminable.

Third, erosion and the deposition of younger rocks have both removed and concealed Precambrian formations, so the accessible geologic record of Precambrian events, especially the earliest ones, is less complete than that of Phanerozoic events.

An interesting, if not surprising, feature of Precambrian rocks is that, except for the changes induced by metamorphism and deformation, they are, by and large, very similar to rocks formed during Phanerozoic and modern times. The primary difference is the relative rarity of Precambrian rocks formed by *biogenic* processes, a feature attributable to the relative scarcity of organisms then. The general resemblance of Precambrian rocks to more recent rocks, whose conditions of formation are determinable, and the intensive application of radiometric dating in lieu of fossils, render Precambrian history more tractable to scientific interpretation.

Precambrian rocks are common. They do not occur in the oceans because the ages of the present ocean basins are less than about 200 Ma, but they do occur on all continents. Rocks of Proterozoic age (0.57 to 2.5 Ga) are the most widespread and occur over more than two-thirds of North and South America, Africa, India, and Greenland, and are common over large areas of Australia, Europe, Asia, and Antarctica (Fig. 4.1). Archean rocks (2.5 to 4.0 Ga) also occur on all continents but over small and roughly equidimensional areas. Although subjected to metamorphism and deformation long ago, these Archean *cratons*, also called *blocks* or *shields*, now constitute very stable parts of the present-day continents and, except for the ravages of erosion, have undergone little internal change for hundreds of millions of years. The typical Archean craton is only about 0.25 to 0.5 × 10^6 km^2 in area (Condie, 1981: 1), but it is likely that Archean rocks underlie many areas of Proterozoic rocks, and these most ancient continental remnants may constitute nearly 50% or more of the present-day continental mass. Terrestrial Priscoan rocks (> 4.0 Ga) are unknown but, as we shall see later in this chapter, there is evidence for minerals of Priscoan age.

The oldest rocks found so far on Earth are about 3.8 to 3.9 Ga. Among the oldest are sedimentary rocks, formed from the debris of still older rocks of which no other trace has yet been found. Dated rocks of this age are not common, but rocks with ages of 3.5 Ga or more are found on nearly all of the continents. It is impractical to describe here every occurrence of early Archean rocks and the evidence for their age. Instead, we shall examine in some detail the geologic setting and evidence for the ages of early Archean rocks in four areas that have been especially well studied: the North Atlantic craton (Greenland–Labrador), the Superior Province (Great Lakes area of North America), the Pilbara Block (Western Australia), and the Barberton Mountain Land (southern Africa). Before we plunge into detailed descriptions of the geology of these four areas, however, a few remarks and generalizations about Archean rocks may help to put things in context. The next few pages, therefore, are devoted to a brief description of some of the principal rock types found in the typical early Archean *terrane*.

Archean Rocks

A typical early Archean terrane can be subdivided into several broad and interrelated categories of rock units, including *gneisses*, *supracrustal sequences* including *greenstone belts*, and *intrusive rocks* (Fig. 4.3). This classification is somewhat of an oversimplification, but it will serve as a reasonable framework for describing the isotopic evidence within the Earth's oldest known rocks.[1]

Some 80 to 90% of Archean rocks are gneisses, which occur as vast "seas" that surround, intrude, and form the basement for both older and younger rocks. There have been various explanations proposed for the origin of these gneisses, but the current consensus is that most or all originated by partial melting of the mantle, were injected into the crust as *granitoids*, and were changed to gneisses by high pressure and temperature deep within the Archean continental crust.[2] Geochemical and isotopic evidence makes it appear highly unlikely that the gneisses and their progenitors are recycled sediments or other crustal material but indicate that they probably constituted new additions to the Archean continental crust. Indeed, the formation of these ancient *plutonic* rocks was probably the major process of crustal growth during the Archean[3] and perhaps throughout geologic time. For example, this process may be similar to that which generated the Mesozoic *batholiths* that form the cores of the great mountain ranges of the west coasts of North and South America, i.e. the Sierra Nevada and the Andes. If we could examine the rocks that form the roots of these magnificent mountain ranges it is likely that we would find not granitoids but gneisses.

1. Literature on the Precambrian is extensive. For general summaries of Archean rocks, I have found the review articles by Moorbath (1975b, 1977a,b, 1980), Sutton and Windley (1974), and Glikson (1979) and the books by Condie (1976, 1981), Salop (1983), and Windley (1984) especially useful.
2. There is considerable evidence from the pressure and temperature necessary to cause the mineralogical changes now evident in Archean metamorphic rocks that the Archean crust reached thicknesses of 25 to 40 km and in a few places may have been as thick as 60 to 80 km (for example, Moorbath, 1980). The thickness of the continental crust today is similar, averaging about 10 to 20 km but reaching as much as 50 to 60 km beneath major mountain ranges (B. C. Burchfield, 1983).
3. There is lively debate about the growth rate of the early continental crust and whether continental growth has been continuous and largely irreversible (e.g., Moorbath, 1975a, 1977a, b, 1980, Glikson, 1979) or whether the continental crust formed early and has been extensively recycled through the mantle (e.g., Hargraves, 1978, Fyfe, 1980).

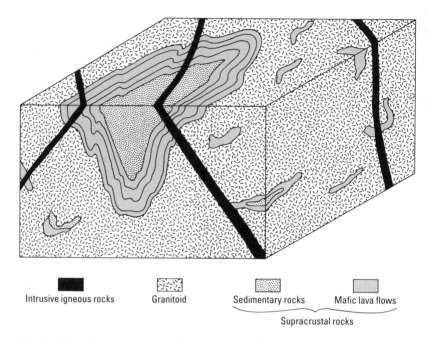

Intrusive igneous rocks   Granitoid   Sedimentary rocks   Mafic lava flows

Supracrustal rocks

Fig. 4.3. Typical sequence in Archean terranes where the Earth's oldest rocks are found. Commonly, the oldest rocks, which occur as fragmented inclusions within the gneiss, are remnants of lava flows and of sedimentary rocks derived from still older rocks of which there is now no trace.

Greenstone belts are stratigraphic accumulations of lava flows and sediments that may reach thicknesses of as much as 20 km. They typically occur in roughly symmetrical basin-like structures from 10 to 50 km in width and 100 to 300 km in length (Condie, 1981: 5). The rocks of greenstone belts were deposited in shallow inland basins or marginal seas, and most subsequently have been subjected to low-grade metamorphism. Archean greenstone belts often contain economically important metal deposits, chiefly of copper, nickel, chromium, silver, and gold, and because of this have been the object of more study than have the more voluminous Archean gneisses. Although many greenstone belts appear to have formed between 2.6 and 2.7 Ga, they range in age from Proterozoic to early Archean (Windley, 1984: 28–29, Condie, 1981: 1–5).

Mafic volcanic rocks, primarily *basalt*, are the primary rock type in greenstone belts and especially predominate near the bottom, where they may be interlayered with *ultramafic* lavas and intrusive rocks (Fig. 4.4). Toward the top of a greenstone belt, *intermediate* and *felsic* volcanic rocks and sedimentary rocks become more common.

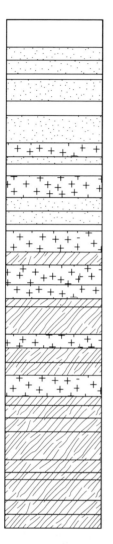

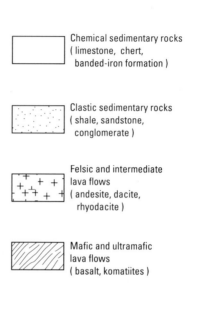

Chemical sedimentary rocks
(limestone, chert,
banded-iron formation)

Clastic sedimentary rocks
(shale, sandstone,
conglomerate)

Felsic and intermediate
lava flows
(andesite, dacite,
rhyodacite)

Mafic and ultramafic
lava flows
(basalt, komatiites)

Fig. 4.4. Hypothetical and simplified stratigraphic section of a typical Archean greenstone belt. The total stratigraphic thickness of a greenstone belt may approach 20 km.

Many of the lava flows are *pillow lavas,* indicating eruption and solidification under water. The sedimentary rocks of greenstone belts are commonly *immature sandstones, shales,* and *conglomerates* derived primarily from the erosion of nearby volcanic rocks. *Chert,* formed either by chemical precipitation or by the accumulation of siliceous microorganisms, is also found in greenstone belts. Other types of sedimentary rocks typical of greenstone belts include limestone and *banded-iron formation.*

There is nothing especially unusual about the types of rocks found in greenstone belts, and all but one have recent analogs. Only banded-iron formation, of which deposits are common in the early Precambrian and are economically important sources of iron worldwide, do not occur in the late Proterozoic and Phanerozoic. These are chemically precipitated sedimentary rocks that consist of alternating layers of iron-rich and iron-poor *silica* (Cloud, 1968, 1973, Sutton and Windley, 1974: 407–08). They are common in sedimentary sequences ranging in age from early Archean to mid-Proterozoic and may constitute nearly 15% of all Precambrian sedimentary rocks. Many workers think that these remarkable deposits precipitated as primitive oxygen-producing organisms evolved and began to flourish in the early Precambrian oceans, providing oxygen that combined with dissolved iron, which originated from volcanic eruptions and weathering of volcanic rocks. Others prefer an inorganic origin. But whether banded-iron formations are biogenic or abiogenic, their formation ceased only when the overabundance of dissolved iron in the ancient waters was exhausted.

What is unique about greenstone belts is the abundance of mafic and ultramafic volcanic rocks and immature sediments in them and the relative paucity of sedimentary rocks derived from siliceous continental sources. Another unique feature of greenstone belts is their overall worldwide similarity in form, composition, and stratigraphic sequence. Apparently shallow seas of limited extent adjacent to areas of high volcanic activity were common during the Archean. Perhaps these were the smaller and more numerous forerunners of our modern oceans. A final note of interest is that greenstone belts contain abundant evidence, in the form of pillow lavas and sediments known to be deposited in shallow seas, that reasonably large bodies of water existed on Earth during the Archean.

A remaining mystery, and the subject of much speculation, is the exact mechanism of formation of greenstone belts. A related question: why are they so common in the Archean? Various explanations of the formation of greenstone belts have been proposed, but none is certain (Sutton and Windley, 1974: 403–04, Moorbath, 1975b: 276–77, Windley, 1984: 48–65, Condie, 1981, 313–81). Perhaps the belts represent remnants of the primordial ocean crust or deposition in basins formed on or near primitive spreading ridges. They may be island-arc deposits, such as are found today around the margins of the Pacific Ocean. It has even been suggested that some may represent meteoritic impact basins, similar to lunar maria but complicated, as the maria are not, by the agents of water and atmosphere (Green, 1972). Most likely, however, greenstone belts were formed in numerous shallow seas on or near the margins of the ancient protocontinents, where volcanic activity, resulting from the rapid and voluminous injection of magma into the growing continental crust, was common. Their abundance in the Archean is probably due to the vigorous processes of volcanism and continental growth in the young and thermally active Earth.

The relation between greenstone belts and the gneisses that dominate the Archean terranes is not always certain. This is because the contacts between these units are often poorly exposed, faulted, or highly deformed. Are greenstone belts deposited on an older basement of granitoid (now gneiss), or do the granitoids intrude the older greenstone belts? Some geologists have argued for a consistent chronologic relationship, but the answer to both of these questions may be "yes." In some localities a rather convincing case can be made that the greenstone sequence was deposited on a pre-existing continental granitoid basement. But there are also numerous examples where the granitoid gneisses clearly intrude the greenstones (Fig. 4.5). This seems to be the case with the oldest of the greenstone belts, some of which are discussed in the following pages. In such instances it seems likely that the greenstones were deposited on oceanic crust, perhaps in basins marginal to a protocontinent, and incorporated by accretion into the continental mass after the injection of massive granitoids. In summary, it seems that the rocks of greenstone belts may be either older or younger than the massive granitoid gneisses and that there is currently no reason to expect the relationship to be everywhere the same.

Archean sedimentary rocks and lava flows that do not occur in the sequence and structure typical of greenstone belts are often called by the broader term *supracrustals* (Fig. 4.3). Supracrustal rocks include sandstones, shales, conglomerates that contain *clasts* of still older granitoids and volcanic rocks, cherts, limestones, and, occasionally, banded-iron formation. The lava flows include mafic and ultramafic types, and some still have recognizable pillow structures, indicating deposition under water. The composition of supracrustal rocks suggests that, as their name implies, they were deposited in shallow seas either on stable continental platforms (Windley, 1984: 13) or perhaps on oceanic crust near continental margins.

Many supracrustals are probably the remnants of greenstone belts whose original form has been largely destroyed by metamorphism, deformation, and the *intrusion* of granitoids. Although some workers have inferred a genetic distinction between greenstone belts and other supracrustal rocks, most view greenstone belts as simply a type of supracrustal sequence with little or no significant genetic differences between the two.

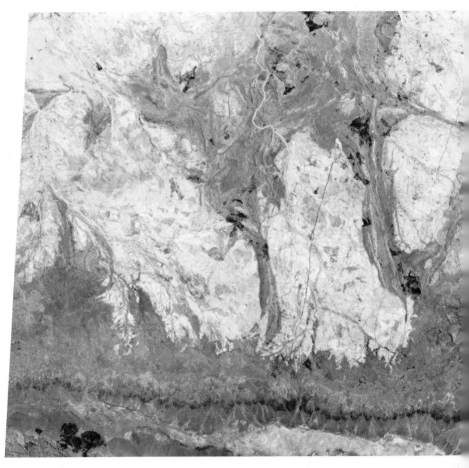

Fig. 4.5. This Landsat photograph of the eastern part of the Pilbara Craton, Western Australia, shows Archean granitoids intruding rocks of an older greenstone belt. U–Pb dating of the Shaw Batholith (lower left) and of volcanic rocks in the greenstone belt shows that these rocks are >3.4 Ga. (Landsat scene ID 84008501301X0, reproduced by permission of EOSAT.)

Of all the Archean rocks studied, the older supracrustals are perhaps the most fascinating to geologists who study the Precambrian. This is because they include the oldest dated rocks found on Earth yet are sedimentary in origin, representing the debris of rocks that must have been older still. Some contain clasts of granitoids, indicating the prior existence of continental crust. More than anything else, the fact that the oldest rocks found so far on Earth are sedimentary inspires geologists to continue the search for remnants of the most primitive crust.

The gneisses, greenstone belts, and other supracrustal rocks are commonly intruded by younger igneous rocks (Fig. 4.3). These range in composition from mafic to felsic and in form from immense batholiths to dikes only a few centimeters thick. These younger intrusive rocks are of interest here primarily because, where dated, they provide a younger limit on the age of the older rocks they intrude. Where these rocks intrude some units but not others they also act as "stratigraphic" markers that are of value in determining the relative sequence of rock units and events in deformed terranes.

In addition to the Archean cratons, which appear to be stable remnants of the earliest continents, there is another important type of terrane called a *mobile belt*. Mobile belts are lenticular portions of the Earth's crust that have undergone more *tectonic* activity than have the more stable crustal blocks on either side of them. They are, accordingly, very highly deformed. There are mobile belts of many geologic ages. The ones of concern here are the Proterozoic mobile belts that occur between some of the Archean cratons. Their origin is uncertain, but they probably formed from the collision of continental fragments, i.e. the Archean cratons on either side. Commonly, these mobile belts contain inliers and fragments of early Archean rock sequences, some more than 3 Ga in age.

North Atlantic Craton

The North Atlantic Archean craton (Fig. 4.6) includes parts of eastern and western Greenland, part of the coast of Labrador in Canada, and small sections of northwestern Scotland and northern Norway (Bridgwater, Watson, and Windley, 1973). Once a single block, this Archean terrane was torn into sections when North and South America were separated from Europe and Africa nearly 200 Ma ago to form the Atlantic Ocean. When rejoined (on paper) the North Atlantic craton consists of a roughly triangular area with sides about 600 km in length. The rocks of the craton are primarily granitoid gneisses, which constitute about 85% of the surface exposures. The remainder con-

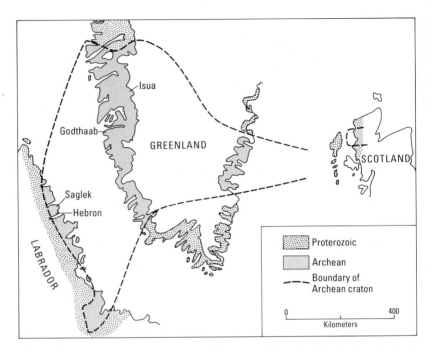

Fig. 4.6. The North Atlantic craton includes parts of Greenland, Labrador, Scotland, and Norway (not shown). The land masses, joined prior to about 200 Ma, are shown closer together than they are now. Precambrian rocks do not occur in the oceans that now divide the fragments of the craton. (After Bridgwater, Watson, and Windley, 1973.)

sists of metamorphosed supracrustal rocks with minor amounts of later intrusive rocks. This Archean terrane is bounded on all sides by Proterozoic mobile belts that consist, in part, of reworked Archean rocks.

Most of the rocks of the North Atlantic craton are 2.6 to 3.0 Ga old (Moorbath, 1977b: 97), but this ancient continental fragment also includes some of the oldest rocks found so far on Earth. Because of their antiquity, the parts of the craton that include these oldest rocks have been especially well studied. A variety of radiometric dating methods has been applied to these rocks, and the data show convincingly that the ages of the oldest are 3.7 to 3.8 Ga. Areas of particular interest are near Godthaab and Isua in southern West Greenland and near Saglek and Hebron on the coast of Labrador (Fig. 4.6).

The area near Godthaab, the capital of Greenland, is dominated by gneiss complexes of two distinct ages; the older complex is called the Amîtsoq gneisses (Fig. 4.7).[4] The Amîtsoq gneisses consist pri-

4. The rocks and geologic history of the Godthaab and Isua areas are described in detail by, among others, McGregor (1973), Bridgwater, McGregor, and Myers (1974),

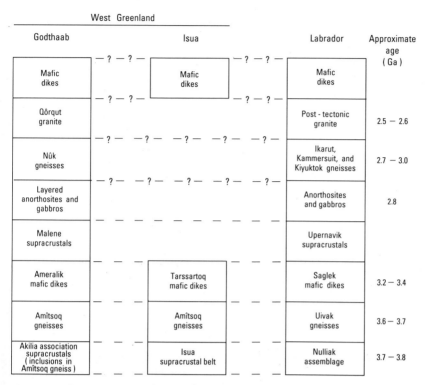

West Greenland

Godthaab	Isua	Labrador	Approximate age (Ga)
— ? — ? —	— ? — ? —		
Mafic dikes	Mafic dikes	Mafic dikes	
— ? — ? —		— ? — ? —	
Qôrqut granite		Post-tectonic granite	2.5 — 2.6
— ? — — ? — — ? — — ? — — ? —			
Nûk gneisses		Ikarut, Kammersuit, and Kiyuktok gneisses	2.7 — 3.0
— ? — — ? — — ? — — ? — — ? —			
Layered anorthosites and gabbros		Anorthosites and gabbros	2.8
Malene supracrustals		Upernavik supracrustals	
Ameralik mafic dikes	Tarssartoq mafic dikes	Saglek mafic dikes	3.2 — 3.4
Amîtsoq gneisses	Amîtsoq gneisses	Uivak gneisses	3.6 — 3.7
Akilia association supracrustals (inclusions in Amîtsoq gneiss)	Isua supracrustal belt	Nulliak assemblage	3.7 — 3.8

Fig. 4.7. Simplified sequence of major rock units, as determined from field relations, in the Godthaab and Isua areas of western Greenland and coastal Labrador. The approximate ages of the dated units are also indicated. (Data for Greenland from McGregor, 1973, Bridgwater, McGregor, and Myers, 1974, Bridgwater, Collerson, and Myers, 1978, and Moorbath, 1975b, 1977b; for Labrador from Bridgwater, Collerson, and Myers, 1978, and Collerson, Kerr, and Compston, 1981.)

marily of *tonalite* and *granodiorite* and contain *rafts* and *inclusions*, as long as several hundred meters, of highly deformed and metamorphosed older rocks (Fig. 4.2b). These inclusions appear to be transformed remnants of mafic lavas, mafic and ultramafic intrusive rocks, clastic sediments, and banded-iron formation. Although their original nature and stratigraphy are somewhat problematical, they are clearly fragments of an older supracrustal sequence and are collectively called the Akilia association, described by McGregor and Mason (1977).

The Amîtsoq gneisses are intruded by the Ameralik dikes. These

Bridgwater, Collerson, and Myers (1978), Allaart (1976), Myers (1976), Gill, Bridgwater, and Allaart (1981), and Nutman et al. (1983, 1984), and the description in the text is taken primarily from these sources. Useful summaries are given by Bridgwater, Watson, and Windley (1973), Moorbath (1975b, 1977a,b, 1980), and Windley (1984), as well as by many of the other references cited in this section.

igneous dikes are mafic in composition and range in thickness from a few tens of centimeters to a few meters. Some were deformed with the Amîtsoq gneisses and are now discontinuous (Fig. 4.2a). The Ameralik dikes provide a valuable stratigraphic marker, first because they cut the Amîtsoq gneisses and its older inclusions but do not intrude younger rocks, and second because they are ubiquitous.

The Malene supracrustals contain a high proportion of mafic volcanic rocks, including some flows with pillow structures, as well as *tuffs* and volcanogenic sediments. Cherts, limestones, and shales occur but are not abundant. All of the rocks are now metamorphosed but still retain some of their original features. The sequence is, in places, as much as a kilometer thick. The Malene supracrustals are not cut by the Ameralik dikes and so are younger than the Amîtsoq gneisses. The supracrustals are, however, intruded by *stratiform anorthosites*, igneous rocks composed almost entirely of *plagioclase feldspar*, that still retain much of their original igneous texture. These anorthosites appear to have been intruded as sheets into the bedding of the Malene supracrustals and along the contact between the supracrustals and the Amîtsoq gneisses.

The Nûk gneisses intrude the Amîtsoq gneisses, the Ameralik dikes, the Malene supracrustals, and the stratiform anorthosites. They cover more area than do the Amîtsoq gneisses and represent a major addition of new material to the continental crust. The Nûk gneisses are similar in composition to the Amîtsoq gneisses and were probably formed in a similar way, by plutonic intrusion. In spite of their similarity in appearance and composition, the Nûk and Amîtsoq gneisses can be easily distinguished by the presence or absence of Ameralik dikes, which are ubiquitous within the Amîtsoq gneisses but nowhere intrude the Nûk gneisses.

The Qôrqut Granite is an elongate plutonic body exposed over an area of some 50 by 18 km near the center of the Godthaab district. The intrusion of the Qôrqut Granite was accompanied by the injection of granitoid dikes that are still undeformed.

The youngest rocks in the Godthaab district are mafic dikes. These postdate most of the deformation undergone by older rocks and are only mildly metamorphosed. They can be distinguished from the Ameralik dikes by their lesser degree of metamorphism and deformation.

Isua is a remote mountainous area 100 km northeast of Godthaab near the edge of the inland ice sheet (Fig. 4.6). Most of the rocks exposed near Isua are gneisses equivalent to the Amîtsoq gneisses in composition and stratigraphic position, and they are commonly given the same name (Fig. 4.7). As in parts of the Godthaab district, these

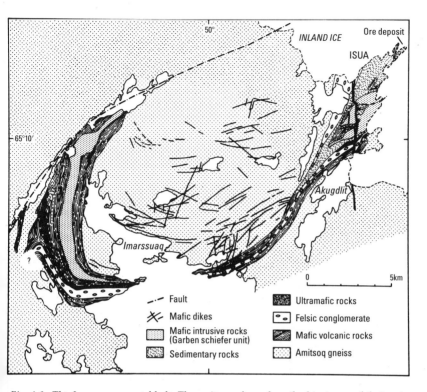

Fig. 4.8. The Isua supracrustal belt. The units are here described in terms of their primary progenitors, but all are now metamorphic rocks. (After Allaart, 1976.)

gneisses are cut by undeformed dikes, the Tarssartoq dikes, which appear to be equivalent to the Ameralik dikes. Rocks younger than the Tarssartoq dikes do not occur at Isua, and both the dikes and the gneisses are much less deformed and metamorphosed than their counterparts at Godthaab. The importance of Isua is that a thick and relatively undisturbed sequence of very ancient supracrustal rocks is exposed there. These remarkable rocks form an incomplete oval approximately 12 by 25 km that is completely enclosed and, in places, intruded by the Amîtsoq gneisses (Fig. 4.8). Field relations show clearly that the Isua supracrustals are older than the gneisses and thus are the oldest known rocks of the North Atlantic Craton.[5]

5. The Isua locality was first visited in 1971 by Stephen Moorbath of Oxford University and V. R. McGregor of the Geological Survey of Greenland at the invitation of a combined Danish/Finnish/U.S.A. mining operation that was exploring the iron deposits. Moorbath (1977a: 159, 1977b: 99–100) described the experience of seeing these ancient rocks for the first time as perhaps the most awe-inspiring of his career.

TABLE 4.1

Radiometric Ages of Some Early Archean
and Related Rocks of the North Atlantic Craton

Unit and locality[a]	Method	Number of samples[b]	Age[c] (Ga)	Source
WEST GREENLAND				
Qôrqut granite				
Godthaab	Rb–Sr	23w	2.53 ± 0.03	Moorbath, Taylor, and Goodwin, 1981
	Rb–Sr	3m	2.52 ± 0.03	Pankhurst et al., 1973
	Pb–Pb	23w	2.58 ± 0.08	Moorbath, Taylor, and Goodwin, 1981
	U–Pb	7z	2.53 ± 0.03	Baadsgaard, 1976
Nûk gneisses (and equivalents)				
Bjorneoen	Pb–Pb	11w	3.02 ± 0.26	P. N. Taylor et al., 1980
	Rb–Sr	11w	2.98 ± 0.05	P. N. Taylor et al., 1980
	Rb–Sr	14w	2.94 ± 0.09	P. N. Taylor et al., 1980; Moorbath and Pankhurst 1976; Pankhurst, Moorbath, and McGregor, 197?
	Rb–Sr	6w	3.08 ± 0.03	Baadsgaard and McGregor, 1981
Buksefjord	Rb–Sr	21w	2.86 ± 0.09	P. N. Taylor et al., 1980; Moorbath and Pankhurst 1976
Faeringehavn	Rb–Sr	8w	2.73 ± 0.06	Moorbath and Pankhurst, 1976
Fiskenaesset	Pb–Pb	9w	2.82 ± 0.07	P. N. Taylor et al., 1980
	U–Pb	5z	2.80 ± 0.10	Pidgeon, Aftalion, and Kalsbeek, 1976
	Rb–Sr	11w	2.84 ± 0.07	P. N. Taylor et al., 1980; Moorbath and Pankhurst 1976
Fredrikshaabs isblink	Rb–Sr	13w	2.60 ± 0.12	Moorbath and Pankhurst, 1976; Pidgeon and Hopgood, 1975
Godthaab	U–Pb	5z	2.82 ± 0.05	Baadsgaard, 1976
	Rb–Sr	11w	2.77 ± 0.19	P. N. Taylor et al., 1980; Moorbath and Pankhurst 1976
Godthaabsfjord	Rb–Sr	11w	2.86 ± 0.30	P. N. Taylor et al., 1980; Moorbath and Pankhurst 1976
Itivinga	Rb–Sr	6w	2.86 ± 0.06	P. N. Taylor et al., 1980
Nordland	Pb–Pb	19w	3.00 ± 0.07	P. N. Taylor et al., 1980
Sermilik	Pb–Pb	11w	3.00 ± 0.09	P. N. Taylor et al., 1980
	Rb–Sr	20w	2.75 ± 0.04	P. N. Taylor et al., 1980; Moorbath and Pankhur? 1976
Layered anorthosite				
Fiskenaesset	Rb–Sr	4w	2.75 ± 0.24	Alexander, Evensen, and Murthy, 1973
	Ar–Ar	1w	2.83 ± 0.08	Alexander, Evensen, and Murthy, 1973
South of Godthaab	Pb–Pb	15w	2.76 ± 0.14	P. N. Taylor et al., 1980
Amîtsoq gneisses				
Godthaab	U–Pb	9z	3.60 ± 0.05	Baadsgaard, 1973, 1976
	U–Pb	8z	3.60	Michard-Vitrac et al., 197?
	Pb–Pb	13w	3.56 ± 0.10	Black et al., 1971

TABLE 4.1 *(continued)*

Unit and locality[a]	Method	Number of samples[b]	Age[c] (Ga)	Source
Isua	Pb–Pb	9w	3.74 ± 0.12	Moorbath, O'Nions, and Pankhurst, 1975
	Rb–Sr	13w	3.64 ± 0.06	Moorbath, O'Nions, and Pankhurst, 1975; Moorbath et al., 1977a
	Rb–Sr	12w	3.62 ± 0.14	Moorbath et al., 1972
Narssaq	Rb–Sr	25w	3.67 ± 0.09	Moorbath et al., 1972
Narssaq, Qîlangarssuit, Isua	Lu–Hf	9w, z	3.55 ± 0.22	Pettingill and Patchett, 1981
Praestefjord	Rb–Sr	18w	3.61 ± 0.22	Moorbath et al., 1972
Qîlangarssuit	Rb–Sr	7w	3.66 ± 0.10	Moorbath et al., 1972
Simiutât	Pb–Pb	7w	3.62 ± 0.13	Griffin et al., 1980
	Rb–Sr	7w	3.56 ± 0.14	Griffin et al., 1980
Early supracrustals				
Various units, Isua	U–Pb	16z	3.81 ± 0.02	Baadsgaard et al., 1984
Conglomerate	U–Pb	8z	3.77 ± 0.01	Michard-Vitrac et al., 1977
	Rb–Sr	8w	3.66 ± 0.06	Moorbath, O'Nions, and Pankhurst, 1975; Moorbath et al., 1977a
	Rb–Sr	5w	3.71 ± 0.07	Jacobsen and Dymek, 1988
Garbenschiefer + conglomerate	Sm–Nd	12w	3.75 ± 0.04	Hamilton et al., 1983
Iron formation	Pb–Pb	11w	3.70 ± 0.07	Moorbath, O'Nions, and Pankhurst, 1973
Akilia association, Akilia	U–Pb	11z	3.59 ± 0.04	Baadsgaard et al., 1984
BRADOR				
Post-tectonic granite				
Saglek	U–Pb	4z	2.52	Baadsgaard, Collerson, and Bridgwater, 1979
Younger gneisses				
Ikarut, Hebron	Rb–Sr	7w	2.77 ± 0.14	Collerson, Kerr, and Compston, 1981
Kammersuit, Nachvak	Rb–Sr	6w	2.69 ± 0.14	Collerson, Kerr, and Compston, 1981
Granitic sheets, Saglek	Rb–Sr	21w	2.81 ± 0.21	Collerson, Kerr, and Compston, 1981
Kiyuktok, Saglek	Pb–Pb	5w	3.50 ± 0.11	Collerson, Kerr, and Compston, 1981
	Rb–Sr	15w	2.75 ± 0.12	Collerson, Kerr, and Compston, 1981
	Rb–Sr	6w	3.06 ± 0.16	Hurst et al., 1975
Uivak gneisses				
Saglek, Maidmonts Is.	U–Pb	3z	3.76 ± 0.15	Wanless, Bridgwater, and Collerson, 1979
Saglek	Rb–Sr	7w	3.55 ± 0.07	Hurst et al., 1975
Hebron	Rb–Sr	8w	3.56 ± 0.08	Barton, 1982
Saglek and Hebron	Rb–Sr	21w	3.61 ± 0.20	M. Cameron et al., 1981
(plus Amîtsoq)[d] various	Sm–Nd	28w	3.56 ± 0.20	Collerson, McCulloch, and Miller, 1981

NOTE: Units within an area, e.g., West Greenland, are listed in stratigraphic order when this order is known. All the ages are based on isochron (Rb–Sr, Sm–Nd, Lu–Hf), concordia–discordia (U–Pb), or age-spectrum (Ar–Ar) techniques.

[a] Localities are approximate as samples are generally collected over a broad area.

[b] w, whole rock; z, zircon; m, various minerals from same rock.

[c] All ages calculated with the decay constants in Table 3.1. Errors are at the 95% confidence level (two standard deviations). Ages underlined are shown as figures.

[d] Combined isochron using 18 Uivak samples and 10 Amîtsoq samples.

The Isua supracrustal sequence consists of metamorphic rocks whose progenitors include fine-grained clastic sediments, a carbonate-bearing conglomerate, cherts, banded-iron formation, mafic and ultramafic lavas and intrusive rocks, and a massive mafic layer called the *garbenschiefer unit*.[6] The sequence was deformed before intrusion of the Amîtsoq gneisses and the rocks are metamorphosed, but they still retain many recognizable features that reflect their supracrustal origin (Allaart, 1976, Nutman et al., 1983). Although not certain, it is probable that these ancient supracrustal rocks are the less deformed and less metamorphosed equivalents of the inclusions of the Akilia association found within the Amîtsoq gneisses near Godthaab.

Because of their antiquity, the early Archean rocks near Godthaab and Isua have been the objects of a great many radiometric age measurements, the effort being led primarily by S. Moorbath of Oxford University, who, with his colleagues, has spent more than a decade studying these rocks. Some of the results of these efforts are summarized in Table 4.1. The radiometric age data confirm the sequence of rock units as determined by field relationships, and show that the Isua supracrustals and the Amîtsoq gneisses are, indeed, some of the oldest rocks found so far on Earth.

The ages of three of the units within the Isua supracrustal sequence have been determined by four independent methods. The conglomerate unit (Fig. 4.8) contains boulders and clasts of felsic volcanic rock embedded in a fine-grained matrix of limestone and volcanic debris. The part of this unit exposed in the western arc of the Isua supracrustal belt is primarily a fine-grained *schist*, but the schist is thought to be merely a textural variation of the conglomerate. The conglomerate unit, including the schist, appears to have been formed by the erosion and redeposition in shallow water of a felsic volcanic rock, most probably a volcanic ash deposit.[7] A Rb–Sr isochron (Fig. 4.9) on five whole-rock samples of boulders and matrix from the conglomerate and three samples of the schist gives an age of 3.66 ± 0.06 Ga (Moorbath, O'Nions, and Pankhurst, 1975, Moorbath et al. 1977a).[8]

6. *Garbenschiefer*, a textural term, describes a platy rock with distinctive spots that resemble caraway seeds.

7. Another interpretation of this unit is that its progenitor was a chert-bearing limestone.

8. The result from one sample was not used by these authors to calculate the age because of its excessive deviation from the isochron. In general, I have attempted not to reinterpret data but to present the researchers' results more or less as they appear in the original publications. The exception is that I have recalculated ages, where necessary, using the currently accepted decay constants so that all ages are comparable. Whether or not the omission of this datum is justified is an arguable point, but the effect of including it in the calculation is negligible—it decreases the age very slightly

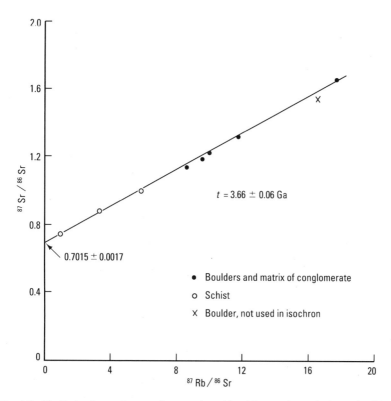

Fig. 4.9. Rb–Sr isochron diagram for samples of boulders and matrix from the felsic volcanogenic conglomerate unit and the related volcanogenic sediment (now a schist) at Isua. (After Moorbath et al., 1977a.)

The inclusion of data from both the conglomeratic and fine-grained phases of this formation on the same isochron is probably justified in view of the likelihood of their common origin. If the data are treated separately the precision is degraded, primarily because of the lack of "spread" along the isochron, but the results are not changed substantially. Further analysis of five additional samples of the conglomerate unit and related rocks gave a Rb–Sr isochron age of 3.71 ± 0.07 Ga (Jacobsen and Dymek, 1988), in agreement with the earlier results.

The Isua conglomerate unit has also been dated by the U–Pb concordia–discordia method from eight single zircon grains separated from two of the volcanic boulders in the conglomerate (Michard-Vitrac et al., 1977). The results of the eight analyses fall on a discordia

and increases the estimated uncertainty a bit—as it has no impact on the important conclusion that the conglomerate unit is more than 3.6 Ga.

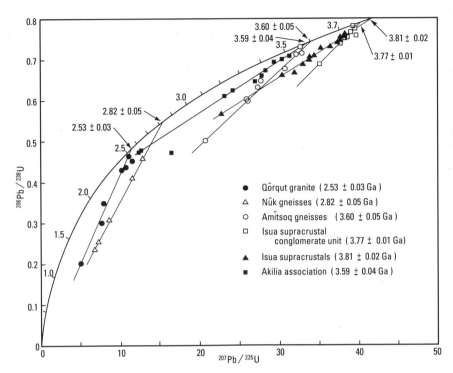

Fig. 4.10. U–Pb discordia diagram for zircons from samples of the Amîtsoq gneisses, the Nûk gneisses, the Qôrqut granite, and the Isua supracrustal conglomerate unit. (After Baadsgaard, 1976. Data on the Amîtsoq gneisses from Baadsgaard, 1973, on the Nûk gneisses and Qôrqut granite from Baadsgaard, 1976, and on the Isua supracrustal unit from Michard-Vitrac et al., 1977, and Baadsgaard, 1984.)

with an upper intercept of 3.77 ± 0.01 Ga (Fig. 4.10). Most of these values lie relatively close to the concordia, which indicates that these particular Isua rocks have not been highly disturbed, i.e. there has been relatively little loss of lead from the zircon grains. The oldest age for the Isua rocks is based on U–Pb analyses of 16 zircon concentrates from 10 samples representing various volcanic and sedimentary units within the Isua supracrustal sequence (Baadsgaard et al., 1984, who did not describe the exact units sampled). The analyses define a discordia with an intercept of 3.81 ± 0.02 Ga (Fig. 4.10).

As is common in metamorphic rocks, the U–Pb age of the zircons appears to be older than the Rb–Sr ages of the whole-rock samples, and an analysis of the uncertainties assigned to the measurements indicates that the two are, indeed, statistically different. When age data are interpreted, however, there are often factors in addition to the assigned errors that need to be considered. First, isochron and discordia errors are calculated from the statistical fit of the line to the data. They are primarily a measure of analytical precision and do not necessarily reflect the true uncertainty in accuracy. For example, they usually do not allow for the uncertainties in the decay constants or the isotopic abundances used in the calculations. They rarely, if ever, include any allowance for possible geological disturbances of the isotopic systems because the effects of such disturbances, even if recognized, are virtually impossible to quantify.

Second, the Rb–Sr and U–Pb systems may be (likely are) measuring different events. If the usual interpretation of U–Pb zircon data holds for the data from the conglomerate unit, then the age is probably near the time of crystallization of the original volcanic rock from which the conglomerate was derived. The Rb–Sr whole-rock isochron age, however, may represent the time of metamorphism of the Isua supracrustal unit. Thus, the two ages found by the two different methods are not necessarily inconsistent. It is not unusual for U–Pb zircon discordia ages of igneous rocks to be slightly older than the isochron ages calculated from other decay schemes. The usual interpretation is that the U–Pb ages reflect the time of crystallization of the zircons, whereas the other decay schemes record some time during the cooling of the rock. In the case of the conglomerate unit, the interpretation is even less certain, because it was originally a sedimentary rock. A conservative interpretation, therefore, is that the U–Pb and Rb–Sr ages represent events that may range from the time of formation of the original volcanic rocks that supplied the material for the conglomerate to the time of metamorphism of the Isua sequence. Thus these are minimum ages for the age of the original source material.

The garbenschiefer unit is a schist that occurs several places within the Isua sequence (Fig. 4.8). Field relations and chemical data indicate that it was probably intruded as a mafic igneous rock into the sedimentary sequence during metamorphism (Allaart, 1976: 183, Nutman et al., 1983: 8). Because of its original composition and ensuing metamorphism, the unit is not amenable to U–Pb, Pb–Pb, or Rb–Sr dating, but it has been dated by the Sm–Nd technique (Hamilton et al., 1978, 1983). Eight samples of the garbenschiefer unit and four of the conglomerate unit fall on a single Sm–Nd isochron with an age of 3.75 ± 0.04 Ga (Fig. 4.11). The data from the garbenschiefer unit alone give an isochron age of 3.73 ± 0.15 Ga. There is insufficient difference in Sm–Nd isotopic composition of the four conglomerate samples to yield a meaningful independent isochron age. The validity of including data from formations with two quite different origins and histories in the same isochron calculation is debatable because it is doubtful that there exists a simple genetic relationship between the two

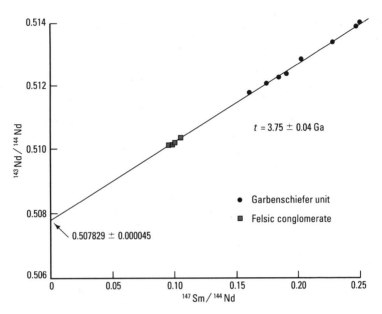

Fig. 4.11. Sm–Nd isochron diagram for samples from the garbenschiefer unit and the felsic conglomerate of the Isua supracrustals. (Data from Hamilton et al., 1978, 1983.)

units. However, the fact that these data fall on a common isochron indicates that they were derived essentially contemporaneously from a source or sources with indistinguishable initial Nd ratios and in that sense are consanguineous (Hamilton et al., 1978: 42). As in the case of the conglomerate unit, it is uncertain whether the Sm–Nd age represents the age of intrusion and crystallization of the garbenschiefer unit or the age of metamorphism of the Isua sequence.

Banded-iron formation, usually consisting of alternating layers of siliceous ironstone, quartz (recrystallized chert), and, sometimes, carbonate, occurs throughout the Isua sequence but is most abundant at the extreme northeast end of the supracrustal arc. Here the ironstones become sufficiently abundant that they constitute an ore body, partly buried by the inland ice sheet, with estimated reserves of some two billion tons (Allaart, 1976: 183–84). The banded-iron formation has been dated by the Pb–Pb method applied to samples of the siliceous ironstone, the interlayered carbonate rock, and both *magnetite* and *silicate* separated from the ironstone. The isochron age (Fig. 4.12) is 3.70 ± 0.07 Ga and is thought to represent the time of metamorphism, although a depositional age cannot be precluded (Moorbath, O'Nions, and Pankhurst, 1973: 139).

Although it is uncertain whether the various radiometric ages (Table 4.1) of formations within the Isua sequence represent ages of

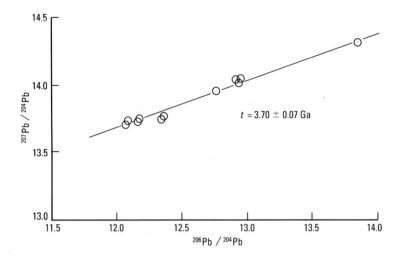

Fig. 4.12. Pb–Pb isochron diagram for samples of the banded-iron formation in the supracrustal sequence at Isua. (After Moorbath, O'Nions, and Pankhurst, 1973.)

metamorphism, deposition, or crystallization of parent igneous material, their consistency is remarkable, especially in view of the variety of units dated and the variety of methods used. The conclusion is inescapable that the Isua supracrustals are at least 3.7 Ga. The age of the source material for the conglomerate, as represented by the U–Pb age of zircons from the volcanic boulders, approaches, and may slightly exceed, 3.8 Ga, while the U–Pb zircon discordia for the mixture of supracrustal units suggests that these rocks are, indeed, in the neighborhood of 3.8 Ga.

Only one age measurement on the rocks of the Akilia association, the presumed equivalent of the Isua supracrustals, has been made (Baadsgaard et al., 1984). Zircons from samples collected near Akilia and Qilangarssuit, south of Godthaab (Fig. 4.13), fall on a ten-point U–Pb discordia with an upper intercept of 3.59 ± 0.04 Ga. This age probably records the intense metamorphism that accompanied the emplacement of the earliest gneisses at 3.6 Ga rather than the initial depositional age. Even though the age is considerably younger than the ages obtained for the Isua rocks, the data are consistent with the observation that the rafts and inclusions that constitute the Akilia association are more highly metamorphosed than the Isua sequence and that the Isua supracrustals generally escaped the intense metamorphism that affected their counterparts on the Greenland coast.

The Amîtsoq gneisses have been dated by four methods applied to samples from a variety of localities near Isua and Godthaab (Table 4.1, Fig. 4.13). Rb–Sr analyses of 82 whole-rock samples from six

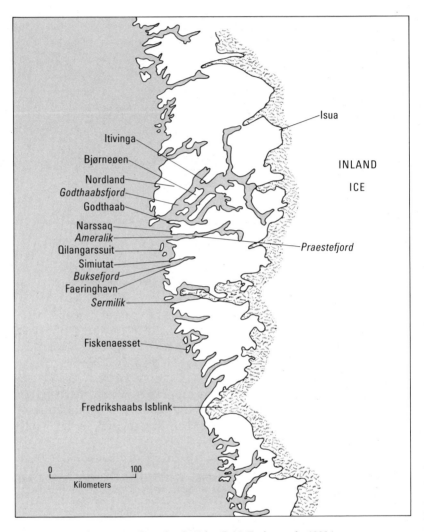

Fig. 4.13. Southern West Greenland. (After P. N. Taylor et al., 1980.)

localities give isochron ages that fall within the range of 3.56 to 3.67 Ga. Of particular interest is the concordance of ages from Isua and the Godthaab district. Moorbath and his colleagues, for example, analyzed 13 samples of Amîtsoq gneisses from the Isua area (Moorbath, O'Nions, and Pankhurst, 1975, Moorbath et al., 1977a). The samples were collected from gneiss veins that cut the supracrustal sequence, from gneisses near the contact with the supracrustals, and from gneisses several kilometers away from the supracrustal belt. The samples fall on an isochron indicating an age of 3.64 ± 0.06 Ga (Fig. 4.14). For comparison, 25 samples from localities near Narssaq, 15 km or so south of Godthaab, give an isochron age of 3.67 ± 0.09 Ga. Rb–Sr analyses of samples of the Amîtsoq gneisses from other localities give similar results (Table 4.1).

Other decay schemes, where applied, confirm the Rb–Sr ages (Table 4.1). Pb–Pb analyses of nine samples of the Amîtsoq gneisses near Isua, for example, give an isochron age of 3.74 ± 0.12 Ga. Lu–Hf analyses of samples from both Isua and the Godthaab district give an isochron age of 3.55 ± 0.22 Ga. Both of these studies were done on samples that were also used for some of the Rb–Sr analyses.

H. Baadsgaard, of the University of Alberta, has analyzed zircons from the Amîtsoq gneisses near Godthaab and Isua (Baads-

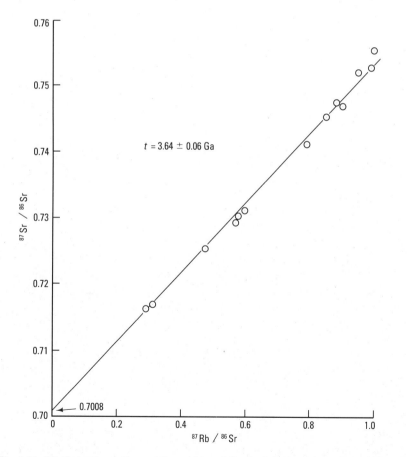

Fig. 4.14. Rb–Sr isochron diagram for samples of the Amîtsoq gneisses at Isua. (After Moorbath et al., 1977a.)

gaard, 1973, 1983, Baadsgaard, Lambert, and Krupicka, 1976). The
Godthaab data fall on a U–Pb discordia with an upper intercept of
3.60 ± 0.05 Ga (Fig. 4.10). These results have been confirmed by
Michard-Vitrac et al. (1977), who analyzed eight single zircon grains
from two samples near Godthaab. These authors did not calculate a
separate discordia age for their results, but commented that most of
their zircon data agreed, within analytical uncertainties, with the
3.60-Ga discordia age found by Baadsgaard. The data from the Isua
zircons, of which Baadsgaard analyzed 22 samples, are somewhat
more scattered than the Amîtsoq data but suggest an age of about 3.7
Ga. Baadsgaard (1983: 41) has speculated that the parent material of
the Amîtsoq gneisses near Isua was intruded more than 3.7 Ga and
that the Isua zircons were not as completely reset during initial metamorphism as the Godthaab zircons, which now record the time that
the original material was first converted to gneiss.

Although isotopic data from zircons and whole-rock samples of
the Amîtsoq gneisses give consistent results that indicate an emplacement age of 3.7 Ga or more and a (probable) age of initial metamorphism of about 3.60–3.65 Ga, other minerals have been highly disturbed by more recent events (Pankhurst, Moorbath, and McGregor,
1973, Baadsgaard, Lambert, and Krupicka, 1976, Baadsgaard, 1983).
For example, individual K–Ar and ^{40}Ar/^{39}Ar ages on the minerals *biotite, hornblende,* and *muscovite* are inconsistent and range from 1.67 Ga
to an impossibly high 4.85 Ga. U–Th–Pb and Rb–Sr measurements
on a variety of separated minerals likewise show a confusing pattern
rather than the straightforward isochron relationships of the whole-
rock samples. The data suggest that the mineral systems were open
isotopic systems at about 2.7 and again at about 1.6 Ga.

The U–Pb discordia method for zircons is known for its ability
to "see through" later metamorphism to the crystallization age of a
rock, so the 3.6-Ga ages from this method are not surprising. But why
do whole-rock samples of the Amîtsoq gneiss give consistent Rb–Sr,
Lu–Hf, and Pb–Pb ages when the minerals (except zircon) from
these same rocks clearly were open systems during later metamorphism? A logical answer is that when the rocks were heated, the isotopes moved, but not very far, so that the isotopes were homogenized
only within small portions of the rock, not throughout the entire rock
body (Moorbath, 1975c, 1977b). During heating a ^{87}Sr atom, for example, will move out of the Rb-bearing mineral where it was created
by radioactive decay but in whose crystal structure it does not fit com-
fortably, only to be captured by an adjacent calcium mineral (e.g. plagioclase) where it can substitute easily for that element. Thus, rock
samples that contain many thousands of individual mineral grains re-

main systems essentially closed to ^{87}Sr even though the individual mineral grains do not. The same argument can be applied to the other isotopes used in dating these rocks. This explanation, or some similar mechanism, must be correct for there can be little doubt that the whole-rock isochrons represent the age of the gneisses—the data in Table 4.1 are far too consistent to be explained as mere coincidence.

Several of the units that postdate the Amîtsoq gneisses have been dated, and the results are consistent with their known stratigraphic positions (Table 4.1). The layered anorthosite unit gives Rb–Sr, Pb–Pb, and ^{40}Ar/^{39}Ar ages of 2.75 to 2.83 Ga. The Nûk gneisses and associated rocks from a variety of localities from both inland and along the coast of Western Greenland give Rb–Sr, Pb–Pb, and U–Pb zircon ages (Fig. 4.10) that range from 2.60 ± 0.12 to 3.08 ± 0.03 Ga. It is likely that the various rock bodies that have been included as Nûk gneisses were emplaced by multiple intrusions over a long period of time, and that the range of ages found for the Nûk gneisses probably represents real differences in age. Finally, the age of the Qôrqut granite has been determined as about 2.5 Ga by Rb–Sr, Pb–Pb, and U–Pb methods.

The Archean rocks of the Labrador coast in the region near Saglek and Hebron (Fig. 4.15) are similar in nearly all respects to those of western Greenland (Bridgwater et al., 1975, Bridgwater, Collerson, and Myers, 1978, Collerson, Jesseau, and Bridgwater, 1976, Collerson, Kerr, and Compston, 1981, Bridgwater and Collerson, 1976, and Baadsgaard, Collerson, and Bridgwater, 1979, from whom the following summary is taken). This is not at all surprising, because the Saglek–Hebron area was very close to the Godthaab area before the breakup of the North Atlantic craton (Fig. 4.6). Like western Greenland, the Saglek–Hebron area is dominated by granitoid gneisses of two general ages. The older gneisses (Uivak gneisses[9]) contain highly metamorphosed rafts and inclusions of supracrustal rocks (Nulliak assemblage) as much as 2 km long by 50 to 100 m wide (Fig. 4.7). These supracrustals include metamorphosed mafic and ultramafic intrusive rocks, volcanic rocks, banded-iron formation, chert, limestones, and clastic sediments derived from volcanic source rocks. The Uivak gneisses are intruded by the Saglek mafic dikes, which, like the Ameralik dikes of western Greenland, do not cut younger rocks.

9. On the basis of texture and mineralogy, the Uivak gneisses of the Saglek area have been subdivided into the Uivak I and Uivak II, but in places they are indistinguishable and probably are very closely related. Near Hebron, the older gneisses have been called the Hebron gneisses, but they are thought to be equivalent to the Uivak gneisses. For this discussion the Uivak I, Uivak II, and Hebron gneisses are collectively called the Uivak gneisses.

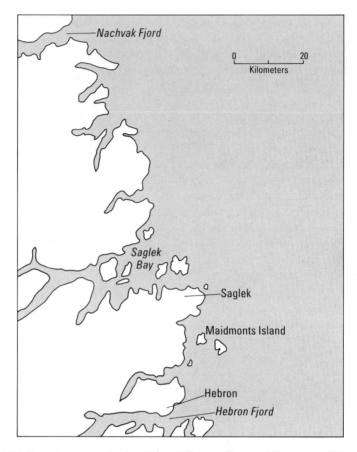

Fig. 4.15. Part of eastern Labrador. (After Collerson, Kerr, and Compston, 1981.)

The Uivak gneisses are tectonically interlayered with the Upernavik supracrustal sequence, which is not cut by the Saglek dikes and is therefore younger than the Uivak gneisses. The Upernavik supracrustals, which occupy the same stratigraphic position as the Malene supracrustals of western Greenland, occur in lenticular bodies as much as 2 km thick and several tens of kilometers long. These rocks are predominantly metamorphosed sediments derived from volcanic sources, but intrusive mafic and ultramafic bodies, pillow lavas, banded-iron formation, and limestones are also present.

The Upernavik supracrustals and the Uivak gneisses are intruded by thin sheets of anorthosite and associated *gabbro* and by younger granitoid gneisses that appear to be equivalent in both age and stratigraphic position to the Nûk gneisses. These younger gneisses, which have been given different names in different areas, can be divided into two types on the basis of their original parent material. The first type, which includes the Kiyuktok and Iterungnek gneisses near Saglek Bay, was formed by *remobilization* of Uivak gneisses. The second type, which includes the Ikarut gneisses near Hebron and the Kammarsuit gneisses near Nachvak Fjord, consists of metamorphosed intrusive rocks and, like the Nûk gneisses, represents the addition of new granitoid material to the crust shortly before its metamorphism to gneisses. The younger gneisses are intruded by post-tectonic granites and, finally, by mafic dikes.

The Uivak gneisses from both the Saglek and Hebron areas give Rb–Sr isochron ages in the range of 3.55 to 3.61 Ga (Table 4.1, Fig. 4.16). There has been, however, some controversy about the Hebron data. Barton (1975) originally obtained a Rb–Sr isochron age of 3.54 ± 0.10 Ga for six whole-rock samples. Collerson et al. (1982), using Barton's samples plus new collections, found much more scatter in the

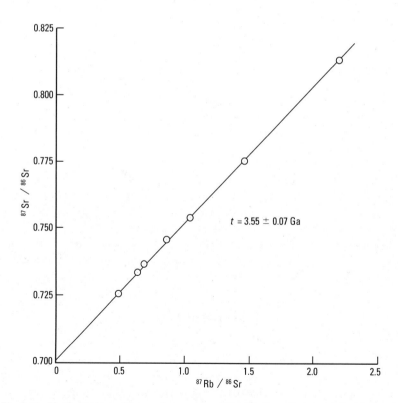

Fig. 4.16. Rb–Sr isochron diagram for samples of the Uivak gneisses near Saglek, eastern Labrador. (After Hurst et al., 1975.)

data than did Barton and claimed that the isochron age, although near 3.6 Ga, had a much larger uncertainty than indicated by Barton. Barton (1982), using improved methods, reanalyzed his samples, confirmed both the original age and the excellent isochron fit, and concluded that the scatter in the data of Collerson and his colleagues was probably due to their use of inhomogeneous samples. In spite of this controversy, the age of the Uivak gneisses is not now in question, having been confirmed by additional Rb–Sr dating and by a U–Pb discordia age of 3.76 ± 0.15 Ga on zircons from samples collected at Maidmonts Island (Table 4.1). Furthermore, a 28-point Sm–Nd isochron for a combined lot of Uivak and Amîtsoq whole-rock samples indicates an age of 3.56 ± 0.20 Ga.

As in western Greenland, mineral analyses of samples from Labrador tend to give younger ages than do whole-rock measurements. U–Th–Pb analyses of apatite and sphene from the Uivak gneisses reflect primarily younger metamorphic events, while even U–Pb zircon data are highly variable, not falling on a convincing discordia (Baadsgaard, Collerson, and Bridgwater, 1979).

Younger gneisses from Saglek, Hebron, and Nachvak that were formed from consanguineous (or nearly so) granitoid intrusions give Rb–Sr isochron ages of 2.69 to 2.81 Ga, which are consistent with the ages of the Nûk gneisses in western Greenland. The Kiyuktok gneisses near Saglek, however, give inconsistent Rb–Sr results and a Pb–Pb age that more nearly reflects the age of the parent Uivak gneisses, from which they were apparently derived, than the age of the younger metamorphism (Table 4.1).

In summary, the early Archean rocks of western Greenland and eastern Labrador are very similar in type, stratigraphy, and geochronology, and there is little doubt that the North Atlantic Craton was once a single entity. The multiplicity of isotopic age measurements on these rocks also leaves little doubt about their antiquity. The Amîtsoq–Uivak gneisses and the earlier Isua supracrustal rocks are 3.5–3.7 and 3.7–3.8 Ga in age, respectively, and are among the oldest and best-dated rocks on Earth.

Superior Province

The Superior Province of North America, with an area of 2.6 × 10^6 km^2, is the world's largest known Archean crustal block (Fig. 4.1). Most of the exposed rocks are late Archean with radiometric ages of 2.6 to 2.8 Ga (Condie, 1981: 12). In the southwest part of the Superior Province, however, there exists a small area where early Archean rocks have been found and dated. Some of these rocks appear to be at least 3.5–3.6 Ga in age.

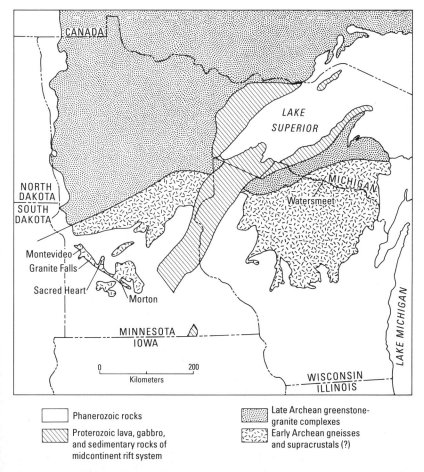

Fig. 4.17. Known and inferred distribution of Archean basement rocks in the southwestern part of the Superior Province. Except for the rocks of the midcontinent rift system, the Proterozoic rocks that cover much of the Archean rocks are not shown. Archean basement probably extends for some distance west and south beneath the Phanerozoic cover. (After Sims and Peterman, 1981.)

The area immediately to the west and south of Lake Superior consists of two distinct terranes (Fig. 4.17), as described by Morey and Sims (1976), Sims (1980), Condie (1981), and Sims and Peterman (1981). The northern terrane resembles large areas of the Superior Province to the north and east and consists of a complex sequence of greenstone belts intruded by granitoids. The greenstones are mildly metamorphosed but deformed mafic and felsic volcanic and intrusive rocks, volcanogenic sediments, and banded-iron formation that were originally deposited in shallow seas either on oceanic crust or within the borderland of an ancient protocontinent. Radiometric dating shows that the age of this greenstone–granitoid terrane is primarily 2.6 to 2.8 Ga.

The basement rock of the southern terrane consists of a highly deformed and metamorphosed gneiss complex, the age of most of which is more than 3.0 Ga, that was intruded by later granitoids. The boundaries of this ancient terrane are not well known, but it is likely that Archean rocks extend for some distance to the west and south beneath the Phanerozoic cover.

The two terranes are separated by a highly deformed tectonic zone some 25–40 km in width. They are thought to have originated as separate terranes and to have been joined in the late Archean, perhaps as the result of a continental collision. Both terranes are rent by the mid-continent rift system, a Proterozoic continental disjunction in which a thick sequence of mafic lavas and sediments were deposited.

The history of the early Archean rocks in the southern terrane has been difficult to decipher for two reasons. First, the gneisses are mostly covered by Proterozoic rocks and are exposed only in small areas. Second, they were formed and have been affected by a complex series of geologic events that included multiple intrusion, metamorphism, deformation, and heating over an interval of nearly 2 Ga. Early attempts to determine the ages of the ancient gneisses were only marginally successful and the results were somewhat controversial.[10] Fortunately, recent field, geochemical, and isotopic work has led to a clearer understanding of the origin of the gneisses, and there has been much progress in deciphering the history and ages of these rocks.

Two areas where the oldest gneisses occur have been studied in some detail. These include an elongate exposure within the Minnesota River Valley in the vicinity of Morton and Granite Falls and a small area of gneiss exposed near Watersmeet in southern Michigan (Fig. 4.17). In both areas, enclaves and inclusions of *amphibolites* occur within the gneisses. Field evidence suggests that some of the amphibolites are older and some are younger than the gneisses. The chemical composition of the older amphibolites suggests that their progenitors were mostly mafic and ultramafic lavas. The younger amphibolites may be remnants of mafic dikes. The amphibolites have not proven amenable to radiometric dating and their original age is unknown. The ages of the old gneisses, however, have been determined and they appear to be at least 3.5 to 3.6 Ga.

10. It would serve no purpose here to recount this history, but readers interested in the matter should read, in order, Catanzero (1963), Goldich, Hedge, and Stern (1970), Goldich and Hedge (1974), Farhat and Wetherill (1975), Goldich and Hedge (1975), and Michard-Vitrac et al. (1977). The matter is summarized and resolved by Goldich et al. (1980) and Goldich and Wooden (1980).

The Morton Gneiss, named after the nearby town of Morton, is a complex rock consisting of multiple rock units that were formed at different times and are now interlayered on a fine scale. The oldest unit is a granitoid gneiss that originated as an igneous rock emplaced into a sequence of volcanic rocks, now represented by the amphibolite inclusions, at least 3.5–3.6 Ga. This gneiss and the amphibolites were intruded by two later granitoid (now gneiss) phases about 3.0 and 2.6 Ga. The three granitoid phases differ somewhat in composition and are usually called by rock names, but they will here be referred to simply as the older, intermediate, and younger granitoid gneisses. The gneisses were intruded by granitoid dikes 2.6 Ga, and were affected by at least two intense periods of metamorphism at 3.0 and 2.6 Ga and by a regional heating associated with the emplacement of mafic dikes about 1.8 Ga. The evidence for this sequence of events is complex and will not be reviewed here, but the principal evidence for the age of the oldest granitoid gneiss and some of the younger associated rocks is of interest and worth brief discussion.

Much of the uncertainty, mentioned above, in early attempts to determine the age of the old gneisses in the Minnesota River Valley was due to the failure to recognize the complexity of the gneisses and, in not doing so, inadvertently mixing rock types of different ages. This problem was resolved by detailed chemical and *petrographic* work that allowed the distinction of the various units and led to more appropriate sampling (Goldich and Hedge, 1975, Goldich and Wooden, 1980, the latter a comprehensive summary of the geochronology of the Morton gneiss and associated rocks). A 26-point Rb–Sr isochron on whole-rock samples from the older granitoid phase of the Morton Gneiss gives an imperfect isochron of 3.48 ± 0.11 Ga (Fig. 4.18, Table 4.2). The scatter of data about the isochron is more than can be accounted for by analytical error, indicating some disturbance of the Rb–Sr system. S. S. Goldich and J. L. Wooden of the U.S. Geological Survey, who, with their colleagues C. E. Hedge and Z. E. Peterman, also of the U.S. Geological Survey, were instrumental in deciphering the geochronology of this area, have proposed that the scatter of data is due to a gain of Rb during a *high-grade metamorphism* at 3.05 Ga and have developed geochemical arguments to support this hypothesis.

U–Pb analyses of zircon from the older gneiss likewise generate scattered values. Results from the zircons analyzed do not all fall precisely on a common discordia (Fig. 4.19), but when two of the data are omitted, the remaining eight fall on a reasonably good discordia with an upper intercept of 3.59 ± 0.12 Ga and a lower intercept of 1.91 ± 0.17 Ga, which is approximately concordant with the latest episode of heating. Although the fit to the discordia is greatly improved, the ex-

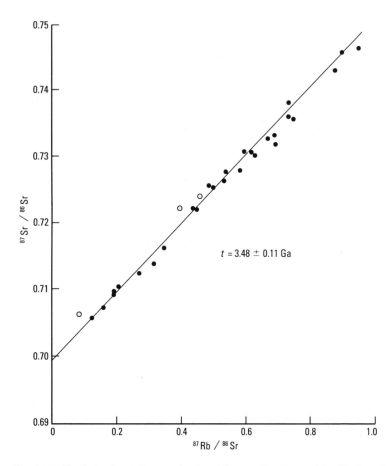

Fig. 4.18. Rb–Sr isochron diagram for the older granitic phase of the Morton Gneiss. The data represented by open circles were not included in the isochron calculation because they are from recrystallized veins within the older granitic gneiss. (After Goldich and Wooden, 1980.)

clusion of the two data, one on technical grounds and the other because of its derivation from a sample unusually low in uranium, is questionable. A more conservative interpretation of these data is provided by the intercept of a line drawn through the origin and the three least discordant points, which represent zircons of different size from the same sample (Goldich and Wooden, 1980: 86, 89). This intercept of 3.3 Ga is, in fact, the [207]Pb/[206]Pb age of the three samples and is a reasonably certain minimum age for the gneiss. The rationale for this interpretation is that the most recent time that lead could be lost is today, in which case the "discordia" must pass through the origin, i.e. $t = 0$; any earlier lead loss would result in an older upper intercept.

TABLE 4.2

*Radiometric Ages of Some Early Archean and
Related Rocks of the Superior Province, North America*

Unit and locality[a]	Method	Number of samples[b]	Age[c] (Ga)	Source
Felsic dikes				
Morton, Minn.	Rb–Sr	6w	2.59 ± 0.04	Goldich and Wooden, 1980
Sacred Heart Granite				
Morton, Minn.	Pb–Pb	6m	2.60 ± 0.01	Doe and Delevaux, 1980
	Rb–Sr	5m	2.64	Goldich, Hedge, and Stern, 1970
	Rb–Sr	4m	2.32 ± 0.10	Tsunakawa and Yanagisawa, 1981
Puritan Quartz Monzonite				
Watersmeet, Mich.	Rb–Sr	8w	2.65 ± 0.14	Sims, Peterman, and Prinz, 1977
Granite near Thayer				
Watersmeet, Mich.	U–Pb	5z	2.74 ± 0.06	Peterman, Zartman, and Sims, 1980
Younger granitoid gneiss				
Morton, Minn.	Rb–Sr	10w	2.56 ± 0.06	Goldich and Wooden, 1980
Intermediate granitoid gneiss				
Morton, Minn.	Rb–Sr	6w	2.64 ± 0.12	Goldich and Wooden, 1980
Older granitoid gneisses				
Morton, Minn.	U–Pb	8z	3.59 ± 0.12	Goldich and Wooden, 1980
	U–Pb	19z	3.54 ± 0.04	Kinney, Williams, and Compston, 1984
	U–Pb	7z	3.66 ± 0.04	Goldich and Fischer, 1986
	Rb–Sr	26w	3.48 ± 0.11	Goldich and Wooden, 1980
Granite Falls, Minn.	U–Pb	1z	>3.23	Goldich et al., 1980
	Rb–Sr	10w	3.68 ± 0.07	Goldich et al., 1980
Watersmeet, Mich.	U–Pb	9z	3.62 ± 0.01	Kinney, Williams, and Compston, 1984

NOTE: Units are listed in stratigraphic order when this order is known. All of the ages are based on isochron (Rb–Sr, Sm–Nd, Lu–Hf) or concordia–discordia (U–Pb) techniques.
[a] Localities are approximate as samples are generally collected over a broad area.
[b] w, whole rock; z, zircon; m, various minerals from same rock.
[c] All ages calculated with the decay constants in Table 3.1. Errors are at the 95% confidence level (two standard deviations). Ages underlined are shown as figures.

Recent isotopic analyses of the innermost growth zones of individual zircon crystals have been obtained by means of an *ion microprobe*. This remarkable new instrument is capable of yielding precise isotopic analyses on tiny (diameter 0.025 mm) portions of single mineral grains. The results provide a precise U–Pb discordia age of 3.54 ± 0.04 Ga for the Morton Gneiss (Kinney, Williams, and Compston, 1984). The ion-microprobe data also show that these inner zones have been overgrown by rims of younger material during subsequent metamorphisms. This has been confirmed by analyses of seven zircon samples whose outer rims were removed by air abrasion, leaving only

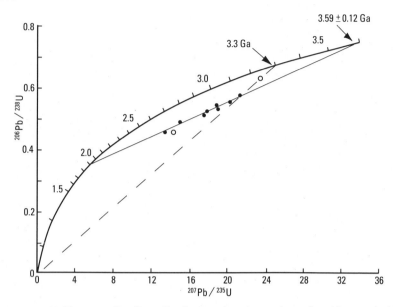

Fig. 4.19. U–Pb concordia–discordia diagram for zircons from the older granitoid phase of the Morton Gneiss. The zircons represented by open circles were not used to determine the solid discordia line. A line (dashed) from the origin through the point representing the least discordant zircon (open circle at right) defines a minimum age of 3.3 Ga. (After Goldich and Wooden, 1980.)

the inner (and older) cores (Goldich and Fischer, 1986). Results from these abraded zircons plot on a discordia with an upper intercept of 3.66 ± 0.04 Ga, which is the oldest precise age for the Morton Gneiss yet measured. Thus, the scatter from the earlier analyses of bulk zircon samples appears to be due to inhomogeneities on a very small scale.

In the vicinity of Granite Falls and Montevideo the gneisses can be grouped into mafic and felsic types, which are interlayered. The mafic gneisses were formed from sediments and from mafic lavas and intrusive rocks. Their age has not been determined radiometrically. The felsic gneisses were formed from igneous rocks of several granitoid compositions and fall into two age groups, the earliest about 3.6–3.7 Ga and a younger group about 3.0–3.1 Ga in age (Goldich et al., 1980). In addition to the events that formed these older igneous rocks, there is evidence of intense metamorphism and deformation accompanied by the intrusion of granitoid bodies about 2.6 Ga, and of a regional heating accompanied by igneous activity about 1.8 Ga. As in the Morton area, the older felsic gneisses contain enclaves and inclusions of amphibolite, but the interlayered mafic gneisses, common in the Granite Falls area, have not been found in the Morton gneiss.

The interlayering of felsic and mafic gneisses near Granite Falls may be due to (1) intrusion of granitoids into an older sequence of volcanic and sedimentary rocks, (2) intrusion of granitoids into an older mafic basement complex, or (3) deformation. Although the choice among these possibilities is not clear, most workers who have studied the area prefer the first explanation (Goldich et al., 1980: 19–20, 40).

As in the Morton area, the isotopic systems in the gneisses near Granite Falls have been disturbed by the many metamorphisms and heatings to which they have been subjected, and much of the radiometric data are difficult to interpret. Rb–Sr analyses of ten whole-rock samples of the older felsic gneisses give a reasonably good isochron indicating an age of 3.68 ± 0.07 Ga (Table 4.2). There are insufficient U–Pb zircon analyses of the older felsic gneisses for a discordia intercept to be determined, but if the discordance is due to recent lead loss (as with the Morton zircon, Fig. 4.19), the least discordant zircon has a minimum age, i.e. a ^{207}Pb/^{206}Pb age, of 3.23 Ga. The Sacred Heart Granite, a plutonic body emplaced during a period of intense deformation and metamorphism, appears to be about 2.6 Ga old on the basis of Rb–Sr and Pb–Pb analyses (Table 4.2).

Near Watersmeet in northern Michigan (Fig. 4.17), in an area within the boundary zone between the late Archean granite–greenstone terrane and the early Archean gneiss terrane, there is an elliptical body of gneiss measuring 25 × 10 km. This gneiss dome, known informally as the "gneiss near Watersmeet," is composed primarily of a felsic granitic gneiss intruded by felsic dikes (Sims and Peterman, 1976, Peterman, Zartman, and Sims, 1980). It is surrounded and overlain by metamorphosed Proterozoic volcanic and sedimentary rocks. The composition of the gneiss suggests that it was originally an igneous rock, but whether volcanic or plutonic is not known. To the west of the gneiss dome there are small exposures of later granitoids, among them the Puritan Quartz Monzonite and several bodies known collectively as the "granite near Thayer."

The Rb–Sr data on samples of the older gneisses near Watersmeet record only the severe episode of metamorphism and deformation that occurred about 1.7–1.8 Ga—evidence of the pre-1.8-Ga history of these rocks has been erased. Results of U–Pb analyses of eight bulk zircon samples from the older gneiss do not all fall precisely on a common discordia. Regression of all of the data gives upper and lower intercepts of 3.78 and 1.74 Ga, respectively, but the failure of the points to lie in a line indicates that the U–Pb systems in the zircons have been adversely affected by the several metamorphisms, and the intercepts do not provide satisfactory ages for the gneiss. Analyses of portions of individual zircon crystals with the ion microprobe, however, show that the zircons are composite grains with several generations of overgrowths. Those inner growth zones lowest in uranium (and, therefore, presumably the oldest) plot on a well-defined discordia with an upper intercept of 3.62 ± 0.01 Ga (Kinney, Williams, and Compston, 1984). Rb–Sr and U–Pb ages of the granite near Thayer and the Puritan Quartz Monzonite show that these younger intrusive bodies are late Archean (Table 4.2).

An additional bit of evidence bearing on the antiquity of the older gneiss near Watersmeet is the Sm–Nd model age calculated by M. T. McCulloch and G. J. Wasserburg (1980) of the California Institute of Technology. This method, the accuracy of which is dependent on the assumptions of the model, is based on the observation that when the initial $^{143}Nd/^{144}Nd$ ratios of crustal rocks and of chondrite meteorites are plotted against their age, the data fall on or very near a straight line (for reviews of this method, see McCulloch and Wasserburg, 1978, O'Nions et al., 1979, and DePaolo, 1981). This line (Fig. 4.20, line $A–B$) is thought to represent the evolution of Nd isotopes with time in the mantle. When magma is separated from the mantle and emplaced as crust, the Sm/Nd ratio of the new rock is changed (lowered) because of chemical fractionation, and the Nd isotopic composition of the new rock will evolve along a different path (Fig. 4.20, line $C–D$). If we presume that this model is correct, and there is evidence that it is, at least to a first approximation, then it is possible to find the time when a crustal rock was derived from the mantle by measuring its present-day $^{143}Nd/^{144}Nd$ and $^{147}Sm/^{144}Nd$ ratios and extrapolating back in time to find the intersection of the two evolution paths. For the gneiss at Watersmeet, this intersection lies at 3.60 ± 0.04 Ga, which is in excellent agreement with the U–Pb discordia age based on ion-probe analyses. How sound is this model age? That question is not easy to answer, but it is safe to say that it is not nearly as convincing as an isochron age. Nd model ages are dependent on the assumptions of the model, in this case a uniform and known mantle reservoir and a simple one-stage history for the rock. In addition, the method has no built-in self-checking features as do isochrons and concordia–discordia diagrams. But the agreement between the Nd model age and the U–Pb discordia age probably is not a coincidence because such concordance has been found elsewhere. For example, the Nd model age of the Amîtsoq gneisses of Greenland is 3.62 ± 0.09 Ga (DePaolo and Wasserburg, 1976, McCulloch and Wasserburg, 1980: 137), which is in impressive agreement with the various isochron ages of that unit (Table 4.1).

The Rb–Sr and U–Pb isotopic data on the older gneisses near Morton, Granite Falls, and Watersmeet indicate that these units are

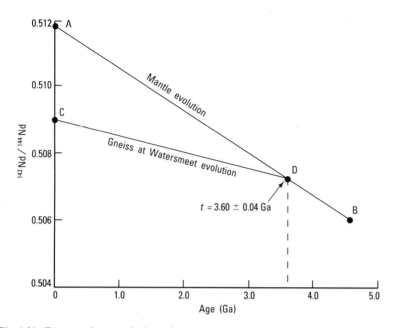

Fig. 4.20. Diagram showing the basis for the Sm–Nd model age of the gneiss at Watersmeet. Line *A–B* is the change in ^{143}Nd/^{144}Nd with time in the mantle. Point *C* is the measured value for the gneiss at Watersmeet. The slope of line *C–D* is determined from the Sm/Nd ratio of the gneiss. The intersection of the two lines, point *D*, defines the time the parent magma of the gneiss was separated from the mantle. (After McCulloch and Wasserburg, 1980.)

early Archean, probably about 3.6 Ga, and thus comparable in age to the Amîtsoq gneisses of western Greenland. This conclusion is strengthened, though not proven, by the 3.60-Ga Sm–Nd model age of the gneiss near Watersmeet. Like the Amîtsoq gneiss and its counterparts in Labrador, the igneous progenitors of the ancient gneisses of the Superior Province appear to have been emplaced into supracrustal lava flows and sediments whose age is unknown. All data considered, the gneisses of the Superior Province qualify as some of the Earth's oldest known crustal rocks.

Pilbara Block

The western part of the Australian continent consists of two Archean cratons surrounded largely by Proterozoic mobile belts (Fig. 4.21). The Yilgarn Block, the more southerly of the two, covers an area of nearly 650,000 km² and extends to the northeast beneath Proterozoic sedimentary cover. The Archean rocks of the Pilbara Block

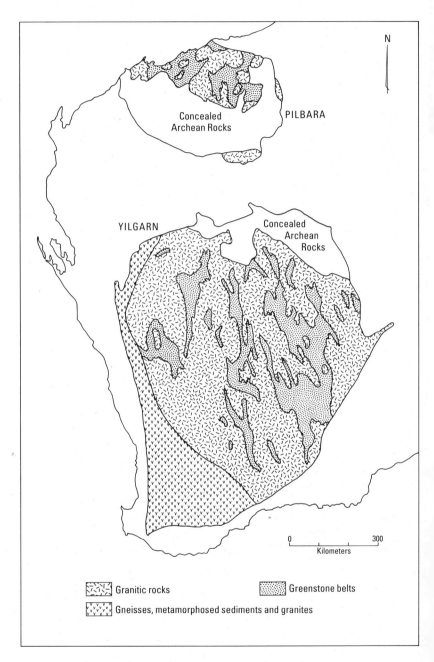

N

PILBARA

Concealed
Archean Rocks

YILGARN

Concealed
Archean
Rocks

0   300
Kilometers

Granitic rocks

Greenstone belts

Gneisses, metamorphosed sediments and granites

Fig. 4.21. Principal rock types of the Archean Pilbara and Yilgarn blocks, Western Australia. (After Rutland, 1981.)

are exposed over an area of only about 56,000 km², but nearly half of the craton is covered on the south by Proterozoic sedimentary rocks, so its total area may exceed 100,000 km². The area between the Yilgarn and Pilbara blocks is underlain by reworked Archean basement, suggesting that at one time the two blocks may have been connected.

Both the Pilbara and the Yilgarn blocks are granitoid–greenstone terranes with thick sequences of mildly metamorphosed volcanic and sedimentary rocks into which several generations of granitoid rocks have been emplaced.[11] There, however, the similarity ends. The Yilgarn Block contains large areas of granitoid gneiss with inclusions and rafts of older supracrustal rocks, whereas extensive areas of gneiss have not been found in the Pilbara Block. There are also differences in the styles of deformation as well as in the ages of the major rock units and tectonic events, leading the geologists who have worked on the history of these cratons to conclude that the two blocks evolved independently.

The Pilbara Block consists of complex, domelike plutons surrounded by deformed, basin-shaped greenstone belts (Figs. 4.22 and 4.5). At one time it was thought that the distribution of the present greenstone belts reflected original depositional basins, but extensive and detailed mapping by A. H. Hickman, S. L. Lipple, and their colleagues of the Geological Survey of Western Australia has shown that the greenstones were deposited as a continuous, tabular stratigraphic sequence and have been deformed into their present distribution and geometry. There is no doubt that the present domes were emplaced into the greenstones—the field relations clearly show that all granitoid—greenstone contacts are either magmatic or tectonic, not depositional—but there are two general hypotheses to explain the age and structural relationships between the greenstones and the granitoids. According to some, as summarized by A. H. Hickman (1981: 57), the greenstones were deposited on oceanic crust and later intruded by granitoid magma, while others surmise that the greenstones were deposited on granitoid basement, remnants of which may still exist within the complex granitoid domes. According to this latter view, for which there is now considerable evidence, the granitoid domes were emplaced in their present form not as liquid magma but by solid-state mobilization and doming of pre-existing granitoid crust, and are probably connected at depth. Detailed mapping of the Shaw Batholith (Fig. 4.22), for example, has shown that its internal structure resembles a gneiss terrane with supracrustal inclusions, incorporated

11. The literature on the Precambrian of Western Australia is voluminous. I have relied primarily on the articles by Blockley (1975), A. H. Hickman (1975, 1981, 1983), Lipple (1975), Gee (1979), Glikson (1979), Rutland (1981), and Hallberg and Glikson (1981).

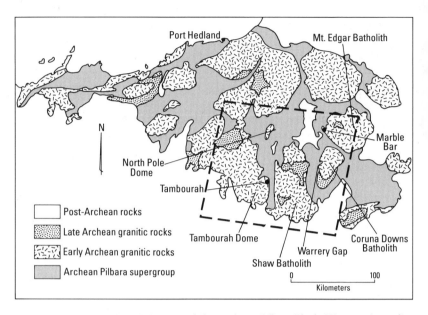

Fig. 4.22. Generalized geologic map of the Archean Pilbara Block, Western Australia. The late Archean granitic rocks were emplaced ca. 3.0 Ga, after the major deformation that deformed the Pilbara Supergroup and accompanied the emplacement of the early Archean granitic domes. Dashed line shows the approximate area shown in the Landsat photograph of Figure 4.5. (After A. H. Hickman, 1981.)

by solid-state deformation, that can be traced into the surrounding greenstone belt (Bickle et al., 1980, 1985).

The formations that compose the Archean greenstones of the Pilbara Block are collectively called the Pilbara Supergroup (Table 4.3). This rock sequence has a cumulative thickness of about 30 km, although its actual thickness in individual areas does not exceed about 10 km. It is clear that part of the Pilbara sequence was deposited in shallow seas, for the rocks include pillow lavas, cherts, banded-iron formation, and shallow-water clastic sediments, but whether on oceanic or continental crust is uncertain.

Not all of the formations of the Pilbara Supergroup occur everywhere within the Pilbara Block. There are some, however, that can be traced over great distances, and it is these "marker" formations that demonstrate that the greenstones were deposited as a tabular unit over the whole of the Pilbara rather than as isolated sequences in separate basins. The Towers Formation, for example, is the principal marker within the Warrawoona Group. It consists of three distinctive chert units separated by volcanic and volcanogenic sedimentary rocks, and can be traced from east to west for 430 km across the whole of the

TABLE 4.3

Stratigraphy of the Archean Pilbara Supergroup, Pilbara Block, Western Australia

Group and subgroup	Formation	Principal rock types	Thickness (km)	Age (Ga)
——	Negri Volcanics	basalt, andesite	0.2	
——	Louden Volcanics	basalt, ultramafic rocks	1.0	>2.6
Whim Creek Group	Rushall Slate	slate	0.2	
	Mons Cupri Volcanics	felsic volcanic rocks	0.5	>2.6
	Warambi Basalt	basalt	0.2	
Gorge Creek Group	Mosquito Creek Formation	sandstone, shale	5.0	>3.0
	Lalla Rookh Sandstone	sandstone, conglomerate	3.0	
	Honeyeater Basalt	basalt	1.0	
Soansville Subgroup	Cleaverville Formation	banded-iron formation, chert	1.0	
	Charteris Basalt	basalt	1.0	
	Corboy Formation	sedimentary rocks	1.5	
Warrawoona Group	Wyman Formation	rhyolite	1.0	
Salgash Subgroup	Euro Basalt	pillow basalt, ultramafic rocks	2.0	
	Panorama Formation	felsic volcanic rocks, chert	1.0	
	Apex Basalt	basalt, ultramafic rocks	2.0	>3.3
	Towers Formation	chert, basalt, felsic volcanic rocks	0.5	
——	Duffer Formation	felsic volcanic rocks	5.0	3.45
Talga Talga Subgroup	Mount Ada Basalt	pillow basalt, chert, banded-iron formation	2.0	
	McPhee Formation	chert, ultramafic rocks	0.1	
	North Star Basalt	basalt, ultramafic rocks, chert	2.0	3.56

SOURCES: Lipple, 1975; A. H. Hickman, 1981; Jahn et al., 1981.

Pilbara Block. Likewise, the McPhee and Panorama formations can be traced for great distances and serve as marker units within the Talga Talga and Salgash subgroups. The Cleaverville Formation, which consists primarily of banded-iron formation and maintains a thickness of nearly a kilometer over most of the Pilbara Block, is a prominent marker within the Gorge Creek Group. The exact ages of most of the formations of the Pilbara Supergroup, particularly the younger units, are unknown, but both the Warrawoona and Gorge Creek groups are intruded by granitoids 2.9–3.3 Ga in age, and the lower part of the Warrawoona, which is relatively well-dated, is 3.5–3.6 Ga.

Like most Archean greenstones, the Pilbara Supergroup is domi-
nantly volcanic near the bottom (Warrawoona Group), with clastic
sedimentary rocks increasing upward (Gorge Creek Group), but the
sequence is not exactly typical. Ultramafic rocks are not concentrated
near the bottom of the section, but are most common in the Salgash
Subgroup, whereas the thickest sequence of felsic volcanic rocks, the
Duffer Formation, occurs near the bottom (compare Fig. 4.4 with
Table 4.3). The occurrence of large volumes of felsic rocks near the
bottom of the Pilbara sequence has been used as an argument that the
Pilbara Supergroup was deposited on a granitoid continental, rather
than a basaltic oceanic, crust (A. H. Hickman, 1981: 62).

The rocks of the Pilbara Block have undergone several periods of
deformation, the principal one resulting in simultaneous *diapiric* intrusion of the granitic domes, folding of the greenstones into their
present *synclinal* forms, and metamorphism. This pervasive deformation, deformation D2 of A. H. Hickman (1975), was not instantaneous
but occurred over the period 3.0–3.3 Ga (Oversby, 1976, Pidgeon,
1978b); it affected the Warrawoona and Gorge Creek groups but not
younger rocks. Subsequent deformations have resulted in (comparatively) minor folding and faulting, sometimes accompanied by the injection of igneous dikes.

The granitoids, which are exposed over about 60% of the Pilbara
Block, are of two principal generations. The older granitoids, which
predominate, are foliated gneissic complexes that form the tectonically emplaced domes. These granitoids are early Archean—some are
nearly 3.5 Ga—and were affected by the 3.0-Ga deformation. Some of
these older granitoids appear to have been emplaced at the time of the
deformation. The younger generation of granitoids comprises late
Archean post-tectonic bodies that form volumetrically minor *plutons*
and were intruded 2.6–2.7 Ga (Fig. 4.22).

Only two of the formations in the Pilbara Supergroup have been
dated by isochron or concordia–discordia methods. These are the
North Star Basalt, the lowest (and, therefore, the oldest) formation in
the Talga Talga Subgroup, and the Duffer Formation (Table 4.3). The
radiometric ages of these units show that the lower part of the Warrawoona Group is about 3.5–3.6 Ga old (Table 4.4).

The North Star Basalt is a 2-km-thick volcanic formation whose
name is misleading because the unit contains a wide variety of rock
types. In addition to basalt flows, which preponderate, the North Star
includes ultramafic, intermediate, and felsic lavas as well as minor
amounts of sedimentary rocks, including chert. All have been mildly
metamorphosed and are now greenstones. The most precise age for
the North Star Basalt was obtained by P. J. Hamilton, of the University
of Cambridge, and his colleagues with the Sm–Nd isochron method.

TABLE 4.4

Radiometric Ages of Some Early Archean and Related Rocks
of the Pilbara Block, Western Australia

Unit	Locality[a]	Method	Number of Samples[b]	Age[c] (Ga)	Source
PILBARA SUPERGROUP					
Duffer Formation	Marble Bar	U–Pb	11z	3.45 ± 0.02	Pidgeon, 1978a
	Marble Bar	Rb–Sr	6w	3.23 ± 0.28	Jahn et al., 1981
	Warrery Gap	Rb–Sr	6w	3.51 ± 0.06	Cooper, James, and Rutland, 1982
North Star Basalt	Marble Bar	Rb–Sr	3w	3.57 ± 0.18	Jahn et al., 1981
	Marble Bar	Sm–Nd	4w	3.56 ± 0.54	Jahn et al., 1981
	Marble Bar	Sm–Nd	6w	3.56 ± 0.03	Hamilton et al., 1981
GRANITOIDS					
Older deformed gneissic granitoids					
Coruna Downs Batholith		Rb–Sr	16w	3.23 ± 0.03	Cooper, James, and Rutland, 1982
Tambourah Dome		Pb–Pb	6w	2.94 ± 0.03	Oversby, 1976
		Pb–Pb	5w	3.07 ± 0.01	Oversby, 1976
		Rb–Sr	4m	2.70 ± 0.01	Oversby, 1976
		Rb–Sr	4m	2.61 ± 0.01	Oversby, 1976
Mt. Edgar Batholith		U–Pb	2z	3.28 ± 0.02	Pidgeon, 1978b
		Rb–Sr	10w	3.06 ± 0.37	deLaeter and Blockley, 1972
Shaw Batholith		U–Pb	2z	3.42 ± 0.04	Pidgeon, 1978b
		U–Pb	52z	3.47 ± 0.02	Williams et al., 1983
		Rb–Sr	10w	2.89 ± 0.08	deLaeter, Lewis, and Blockley, 1975
		Rb–Sr	13w	3.09 ± 0.05	Cooper, James, and Rutland, 1982
Satellite dome and dikes	Warrery Gap	Rb–Sr	12w	3.27 ± 0.02	Cooper, James, and Rutland, 1982
Post-deformation dikes	Warrery Gap	Rb–Sr	3w	2.99 ± 0.03	Cooper, James, and Rutland, 1982

NOTE: Stratigraphic order within the Pilbara Supergroup is shown in Table 4.3. All of the ages are based on isochron (Rb–Sr, Sm–Nd) or concordia–discordia (U–Pb) techniques.
[a] Localities are approximate, for samples are generally collected over a broad area.
[b] w, whole rock; z, zircon; m, various minerals from same rock.
[c] All ages calculated with the decay constants listed in Table 3.1. Errors are at the 95% confidence level (two standard deviations). Ages underlined are shown as figures.

They selected samples from six volcanic units that ranged in composition from ultramafic to felsic in order to obtain as much variation in samarium and neodymium concentrations as possible. The data form a good linear array indicating a Sm–Nd isochron age of 3.56 ± 0.03 Ga (Fig. 4.23). As Hamilton and his co-workers pointed out, it is not known with certainty that the samples are from the same igneous suite, but the fact that the data fit an isochron indicates that the six volcanic units originated at essentially the same time with the same initial neodymium isotopic ratio and have had the same Sm–Nd iso-

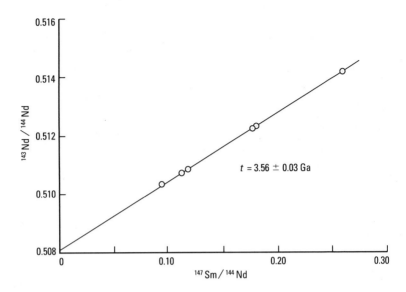

Fig. 4.23. Sm–Nd isochron diagram for samples of volcanic rocks from the North Star Basalt, the lowest formation in the Pilbara Supergroup, Pilbara Block, Western Australia. (After Hamilton et al., 1981.)

topic "history" since formation. Thus, the age is interpreted as the age of volcanism.

Two other attempts to date the North Star Basalt, while confirming the 3.56-Ga age, have yielded less precise ages (Jahn et al., 1981). Samples of four volcanic rocks, consisting of three *dacites* and one *andesite*, gave a Sm–Nd isochron age of 3.56 ± 0.54 Ga. The large error is not due to any imprecision in the isotopic measurement but rather to the samples' similarity in Sm and Nd isotopic composition and the resulting lack of "spread" on the isochron diagram. The Sm–Nd model ages (see Fig. 4.20 for basis) of the four samples are 3.49, 3.59, 3.59 (dacites), and 3.46 (andesite) Ga. Rb and Sr measurements were made on these same four samples. The three dacites gave an isochron age of 3.57 ± 0.18 Ga, but the result from the andesite sample did not fall on the isochron. It may seem curious that the andesite sample behaved as a closed system for Sm and Nd but not for Rb and Sr, but this is probably due to the higher resistance of the Sm–Nd system to postformation metamorphism.

The most precise age for the Duffer Formation was obtained by R. T. Pidgeon (1978a), of the Australian National University, using the U–Pb concordia–discordia method on zircons from a dacite lava flow (or possibly an intrusive *sill*). The zircons were separated on the basis of size and magnetic properties into seven fractions and analyzed, four of the fractions in duplicate. The 11 data show a slight discor-

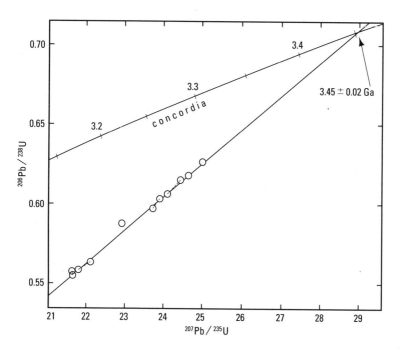

Fig. 4.24. U–Pb concordia–discordia diagram for zircons from the Duffer Formation of the Pilbara Supergroup, Pilbara Block, Western Australia. Points connected by a dashed line represent replicate analyses of the same zircon fraction. (After Pidgeon, 1978a.)

dance, i.e. they do not fall far from concordia, but all fit a common chord with an upper intercept of 3.45 ± 0.02 Ga (Fig. 4.24). An additional fraction, consisting of the smallest zircons, was highly discordant and was not included in the age determination.[12] A Sm–Nd model age of 3.40 Ga for the dacite flow is in good agreement with the U–Pb zircon age (Collerson and McCulloch, 1982).[13]

Three attempts to date the Duffer Formation by using the Rb–Sr method have yielded disappointing results. Pidgeon, in the study cited above, analyzed six samples from the dacite lava flow that he had dated by the U–Pb method and found that the data did not fall on a single isochron, from which he concluded that the samples had not behaved as closed systems for rubidium and strontium since formation. B. M. Jahn, of the University of Rennes, and his co-workers (1981) studied six samples of felsic lava flows and two of *tuffaceous*

12. It is not uncommon for very small zircons in metamorphic rocks to be highly discordant.

13. There are also a number of Pb isotopic model ages for the Duffer Formation (3.44 Ga) and related Archean rocks of the Pilbara Block. Pb models tend to be more complex than Sm–Nd models and I have chosen not to discuss them here, except as they are directly related to age-of-Earth calculations (Chapter 7). For a summary of Pb isotopic model ages for the Pilbara, see Richards, Fletcher, and Blockley (1981).

units. The data from the six lava flows fall on an isochron indicating an age of 3.23 ± 0.28 Ga, but the two data from the tuffaceous samples fall off of the isochron. While this isochron is poorly defined, the age is consistent with the U–Pb zircon age, within analytical error.

The most recent attempt to date the Duffer Formation by the Rb–Sr method was by a team from the University of Adelaide (Cooper, James, and Rutland, 1982). They analyzed 15 samples from various felsic volcanic rocks but found no isochron relationship. Five of the low-rubidium samples, however, fall on a line whose slope indicates an age of 3.51 ± 0.06 Ga, in agreement with Pidgeon's U–Pb zircon age. While the authors admit that the preferential selection of five of fifteen points is open to criticism, they defend the validity of the age on the basis that the low-rubidium samples are the least likely to be disturbed by later metamorphism, the initial $^{87}Sr/^{86}Sr$ ratio (0.7006 ± 0.0011) is reasonable for these rocks, and the isochron fit to the five data is very good. While Cooper and his colleagues may be correct in asserting that their 3.51-Ga age for the Duffer Formation is accurate, the age is of minimum value because its acceptability is based primarily on agreement with Pidgeon's U–Pb zircon age and not on the internal criteria that are the strength of the isochron method.

Even though the Rb–Sr method has been only marginally successful in dating the Warrawoona Group, the ages of the North Star Basalt and the Duffer Formation are precisely determined by the Nd–Sm and U–Pb zircon methods at 3.56 ± 0.03 and 3.45 ± 0.02 Ga, respectively (Table 4.4). These ages satisfy the internal validity criteria of their respective isochron methods and agree with the stratigraphic order of the rock units as determined independently by field mapping. The ages also differ significantly, and indicate that approximately 100 Ma elapsed between the deposition of these two formations.

Although the North Star Basalt and the Duffer Formation are the only two units in the Pilbara Supergroup for which there are good radiometric ages based on either isochron or concordia–discordia data, several interesting and important ages have been determined for rocks that form the granitoid domes. Some of these data provided minimum ages for parts of the Pilbara Supergroup, whereas others are interesting because they show that parts of the granitoid domes are composed of very old rocks. A few of these data are listed in Table 4.4; more comprehensive compilations may be found in Compston and Arriens (1968), Arriens (1971), de Laeter, Libby, and Trendall (1981), Hamilton et al. (1981), and Cooper, James, and Rutland (1982).

V. M. Oversby, at the time at the Australian National University, determined Pb–Pb and Rb–Sr ages for a series of samples of the older gneissic granitoids from seven sites between Port Hedland and Tambourah (Fig. 4.22). Samples from four of the sites recorded evidence of a pervasive metamorphism at about 2.95 Ga (for example, Pb–Pb ages from Tambourah Dome, Table 4.4), which, she suggested, represented the time of emplacement of the granitoid domes (Oversby, 1976; Pidgeon, 1978b, reached a similar conclusion). Since field evidence indicates that the emplacement of the domes was coincident with the widespread deformation that affected the Mosquito Creek Formation and all of the underlying units, the minimum age of the Gorge Creek Group is about 3.0 Ga.

More recent work by Cooper, James, and Rutland (1982) has shown that the regional deformation dated by Oversby probably occurred over a period of several hundred million years. Granitoids near Warrery Gap, emplaced after the deformation, and others from within the Coruna Dome that are only slightly deformed, have Rb–Sr isochron ages of 3.27 ± 0.02 and 3.23 ± 0.03 Ga, respectively. These workers also determined a Rb–Sr age of 3.09 ± 0.05 Ga for a deformed granitoid of the Shaw Batholith near Tambourah, and a Rb–Sr age of 2.99 ± 0.03 Ga for postdeformation granitoid dikes near Warrery Gap. Thus, deformation apparently occurred over a period of about 300 Ma ending only slightly less than 3.0 Ga.

Some of the dated granitoids intrude parts of the Warrawoona Group and thus provide information on the minimum ages of the stratigraphic units (A. H. Hickman, 1981: 59). For example, field evidence indicates that several units were emplaced after deposition of the Apex Basalt. These include granitoids and granitoid gneisses with Rb–Sr isochron ages (Table 4.4) of 3.27 ± 0.02, 3.23 ± 0.03, and 2.89 ± 0.08 Ga, and Pb–Pb isochron ages of 2.94 ± 0.03 and 3.07 ± 0.01 Ga. Thus, the Apex Basalt, and probably the entire Salgash Subgroup, was deposited more than 3.3 Ga. Granitoid gneisses that are known to intrude the Duffer Formation include two units from the Mt. Edgar Batholith with Rb–Sr and U–Pb ages of 3.06 ± 0.37 and 3.28 ± 0.02 Ga, respectively. These minimum ages are consistent with the age of 3.45 Ga for the Duffer Formation.

The oldest granitoids found so far in the Pilbara Block are gneisses from the Shaw Batholith. U–Pb zircon analyses of samples from two localities in the Shaw resulted in ages of 3.42 ± 0.04 and 3.47 ± 0.02 Ga (Collerson and McCulloch, 1982: 173, Williams et al., 1983: 203). The latter age (Fig. 4.25) is the result of measurements on the inner parts, or cores, of more than 50 individual zircon grains with the ion microprobe. The rims of the grains, which apparently are the result of overgrowths during later metamorphisms, are somewhat younger (about 2.8 Ga). The validity of the 3.47-Ga U–Pb zircon age is confirmed by Sm–Nd model ages of 3.43 ± 0.03 Ga for the same sample, and of 3.36 to 3.46 Ga for nearby samples.

In summary, the radiometric evidence clearly shows that most of

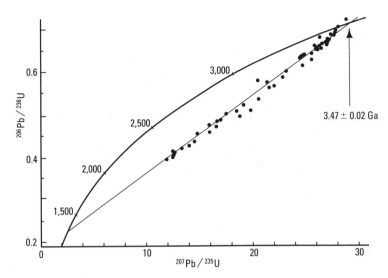

Fig. 4.25. U–Pb concordia–discordia diagram for the cores and innermost rims of zircon crystals from a granitic gneiss of the Shaw Batholith, Pilbara Block, Western Australia. The analyses were done with the ion microprobe. (After Williams et al., 1983.)

the Pilbara Supergroup is older than 3 Ga, and that the lowest part of
the stratigraphic sequence, i.e. the North Star Basalt, is 3.56 Ga in
age. The granitoid domes that were emplaced into the Gorge Creek
and Warrawoona groups appear to be mostly older than 3 Ga, but the
oldest granitoids found so far are nearly 100 Ma younger than the
North Star Basalt.

Barberton Mountain Land

Perhaps the best preserved and most complete greenstone succession known is the Barberton greenstone belt. This remarkable sequence of Archean volcanic and sedimentary rocks is exposed in the
Barberton Mountain Land in northern Swaziland and the southeastern part of the Transvaal Province of South Africa (Fig. 4.26). It is one
of several greenstone belts in the Kaapvaal Craton, an Archean continental fragment with a total area of about 585,000 km². Approximately
86% of the Kaapvaal Craton is concealed beneath Proterozoic sedimentary and volcanic rocks or obscured by the Bushveld Complex, a
2.1-Ga-old layered mafic intrusive body with an area of some 67,000
km². Of the Archean rocks exposed in the Kaapvaal Craton, more
than 91% are granitoids and granitoid gneisses, and the remainder
are greenstones (Anhaeusser and Wilson, 1981: 421).

The rocks of the Barberton greenstone belt constitute a 20-km-
thick volcanic and sedimentary sequence, known as the Swaziland

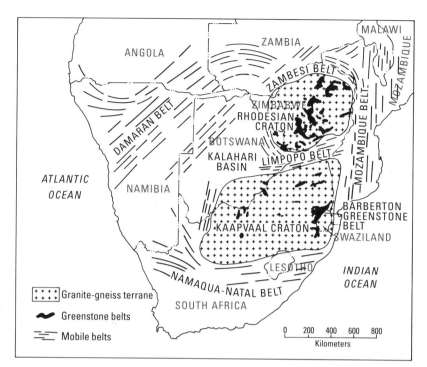

Fig. 4.26. The Archean cratons of southern Africa. The boundaries of the Kaapvaal and Rhodesian cratons are uncertain because large areas are concealed beneath younger rocks. (After R. Mason, 1973.)

Supergroup, that is now highly deformed and enveloped by granitoids and granitoid gneisses (Fig. 4.27).[14]

An interesting and instructive feature of the Swaziland Supergroup is the progressive change in rock type and composition from bottom to top (Table 4.5). The lowest units of the Swaziland Supergroup are predominantly ultramafic lava flows. Sedimentary rocks are negligible in volume and consist of occasional chemical precipitates. With time, i.e. upward in the stratigraphic column, the volcanic rocks become more felsic, and pyroclastic rocks increase in volume at the expense of lava flows. In addition, both the volume and maturity of clastic sedimentary rocks increases toward the top of the sequence, while volcanic rocks decrease in abundance.

The Onverwacht Group, which constitutes the lower three-

14. The geology of the Barberton greenstone belt and the related granite–gneiss terrane is summarized by Viljöen and Viljöen (1969a) (and the nine companion papers by these same two authors in the volume), Anhaeusser (1973: 361–69, 1978), Hunter (1974), Anhaeusser and Wilson (1981: 424–40), Condie (1981: 26–28, 45–49), Tankard et al. (1982: 21–86), and the papers in Anhaeusser (1983), which also includes a geologic map in color.

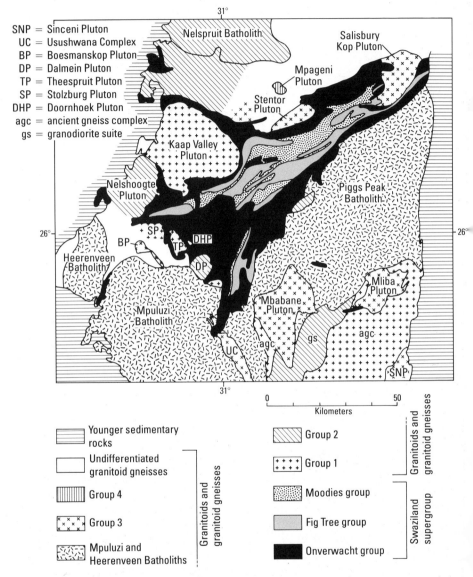

Fig. 4.27. Simplified geologic map of the early Archean Barberton greenstone belt (Swaziland Supergroup) and related granitoids of the Barberton Mountain Land, southern Africa. The relative ages of units in the granitoid–gneiss terrane are based primarily on radiometric age data. (After Viljöen et al., 1983, and Barton, 1983.)

TABLE 4.5

Stratigraphy of the Archean Swaziland Supergroup,
Barberton Mountain Land, Southern Africa

Group and subgroup	Formation	Principal rock types[a]	Thickness (km)	Age (Ga)
Moodies Group	Baviaanskop Formation	sandstone, shale, conglomerate	0.7	
	Joe's Luck Formation	sandstone, shale, conglomerate, BIF, chert	0.7	
	Clutha Formation	sandstone, shale, conglomerate, BIF, chert	1.6	
Fig Tree Group	Schoongezicht Formation	felsic tuff, breccia, lava	0.6	>2.92
	Belvue Road Formation	shale, felsic tuff, chert	0.6	
	Sheba Formation	sandstone, shale, chert	1.0	
Onverwacht Group Geluk Subgroup	Swartkoppie Formation	mafic and felsic volcanic rocks, chert	0.9	
	Kromberg Formation	mafic lava and breccia, felsic volcanic rocks, chert	1.9	
	Hoogenoeg Formation	mafic lava, felsic volcanic rocks, chert	4.8	
	[Middle Marker]	chert	0.006	>3.28
Tjakastad Subgroup	Komati Formation	mafic and ultramafic lavas (komatiites)	3.5	3.53
	Theespruit Formation	mafic and ultramafic lavas, felsic tuff, chert	1.9	
	Sandspruit Formation	ultramafic and mafic lavas	2.1	

SOURCES: Anhaeusser, 1973: 36; Anhaeusser and Wilson, 1981: 433.
[a]Rock types listed in approximate order of abundance. BIF, banded-iron formation.

fourths of the Swaziland Supergroup, is predominantly volcanic and
has been divided into a lower ultramafic unit (Tjakastad Subgroup)
and an upper mafic to felsic unit (Geluk Subgroup), which are separated by a thin (6 meters thick) but persistent black and white chert
unit known as the Middle Marker. The lowest unit in the Tjakastad
Subgroup, the Sandspruit Formation, consists of about two-thirds ultramafic and one-third mafic volcanic rocks and (some) intrusive sills.
The composition of the rocks changes upward in the section, and in
the Komati Formation[15] the relative proportions of ultramafic and
mafic rocks are reversed. Ultramafic rocks are negligible above the
Middle Marker, and the Geluk Subgroup is composed primarily of

15. The Komati Formation is the type section of komatiites, the name given to
ultramafic and mafic volcanic rocks that are unusually high in MgO, low in K_2O, and

mafic and felsic volcanic rocks and chemical sedimentary rocks. Much of the Geluk Subgroup appears to be the result of multiple volcanic cycles, each cycle consisting of mafic volcanic rocks followed by felsic lavas and pyroclastic rocks and ending with the deposition of chert. Pillow structures are common in the lavas of the Onverwacht Group, and many of the pyroclastic rocks have features indicating submarine deposition. It is likely that the entire Onverwacht Group was deposited in water.

The Fig Tree and Moodies groups consist mostly of clastic sedimentary rocks, which become progressively coarser and more mature toward the top. Felsic volcanic rocks, primarily tuffs and *breccias*, occur in the Fig Tree Group but are minor constituents of the Moodies Group. The rocks of the Fig Tree Group, which are largely shales, immature sandstones, cherts, and felsic volcanic rocks, have features that indicate deposition in relatively deep water. In contrast, the rocks of the Moodies Group, which represent the cyclic deposition of conglomerate, mature sandstones and shales with minor amounts of volcanic rocks, chert, and banded-iron formation, appear to be the result of near-shore deposition in shallow water. Thus, the rocks of the Swaziland Supergroup appear to record the progressive development and infilling of an ancient volcanic and sedimentary basin. Isotopic ages of the rocks of the Onverwacht Group indicate that deposition began more than 3.5 Ga.

The granitoids surrounding and enveloping the Barberton greenstone belt have been divided into several groups based on their compositions and isotopic age relationships (Anhaeusser, 1973: 365–69, Barton, 1981: 25–27, 1983) (Fig. 4.27). The oldest group (group 1, 3.4–3.5 Ga) includes both foliated granitoid gneisses and diapiric plutons. The gneisses, which are more voluminous some distance south of the Barberton greenstone belt, are collectively and informally called the Ancient Gneiss Complex (Hunter, 1970, Hunter, Barker, and Millard, 1978). The precursors of these gneisses are not entirely certain, but at least some appear to have been derived from mafic or ultramafic rocks. Nowhere are these ancient gneisses in direct contact with the rocks of the Barberton greenstone belt, so their exact field relation to the volcanic–sedimentary sequence is uncertain. The diapiric plutons contain inclusions and fragments of greenstones. Some of the plutons are in contact with rocks of the Onverwacht Group, but the contacts are tectonic rather than intrusive. In several places, however, the emplacement of the plutons (e.g. the Kaap Valley Pluton) appears to have resulted in extensive deformation of the Onverwacht, Fig Tree,

---

have CaO/Al_2O_3 ratios greater than 1. Komatiites are unique to the Precambrian (Viljöen and Viljöen, 1969b, Viljöen et al., 1983).

and Moodies groups, suggesting that the plutons are younger than the Swaziland Supergroup.

Group 2 crystalline rocks (3.1–3.4 Ga) consist of a complex assemblage of gneisses and granitoids that have radiometric ages younger than do the group 1 rocks. The Mpuluzi, Piggs Peak, and Heerenveen batholiths are relatively homogeneous granitoids with radiometric ages intermediate between those of groups 2 and 3. The rocks of groups 3 (2.7–3.0 Ga) and 4 (2.2–2.6 Ga) represent the most recent Archean igneous events and consist of medium- to coarse-grained granitoid plutons with sharply defined intrusive contacts.

There has been considerable debate about the exact relationship between the Barberton greenstones and the ancient gneisses and about the nature of the crust on which the Swaziland Supergroup was deposited. One hypothesis is that the greenstones were deposited on a granitoid crust, remnants of which may be represented by the ancient gneisses (Hunter, 1974). Another is that the greenstones were deposited on oceanic crust and the granitoid gneisses are the result of partial melting and differentiation below the subsiding sedimentary basin (Viljöen and Viljöen, 1969a, Anhaeusser, 1973, 1978). Indeed, it has been proposed that the ultramafic and mafic rocks of the Tjaka-stad Subgroup may actually be a part of the ancient oceanic crust on which the younger units of the Swaziland Supergroup were deposited. Although the matter is far from settled, several lines of evidence currently favor some form of the latter hypothesis. The first is that the inclusions in the ancient gneisses are similar to the rocks of the Tjaka-stad Subgroup, suggesting that the gneisses incorporated fragments of the pre-existing ultramafic rocks during emplacement. The second is the absence of granitoid debris in the lower parts of the Swaziland Supergroup. This indicates either that granitoids did not exist during deposition of the lowest units of the Swaziland Supergroup or at least that they were not available at the surface as a source of sediment. Finally, the radiometric ages obtained thus far seem to indicate that the ancient gneisses and diapiric plutons of group 1 are some 100 Ma younger than the rocks of the Tjakastad Subgroup.

The emplacement of the various generations of gneissic and granitoid rocks has resulted in the metamorphism and deformation of the Barberton greenstones, processes that have taken their toll on the isotopic systems and reset or disturbed most of the radiometric geo-chronometers. As a result, many of the isotopic ages obtained for units of the Swaziland Supergroup appear to reflect times of recrys-tallization due to metamorphism rather than times of initial deposition. So far, only the Komati Formation has yielded radiometric ages that can be reasonably interpreted as representing the age of initial deposition and crystallization of these ancient rocks (Table 4.6).

TABLE 4.6
Radiometric Ages of Some Early Archean
and Related Rocks of the Barberton Mountain Land, Southern Africa

Unit and locality[a]	Method	Number of samples[b]	Age[c] (Ga)	Source
SWAZILAND SUPERGROUP				
Fig Tree Group	Rb–Sr	11w	2.92 ± 0.04	Allsopp, Ulrych, and Nicolaysen, 1968
Onverwacht Group				
Uppermost Hoogenoeg or lowermost Kromberg formations	Pb–Pb	4s	3.35 ± 0.20	Van Niekerk and Burger, 1969
Middle Marker	Rb–Sr	5w	3.28 ± 0.07	Hurley et al., 1972
Komati and Theespruit formations	Sm–Nd	10w	3.53 ± 0.05	Hamilton et al., 1979, 1983
Komati Formation	Rb–Sr	6m	3.43 ± 0.20	Jahn and Shih, 1974
	Ar–Ar	1w	3.49 ± 0.01	Lopez Martinez et al., 1984
	Ar–Ar	1w	3.41 ± 0.03	Lopez Martinez et al., 1984
	Ar–Ar	1w	3.42 ± 0.01	Lopez Martinez et al., 1984
	Ar–Ar	1w	3.30 ± 0.02	Lopez Martinez et al., 1984
	Ar–Ar	1w	3.45 ± 0.01	Lopez Martinez et al., 1984
	Pb–Pb	12w	3.46 ± 0.07	Brèvart, Dupré, and Allègre, 1986
GRANITOIDS				
Group 3				
Stentor Pluton	Rb–Sr	9w	2.76 ± 0.05	Barton et al., 1983
Felsic gneiss (1)	Rb–Sr	8w	2.73 ± 0.23	Barton et al., 1983a
Batavia Pluton phase A (4)	Rb–Sr	5w	2.77 ± 0.05	Barton et al., 1983b
Batavia Pluton phase B (4)	Rb–Sr	4w	2.76 ± 0.07	Barton et al., 1983b
Felsic granite, Ushawana Complex	Rb–Sr	7w	2.81 ± 0.06	Davies et al., 1970
Boesmankop Pluton	Rb–Sr	11w	2.85 ± 0.03	Barton et al., 1983b
Mafic dike, Farm Weergevonden (3)	Rb–Sr	10w	2.87 ± 0.07	Barton et al., 1983b
Felsic gneiss B, Farm Weergevonden (3)	Rb–Sr	10w	2.92 ± 0.03	Barton et al., 1983b
Felsic gneiss B, Farm Theeboom (2)	Rb–Sr	10w	2.94 ± 0.08	Barton et al., 1983b
Mpuluzi Batholith	Rb–Sr	4w	3.00 ± 0.12	Allsopp et al., 1962
Mpuluzi and Heerenveen batholiths	Rb–Sr	16w	3.03 ± 0.03	Barton et al., 1983b
Group 2				
Nelshoogte Pluton	Rb–Sr	11w	3.18 ± 0.08	Barton et al., 1983b
Felsic gneiss A, Farm Weergevonden (3)	Rb–Sr	9w	3.15 ± 0.02	Barton et al., 1983b
Felsic gneiss A, Farm Theeboom (2)	Rb–Sr	11w	3.19 ± 0.07	Barton et al., 1983b
Doornhoek Pluton, medium-grained phase	Rb–Sr	7w	3.18 ± 0.20	Barton et al., 1983b
Doornhoek Pluton, fine-grained phase	Rb–Sr	3w	3.19 ± 0.05	Barton et al., 1983b
Granodiorite suite	Rb–Sr	9w	3.35 ± 0.06	Barton et al., 1983a

TABLE 4.6 (*continued*)

Unit and locality[a]	Method	Number of samples[b]	Age[c] (Ga)	Source
Group 1				
Theespruit Pluton	Rb–Sr	10w	3.43 ± 0.14	Barton et al., 1983b
Stolzburg Pluton	Rb–Sr	10w	3.48 ± 0.09	Barton et al., 1983b
Kaap Valley Pluton	Rb–Sr	13w	3.49 ± 0.17	Barton et al., 1983b
Bimodal suite, Ancient Gneiss Complex (1)	Sm–Nd	6w	3.42 ± 0.03	Carlson, 1983; Carlson, Hunter, and Barker, 1983
	Rb–Sr	11w	3.56 ± 0.11	Barton et al., 1980

NOTE: Stratigraphic order within the Swaziland Supergroup is shown in Table 4.5. All of the ages are based on isochron (Rb–Sr, Sm–Nd, Pb–Pb) or age-spectrum (Ar–Ar) techniques.
[a] Numbers refer to localities shown in Figure 4.33.
[b] w, whole rock; s, sulfides; m, various minerals from same rock.
[c] All ages calculated with the decay constants in Table 3.1. Errors are at the 95% confidence level (two standard deviations). Ages underlined are shown as figures.

The most precise age for the Komati Formation has been obtained by P. J. Hamilton of Cambridge University and his colleagues using the Sm–Nd method (basic data from Hamilton et al., 1979, recalculated by Hamilton et al., 1983). Nine samples from the Komati Formation and one sample from the Theespruit Formation fall on an excellent isochron indicating an age of 3.53 ± 0.05 Ga (Fig. 4.28). The samples were selected to provide an optimum spread in the Sm/Nd ratio and were collected from seven mafic and ultramafic lava flows and two felsic lava flows from the Komati Formation and from one felsic lava from the Theespruit Formation. If the three analyses of felsic lavas are excluded from the isochron calculation, the resulting age is 3.51 ± 0.06 Ga. Two samples of lava flows from the Hoogenoeg Formation also fall on the isochron shown in Figure 4.28. If these samples are included in the age calculation with the other ten, the result is 3.57 ± 0.05 Ga.

The Rb–Sr method has proved less successful at yielding precise emplacement ages of the Barberton greenstones than has the Sm–Nd method, apparently because it is more susceptible to partial or complete resetting during postdepositional metamorphism. J. M. Barton, Jr., of the Bernard Price Institute for Geophysical Research, University of the Witwatersrand, for example, failed to obtain an isochron for 12 samples of the Komati Formation (Barton et al., 1980). B.-M. Jahn and C.-Y. Shih (1974), then of NASA's Johnson Space Center, made Rb–Sr analyses of a variety of ultramafic, mafic, and felsic volcanic rocks and of sedimentary rocks from the Theespruit, Komati, Hoogenoeg, and Kromberg formations as well as the Middle Marker. Apparently the samples they selected have not remained systems

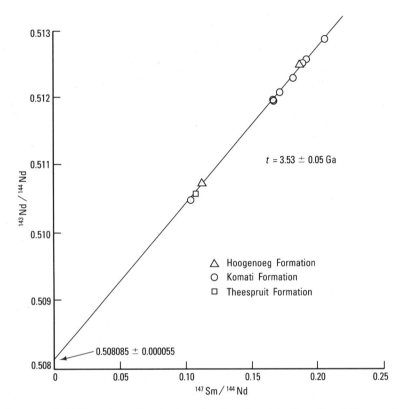

Fig. 4.28. Sm–Nd isochron diagram for volcanic rocks of the Komati and Theespruit formations, Onverwacht Group, Barberton Mountain Land, southern Africa. Two samples from the Hoogenoeg Formation are shown for comparison but were not included in the isochron calculation. (After Hamilton et al., 1979.)

closed to rubidium and strontium since initial crystallization for none of the data formed linear arrays that can be interpreted as isochrons. Five mineral fractions of different sizes and densities from a single basaltic komatiite flow of the Komati Formation, however, did define an isochron with an age of 3.43 ± 0.20 Ga. Because of the obvious open-system behavior of their other (whole rock) samples, these authors interpreted this mineral isochron as a probable metamorphic age, but it is consistent, within the analytical uncertainties, with the Sm–Nd age of 3.53 Ga.

Brévart, Dupré, and Allègre (1986), of the University of Paris, have obtained a relatively precise Pb–Pb isochron for 11 komatiitic lava flows and one tholeiitic basalt from the Komati Formation. The resulting value of 3.46 ± 0.07 Ga is also in good agreement with the Sm–Nd age for the Komati Formation.

A somewhat surprising result was recently obtained for lava flows of the Komati Formation with the $^{40}Ar/^{39}Ar$ incremental heating method. Normally, the K–Ar method is of limited use in determining the initial emplacement ages of ancient metamorphic rocks because of the ease with which argon escapes from rocks during postcrystallization heating. M. Lopez Martinez and her colleagues at the University of Toronto, however, have recently discovered that the Barberton komatiites have remarkable retention of argon and give $^{40}Ar/^{39}Ar$ incremental heating plateau ages that approach the crystallization ages of the rocks (Table 4.6). Sixteen incremental heating analyses of five komatiites and two komatiitic basalts gave plateau ages ranging from 3.11 ± 0.01 to 3.49 ± 0.01 Ga (Fig. 4.29). Because the rocks of the Komati Formation were extensively metamorphosed early in their history and by comparison with the Sm–Nd age, the authors concluded that the oldest of the $^{40}Ar/^{39}Ar$ plateau ages probably reflect the age of initial metamorphism, which occurred within 100 million years of emplacement or about 3.45 to 3.49 Ga (Lopez Martinez et al., 1984).

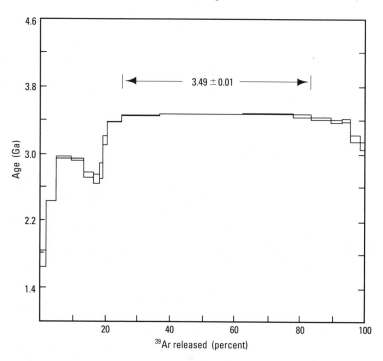

Fig. 4.29. $^{40}Ar/^{39}Ar$ age spectrum for a sample of komatiite from the Komati Formation, Onverwacht Group, Barberton Mountain Land, southern Africa. The vertical thickness of the boxes indicates the standard deviation of the value for each gas increment. Error in the plateau age indicates two standard deviations. (After M. Lopez Martinez et al., 1984.)

There have been attempts to date sedimentary rocks from both the Middle Marker and the Fig Tree Group by the Rb–Sr method, but the ages so obtained may reflect the times of later metamorphisms rather than either deposition (ca. 3.5 Ga) or the earliest postdepositional metamorphism (ca. 3.45–3.5 Ga). Five samples of metamorphosed shale from within the Middle Marker show an excellent isochron relationship with a calculated age of 3.28 ± 0.07 Ga, whereas 11 whole-rock shale and sandstone samples from the Fig Tree Group fall on a precise isochron of 2.92 ± 0.04 Ga (Table 4.6). Even though these ages are in accord with the known stratigraphic relations within the Swaziland Supergroup, we will see later that there is good reason to think that the ages do not represent the ages of either initial formation or earliest metamorphism. Likewise, four *sulfide* samples from a felsic lava flow in either the uppermost Hoogenoeg or lowermost Kromberg formations gave a Pb–Pb isotopic age of 3.35 ± 0.20 Ga (Van Niekerk and Burger, 1969),[16] which probably is the age of sulfide mineralization induced by metamorphism some time after the host volcanic rocks were deposited.[17]

The ages of the granitoids that surround and encompass the Barberton greenstones are not only of interest in themselves—some are very early Archean—but they provide a possible upper limit for the age of the Swaziland Supergroup.[18] Of particular interest are the rocks of the Ancient Gneiss Complex and the diapiric plutons that have been assigned to group 1.

The Ancient Gneiss Complex, which is exposed most extensively in Swaziland, south of the Barberton greenstone belt, is a complexly deformed suite of gneisses that has been divided into three major units: (1) the "Bimodal Suite," which consists of interlayered felsic and mafic gneisses, (2) homogeneous granitoid gneiss, dated at about 3.3 Ga, that intrudes the Bimodal Suite, and (3) inclusions of

16. The authors described the unit only as a felsic lava within the Onverwacht Group, but a comparison of the geographic coordinates of the sample locality with the geologic map of Anhaeusser, Robb, and Viljöen (1983) indicates that the flow must be from either the uppermost Hooggenoeg or lowermost Kromberg formations.

17. Saager and Koppel (1976) have calculated minimum and maximum ages of 3.45 ± 0.02 and 3.81 ± 0.07 Ga for sulfide minerals associated with gold mineralization in the Onverwacht Group. These are two-stage Pb isotope model ages based on assumptions about the history of lead mineralization. Such lead model ages will not be discussed here because of their complexity and the uncertainty of the underlying assumptions, but the results are interesting and informative.

18. There are considerable geochronologic data for the granitoid rocks that intrude and surround the Barberton greenstone belt, especially from localities to the south. The data discussed here and included in Table 4.6 include only those within the area shown in Figure 4.27. Data from dissertations, which tend to be difficult to obtain, also have been excluded as have other unpublished data. For more complete summaries see Barton (1981, 1983).

supracrustal rocks that may be disseminated remnants of the Swaziland Supergroup (Hunter, Barker, and Millard, 1978, Barton et al.,
1980). Field relations suggest that the felsic phase of the Bimodal Suite
may be among the oldest of the granitoid rocks, although it does not
provide any conclusive information on the felsic phase's relationship
with the Swaziland Supergroup. One possibility is that the Bimodal
Suite represents the basement on which the Swaziland Supergroup
was deposited. Another is that the felsic gneiss is younger than the
Swaziland Supergroup and that the mafic phase was derived from inclusions of the older greenstones.

Recent dating of the Bimodal Suite by R. W. Carlson, of the Carnegie Institution of Washington, and his colleagues may have solved
this dilemma. Results from six samples of the granitoid gneiss fall
on a precise Sm–Nd isochron indicating an age of 3.42 ± 0.03 Ga
(Fig. 4.30). Comparison of this age with the Sm–Nd age of 3.53 ± 0.05

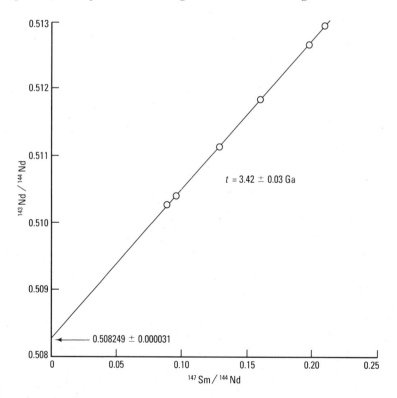

Fig. 4.30. Sm–Nd isochron diagram for samples of the Bimodal Suite of the Ancient
Gneiss Complex, Swaziland, southern Africa. These rocks are probably the oldest in
the granitoid–gneiss terrane that envelops the Barberton greenstone belt. (After Carlson, 1983; see also Carlson, Hunter, and Barker, 1983.)

Ga for the Komati and Theespruit formations (Table 4.6) indicates that the felsic gneiss is about 100 Ma younger than the lower part of the Onverwacht Group. A Rb–Sr isochron age of 3.56 ± 0.11 Ga for ten samples of the felsic gneiss is in accord with the Sm–Nd age but is insufficiently precise to provide conclusive information about the age relationship between the gneiss and the Barberton greenstones.

The Kaap Valley, Theespruit, and Stolzburg plutons intrude and deform the rocks of the Swaziland Supergroup (Fig. 4.27) and therefore were emplaced, either by magmatic (liquid) or diapiric (semi-solid or solid state) intrusion, after the greenstones were deposited (Barton, 1981, 1982). Rb–Sr isochron ages ranging from 3.42 ± 0.14 to 3.49 ± 0.17 Ga have been obtained for these plutons (Table 4.6; Fig. 4.31). It is not known with certainty whether these ages are the ages of emplacement or metamorphism, but the ages agree with the known field relations and, therefore, are probably emplacement ages (Barton et al., 1983b: 71).

Over the past several years a considerable amount of Rb–Sr ra-

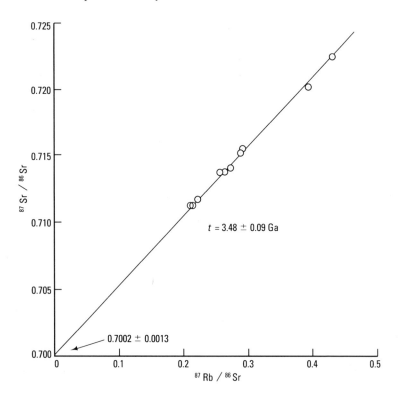

Fig. 4.31. Rb–Sr isochron for whole-rock samples from the Stolzburg Pluton, South Africa. (After Barton et al., 1983b.)

diometric data on the ages of the granitoids and gneisses that surround and envelop the Barberton greenstone belt has been produced.
Most of these new data are due to the efforts of J. M. Barton and his
colleagues, and some still await publication, though many, and some
of the mineral data, were summarized by Barton (1983: 76). Those
data accessible in the literature at the time this chapter was written are
summarized in Table 4.6. The data, published and unpublished,
show that the crust surrounding the Barberton greenstones was the
scene of more or less continuous igneous intrusive activity and repeated metamorphism throughout the Archean and into the early
Proterozoic at least until 2.2 Ga. As we have seen with other Archean
rocks, the most consistent Rb–Sr data seem to come from analyses of
whole-rock samples, whereas analyses of individual minerals frequently either do not provide isochrons or appear to record times of
later metamorphisms.

   In summary, the data currently available show that the rocks of
the Onverwacht Group were deposited more than 3.5 Ga, the most
precise age being the Sm–Nd isochron age of 3.53 ± 0.05 Ga. This
was followed within 100 Ma or less by the emplacement of granitoids
now represented by the felsic phase of the Bimodal Suite of the Ancient Gneiss Complex, which has a Sm–Nd isochron age of 3.42 ±
0.03 Ga, and by the diapiric plutons, which have Rb–Sr ages of 3.43 ±
0.14 to 3.49 ± 0.17 Ga. This earliest igneous and tectonic activity
strongly deformed and metamorphosed the volcanic and sedimentary
rocks of the Swaziland Supergroup, and the metamorphism appears
to be reflected in the $^{40}Ar/^{39}Ar$, Rb–Sr, and Pb–Pb ages of the Komati
Formation. Because the diapiric plutons, in particular the Kaap Valley
Pluton, appear to have deformed the entire Swaziland Supergroup, it
is likely that the sedimentary rocks of both the Moodies and Fig Tree
groups are older than the diapiric plutons, i.e. older than about 3.45
Ga. If this is so, then the Rb–Sr ages of 3.3 and 2.9 Ga for the Middle
Marker and the Fig Tree Group, respectively, represent ages of crystallization due to metamorphism induced or accompanied by subsequent igneous activity (probably the emplacement of the granitoid
rocks of groups 2 and 3) rather than ages of initial deposition.

Other Ancient Rocks

   In the preceding sections we have examined the geology and radiometric data from four well-studied areas in which rocks approaching or exceeding 3.5 Ga in age have been found. These are not the
only places, however, where early Archean rocks occur, although the
other localities have not been as extensively studied. For example,

THE AGE OF THE EARTH

samples of the ancient gneisses of the Imataca Complex of Venezuela have yielded a Pb–Pb isochron age of 3.77 ± 0.02 Ga (Montgomery, 1979). The Imataca Complex has been correlated with similar Archean gneisses in Liberia and Sierra Leone,[19] from which whole-rock Rb–Sr ages of 3.1–3.7 Ga have been obtained (Hurley, Fairbairn, and Gaudette, 1976).

Rb–Sr isochrons ranging from 3.42 ± 0.06 Ga for granitoid that intrudes greenstones to 3.52 ± 0.06 Ga for granitoid gneisses have been reported from the Rhodesian Craton in Zimbabwe (M. H. Hickman, 1974, Hawkesworth et al., 1975, Moorbath, Wilson, and Cotterill, 1976, Moorbath et al., 1977b). To the south of the Rhodesian Craton, the Limpopo mobile belt (Fig. 4.26) contains well-documented early Archean gneisses near the Sand River (Barton, Ryan, and Fripp, 1978, 1983, Barton, 1981). Gray gneisses and the more felsic gneisses that enclose them give Rb–Sr isochron ages of 3.73 ± 0.06 and 3.78 ± 0.10 Ga, respectively, while basaltic dikes that intrude the gneisses give a Rb–Sr isochron age of 3.57 ± 0.10 Ga.

The Slave Province, in the northwestern part of the Canadian shield, consists predominantly of late Archean supracrustal rocks intruded by granitoid gneisses. In the western part of the Slave Province, however, is a small area of much older rocks, the Acasta gneisses, that range from granitic to tonolitic in composition. Remnants of supracrustal rocks are also present, but their relation to the gneisses is uncertain. U–Pb ion probe analyses of 53 zircon grains from two samples of the Acasta gneisses indicate that the original granitoids from which the gneisses formed crystallized at 3.962 ± 0.003 Ga (Bowring, Williams, and Compston, 1989). Even though these rocks have not yet been studied as extensively as those at Isua, they currently appear to be the oldest known rocks on Earth.

Six whole-rock samples of amphibolite that occurs as enclaves within granitoid gneiss in Qinan County, Hebei Province, China, give a well-defined Sm–Nd isochron of 3.50 ± 0.08 Ga (Xuan, Ziwei, and DePaolo, 1986). Ion-microprobe analyses of cores of zircon crystals in gneiss from Mount Sones, Enderby Land, Antarctica, show that the granitoid progenitor of the gneiss crystallized 3.93 ± 0.01 Ga (Black, Williams, and Compston, 1986). Rocks exceeding 3 Ga in age have also been reported from India, the USSR, and elsewhere in China, but the data are highly disputable (Kratz and Mitrofanov, 1980, Kazansky and Moralev, 1981, Salop, 1983, Bibikova, 1984, Shcherbak et al., 1984, Sun and Wu, 1981, Jahn and Zhang, 1984, Basu et al., 1981, Baksi et al., 1984).

19. These now-distant areas were joined before Africa and South America separated in early Jurassic time.

Recent evidence suggests the exciting possibility that rocks exceeding 4.0 Ga in age eventually may be found in the Yilgarn Block of Western Australia, a terrane where undisputed early Archean rocks are abundant (de Laeter et al., 1981, McCulloch, Collerson, and Compston, 1983, Claoue-Long, Thirwall, and Nesbitt, 1984). In the northwestern part of the Yilgarn Block, near Mount Narryer, there occurs a sequence of metamorphosed sedimentary rocks surrounded by a banded gneiss. This banded gneiss has a Sm–Nd model age of 3.63 ± 0.04 Ga, and U–Pb analyses of individual zircon grains by the ion microprobe shows that some of the zircon cores are as old as 3.69 Ga (de Laeter et al., 1981, Compston et al., 1983). The banded gneiss is intruded by a granitoid gneiss that has a Sm–Nd model age of 3.51 ± 0.05 Ga and a Rb–Sr whole-rock isochron age of 3.35 ± 0.04 Ga. The latter age is the same as the U–Pb ion-microprobe ages of the youngest overgrown rims of zircons from the banded gneiss and probably reflects a major metamorphic event at about 3.3 Ga.

Within the sedimentary sequence there occurs a *quartzite* (metamorphosed sandstone) composed of mineral grains derived from preexisting crust and redeposited in Archean seas 3.3–3.5 Ga. The quartzite contains grains of zircon that have been analyzed with the ion microprobe by D. O. Froude and his colleagues (1983) at the Australian National University. These analyses were made as part of a systematic search for zircons with ages in the range 4.0–4.5 Ga, the period in Earth's history from which no rocks have yet been found. Most of the zircon grains are 3.5–3.7 Ga old (Fig. 4.32), but four grains appear to exceed 4.0 Ga.[20] The authors offer two interpretations for these data. One is that two of the grains (12 and 31) are about 4.18 Ga, whereas the other two are about 4.11 Ga. The second interpretation, preferred by the authors, is that all four zircons are about 4.15 Ga and have suffered some lead redistribution during subsequent metamorphism. Regardless of which interpretation is accepted, these four mineral grains appear to have been derived from pre-existing granitoid crust that crystallized more than 4.1 Ga. In a subsequent ion-microprobe study of 140 zircon grains from a conglomerate in the Jack Hills Metasedimentary Belt about 60 km northeast of Mt. Narryer, 17 grains with U–Pb ages exceeding 4.0 Ga were found, including one grain with the remarkably old age of 4.28 ± 0.06 Ga (Compston and Pidgeon, 1986). Thus, although not rocks, the zircon crystals from the Mt. Narryer and Jack Hills areas represent the oldest materials

20. Schärer and Allègre (1985) analyzed 32 zircons from the quartzite by conventional techniques and failed to find any whose age exceeds 3.8 Ga. Compston et al. (1985) argued, however, that Schärer and Allègre were simply unlucky and that statistically there was a 27% chance of such a result.

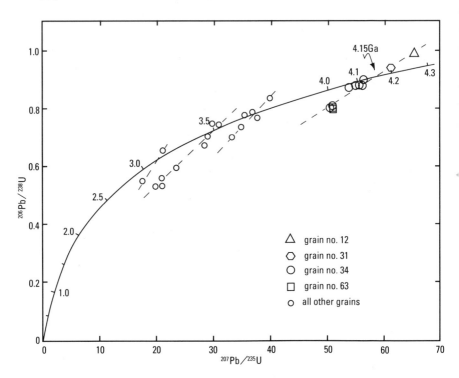

Fig. 4.32. U–Pb concordia–discordia diagram for detrital zircon grains from quartzite near Mt. Narryer, Yilgarn Block, Western Australia. The four older grains appear to have been derived from a crustal source with age about 4.15 Ga and represent the oldest terrestrial materials yet found. (After Froude et al., 1983.)

found so far on Earth and offer hope that remnants of the Earth's earliest crust may yet be found.

The Record and the Missing Record

As we have seen, the oldest well-studied rocks on Earth are metamorphosed supracrustal rocks that are intruded and enveloped by only slightly younger granitoids. The oldest of these are found at Isua in western Greenland, where two sedimentary units and a mafic intrusive body have been dated by Rb–Sr, U–Pb, Pb–Pb, and Sm–Nd methods at 3.7–3.8 Ga. The Isua supracrustal sequence is surrounded and enclosed by the Amîtsoq granitoid gneisses, which have been extensively dated at a number of localities by Rb–Sr, Lu–Hf, U–Pb, and Pb–Pb techniques, with results ranging from 3.5 to 3.7 Ga. On the Labrador coast, the Uivak gneisses, which are equivalent to the Amîtsoq gneisses, also have radiometric ages of 3.6–3.7 Ga.

In the Minnesota River Valley and in northern Michigan, granitoid gneisses near Morton, Granite Falls, and Watersmeet have U–Pb and Rb–Sr radiometric ages of 3.5 to 3.7 Ga.

In the Pilbara Block of Western Australia, the Warrawoona Group is intruded by granitoid gneisses. Felsic (Duffer Formation) and mafic (North Star Basalt) volcanic rocks within this supracrustal sequence have been dated at 3.4–3.6 Ga by Rb–Sr, Sm–Nd, and U–Pb methods. The Shaw Batholith, a granitoid gneiss that intrudes the Warrawoona, has a U–Pb age of 3.5 Ga.

In Swaziland, the sedimentary Theespruit Formation and the volcanic Komati Formation, which occur within the lower part of the Onverwacht Group, the lowermost unit of the Barberton greenstone belt, have Sm–Nd, Rb–Sr, Pb–Pb, and ^{40}Ar/^{39}Ar radiometric ages of 3.4–3.5 Ga. These supracrustal rocks are enveloped by gneisses, one of which, known as the Bimodal Suite of the Ancient Gneiss Complex, has Rb–Sr and Sm–Nd ages of 3.4–3.5 Ga.

In the Slave Province, in the northwestern part of the Canadian shield, U–Pb analyses of zircon crystals indicate that the granitoids from which the Acasta gneisses formed crystallized at 3.962 ± 0.003 Ga. These rocks are currently the oldest known on Earth.

Rocks exceeding 3.5 Ga in age have also been found in Venezuela, China, Zimbabwe, and Antarctica. On Antarctica, zircons within a gneiss near Mount Sones, Enderby Land, have a U–Pb age of 3.93 ± 0.01 Ga.

Some zircon crystals from two sedimentary units near Mt. Narryer and Jack Hills in the Yilgarn Archean Block of Western Australia have U–Pb ages of from 4.0 to 4.3 Ga. Their source rocks have not been found, but the zircons are the oldest materials found so far on Earth.

It is known from other evidence to be discussed that the Earth's age is most probably between 4.5 and 4.6 Ga, yet the oldest rocks found on Earth are only about 3.8–3.9 Ga. What happened to the rocks that represent the first two-thirds to three-fourths of a billion years of Earth's history? The answer to this question is not really known—there are only speculations and possibilities. One possibility is that during that period of Earth's history not only was the first continental crust forming, it was also being vigorously recycled and regenerated. Thus, the earliest crustal rocks may have been consumed by recycling into the primitive mantle almost as fast as they were generated. A second possibility arises from the observation that the Moon and, by inference, the Earth were subjected to intense bombardment by large meteorites from the time of their initial formation to about 3.8 Ga. This bombardment occurred because the planets were still sweeping up huge masses of material from their orbital paths. Perhaps the bombardment was sufficiently intense to obliterate the first crustal rocks. A third possibility is that the record of the Earth's early history exists somewhere but has not been found. The discovery of zircon grains 4.0–4.3 Ga old in sedimentary rocks of the Yilgarn Block of Australia offers the hope that some rocks from Earth's earliest history may yet be discovered. The correct reason for the absence of the most ancient rocks may well be some combination of the above.

The absence of known rocks that represent the first two-thirds to three-fourths of a billion years of Earth's history is probably due to destruction owing to vigorous crustal recycling, intense meteoritic bombardment, lack of discovery, or some combination thereof. But whatever the reason for the missing record on Earth may be, we can learn much about the history and age of the Earth by examining the evidence from more primitive bodies in the Solar System, in particular the Moon and meteorites.

CHAPTER FIVE

Moon Rocks: Samples from a Sister Planet

We may hope therefore to find some profit in contemplating for a few moments this land of the skies: and although we may not look for very speedy "annexation," we may possibly gather some facts and ideas which the decree of Truth will annex to the domain of Science.

J. D. DANA (1846: 336)

The trips to the Moon by the Apollo astronauts were surely the greatest feats of engineering and exploration in the history of humankind. In addition to their technical and spiritual benefits, the manned lunar missions had significant scientific worth, for they gave scientists, for the first time, an exciting opportunity to study rock samples collected from another planet.[1] The Moon, a body much smaller than the Earth, lost its internal heat relatively early in its history. As a result, it ceased to be an internally active planet about a billion years or more ago. Thus, lunar rocks, unlike their counterparts on Earth, primarily record events early in the Moon's history.

But what can we learn about the age of the Earth by studying the Moon? The answer lies in the fact that the Moon and Earth, along

1. S. R. Taylor (1975: xiii) estimated that the scientific literature on the returned lunar samples exceeded 30,000 pages, and that was more than a decade ago! In contrast, there are surprisingly few current books summarizing the results of the studies of lunar samples and the Apollo lunar surface experiments. The books by S. R. Taylor (1975) and Cadogan (1981) are excellent, comprehensive, and admirably readable. Anyone interested in the Moon should read them both. The book by French (1977) is lighter fare and suitable for readers who prefer less detail in a few evenings of recreational reading. Guest and Greeley (1977) is a good introductory text focusing on the surface features and stratigraphy, while Kopal (1974) emphasizes lunar geophysics. An excellent intermediate summary is the one by Wilhelms (1984). For capsule views, see the short articles by J. A. Wood (1975) and S. R. Taylor (1979). I have relied heavily on these books and articles for the content of the first half of this chapter.

with the entire Solar System, formed more or less simultaneously (see Chapter 1). Both the theoretical and experimental evidence for that is quite good—in fact there is no evidence to the contrary—and the age of the Moon therefore has a direct bearing on the age of the Earth. There is also very good evidence that the Moon formed relatively near the Earth and may, in fact, be the product of a collision between the Earth and a large body.

Hundreds of radiometric age measurements on lunar rocks have now been made. The majority fall within the range 3.2–4.0 Ga and the oldest approach 4.5 Ga. There are many rocks from the moon older than the very oldest rocks found on Earth, and there are few lunar rocks with formation ages younger than Earth's early Archean. Thus, compared to Earth rocks, Moon rocks are exceedingly old and record events early in the history of the Moon. Some of these events, particularly the heavy meteoritic bombardment that left deep scars on the Moon's surface, probably affected the Earth as well as the Moon, but all Earthly record of them appears to have been erased by the vigorous crustal recycling and erosion that continually reshapes the face of our planet. In a sense, then, the rock record of the Moon and Earth are complementary, each providing detailed records of quite different times in the evolution of our region of the Solar System.

Like radiometric ages of terrestrial rocks, lunar age data and their significance are best understood in their geological context. Before we examine some of the evidence for the ages of the Moon rocks and of the Moon itself, therefore, it will be worthwhile to inquire into the origin of the Moon and its relationship to Earth, the principal features of the lunar surface and the processes that have sculpted the face of the Moon, the various types of Moon rocks and how they formed, and what is known about the history and evolution of the Moon.

Hypotheses for the Origin of the Moon

Virtually all hypotheses for formation of the Moon fall into one of four groups: fission, capture, formation with Earth as a double planet, and collision.[2]

The *fission hypothesis*, in which the Moon is presumed to have separated from the Earth, was proposed by George Darwin in 1879 (see Chapter 2). According to his proposal the rapid rotation of the molten Earth resulted in a tidal bulge that, because of resonance with

2. The references cited in note 1 summarize the first three hypotheses for the Moon's origin. In addition, see Brush (1982), A. E. Rubin (1984), and G. J. Taylor (1985). The collision hypothesis is summarized by Stevenson (1987).

the Sun and the solar tides, threw off a large "droplet" that solidified in orbit and became the Moon. Three years later, Osmond Fisher proposed that the Pacific Ocean represents the unhealed scar where the Moon used to be. It is now known, however, that the Pacific Ocean, whose age is less than 200 Ma, did not exist early enough to have contributed material for the Moon.

Aside from the near identity of the Moon's density and the uncompressed density of Earth's mantle, which may be simply a coincidence, there are few arguments in favor of the fission hypothesis and considerable evidence, both dynamic and chemical, against it. The principal dynamic objections to the fission hypothesis are six. First, the angular momentum in the Earth–Moon system is insufficient by a factor of three for the proposed primordial Earth to have formed the tidal bulge necessary for fission. Some of the missing angular momentum might have been lost through time by the escape of atmospheric gases from the Earth and by the heat generated from tidal friction, but these mechanisms can account for only a small fraction of the deficiency. Second, tidal friction in the Earth would dampen a tidal bulge before it could become large enough to throw off part of its mass. Third, the mass of the daughter planet produced by fission should have been about one-fifth of the mass of the Earth. The Moon's mass, however, is only 1/81.3 of the Earth's mass. Some of the Moon's mass, largely volatiles, might have been lost to the gravitational attraction of the Earth, but it is virtually impossible to account satisfactorily for such a large discrepancy. Fourth, a lunar fission fragment initially would have been within the *Roche limit*[3] and would have been destroyed by tidal forces. Fifth, it is highly improbable that a fission fragment would have had sufficient orbital velocity to stay in orbit around the Earth. Instead, it would have either escaped entirely to orbit the Sun or fallen back to Earth. Finally, if born by fission the Moon should orbit within a few degrees of Earth's equatorial plane. Instead, the inclination of the Moon's orbit varies from 18.5° to 28.5°.

In addition to the physical difficulties, there are serious chemical problems with the fission hypothesis. Compared to the Earth, the Moon is depleted in iron (Fe), many of the *siderophile* elements,[4] and the volatile elements like potassium, and slightly enriched in the refractory elements like uranium and the rare earths; its composition more nearly matches the composition of the Earth's mantle than that

3. A body inside of Earth's Roche limit, which depends on the density of the body but is about three Earth radii for objects the density of the Moon and asteroids, will be torn apart because tidal forces exceed the gravitational forces that hold the body together.
4. Including nickel, osmium, iridium, rhenium, molybdenum, palladium, and cobalt.

of the bulk Earth. The depletion of iron and the siderophile elements in the Earth's mantle is due to their segregation and concentration in the metallic core. The Moon, however, has either a very small core with a maximum radius of about 500 km or no core at all—the evidence is inconclusive—and it seems impossible to account for the depletion of iron and the siderophile elements by purely internal mechanisms (Toksöz et al., 1974, S. R. Taylor, 1979, Newson, 1984). Thus, the material of the Moon must have been differentiated from a more primitive body before the Moon formed. So far, this might sound like an argument in support of the fission hypothesis, but it turns out not to be so. In spite of the similarities in the compositions of the Moon and the Earth's mantle, there are also significant differences that nearly preclude the Moon from having an Earthly origin.

Although both the Earth's mantle and the Moon are depleted in Fe and siderophile elements relative to the rest of the Solar System, the Fe content of the Moon (ca. 13%) is nearly twice that of Earth's mantle (ca. 7.5%). At the same time, the concentrations of siderophile elements are much lower in the Moon than in the mantle of the Earth. Mo and Re, for example, are depleted in the Moon by factors of more than 10 and 100, respectively. How can the siderophile elements be lower in the Moon while Fe is substantially higher? There are other differences as well. Au (gold), Sb (antimony), and Ge (germanium) are lower by a factor of 100 in the Moon; Bi (bismuth), Th (thorium), and In (indium) are depleted by a factor of about 50; the alkali metals K, Rb, and Cs (cesium) are deficient by about a factor of two; and many of the elemental ratios, e.g. Mo/Nd, are not at all the same. Various means of accomplishing these variations between the Moon and Earth's mantle have been proposed, as summarized by Newson (1984), but the mechanisms are complicated and, as yet, unconvincing.

A variation of the fission hypothesis, proposed by A. E. Ringwood of the Australian National University, is that the early Earth differentiated and then became so hot that metals and oxides were vaporized to form an "atmosphere" that collected in the equatorial plane of the Earth and finally condensed to become the Moon. It is highly improbable, however, that the proposed "atmosphere" would have sufficient energy to attain orbit about the Earth.

The *capture hypothesis* envisions a Moon formed elsewhere and gravitationally captured when it passed near the Earth. This hypothesis has one attractive feature—it neatly solves the problem of the chemical differences between the Earth and Moon. If the Moon formed elsewhere, then these differences are unimportant.

Most capture models call for formation of the Moon somewhere in the inner part of the Solar System, but one turn-of-the-century advocate of capture, the astronomer T. J. J. See, proposed that the Moon formed in the outer part of the Solar System near the orbit of Neptune. It has even been proposed that the Moon originated outside of the Solar System entirely, but the coincidence of the Moon's age and the age of the Solar System indicates that formation around another star is highly improbable.

Capture of a body formed in some distant and unexplored part of the Solar System is an aesthetically unsatisfying solution because it simply separates the problem of composition from the available data. In addition, there are some limits on the region of the Solar System in which the Moon can have formed. Analyses of Earth, Moon, and meteoritic materials, backed by theoretical considerations, have shown that the ratios of certain light isotopes, particularly of oxygen, vary from region to region within the Solar System because of the light-isotope fractionation and nuclear processes in the condensing Solar nebula. The compositions of oxygen isotopes in lunar rocks are identical to those in the Earth and the differentiated meteorites but different from those in the more primitive meteorites (R. N. Clayton and Mayeda, 1975, R. N. Clayton, 1978). This appears to restrict formation of the Moon at least to the same region in space, i.e. the same distance from the Sun, as the Earth. But if this is so, then how can the chemical differences between the Moon and the bulk Earth be explained?

The dynamic difficulties with capture are also severe. Unless there is some mechanism for dissipating its energy and decreasing its velocity, an approaching body will not go into orbit around the Earth but will merely be deflected and return to an orbit about the Sun. Spacecraft accomplish orbital injection by firing rocket engines to brake their approach, but it is a far more difficult feat to slow a planet just the right amount to ensure orbit. In addition, the orbit of the Moon around the Earth is nearly circular, which requires an even more exacting set of circumstances. Capture of a body from without the Solar System adds the additional requirement that the body escape the gravitational field of its own star and enter the Solar System on a path very nearly in the *ecliptic plane*. These requirements make the odds of capture exceedingly small, but not altogether impossible.[5]

Several variations on the capture hypothesis have been advanced to overcome some of the dynamic and chemical problems with

5. Calculating the odds of an event, especially after it has happened, can be misleading. As an illustrative experiment, deal yourself a "hand" of 52 cards from a shuffled deck and lay them out on a table in the order dealt. The odds of dealing that particular sequence of cards is $1/52!$ or 1.2×10^{-68}. From this exceedingly small probability you might conclude that the hand you just dealt was impossible, yet there it is before you. The same might be true of the capture of the Moon.

the capture of a fully formed planet. Mostly they call for the Moon to be aggregated from one or more moon-sized bodies that were already chemically differentiated when captured by the Earth. This sort of origin is consistent with the similarity of oxygen isotope compositions of lunar rocks and differentiated meteorites. According to one version, proposed by J. A. Wood and H. E. Mitler of the Smithsonian Astrophysical Observatory, a number of smaller bodies passed within the Roche limit of the Earth, where tidal forces caused them to disintegrate. The silicate mantles of the objects remained in orbit to form the Moon by aggregation, while the iron cores either fell to the Earth or returned to orbit around the Sun as iron meteoroids. Although this type of mechanism may satisfy the chemical constraints, there is very little physical basis to suppose that the iron-rich fragments would escape Earth's orbit. Most likely, they would simply reaggregate with the silicates and form a moon of composition similar to that of the original bodies.

In the *double-planet hypothesis*, the Moon and Earth formed together as a planetary pair. This satisfies the proximity requirement of the oxygen isotope data in a straightforward way and avoids the severe dynamic problems of capture and fission, but the chemical differences between the two planets remain to be explained. A variation of this hypothesis, proposed by G. K. Gilbert (1893: 289–90) is that the Moon accreted from a ring of planetesimals that formed around the Earth shortly after the Earth formed. Another variation, advanced by A. Harris and W. Kaula of the University of California at Los Angeles, is that during the early stages of formation the Earth captured a small moon and then both the Earth and Moon grew by accretion as a double planet. All of the double-planet hypotheses require some fractionation processes during accretion to account for the difference in composition between the two planets, but the processes that might lead to such fractionation are not well understood.

The *collision hypothesis*, originated by W. K. Hartmann and D. R. Davis (1975) of the Planetary Science Institute in Tucson, Arizona, calls for material from the proto-Earth to be injected into orbit by the impact of a large body, perhaps the size of Mars or Mercury, where it would later aggregate as the Moon. Although initially considered unlikely, this hypothesis has enjoyed renewed interest. Preliminary dynamic and geochemical modeling suggests that such an impact might well result in the formation of a proto-Moon with a mass and composition similar to the contemporary Moon's (Hood, 1986, Stevenson, 1987).

Theoretical considerations indicate that numerous bodies the size of Mars or larger probably struck the proto-Earth within a 20–30-

Ma period during the early stages of Solar System formation (Wetherill, 1985b). The dynamics and chemical consequences of large impacts, however, are poorly understood although currently under study. The enormous energy of large impacts would have generated a silicate-vapor atmosphere and melted the surface of the proto-Earth (Stevenson, 1987) and the Moon. The collision hypothesis, which would apply to all of the planets, might also account for the large angle of inclination of Earth's equator from the plane of the ecliptic and the retrograde spins of Venus and Uranus. There are, however, many uncertainties. What are the implications of a magma ocean for early Earth history? How many impacts occurred and what was the size and composition of the impactors? What relative proportions of proto-Earth material versus impactor material were injected into Earth orbit? Did the ejecta form a smooth disk orbiting the Earth or were they "clumped"? How many proto-Moons formed before the large impacts ceased?

Although many questions remain, the uncertainties in the collision hypothesis are due primarily to the recency of attention accorded the collision hypothesis and the corresponding lack of research results, a situation that is rapidly changing. Unlike the other hypotheses for lunar origin, there is as yet no substantial negative evidence against the collision hypothesis, and so it is currently, if tentatively, favored.

As should be obvious from this brief summary, we do not know how the Moon originated or its exact genetic relationship to the Earth. At the heart of this dilemma is the lack of chemical data from other parts of the Solar System and the extreme complexity of the dynamic and chemical processes, also imperfectly understood, that lead from the debris of star formation to the formation of planets. It is important to note, however, that unless the Moon was captured from another star system, a most unlikely scenario, all reasonable hypotheses for the formation of the Earth, Moon, and Solar System require them to form at very nearly the same time. Thus, there is much to learn about Earth's age from studies of the ages of lunar rocks.

The Lunar Landscape

Anyone who has looked at the face of the Moon has noticed distinctive light and dark areas (Fig. 5.1). The difference in brightness is caused by a dissimilarity in the ability of the two fundamental types of lunar terrain to reflect sunlight; it is due partly to composition and partly to surface roughness. The dark areas, the *maria* (singular, *mare*) or "seas," are relatively smooth lowland plains. The light areas are

North

Fig. 5.1. The near side of the Moon, photograph (above) and diagram (opposite), showing some of the principal named features and the locations of the six Apollo (A) and three Luna (L) landing sites from which samples were returned to Earth. The dark smooth areas are maria; the brighter rugged areas are terrae (highlands). (Lick Observatory Photograph L-9, reproduced with permission.)

the *lunar highlands*, also called *terrae*. They are the rougher, more heavily cratered upland areas.

The maria account for about 17% of the lunar surface, but they are not distributed uniformly. Nearly all maria are on the Moon's near side; they are comparatively rare on the far side. Even before the first Apollo missions provided samples of material from the maria, photogeologists had deduced from the presence of small volcanic domes and vents, flow fronts, lava tubes, and compression ridges that the

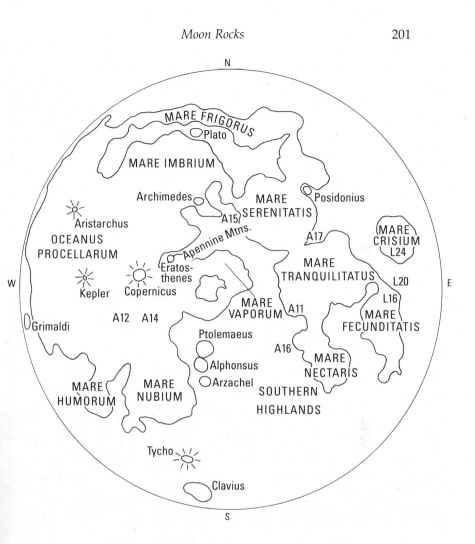

N

MARE FRIGORUS
⃝Plato

MARE IMBRIUM

Archimedes⃝   MARE ⃝Posidonius
              SERENITATIS
          A15
Aristarchus   ⟋Apennine Mtns.
OCEANUS   A17   MARE
PROCELLARUM   CRISIUM
                                            L24
      ⟋Eratos-   MARE
Kepler thenes   TRANQUILITATUS   L20
     Copernicus
                                         L16
W ⃝Grimaldi   A12 A14   MARE   A11 MARE   E
                           VAPORUM   FECUNDITATIS
                  Ptolemaeus
                                  A16
                  ⃝Alphonsus   MARE
              ⃝Arzachel   NECTARIS
   MARE   MARE   SOUTHERN
   HUMORUM NUBIUM
                            HIGHLANDS

        Tycho

         ⃝Clavius

S

maria were formed by the filling of low-lying areas with voluminous
lava flows. The lack of major volcanoes within or near the maria indicates that most of these flows erupted from fissures in a manner similar to the flood basalts of the Columbia River Plateau of the northwestern United States and the Deccan Traps of India. Compared to
the highlands, the maria are very flat. Slopes of only 1:500 to 1:2,000
are common, and elevations typically change less than 150 m over distances of 500 km or so (S. R. Taylor, 1975: 27).

Except for a few large and spectacular craters, the maria are the
Moon's youngest major features. Where they contact the surrounding
highlands the basaltic lava flows of the maria can be seen to fill em-
bayments in the older highland topography (Fig. 5.2). Another indication of their relative youth is that the maria are cratered much less

Fig. 5.2. The eastern edge of Mare Serenitatis as photographed by the Apollo 17 mapping camera. The relatively smooth, dark material of the mare is formed of lava flows that filled a large circular basin. There are fewer craters in the mare than in the rugged highlands to the right because of the difference in age. The partly flooded crater Posidonius, which is about 100 km in diameter, is near the top of the photograph. Just to the south of Posidonius and similar in size is the older flooded crater Le Monnier. Both craters are visible in Figure 5.1. The Apollo 17 landing site in the Taurus–Littrow Valley is shown by the arrow in the lower right. (NASA photograph AS17-0940.)

heavily than are the highlands. Though the maria are youthful lunar features, they are ancient by Earth standards. Radiometric ages of basalt samples from maria are all within the range 3.0–3.9 Ga. Although samples are available from only a few of the maria, crater density and other geologic considerations suggest that mare volcanism may have continued until 1 Ga or so ago, although at a much diminished rate (Schultz and Spudis, 1983).

As their name implies, the lunar highlands are upland areas. Typically, their mean elevations are 3 to 4 km above the maria, although Oceanus Procellarum (Fig. 5.1) is only about a kilometer below the adjacent highlands. The highlands represent the top surface of a lunar crust several tens of kilometers thick and are the oldest regions of the Moon. Crater densities show that there are two types of highland surfaces, an older one that is heavily cratered and a younger one, the so-called light plains, that is less cratered. The older highland surface is very heavily cratered by both large and small craters, and is extremely rough on a fine scale, which contributes to its higher reflectivity and light appearance. Highland rocks differ in composition from mare basalts in, among other things, their content of lighter-colored minerals. Thus, their compositions also tend to make highland rocks more reflective than mare basalts.

Even before the Apollo landings it was known from the difference in crater densities and from stratigraphic relations that the highlands were older than the maria, and radiometric dating has confirmed that conclusion. The majority of highland rock ages cluster within the range 3.8–4.0 Ga, but some are older, and a few have radiometric ages near 4.5 Ga.

If any single process can be said to have dominated the evolution of the lunar landscape, it is cratering due to the impact of rock bodies striking the Moon.[6] Lunar impact craters are ubiquitous and come in all sizes, from the largest of the multi-ring *basins* several thousand kilometers in diameter to tiny micro-pits less than 0.001 mm in diameter that are found on the surfaces of individual mineral grains and rock surfaces. Many of the larger craters can be seen easily from Earth with the naked eye or with binoculars.

The large multi-ring basins consist of a central depression surrounded by one or more concentric "rings" of mountains (Fig. 5.3). There are 43 basins with diameters greater than 220 km distributed more or less uniformly over both the near and far sides.[7] They are enormous features that are responsible for most of the Moon's major topography. The largest basin, with a diameter of 2,000 km and a relief of some 8 km, is on the far side in the southern hemisphere. The Imbrium Basin, a prominent feature of the northwestern near side (Fig. 5.1), is approximately the size of Texas. It has a diameter of 1,340

6. For more than a century and a half there was lively debate about the volcanic (e.g., Dana, 1846) vs impact (e.g., Gilbert, 1893) origin of the Moon's craters. The matter was settled in the 1960's and it is now clear that, with only a few small and rare exceptions, virtually all lunar craters are due to impact. The debate and evidence are summarized by S. R. Taylor (1975: 27–30), Guest and Greeley (1977: 30–33, 98–116), and Cadogan (1981: 48–56).

7. Hartmann and Wood (1971) and Howard, Wilhelms, and Scott (1974) discussed multi-ring basins in detail.

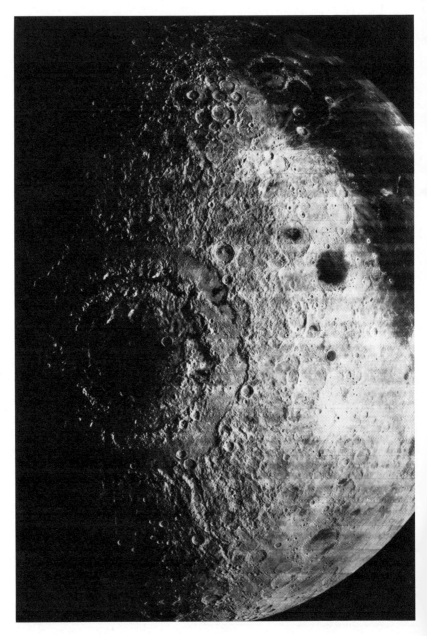

Fig. 5.3. The multi-ring Orientale Basin, located on Moon's far side, is 900 km in diameter and was formed by the impact of an asteroid-sized body about 3.8 Ga. It is the youngest of the ringed basins. Unlike the basins on the near side, Orientale was only partly filled with mare basalts. Concentric rings of mountains are an inherent feature of very large impact structures. Oceanus Procellarum can be seen in the top right. The highlands surrounding Mare Orientale are saturated with craters. (NASA lunar orbiter photograph IV-187M.)

km and its ringed mountain ranges rise more than 7 km above its floor. Now flooded with basalt lava and the site of Mare Imbrium, its circular form is still plainly visible.[8]

The lunar basins were caused by the impact of planetesimals (large asteroids) typically some tens of kilometers in diameter. The amount of energy involved in these impacts was enormous. It has been estimated that the excavation of the Imbrium Basin, for example, required an energy of about 3×10^{26} joules (S. R. Taylor, 1979: 106). This is about 10^7 times the energy released in all of the earthquakes in the Earth in one year or enough to energize 10^{10} 100-watt light bulbs for 10 Ma. This amount of energy corresponds to the collision of an asteroid some 100 km in diameter traveling at a velocity of 16 km/s.

There are no impact features on Earth even remotely comparable in size to the lunar basins, although equally large ringed basins are found on both Mars and Mercury. The absence of such features on Earth is due to both the antiquity of basin formation and the active nature of the Earth's crust and upper mantle. Radiometric dating indicates that the lunar basins were formed between 3.8 and 4.5 Ga ago while the young planets were collecting the last of the larger bodies in their orbital paths. Probably the Earth was also subjected to a similar bombardment and basin-forming at the same time. On the Earth, however, all evidence of pre-existing basins has been destroyed by the constant crustal recycling that is absent on the smaller, now-inactive planets where basins are still preserved.

The effects of basin formation on the lunar surface are profound. Not only do multi-ring basins account for most of the Moon's major topography and act as the principal sites for maria, but the frequent and violent collisions pulverized and fractured virtually the entire early lunar crust to depths of 20 km or so. As a result, there appear to be no "original" highland igneous rocks, only impact breccias and fragments. In addition, the ejecta from basin formation cover large areas of the lunar highlands to depths of hundreds of meters. These ejecta "blankets" constitute traceable formations that are the basis for lunar stratigraphy. The ejecta from the youngest basins, principally Imbium and Orientale, may form the light highland plains. Even these youngest basin ejecta do not occur on the maria, indicating that basin formation preceded the eruption of maria lavas. At one time it was thought that maria lavas were generated by melting due to the basin-forming impacts, but the lack of basin ejecta on any of the maria and the great age differences (typically several hundred Ma) between basin formation and the maria lavas preclude that possibility. Basin

8. The terms *basin* and *mare* are not synonymous. The basin is the impact feature, whereas the mare is the lava plain flooding the basin. A basin need not have an accompanying mare, though nearly all do.

formation also resulted in secondary impact craters, some as much as 25 km in diameter. These are due to the impact of large blocks of lunar crustal material ejected from the basin site at low angles and at velocities less than that required to escape from the Moon.

In addition to the ringed basins, the lunar surface is peppered with smaller impact craters whose number increases rapidly with decreasing diameter. There are, for example, about 80 craters with diameters between 100 and 220 km and thousands with diameters between 1 and 100 km. Even though smaller than basins, the larger of the craters are enormous features.[9] Copernicus, for example, with a diameter of 91 km and visible in Figure 5.1, is 25% larger than the state of Delaware and 3.4 km deep. As is the case with basins, the impacts that formed the larger craters not only reshaped the lunar landscape but pulverized the upper parts of the crust and redistributed material over the lunar surface.

The numerous smaller craters, i.e. those with diameters less than one kilometer, are of both primary and secondary origin. The primary craters are excavated by the impact of extralunar bodies whereas secondary craters are the result of impacts by blocks of lunar rock ejected from the primary craters. Many of the impacts that formed these smaller craters did not disturb bedrock but expended their energy in stirring and mixing the material on the lunar surface and in forming new rocks (breccias) from the particles of the lunar soil.

The microcraters, which range in diameter from about one centimeter to less than 10^{-3} mm, are caused by the impact of cosmic dust and, at the smaller end of the range, of high-velocity ions from the Sun, i.e. the solar wind. The dust particles, which commonly range from about 10^{-4} to 10^{-7} g in mass, are primarily material shed by comets as they pass through the inner reaches of the Solar System, although some dust may be generated by collisions in the asteroid belt between Mars and Jupiter. These tiny dust grains, composed primarily of silicates and metallic iron, impact the lunar surface with an average velocity of about 20 km/s, causing erosion through mechanical abrasion, melting, vaporization, ionization, and lateral transport of material. The production rates of microcraters are very high. It has been estimated, for example, that the present production rate of microcraters with diameters greater than 0.5 mm is about 5 craters per Ma for each square centimeter of lunar surface, and that a fresh surface is saturated[10] with such craters within about 1 Ma (S. R. Taylor,

9. The distinction between large craters and small basins is based on the presence or absence of multiple rings, which are an intrinsic product of larger impacts. Otherwise, basins and craters are a continuum of the same phenomenon.

10. Saturation means that the density is at maximum so that, on the average, the formation of a new crater results in the destruction of an older one of similar size and the total number is unchanged by continued bombardment.

1975: 89). These micrometeorite impacts erode the lunar surface at a rate of about 1 mm/Ma, while ions from the solar wind account for an additional 0.1 mm/Ma of wear and tear on the exposed surfaces of lunar rocks.

Because the Moon has been subjected to cratering throughout its history, it is not surprising that the density of primary lunar craters, i.e. those with diameters of 1 to 100 km, is a reasonably reliable indicator of the relative ages of lunar formations. Crater counting is, in fact, one of the most important tools of lunar stratigraphy. Cratering rates, however, have not been uniform for the past 4.5 Ga. Comparison of the densities of large craters on lunar surfaces of different age, as determined by radiometric dating of the returned lunar samples, has shown that there was a sharp decrease in impact rates between about 4.0–3.5 Ga followed by a gradual decline in cratering since then (Fig. 5.4). The sharp decline represents the final phase in the accretional process and explains why the oldest craters are not much older than 4.0 Ga. Prior to 4.0 Ga the lunar surface was subjected to saturation bombardment with the result that craters that formed much before the sudden decline in impact rates were quickly obliterated by the formation of younger craters.

Although present cratering rates are about an order of magnitude less than the average rate for the past 3.5 Ga, current observations show that they are not zero. For example, seismic detectors left on the Moon by the Apollo astronauts record about 70–150 moonquakes each year caused by the impact of meteorites with masses between 0.1 and 1000 kg. In addition, 15 microcraters, caused by the impact of cosmic dust grains, were found on the glass windows of the Apollo spacecraft that made the journey to the Moon and back (Cadogan, 1981: 233).

While the heavily cratered lunar landscape may seem exotic to Earthbound inhabitants, impact cratering is not unique to the Moon. Parts of Mars and Mercury are heavily cratered, and there is no doubt that the Earth has been (and continues to be) subjected to a similar bombardment. The Earth's atmosphere, however, restricts the size and velocity of objects reaching the Earth's surface from space, thereby minimizing the frequency and intensity of impacts. In addition, Earth's continually active crust and vigorous agents of erosion work against the preservation of what few craters do form. Extraterrestrial objects between about 10^3 and 10^{-7} g never impact Earth's surface because they "burn up" in the atmosphere due to frictional heating. Smaller particles do not burn up but instead are slowed and drift downward through the atmosphere to eventually settle on the surface as dust. It is estimated from satellite penetration data that some 88×10^4 kg of meteoritic dust is deposited on the Earth each year, and such

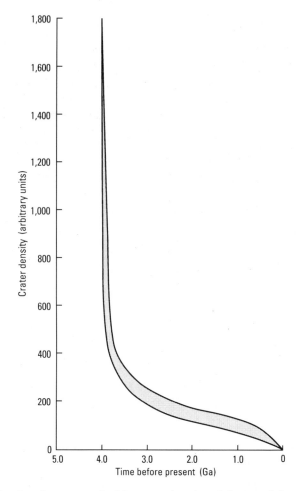

Fig. 5.4. Density of craters on the Moon as a function of the age of the surfaces on which they occur, as determined by dating of Apollo samples. The shaded band encompasses the errors in the crater density measurements. The rapid decrease at about 4 Ga probably represents the final "collection" of accretional debris. (After Soderblom et al., 1974.)

material has been found in deep-sea sediments (Barker and Anders, 1968, Dohnanyi, 1972, Parkin, Sullivan, and Andrews, 1980). Bodies larger than 10^3 g occasionally penetrate the atmosphere and impact the Earth's surface (Chapter 6). The larger of these impacts create substantial craters that may be preserved for considerable lengths of time. One hundred sixteen known and probable impact craters, ranging in diameter from a few tens of meters to as much as 140 km, have been identified on the Earth (Grieve, 1982, 1987). The Pleistocene-age Meteor (or Barringer) Crater in Arizona, the 15-Ma-old Ries Basin in Germany, and the two late Paleozoic craters that form Clearwater Lakes in Quebec, Canada, are well-known examples. Although no associated crater has been identified, there is now good evidence that an asteroid some 10 km or so in diameter struck the Earth at the end of the Cretaceous, possibly causing the *mass extinctions,* including the disappearance of the dinosaurs, that mark the close of that geologic period.[11]

The incessant bombardment of the lunar surface by extralunar objects and particles has resulted in a layer of comminuted rock called the *regolith,* discussed by Heiken (1975), S. R. Taylor (1975: 57–78), Guest and Greeley (1977: 136–50), Langevin and Arnold (1977), and Cadogan (1981: 194–250). This layer of "soil" covers virtually the entire surface of the Moon (Fig. 5.5). The lunar regolith differs substantially from the soils on Earth because the principal processes active on the surfaces of the two planets are very dissimilar. On Earth, soils are the result of mechanical, chemical, and biological activity. Chemical alteration due to the interaction of the atmosphere, the hydrosphere, and living organisms with rocks and rock particles is especially important, whereas mechanical processes, principally gravitational transport, play a much lesser role. On the Moon, which lacks an atmosphere, water, and living organisms, soil-forming processes are almost entirely mechanical. In addition to gravitational transport and expansion and contraction due to heating and cooling, both of which play important but minor roles, the principal soil-forming agent on the Moon is the impact of meteorites, cosmic rays, and the ions of the solar wind—chemical alteration is almost entirely absent.

Although the thickness of the regolith averages about 4–5 m over the entire lunar surface, it is highly variable, being primarily a function of the age of the surface. It varies from 10–15 m in the highlands and the older maria to a few centimeters on the rims of young craters. On the younger maria, the regolith is generally only a meter or two thick.

The dominant component of the regolith is rock and mineral fragments. The fragments vary in size from a few micrometers to tens of meters, but the bulk of the material is smaller than 1 mm. Among particles less than 1 mm in size, glass, one of the more interesting components of the lunar soil, is abundant. Particles greater than 1 mm are primarily of rock and mineral fragments, with impact glass

11. The evidence for this event is a thin layer of sediment, found at the Cretaceous–Tertiary boundary, that contains an unusually high concentration of the platinum-group metals, which are common only in meteorites, and of shocked minerals. This layer has been found in more than 75 places worldwide. It is hypothesized that the layer is the debris from the impact of an asteroid and further that the impact put enough dust into the atmosphere to darken the sky for several years, thereby disrupting the food supply and leading to mass faunal extinctions. The dinosaurs may have been victims. For more information on this intriguing hypothesis see Silver and Schultz (1982) or Alvarez (1987).

Fig. 5.5. Geologist Harrison H. Schmidt, the only scientist to land on the Moon, examines the large boulder at Station 6 near the Apollo 17 landing site in the Taurus–Littrow Valley. The complex breccia boulder rolled downhill from the adjacent highlands. The ubiquitous lunar regolith, which is composed of comminuted rock, glass, and rock fragments, is the product of repeated meteorite impacts and radiation. None of the rocks visible in the photograph are "in place" where they originally formed. Note the numerous small craters in the regolith. (NASA photograph AS17-140-21497.)

being scarce. Much of the glass occurs as small, spherical glass beads of various colors. Most of the glass is *impact glass* and is formed by impact melting of the soil itself, and glass was found lining many of the small craters visited by the Apollo astronauts. Some of the glass spheres, however, especially those devoid of rock fragments and with ages and compositions identical to those of mare basalt, are probably volcanic, representing liquid droplets scattered by the fountains of lava that must have accompanied the copious eruptions of mare basalt. About 1.5% of the regolith is meteoritic material and represents the disintegrated remains of extralunar objects.[12]

Although impact cratering causes lateral transport and may displace material thousands of kilometers, most of the regolith is derived locally. It has been estimated that more than 50% of the regolith at any given spot on the Moon represents material from within 3 km of the site, while only 5% and 0.5% has traveled more than 100 and 1000 km, respectively (S. R. Taylor, 1975: 61–62).

The regolith represents the Moon's active surface layer, but its turnover rate is not particularly high. On the basis of current impact rates, the upper 0.5 mm of the regolith is completely "stirred" only once every 10,000 years or so. Mixing times for depths of 1 cm and 1 m are 10 Ma and 1 Ga, respectively. Efforts on the Apollo missions to penetrate the regolith with a coring apparatus were only marginally successful because, owing to compaction, the density of the regolith increases rapidly with depth. The longest core, 242 cm, was collected on the Apollo 15 mission and contained 42 distinct layers, each representing deposits of ejecta from successive and nearby impacts (Guest and Greeley, 1977: 141). The preservation of such layering is consistent with the very low turnover rate of the regolith.

Beneath the regolith is the *megaregolith*, a thicker layer of ejecta from basin-forming impacts. Its thickness has not been measured, but theory indicates that it must vary considerably from place to place as a function of distance from the lunar basins. For example, the calculated thickness of the megaregolith at the Apollo 14 landing site is only about 280 m, whereas the thickness at the Apollo 17 site is nearly 2 km (McGetchin, Settle, and Head, 1973). Since the megaregolith consists of basin ejecta, it occurs beneath, rather than on top of, the mare lavas. The nature of the material beneath the megaregolith is uncertain, but it is thought to be brecciated and fractured crustal material, disrupted to depths of as much as 20–30 km during the very intense early lunar bombardment. It is probable that the contacts between the regolith, the megaregolith, and the fractured crust are gradational rather than sharp.

The Nature of Moon Rocks

One of the most significant scientific benefits of the Apollo program was the return of samples of rock and soil for study by Earth-

12. Meteorites reach the Moon's surface at full velocity and disintegrate on impact. In contrast, those reaching the Earth's surface have been greatly slowed by the atmosphere and so larger fragments are more common.

TABLE 5.1

Nine Lunar Missions Returned Samples to Earth for Study

Mission	Date	Location	Geologic setting	Returned samples Weight (kg)	Types[a]
Apollo 11	Jul. 1969	Mare Tranquillitatis	Open mare plain	21.7	C, B, S
Apollo 12	Nov. 1969	Oceanus Procellarum	Open mare plain	34.4	C, B, S
Luna 16	Sep. 1970	Mare Foecunditatis	Open mare plain	0.101	S
Apollo 14	Feb. 1971	Fra Mauro region	Imbrium ejecta blanket (Fra Mauro Formation)	42.9	B, S, C
Apollo 15	Jul. 1971	Palus Putredinis	Mare at highland border, inside third (Apennine) ring of Imbrium Basin near Hadley Rille	76.8	C, B, S
Luna 20	Feb. 1972	Apollonius Highlands	Highlands near mare border, on ejecta from Crisium and Fecunditatus basins	0.030	S
Apollo 16	Apr. 1972	Southern Highlands	Highland Light Plains (Cayley Formation) and hummocky terrain (Descartes Formation)	94.7	B, S
Apollo 17	Dec. 1972	Taurus–Littrow Valley	Mare (Serenitatis) at base of highlands	110.5	C, B, S
Luna 24	Aug. 1976	Mare Crisium	Open mare plain	0.170	S

SOURCES: S. R. Taylor, 1975: 4–9; Cadogan, 1981: 89–123, 135–40.
[a]C, crystalline rocks; B, breccias; S, soil, which includes small fragments of crystalline rocks and breccias as large as a centimeter or so. Listed in approximate order of abundance.

bound scientists. Nine missions, six from the United States and three from the USSR, returned a total of nearly 382 kg of samples (Table 5.1).[13] This priceless material consists of crystalline rocks, breccias, and soil, the latter in the form of both scooped samples and cores, from a variety of geological environments. It is important to realize that because of the ubiquity and thickness of the regolith, outcrops of undisturbed rock are rare on the Moon. In fact, none of the lunar samples were collected *in situ* at their exact places of origin, although many probably came from nearby. All were excavated by meteoritic impacts and many have been moved repeatedly, some for considerable distances. Thus, pieces of highland rocks have been found even at the landing sites in the open maria many tens of kilometers from the nearest highlands.

13. As S. R. Taylor (1975: 8) pointed out, if the entire Apollo program is charged to the returned samples the unit cost is $28,500 per pound. But, as he also emphasizes, this is faulty bookkeeping because it does not include the returns from other aspects of the Apollo program (e.g., surface and orbital experiments) and assumes that the total rationale for the program was scientific—a wishful thought. The Russian samples are far more valuable than their total weight of 0.3 kg might suggest because they broaden and enrich the coverage of geological localities and environments greatly.

Fig. 5.6. Rock 15556, returned by the Apollo 15 mission, is a vesicular mare basalt. (NASA photograph AS17-71-643328.)

Moon rocks are not exactly like Earth rocks and much has been made of the differences (S. R. Taylor, 1975, Papike et al., 1976, Ridley, 1976, Head, 1976, Basaltic Volcanism Study Project, 1981, Cadogan, 1981). Although these differences are important, the overall similarity of Earth and Moon rocks is equally worthy of note. Contrary to the impression conveyed by many pre-Apollo films and television series, there are no totally new or weird types of rocks in the lunar sample collection. The lunar rocks include both crystalline igneous rocks and impact breccias. Virtually all of the lunar rock types have their terrestrial analogs, albeit not necessarily in the same abundances. For example, basalt is a common rock type on both Earth and Moon and the lunar mare basalts look very much like Earth basalts. Lunar basalts contain the same basic minerals as terrestrial basalts and many are vesicular, just like their Earthly cousins (Fig. 5.6).[14] Similarly, impact and volcanic breccias are found on both the Earth and the Moon. On

14. Several new minerals have been found in lunar rocks, but they are not major constituents and do not change the basic rock types.

TABLE 5.2

Simplified Classification of Mare Basalts and Their Approximate Contents of Four of the Major Oxides That Distinguish Them

Type	Mission	SiO$_2$	Al$_2$O$_3$	TiO$_2$	K$_2$O	Source
Mare basalts						
Olivine basalt	Apollo 12, 15	45	8.5	2.5	0.05	S. R. Taylor, 1975: 136
Quartz basalt	Apollo 12, 15	47	10	2.5	0.05	
High-Ti basalt						
High-K	Apollo 11	40	8.5	11	0.3	
Low-K	Apollo 11, 17	39	9	12	0.1	
Aluminous basalt	Apollo 12, Luna 16	46	13	4	0.1	
Hawaiian basalt		49	14	2.5	0.4	Macdonald, 1968: 502
Sea-floor basalt		49	17	1.5	0.2	Hughes, 1982: 273
Columbia River basalt		50	16	1.6	0.6	Hughes, 1982: 346

NOTE: The basalt types have a range of compositions, but the values shown are representative. The average compositions of Hawaiian, sea-floor, and Columbia River tholeiitic basalts are shown for comparison.

the Moon impact breccias are very common while volcanic breccias are much less so, whereas the reverse is true on the Earth. The overall similarity of Earth and Moon rocks is not especially surprising because it was known before the Apollo landings that the entire Solar System had a common origin and its parts must be compositionally linked.

There have been a number of classifications of mare basalts based on chemistry, mineralogy, and texture, each classification varying according to which differences and which similarities the classifier wished to emphasize. For the purpose of this chapter, I will use a simplified version of a widely used chemical classification based in large part on the relative percentages of silicon, aluminum, titanium, and potassium (Table 5.2).

The principal minerals in the lunar basalts, as in terrestrial basalts, are plagioclase feldspar and *pyroxene*, with the latter mineral being about twice as abundant as the former. In addition, all consist of from about 5 to 25% opaque Fe–Ti oxides, and some consist of as much as 20% or so olivine. Many of the lunar basalts contain a few percent of pure SiO$_2$ as tridymite and cristobalite. One of the more distinctive features of mare basalts, and one that was not anticipated, is their titanium content, which is much greater than that of their nearest terrestrial analogs. Those highest in titanium, collected from the Apollo 11 and 17 sites, constitute the *high-Ti basalts*, which can be subdivided into high- and low-potassium varieties. The *olivine* and *quartz basalts* are so named because they contain these minerals in their *norms*. The olivine basalts also have significant (5–20%) amounts of olivine as actual constituents, whereas the quartz basalts do not actually contain quartz and are devoid of olivine. The so-called *aluminous basalts* are probably nearer in composition to terrestrial basalts than the other types, but still contain more titanium than do the terrestrial varieties (Table 5.2). They are higher in aluminum than other lunar basalts because they contain more plagioclase feldspar, less opaque Fe–Ti oxides, and no olivine.

Most of the lava flows of the maria were emplaced after the early intense lunar bombardment and, as a result, relatively undisturbed samples of maria basalt, with their original igneous textures and mineralogy intact, are common. In contrast, the early lunar crust that now forms the highlands was pulverized and thoroughly mixed to a depth of 10–20 km or so, and the original crustal layering and lateral distribution of highland rock types has probably been destroyed. All of the larger rocks returned from the highlands are intensely brecciated (Fig. 5.7). Many of the breccias represent multiple events, with breccia fragments enclosed within breccia fragments. In the process of breccia formation the rocks were subjected to temperatures, generated upon impact, of as much as 1,100°C or more, resulting in melting

Fig. 5.7. Rock 15459, a breccia from the rim of Spur Crater, near the Apollo 15 landing site in the eastern Mare Imbrium. The breccias, composed of a glassy matrix enclosing angular fragments of rock, are formed by impact. The largest visible clast is 8 cm across. (NASA photograph AS15-71-44181.)

to various degrees followed by recrystallization. In addition, the shock of impact frequently produced mineralogic and textural changes. Thus, lunar breccias are metamorphic rocks, having endured multiple metamorphisms, and are extremely complicated. Various classifications, based on texture and mineralogy and correlated with temperature of formation and metamorphic grade, have been proposed for the breccias. The details of breccia classification, however, are not essential for the purpose of this chapter and will not be discussed further (for a summary, see S. R. Taylor, 1975: 211–30).

Clearly, the composition of the original lunar crust, which has been thoroughly pulverized and mixed, is of considerable interest but difficult to determine. Samples of the highland rock types occur primarily as clasts in the breccias and as small fragments in the regolith, although larger samples representing all of the major rock types have been found. Only rarely do highland samples still retain their original igneous textures—most have been modified by impact processes. Nevertheless, by careful study of the fragments in the soils and the breccias, supplemented by chemical analyses of the various types of glasses found in the soils and by spectrographic data from experiments flown in lunar orbit, scientists have been able to piece together a reasonable picture of the composition of the highland crust.[15]

Highland igneous rocks can be grouped into two general categories—basalts and *cumulates* (Table 5.3). As their name implies, the basalts formed by the cooling and crystallization of a lava. The cumulates, however, represent rocks formed primarily by the density segregation of crystals within cooling magmas and so their origin is in part due to mechanical processes.

The basalts, also called *Fra Mauro basalts* because of their abundance in the Fra Mauro Formation sampled on the Apollo 14 mission, are primarily of two types that are very similar in composition. The major difference between them is that one is higher in K, the rare earth elements (REE), P (phosphorus), and such elements as U and Th, and somewhat lower in Al (aluminum) than the other. The former has been given the humorous acronym *KREEP*; the latter is called aluminous basalt or *low-K Fra Mauro basalt*. The aluminous basalt appears to be the second most abundant of the highland rock types. It is a common constituent of the highland soils sampled on the Apollo 16 and Luna 20 missions and also occurs in the soils collected at the Apollo 15 site near the Apennine Mountains. KREEP basalt is com-

15. The soil surveys, which looked for natural groupings in glass compositions as well as in the compositions of small crystalline particles, have been especially helpful in determining the compositions of highland rock types and their probable relative abundances, for example, Apollo Soil Survey (1971) and Reid (1974). The major rock types deduced from the soil surveys have also been found as larger rocks.

TABLE 5.3

Simplified Classification of Lunar Highland Rock Types and Their Partial Compositions

Rock type[a]	Mineralogical (volume %)			Chemical (weight %)			
	Plagio-clase	Pyrox-ene	Olivine	SiO_2	Al_2O_3	TiO_2	K_2O
Basalts (Fra Mauro basalts)							
KREEP (medium-K) [4]	50	35	10	48	18	2	0.5
Low-K, high-Al (aluminous) [2]	50	35	10	47	19	1	0.1
Cumulates							
Anorthosite [3]	95	4	0	44	35	0.05	0.02
Anorthositic gabbro (highland basalt) [1]	70	20	9	45	26	0.4	0.09
Norite [5]	40	40	5	50	18	0.08	0.05
Troctolite [6]	35	5	60	43	23	0.2	0.04
Dunite [7—very rare]	2	2	95	40	1.5	0.03	0.002

SOURCES: S. R. Taylor, 1975: 233–38; Ridley, 1976: 34–42; Cadogan, 1981: 161–71.
[a]Numbers in brackets represent approximate relative order of abundance with [1] being the most abundant. Some common alternate names are in parentheses. Each type has a range of compositions, but those shown are typical. The anorthosite–norite–troctolite group is collectively called the ANT suite.

mon at the Apollo 12, 14, and 15 sites but is not abundant at the high-land sites. Its distribution suggests that KREEP may be peculiar to the Imbrium Basin, but whether it originated as lava flows due to partial melting of the Moon's interior or was formed by large-scale impact melting is unknown (Cadogan, 1981: 189).

The cumulates have a range of igneous compositions that grade one into the other, so the simple classification in Table 5.3 is some-what artificial but convenient. *Anorthositic gabbro* appears to be the most common highland rock type. It is present at all Apollo and Luna sites, either as small fragments in the soils and breccias or as a common type of impact glass, and its composition resembles the average composition of the highlands as determined from orbital spectrographic data.

Anorthosite is found mainly as soil and breccia fragments and is sufficiently abundant that it is thought to form a significant percentage of the lunar crust. The existence of anorthositic magma is considered highly improbable because of the unusual composition and extremely high temperatures that would be required to form it, so the anorthosites are thought to be cumulates, formed by the aggregation of feldspar crystals from magma even though their textures are inconclusive on this point.

With an increase in pyroxene and olivine at the expense of feldspar, anorthosites grade into *norites*, *troctolites*, and *dunites*. The textures of these rocks indicate that they are cumulates, similar in many respects to those found in terrestrial layered gabbroic intrusions like the Stillwater Complex of Montana. Troctolites have been found as small fragments in the soil from most highland sites and as larger rocks at the Apollo 16 and 17 sites. Norites, which are mineralogically similar to basalts, are found as small fragments and constituents of glass but are rarely found as large rocks. Dunite was found only as small, highly crushed fragments at the Apollo 17 site. The Apollo 17 dunite is one of the oldest rocks yet found on the Moon (4.47 Ga) and will be discussed later.

The Geologic History of the Moon

Having examined some of the principal lunar surface features and the processes that formed them, as well as the major types of lunar rocks, we turn now to the geologic history of the Moon. This will put the highland rocks, basins, craters, and mare lava flows into temporal perspective and provide the framework for outlining the principal radiometric evidence for the ages of lunar rocks and the time of origin of the Moon.

There is a geologic time scale for the Moon (Fig. 5.8) that is similar to the one for the Earth (Table 2.2) except that it is not nearly so detailed. The lunar time scale consists of only five geologic periods, none of which has been further subdivided. The difference in detail between the lunar and terrestrial geologic time scales is due to a number of factors, the most important of which is the degree of access that geologists have to the rock formations of the two planets. Geologists studying the Earth can walk at will on its surface, map the distribution and sequence of its rock types, and collect large numbers of samples for detailed examination and analysis—and have been doing so for more than two centuries. In contrast, geologic mapping of the Moon's formations began in earnest in the 1960's and was done from telescopic photographs, later supplemented by orbital photographs and a few precious samples from nine landing sites. Tens of thousands of geologists have trod the Earth's surface in their effort to decipher its history; only one has ever set foot on the Moon.[16] Nonetheless, the geologic map of the Moon is remarkable in its detail, considering the way in which it was made (for example, Wilhelms and McCauley, 1971).

Except for the fact that virtually all of the observations were made from a distance with telescope and camera, the temporal rela-

16. Several hundred scientists examined the lunar samples. See S. R. Taylor (1975: 6–7) for a summary.

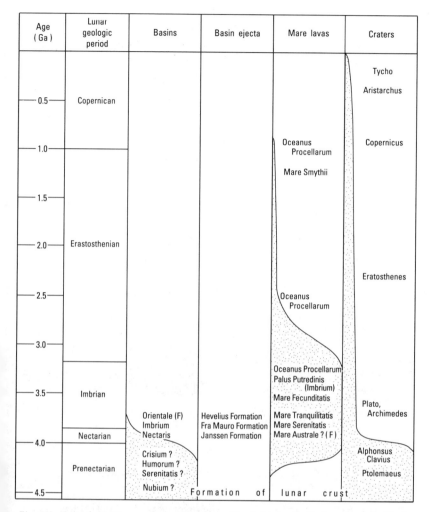

Age (Ga)	Lunar geologic period	Basins	Basin ejecta	Mare lavas	Craters
0.5	Copernican				Tycho / Aristarchus
1.0				Oceanus Procellarum / Mare Smythii	Copernicus
1.5					
2.0	Erastosthenian				
2.5				Oceanus Procellarum	Eratosthenes
3.0					
3.5	Imbrian			Oceanus Procellarum / Palus Putredinis (Imbrium) / Mare Fecunditatis	Plato, Archimedes
		Orientale (F) / Imbrium	Hevelius Formation / Fra Mauro Formation	Mare Tranquilitatis / Mare Serenitatis	
4.0	Nectarian	Nectaris	Janssen Formation	Mare Australe ? (F)	
	Prenectarian	Crisium ? / Humorum ? / Serenitatis ?			Alphonsus / Clavius / Ptolemaeus
4.5		Nubium ?	Formation of lunar crust		

Fig. 5.8. A geologic time scale for the Moon. Shaded areas indicate schematically the duration and intensity of events forming basins, maria, and craters. The locations of the near-side features are shown in Figure 5.1b. (F) indicates a feature on the far side. (After Wilhelms and McCauley, 1971, S. R. Taylor, 1975; 21–27, Guest and Greeley, 1977: 8, El-Baz, 1978, Masursky, Colton, and El-Baz, 1978: 22, Cadogan, 1981; 46–48, 360–63, and Wilhelms, 1984: 113.)

tionships between the various rock units of the Moon were deciphered by means of, by and large, the same general principles used to study the Earth. Foremost among these are the laws of superposition and cross-cutting relationships, which permit an observant geologist to determine the order of emplacement of rock formations at a given locality, though not their ages in years (Fig. 5.9).

Oceanus
Procellarum
→

N

Fig. 5.9. A variety of features of different relative ages can be seen in this photograph taken by the Apollo 15 mapping camera east of the Aristarchus Plateau. Ridges of older highlands peek through the lava flows of the Oceanus Procellarum. The flows have flooded the crater Prinz (at center). In contrast, the crater Aristarchus (at right), whose bright "rays" of ejecta form streaks across the mare, is younger than the lava flows. Aristarchus, 43 km in diameter and one of the youngest and brightest of Moon's impact craters, is also visible in Figure 5.1. The sinuous rills, most of which originate in small craters, are probably lava channels that may have contributed some of the lava that fills the mare. (NASA photograph AS15-2606.)

On the Earth, fossils have permitted planet-wide correlation of rock units and have made possible an especially detailed time scale for the past 570 Ma, the time over which organisms with easily preservable parts (e.g. bones, shells, teeth) have been plentiful. On the Moon, fossils are entirely lacking and so the valuable stratigraphic tool of paleontology is unavailable. As we have seen, however, the Moon is not entirely devoid of regional and planet-wide time–stratigraphic markers. The formation of the multi-ring basins early in the Moon's history, for example, had planet-wide effects through the wholesale modification of topography and the deposition of extensive deposits of ejecta, some of which cover large fractions of the Moon's surface. The blankets of ejecta, in particular, provide the basis for the subdivision of lunar geologic time. Likewise, the flooding of basins by extensive emissions of lava, forming the maria, provides regional time markers. One of the most useful stratigraphic tools available to the lunar geologist is the density of craters on the various surfaces (Figs. 5.2 and 5.4, Table 5.4). Although not linear with time, the density of craters is the only quantitative method of determining the relative age of lunar surfaces and their underlying formations on a planet-wide basis. In a sense, crater density plays a role in lunar geology similar to

TABLE 5.4

Relative and Radiometric Ages of Lunar Basin Ejecta and Mare Lavas as Determined from the Densities of Large Craters and from Radiometric Age Measurements

Basin[a]	Diameter[b] (km)	Relative crater density[c] Ejecta	Mare	Radiometric age (Ga) Ejecta	Mare
Orientale (F)	620 (1,300)	2.4	1.0	3.75–3.81	
Imbrium	1,340	2.5	0.9	3.82–3.83	3.3
Schrödinger (F)	320	5	2.8		
Milne (F)	240	10			
Apollo (F)	435	12			
Moscoviense (F)	410 (700?)	14	4?		
Humboldtianum	600	15	1.5		
Nectaris	840	16	0.9	<3.92	
Grimaldi	220 (410)	16	0.8		
Crisium	450 (1,060)	17	0.5	3.8–4.1	
Janssen	160 (540)	22	0.8		
Humorum	410 (700)	25	0.85	4.1	
Smythii	450 (810)	27	1.9		
Serenitatis	(680)	28	0.6	3.9–4.4	3.8
Southwest Tranquillitatis	(350)	30	1.6	4.5	3.6–3.7
Oceanus Procellarum			0.8	4.4–4.5	3.2
West Nubium	(425)	30		>4.5	
Foecunditatis	(520)	30	0.6	4.4–4.5	3.4
Tycho		0.6		0.1 (est.)	
Copernicus		0.3		0.9	

NOTE: Except for that of the Imbrium Basin, the radiometric ages of basin formation are uncertain, because the relations of dated samples to particular basins are highly inferential. The highlands are saturated with large craters and have a relative crater density of 32. The values for the young rayed craters Tycho and Copernicus are shown for comparison.

SOURCES: Hartmann, 1970, 1972; Hartmann and Wood, 1971; Turner and Cadogan, 1975; Turner, Cadogan, and Yonge, 1973; Nunes, Tatsumoto, and Unruh, 1974; O. A. Schaeffer and Husain, 1974a,b; O. A. Schaeffer, Husain, and Schaeffer, 1976; Wilhelms, 1984.

[a] (F), basin on far side; others, near side.
[b] Most prominent outer ring (outermost ring).
[c] Average near side of mare, 1.0.

that of paleontology in terrestrial geology, even though it does not provide nearly the resolution that fossils do.

A stratigraphic column for the Moon was first proposed by E. M. Shoemaker and R. J. Hackman (1962) of the U.S. Geological Survey. Their original stratigraphy, based on deposits near the southern part of the Imbrium basin, was later modified and applied to the entire near side and much of the far side (Fig. 5.8). As is the case for the terrestrial geologic time scale, the lunar time scale is a relative scale based on the order of deposition of mappable geologic units. Since the Apollo and Luna missions, however, it has been possible to calibrate this time scale by radiometric dating of lunar samples so that the ages of the period boundaries are approximately known.

The Copernican System includes the deposits of the fresh, brightly rayed craters such as Tycho and Aristarchus (Fig. 5.1). The relative times of formation of many of these young craters have been determined by the densities of smaller craters on their floors and ejecta and by the "freshness" of their rays. Copernicus, for which the period is named and one of the oldest of the brightly rayed craters, is thought to be about 800 Ma old (Eberhardt et al., 1973).[17]

The Eratosthenian System refers to the deposits of craters like Eratosthenes, whose once-bright rays have been "erased" by mixing of the regolith and by radiation-induced changes over time. This system also includes a third of the near-side mare lavas, primarily those in Oceanus Procellarum and Mare Imbrium.

The Imbrian System consists of the ejecta from the Orientale and Imbrium basins. These deposits cover a large part of the lunar surface and provide major marker horizons for lunar stratigraphy. About two-thirds of the mare lavas are also included within the Imbrian.

The Nectarian System includes all deposits that predate the formation of the Imbrium Basin down through and including the deposits of the Nectaris Basin. The deposits of the Prenectarian are mostly on the lunar far side and include all materials older than Nectarian. The Nectarian and Prenectarian systems, sometimes collectively called Preimbrian, include most of the materials of lunar basin formation.

An interesting feature of the lunar time scale is that its subdivisions are not uniform in length. This is also true of the terrestrial time scale because both were constructed by dividing the exposed deposits, not time, into units practical for the purpose of mapping and dis-

17. The age of 800 ± 40 Ma is tenuous. It is based on the identification of KREEP glass from an Apollo 12 soil sample as Copernicus ray material and the interpretation of low-temperature $^{40}Ar/^{39}Ar$ age spectrum "plateaus" as due to heating and partial outgasing caused by the Copernicus impact.

tinguishing among the significant events of the geologic history as recorded by the existing rocks. Because of the erosion, nondeposition, burial by younger deposits, and the differences in magnitude between various rock-forming events the result is inevitably unequal division. It is also important to remember that both the lunar and the terrestrial time scales were devised before it was even possible to calibrate them with radiometric dating methods.[18]

Another interesting feature of the lunar time scale is that it is "bottom heavy," i.e. the majority of the significant events in lunar history occurred within the first billion years or so. This is one of the things that makes the Moon such a fascinating object for geologists to study—much of its decipherable history took place precisely in that period of time for which little or no terrestrial geologic record has been preserved.

There is still much uncertainty about the earliest history of the Moon. How, for example, did the Moon acquire a planet-wide crust of anorthositic composition? Where did KREEP come from and why does it seem to be restricted to the vicinity of the Imbrium and Procellarum basins? There are few certain answers to such questions, but there are clues in the compositions of the highland rocks that invite reasonable inferences about the events that led to the formation of the principal rock types.[19]

To review what is known about the highland crust, anorthosite and anorthositic gabbro are the two most common highland igneous rock types found in the returned lunar samples. X-ray spectrographic data from orbiting instruments indicate that the highlands are composed of rocks with unusually high Al/Si ratios, which is consistent with their being anorthositic. In addition, data relayed to Earth by the lunar seismometers show that the upper 60 km or so of the lunar crust has a density consistent with an anorthositic composition. From these facts, it is reasonable to conclude that the uppermost lunar crust, including those parts now covered by younger lavas and ejecta, is composed largely of anorthosite and anorthositic gabbro.

The temperatures required to form anorthositic magma are unreasonably high, and it is improbable that the crust was formed from that sort of melt. It is far more likely that the anorthosite formed as a cumulate from a magma of more normal composition. This could happen by the gravitational sinking of the denser iron and magnesium minerals in a large magma body as it cooled and crystallized, leaving

18. The terrestrial geologic time scale was defined before radioactivity was discovered. The lunar geologic time scale was formulated before lunar samples had been collected.

19. A concise description of one model of the earliest history of the Moon is given by J. A. Wood, 1975. It is as good as any and I have borrowed it for use here.

the lighter feldspar to concentrate near the top. Thus, the most likely scenario for the formation of an anorthositic crust requires a Moon-wide magma ocean a few hundred kilometers deep. This is not as fanciful as it might sound at first. What is required is sufficiently rapid accretion. Given that, the heat from energy released by the impacts of bodies striking the lunar surface during the final stages of accretion would accumulate faster than the heat could be radiated into space or conducted to the Moon's interior. This would result in the melting of the entire lunar surface to a depth of several hundred kilometers—a magma ocean from which an anorthositic upper crust and a more mafic lower crust could then develop by cooling, crystallization, and gravitational processes.

The unusually high contents of aluminum, potassium, phosphorus, uranium, thorium, and rare earth elements in the Fra Mauro basalts can be most easily explained by partial melting of the anorthositic crust (Shirley, 1983). The chemistry of these "incompatible" elements dictates that they would have been concentrated in the anorthositic upper crust and would also be enriched in the liquid formed during partial melting of this crust. This liquid then migrated upward and flowed onto the surface as basaltic lava. It is not known why these basalts are restricted to the vicinity of the Imbrium Basin or where the heat to melt the anorthositic crust partially and generate the lava came from.

Basin formation probably began as soon as the crust had solidified sufficiently to preserve topography and continued until the Moon had acquired all of the large planetesimals in its orbital path, a period of some 600–700 Ma. The chronology of basin formation and the time over which it occurred is reasonably well known from crater counting and isotopic dating (Table 5.4), but there is some uncertainty about the frequency distribution of basin-forming events. The prevailing view is that basin formation was continuous but declining from the time the crust solidified to the final collision that formed the Orientale Basin. This hypothesis is appealing because it is congruent with the continuous decline in cratering (Fig. 5.4), calls for no unusual events, and seems to be supported by the tentative basin chronology shown in Table 5.4. An alternative hypothesis, based primarily on lead-isotopic data to be discussed later, is that there was a final burst of impacts of both large and small bodies that resulted in what has been called the "terminal lunar cataclysm" (Tera, Papanastassiou, and Wasserburg, 1974). This might explain that predominance of ages of 3.8–3.9 Ga for highland rocks, but it requires that a new supply of impacting bodies be available to ravage the Moon some 600 Ma after accretion. This might occur either by chance or as the Moon receded from the Earth after capture, collecting material in different orbits (Cadogan, 1981: 363).

Isotopic dating of mare lavas shows that mare volcanism extended over a period of at least 1 Ga, but morphology, crater density, and relationship to several young, brightly rayed craters in unsampled regions of Oceanus Procellarum and Mare Smythii indicate that some mare lavas may be as young as 1 Ga (Schultz and Spudis, 1983, Head, 1976: 289). At the other end of the scale, a fragment of basalt resembling mare lava found in the Fra Mauro Formation has been dated at more than 4.1 Ga (L. A. Taylor et al., 1983). Although mare volcanism and basin formation overlapped in time, the great age differences between basins and their filling by the maria, typically several hundred million years, indicate that mare lavas were not generated by basin impacts. The chemistry of the mare lavas suggests that they were generated by partial melting of the lunar mantle at depths of 200–350 km (Kesson and Lindsley, 1976: 370–71, Binder, 1982). There is some suggestion that the source region for mare lavas may have deepened with time. Although uncertain, a downward migration of the source is consistent with the expected thermal history of the Moon. The surface would cool as heat was radiated into space, whereas the interior would grow hotter because of the radioactivity of uranium, thorium, and potassium; the combined effect would be a progressive deepening of the conditions necessary for partial melting. Regardless of the depths of mare lava generation, downward cooling and waning volcanism were inevitable. Eventually the lunar interior was no longer able to supply magma, volcanism ceased, and the planet became thermally inactive.

For the past billion years or so the Moon has been quiescent, its surface modified only by the occasional impact of a meteorite and by the relentless assaults of cosmic dust, radiation, and gravity.

The Ages of Moon Rocks

There are many hundreds of radiometric ages for the returned lunar samples, and it would be pointless, in view of the purpose of this book, to review them all. Of primary interest is the isotopic evidence for the oldest rocks and the age of formation of the Moon. But a survey of the isotopic ages of mare lavas and highland rocks is useful if only to demonstrate the overwhelming evidence for the great antiquity of the principal geologic events that have affected the lunar surface.

Radiometric ages for the lunar samples have been obtained primarily by the $^{40}Ar/^{39}Ar$ incremental heating method and by the Rb–Sr and Sm–Nd isochron methods. In contrast to the ^{40}Ar/^{39}Ar method, the conventional K–Ar method proved to be of little use for lunar chronometry because most of the lunar samples have been subjected to impact heating and have experienced some Ar loss. As a result, conventional K–Ar ages are commonly 1 Ga or more younger than ^{40}Ar/^{39}Ar, Rb–Sr, and Sm–Nd ages for the same rocks (for example, Turner, 1970: 1680, table 10). U–Th–Pb methods, while useful as tracers of petrogenic processes and for modeling of lunar evolution, have proven incapable of yielding direct radiometric ages from the lunar samples except in rare circumstances. This is primarily because of the analytical difficulties with the very small, low-lead samples available and because most lunar rocks are devoid of minerals that are high in U and free of initial Pb, like zircon. As a result of these circumstances, only three U–Pb and Pb–Pb internal isochrons have been generated from lunar rocks (Basaltic Volcanism Study Project, 1981: 949, Compston, Williams, and Meyer, 1984).

The majority of the ages of lunar rocks are the result of ^{40}Ar/^{39}Ar incremental heating measurements. This is because an age can be determined by this method on a single small chip of rock weighing 0.1 g or less. In contrast, the Rb–Sr and Sm–Nd isochron methods involve the separation of minerals, a time-consuming process that may require a sample weighing several grams. Thus, the ^{40}Ar/^{39}Ar method is somewhat less laborious and much more economical of sample and so can be applied to more samples than can the Rb–Sr and Sm–Nd isochron techniques.

Many, but not all, lunar samples have been dated by two or more methods or by two or more laboratories using the same method. The number and type of age measurements tend to be a function of sample size. Large rocks, from which material is plentiful, typically have been dated by multiple measurements and by several methods, whereas very small rocks may have been dated by a single measurement by only one technique. Some lunar samples have not been dated but have been reserved for other types of studies or have been set aside in their entirety to ensure that an adequate and representative supply of lunar rocks is available for future use, both scientific and display.

Mare Basalts

The results of 114 age measurements on 52 samples of mare basalt from six mare landing sites are shown in Figure 5.10. The figure is not comprehensive but does include most of the published radiometric ages for mare basalts and is representative of the age data for mare lavas. One obvious feature of the results is that all of the mare basalt samples are ancient by terrestrial standards even though the

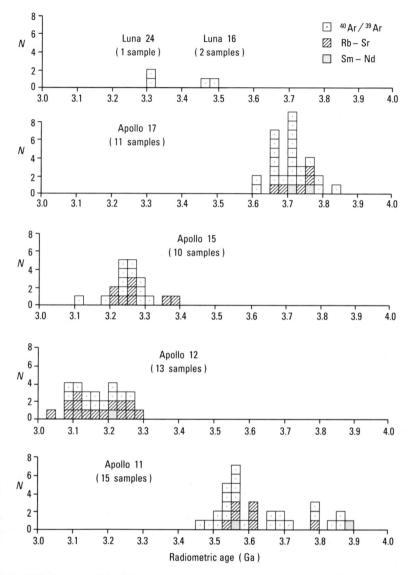

Fig. 5.10. Data on 114 radiometric age measurements on 52 samples of mare basalt. Included are ages whose reported analytical uncertainties at the 95% confidence level (two standard deviations) are 0.1 Ga or less.

flooding of basins by mare lavas was the most recent of the major events in lunar history. The youngest dated mare basalt, collected on the Apollo 12 mission, is slightly greater than 3 Ga in age, but it is highly likely that there are younger mare lavas elsewhere.

The oldest dated mare basalts represented in Figure 5.10 are between 3.8 and 3.9 Ga in age and were collected by the Apollo 11 and 17 astronauts. There is evidence, however, of earlier mare volcanism. Several small fragments of aluminous basalt from the Fra Mauro Formation closely resemble mare basalts in composition and have ages near 3.9 Ga. Six age measurements by $^{40}Ar/^{39}Ar$ and Rb–Sr on basalt sample 14053, for example, fall within the range 3.86–3.90 Ga (Stettler et al., 1973: 878). An even older mare basalt candidate has been dated. A mare-type basalt clast in Apollo 14 breccia 14305 has a Rb–Sr isochron age of 4.14 ± 0.05 Ga (L. A. Taylor, 1983). These older mare-type basalts suggest that the earliest mare volcanism may have preceded the most recent basin-forming events (Imbrium, Orientale) by as much as 300 Ma. Virtually nothing is known, however, about the mode of origin of these intriguing fragments of mare-type basalt. Although their compositions are similar to those of mare lavas, it is by no means certain that these basalts originated in the same type of extensive lava "floods" that formed the extant maria.

The consistency among age measurements on samples from the same landing site is impressive, but becomes all the more so when the results from individual samples are examined. A number of mare basalts have been dated several times by different laboratories and by different techniques (Table 5.5). Multiple results for single samples usually agree to within about ± 2% or better, especially when the few measurements with analytical uncertainties of greater than 0.1 Ga are excluded from the comparison. When data are carefully selected to include only results of the highest precision and to minimize interlaboratory biases, it is possible to resolve age differences of 50 Ma or less and to determine the age sequence of basalts of different chemical type (e.g., Geiss et al., 1977, Guggisberg et al., 1979: 29–36).

Most impressive is the concordance of ages measured by different methods and from different phases separated from the same sample. An example is 10072, a large sample of high-K mare basalt collected by the Apollo 11 astronauts (Fig. 5.11, Table 5.5). A seven-point Rb–Sr isochron and a six-point Sm–Nd isochron both give ages of 3.57 Ga. $^{40}Ar/^{39}Ar$ age spectrum dating of this sample has been done on the whole rock, plagioclase feldspar, ilmenite, and pyroxene. The two analyses of the plagioclase, which yielded the most convincing plateaus and most likely provide the best $^{40}Ar/^{39}Ar$ values for the crystallization age of the sample, agree exactly with the Rb–Sr and Sm–Nd ages. The age-spectrum plateaus for the whole rock, *ilmenite*, and pyroxene are discontinuous through the highest-temperature increments and have only medium-temperature age "plateaus." Nevertheless, the ages found from these imperfect plateaus agree with the other results within the analytical uncertainties. Although sample 10072 is one of the best examples because it has been the object of much study, it is not entirely extraordinary and there are many other samples whose radiometric ages show equally impressive concordance. Such consistency among methods indicates clearly that the $^{40}Ar/^{39}Ar$, Rb–Sr, and Sm–Nd isotopic "clocks" in the lunar rocks are accurate and being properly read.

The age results on the mare basalt samples show that mare volcanism was not simultaneous at all localities but occurred over a period of at least a billion years, probably longer, and may have begun before the most recent basin-forming impacts. For samples from an individual landing site, the age results are distributed over a range of about 200 Ma. This spread in ages is real and primarily reflects differences in the crystallization ages of the samples. It is clear from Figure 5.10 that the filling of individual basins was not instantaneous but occurred over a protracted period of time. Volcanism in Mare Tranquillitatis (Apollo 11) apparently lasted even longer—at least 400 Ma. None of the maria have been sampled at more than a single site or at any substantial depth below the surface, and so the age ranges reflected in Figure 5.10 must be regarded as the minimum ranges for the duration of volcanism in the individual maria.

Highland Rocks

It is no surprise that the radiometric ages of highland rocks indicate that, as a group, they are older than mare lavas (Fig. 5.12). Highland samples, however, are mostly breccias. They tend to be very complicated rocks and the interpretation of their radiometric ages is often equivocal. This is because the breccias are at the same time fragmental rocks, composed of bits and pieces of pre-existing rocks, and metamorphic rocks, reconstituted and modified by the heat and fusion of one or more impacts. It is not uncommon for lunar breccias to contain fragments of earlier breccias, and as many as five generations of breccia within breccia have been found in a single sample. Thus, the radiometric age of a breccia sample may represent the time(s) of breccia formation, the time(s) of crystallization of the rock fragments it contains, or something in between.

As an example, consider the diversity of the ages found by Jessberger, Kirsten, and Staudacher (1977) at the Max Planck Institute for Nuclear Physics in Heidelberg for the various components of breccia 73215. This rock is a light gray breccia that is thought to have formed as an aggregate of clasts and melt during the impact that created the basin of Mare Serenitatis.[20] Some of the clasts have $^{40}Ar/^{39}Ar$

20. Except for Imbrium ejecta, the association of individual samples with basin-forming impact events is highly uncertain and the opinions of different workers on the ages of formation of individual basins vary considerably. This accounts for what may appear as some inconsistency in basin ages.

TABLE 5.5
Radiometric Ages of Some Mare Basalts Dated by Two or More Methods

Mission	Sample no.	Basalt type[a]	Material dated[b]	Method	Age (Ga)[c]	Source
Apollo 11	10044	low-K	wr	Ar–Ar	3.69 ± 0.05	Turner, 1970
			wr	Ar–Ar	3.66 ± 0.04	Guggisberg et al., 1979
			wr	Ar–Ar	3.66 ± 0.03	Geiss et al., 1977
			7	Rb–Sr	3.62 ± 0.07	Lunatic Asylum, 1970
	10062	low-K	wr	Ar–Ar	3.78 ± 0.06	Turner, 1970
			wr	Ar–Ar	3.79 ± 0.04	Guggisberg et al., 1979
			5	Sm–Nd	3.88 ± 0.06	Papanastassiou, DePaolo, and Wasserburg, 1977
			7	Rb–Sr	3.92 ± 0.11	Papanastassiou, DePaolo, and Wasserburg, 1977
	10072	high-K	wr	Ar–Ar	3.49 ± 0.05	Turner, 1970
			wr	Ar–Ar	3.52 ± 0.04	Guggisberg et al., 1979
			pl	Ar–Ar	3.57 ± 0.05	Guggisberg et al., 1979
			pl	Ar–Ar	3.56 ± 0.06	Guggisberg et al., 1979
			il	Ar–Ar	3.58 ± 0.05	Guggisberg et al., 1979
			px	Ar–Ar	3.55 ± 0.05	Guggisberg et al., 1979
			7	Rb–Sr	3.57 ± 0.05	Papanastassiou, DePaolo, and Wasserburg, 1977
			6	Sm–Nd	3.57 ± 0.03	Papanastassiou, DePaolo, and Wasserburg, 1977
Apollo 12	12002	olivine	wr	Ar–Ar	3.21 ± 0.05	Turner, 1971
			wr	Ar–Ar	3.18 ± 0.04	Alexander, Davis, and Reynolds, 1972
			7	Rb–Sr	3.29 ± 0.10	Papanastassiou and Wasserburg, 1971a
	12039	quartz	6	Rb–Sr	3.12 ± 0.06	Nyquist et al., 1977
			3	Sm–Nd	3.20 ± 0.05	Nyquist et al., 1979a
	12051	quartz	wr	Ar–Ar	3.13 ± 0.05	Stettler et al., 1973
			wr	Ar–Ar	3.12 ± 0.07	Stettler et al., 1973
			wr	Ar–Ar	3.23 ± 0.05	Turner, 1971
			wr	Ar–Ar	3.25 ± 0.06	Alexander, Davis, and Reynolds, 1972
			7	Rb–Sr	3.19 ± 0.10	Papanastassiou and Wasserburg, 1971a
			4	Rb–Sr	3.09 ± 0.04	Nyquist et al., 1977
	12064	quartz	wr	Ar–Ar	3.15 ± 0.01	Horn et al., 1975
			wr	Ar–Ar	3.11 ± 0.09	Horn et al., 1975
			5	Rb–Sr	3.11 ± 0.09	Papanastassiou and Wasserburg, 1971a
Apollo 15	15076	quartz	wr	Ar–Ar	3.31 ± 0.04	Stettler et al., 1973
			4	Rb–Sr	3.26 ± 0.08	Papanastassiou and Wasserburg, 1973
	15555	olivine	wr	Ar–Ar	3.29 ± 0.05	Alexander, Davis, and Lewis, 1972
			wr	Ar–Ar	3.25 ± 0.06	York, Kenyon, and Doyl 1972
			wr	Ar–Ar	3.28 ± 0.06	York, Kenyon, and Doyl 1972
			wr	Ar–Ar	3.24 ± 0.06	Husain et al., 1972
			wr	Ar–Ar	3.19 ± 0.02	Podosek, Huneke, and Wasserburg, 1972

TABLE 5.5 *(continued)*

Mission	Sample no.	Basalt type[a]	Material dated[b]	Method	Age (Ga)[c]	Source
			pl	Ar–Ar	3.27 ± 0.02	Podosek, Huneke, and Wasserburg, 1972
			px	Ar–Ar	3.24 ± 0.09	Podosek, Huneke, and Wasserburg, 1972
			6	Rb–Sr	3.23 ± 0.08	Murthy et al., 1972
			6	Rb–Sr	3.27 ± 0.09	Birck, Fourcade, and Allègre, 1975
			7	Rb–Sr	3.25 ± 0.04	Papanastassiou and Wasserburg, 1973
Apollo 17	70035	high-Ti	wr	Ar–Ar	3.67 ± 0.07	Stettler et al., 1973
			wr	Ar–Ar	3.70 ± 0.07	Stettler et al., 1973
			?	Rb–Sr	3.75 ± 0.10	Chappel et al., 1973, cited in Stettler et al., 1973
			7	Rb–Sr	3.74 ± 0.06	Evensen, Murthy, and Coscio, 1973
	70135	high-Ti	7	Rb–Sr	3.67 ± 0.09	Alexander, Davis, and Reynolds, 1972
			3	Sm–Nd	3.77 ± 0.06	Alexander, Davis, and Reynolds, 1972
	75055	high-Ti	wr	Ar–Ar	3.71 ± 0.05	Turner, Cadogan, and Yonge, 1973
			wr	Ar–Ar	3.73 ± 0.04	Huneke et al., 1973
			wr	Ar–Ar	3.77 ± 0.05	Kirsten and Horn, 1974
			5	Rb–Sr	3.75 ± 0.10	Tatsumoto et al., 1973
			6	Rb–Sr	3.69 ± 0.06	Tera, Papanastassiou, and Wasserburg, 1974
	75075	high-Ti	wr	Ar–Ar	3.71 ± 0.04	Jessberger, Horn, and Kirsten, 1975
			wr	Ar–Ar	3.66 ± 0.05	Jessberger, Horn, and Kirsten, 1975
			wr	Ar–Ar	3.61 ± 0.04	Jessberger, Horn, and Kirsten, 1975
			pl	Ar–Ar	3.69 ± 0.02	Jessberger, Horn, and Kirsten, 1975
			pl	Ar–Ar	3.67 ± 0.02	Jessberger, Horn, and Kirsten, 1975
			8	Rb–Sr	3.76 ± 0.12	Nyquist, Bansal, and Wiesmann, 1975
			4	Sm–Nd	3.70 ± 0.07	Lugmair, Scheinin, and Marti, 1975
Luna 16	B-1	aluminous	wr	Ar–Ar	3.45 ± 0.04	Huneke, Podosek, and Wasserburg, 1972
			5	Rb–Sr	3.42 ± 0.18	Papanastassiou and Wasserburg, 1972b

[a] Table 5.2.
[b] Numbers represent number of points in isochron; wr, whole rock; pl, plagioclase; px, pyroxene; il, ilmenite.
[c] All ages calculated with the decay constants in Table 3.1. Errors are at the 95% confidence level (two standard deviations). Ages underlined are shown as figures. Ages are based on either isochron (Rb–Sr, Sm–Nd) or $^{40}Ar/^{39}Ar$ release-spectrum (Ar–Ar) methods. Included are data with analytical uncertainties of 0.1 Ga (two standard deviations) or less, except for three results that are included to maximize the number of methods for the sample.

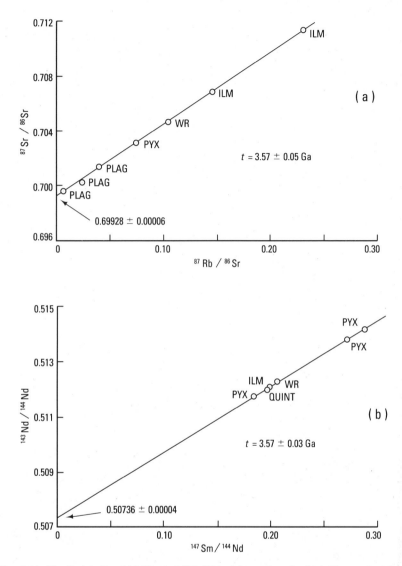

Fig. 5.11. Rb–Sr (a), Sm–Nd (b), and $^{40}Ar/^{39}Ar$ (c) age data for high-K mare basalt 10072. PLAG, plagioclase feldspar; PYX, pyroxene; WR, whole rock; ILM, ilmenite; QUINT, "quintessence," a mixture of fine-grained mineral phases. [After Papanastassiou, DePaolo, and Wasserburg, 1977 (a and b), and Guggisberg et al., 1979 (c).]

plateau ages exceeding 4.2 Ga, whereas the apparent ages of the microcrystalline matrix range from 3.93 to 4.09 Ga (Fig. 5.13). Even though a large proportion of the matrix was formed from impact melt whose radiometric "clock" was completely reset, the matrix also contains fragments of older rocks and minerals whose clocks apparently

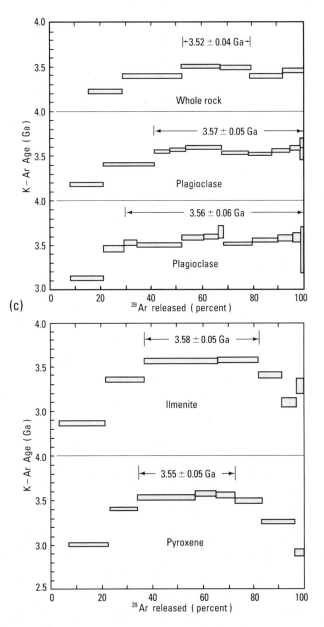

(c)

were not completely reset by the impact heating. The formation age of the breccia, 3.87 ± 0.04 Ga, is recorded only in the felsite clast, a rock consisting of glass and fine-grained feldspar. Petrographic and mineralogic studies indicate that the felsite was incorporated into the breccia while still molten and has not been reheated since. The Rb–Sr iso-

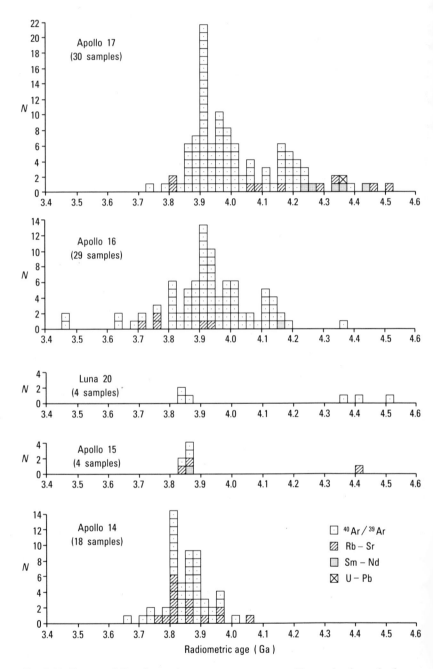

Fig. 5.12. Data on 260 radiometric age measurements on 85 samples from the lunar highlands. The ages included have analytical uncertainties of 0.1 Ga or less at the 95% level of confidence (two standard deviations).

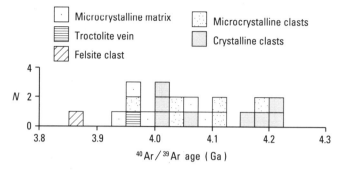

Fig. 5.13. ^{40}Ar/^{39}Ar plateau ages for various components of breccia 73215. (Data from Jessberger, Kirsten, and Staudacher, 1977, and Jessberger et al., 1978.)

chron age of the felsite is 3.82 ± 0.05 Ga (Compston, Foster, and Grady, 1977), which agrees with the ^{40}Ar/^{39}Ar age within the analytical uncertainties. In contrast to the felsite age, the ages measured for the matrix and the other clasts are hybrid apparent ages, each one reflecting some degree of resetting following the original crystallization. Thus, the only certain interpretation is that these ages lie somewhere between the age of the oldest component in the sample and the formation age of the breccia. Even though difficult to interpret, however, such information is valuable for it tells us that two of the clasts in the breccia are at least 4.2 Ga in age.

With these complexities of breccia chronology in mind, let us briefly examine the ages of highland rocks and their implication for the times of formation of some of the major lunar basins. Figure 5.12 includes 260 radiometric ages for 85 lunar samples. Each sample represents one numbered rock or soil collection. Some of the ages in the figure were calculated from different clasts from the same breccia and some on different rock fragments from the same soil sample. Thus the number of rocks represented in Figure 5.12 is considerably greater than 85. The data in the figure are not comprehensive but include most of the published values and are representative of the ages found for highland samples.

The Apollo 14 samples are from the Fra Mauro Formation, and the majority represent ejecta from the Imbrium impact. The few highland samples in the Apollo 15 collection are likewise thought to be Imbrium ejecta. Although it is not possible to date the Imbrium impact directly, the ages of Apollo 14 and 15 samples can provide relatively narrow limits on the time of this event. It is axiomatic that the age of the Fra Mauro Formation, and hence of the Imbrium event, must be equal to or younger than the ages of its constituents. Exceptions to this relationship would be those samples that may have been disturbed by later, smaller impacts or that represent ejecta from the impact responsible for Mare Orientale, a thin layer of which may cover the Apollo 14 site (McGetchin, Settle, and Head, 1973). Most of the Apollo 14 and 15 samples have ages between 3.8 and 3.9 Ga, and there is a pronounced mode at about 3.82 Ga (Fig. 5.12), which provides a reasonable upper limit for the age of the Fra Mauro Formation.

Specific samples also yield some interesting information. Sample 14310, for example, is a crystalline rock, the largest collected on Apollo 14, that appears to have been formed by the heat of impact (James, 1973). The mean of 13 ^{40}Ar/^{39}Ar and Rb–Sr isochron ages on this sample is 3.83 Ga. Although likely, there is no assurance that this rock was produced by the Imbrium impact—it could have been formed earlier—so its age, too, is an upper limit for the Imbrium event. We also know that the Fra Mauro Formation must be older than the mare lava flows that cover it. These flows have mean ages of 3.82–3.83 Ga (Apollo 11) and 3.78–3.84 Ga (Apollo 17). Thus, the age of the Imbrium impact can be narrowed to about 3.82–3.83 Ga.

L. E. Nyquist, of NASA's Manned Spacecraft Center in Houston, and his colleagues have provided additional evidence for the age of the Imbrium event in their Rb–Sr study of Apollo 14 KREEP-rich breccias (Nyquist et al., 1972). They found that values for four high-grade breccias, i.e. those that had been the most severely affected by impact heating, fell on a Rb–Sr isochron of 3.86 ± 0.12 Ga (Fig. 5.14). Results from other Apollo 14 KREEP breccias plot near this same iso-

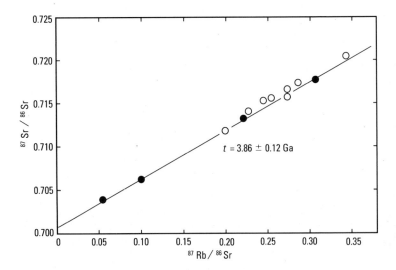

Fig. 5.14. Rb–Sr isochron diagram for Apollo 14 breccias that are high in KREEP. The isochron is drawn through the four breccias of highest metamorphic grade (solid symbols). (After Nyquist et al., 1972.)

chron. Although the interpretation of these whole-rock data is less certain than the interpretation of internal mineral isochrons, the results suggest a redistribution of strontium during impact heating and an age for the Imbrium event that is consistent, within the analytical uncertainties, with the age of 3.82–3.83 Ga deduced from the younger and older limits discussed above.

There is little or no direct evidence for the age of formation of the Orientale Basin, but on the basis of the probable presence of a thin layer of Orientale ejecta at the Apollo 14 site, the few breccias from there with post-Imbrium ages may be Orientale ejecta. If so, then the Orientale impact may have occurred between about 3.75 and 3.81 Ga.

Very few ages have been calculated for the Luna 20 material from the Apollonius highlands because of the small amount of sample returned to Earth by the unmanned spacecraft. There is a cluster of ^{40}Ar/^{39}Ar ages at 3.84–3.85 Ga and some older values discussed below. Geologic mapping indicates that the Luna 20 site is probably dominated by material from the impact that formed Mare Crisium. If the Luna 20 samples represent Crisium ejecta then the Crisium basin must be either equal to or less than 3.84 Ga in age. This age is consistent with the hypothesis that the Crisium impact immediately preceded the Imbrium impact. O. A. Schaeffer and L. Husain (1974a: 1552–54), then of the State University of New York at Stony Brook, however, have speculated that the ages of certain breccias from samples excavated by North Ray Crater near the Apollo 16 site reflect the Crisium and Humorum impacts. If they are correct, then the Crisium Basin formed sometime between about 3.99 and 4.13 Ga and the younger Luna 20 ages record either the Imbrium event or local impacts.

The Apollo 16 site is dominated by samples with ages of about 3.90–3.92 Ga. The geology of this site is complicated and the interpretation of the age results is problematic, primarily because the origin of the Caley and Descartes formations, both of which underlie the Apollo 16 site, is unclear. O. B. James (1981) of the U.S. Geological Survey, however, has argued that all of the recent hypotheses for the origin of these two formations lead to the conclusion that both are composed of material of Nectarian or Prenectarian age. On the basis of a careful examination of the relationships between petrographic type and age results for Apollo 16 breccias and melt rocks, she has concluded that the major subsurface units at the site were deposited no earlier than 3.92 Ga and that this is an approximate upper limit for the age of the Nectaris impact. Younger materials at the Apollo 16 site are most likely Imbrium ejecta and melt rocks from pre-Imbrian, post-Nectarian impacts.

The oldest dated lunar samples were collected at the Apollo 17 site in the Taurus–Littrow Valley on the southeastern border of Mare Serenitatis (Fig. 5.1). The ages of some of these oldest rocks, one of which exceeds 4.5 Ga, are discussed below. Most of the Apollo 17 highlands samples, however, have ages between about 3.85 and 4.0 Ga. According to E. K. Jessberger and his colleagues, most of the Apollo 17 breccia ages do not reflect the times of breccia formation, i.e. impact, but instead reflect varying degrees of resetting of the crystallization ages of their constituents (Jessberger, Kirsten, and Staudacher, 1977, Jessberger et al., 1978). Thus, ages of most breccias tend to be older than the event that produced them. A few breccias, however, contain certain constituents whose ages do represent the time of breccia formation. One example is a felsite clast from breccia 73215, which is discussed above (also Fig. 5.13). Another is the degassed matrix from breccia 73255 (Jessberger et al., 1978, Staudacher et al., 1979). The ^{40}Ar/^{39}Ar plateau ages of these two subsamples, 3.87 ± 0.04 and 3.88 ± 0.02 Ga, respectively, record the age of the impact that formed them, perhaps the formation of the Serenitatis Basin.

The Oldest Rocks

Lunar rocks with radiometric ages greater than 4.2 Ga number only slightly more than a dozen. All except one, a glassy clast of granitic composition (73217), appear to be cumulates, primarily members of the *ANT suite*, an acronym for the anorthosite, norite, troctolite suite of rocks. Most have been found as clasts in breccias and probably all have complex histories involving one or more episodes of postcrystallization impact heating. The radiometric ages for these oldest rocks (Table 5.6) range to as much as 4.5 Ga, but because of their complex histories the interpretation of the data is not always straightforward.

Sample 77215 is a clast of brecciated norite that was enclosed within a large breccia boulder at Apollo 17 Station 7 near the base of North Massif. The norite breccia is enclosed within and cut by melt and is the oldest unit sampled from the complex breccia boulder. The norite has been dated by both Sm–Nd and Rb–Sr isochron techniques, which give concordant ages of 4.37 and 4.33 Ga, respectively (Fig. 5.15). Note that two different chips of sample were used for the Rb–Sr isochron measurements. All of the data fall on a well-defined isochron except for the two pyroxene measurements from chip 145. Why these two differ so is unknown, but it may be due to disturbance of the Rb–Sr system by the breccia-forming event or by later impact heating. If only the data from chip 37 are used, the Rb–Sr isochron age is 4.34 ± 0.05 Ga. ^{40}Ar/^{39}Ar measurements on whole-rock and plagioclase samples from 77215 give plateau ages ranging from 3.90 to

TABLE 5.6

Radiometric Ages of the Oldest Lunar Rocks

Mission	Sample no.	Rock type[a]	Material dated[b]	Method	Age (Ga)[c]	Source
Apollo 15	15455	anorthosite clast	3	Rb–Sr	4.42 ± 0.10	Nyquist et al., 1979a
Apollo 16	67435	plagioclase clast	pl	Ar–Ar	4.35 ± 0.05	Dominik and Jessberger, 1978
Apollo 17	72417	dunite clast	11	Rb–Sr	<u>4.47 ± 0.10</u>	Papanastassiou and Wasserburg, 1975
	73215	feldspathic clast	wr	Ar–Ar	4.22 ± 0.03	Jessberger, Kirsten, and Staudacher, 1977
				Ar–Ar	4.05 ± 0.05	Jessberger, Kirsten, and Staudacher, 1977
	73217	granitoid clast	4z	U–Pb	4.36 ± 0.02	Compston, Williams, and Meyer, 1984
	73255	norite clast	3	Sm–Nd	4.23 ± 0.05	Carlson and Lugmair, 1981
		anorthositic gab-bro clast	wr	Ar–Ar	4.20 ± 0.01	Staudacher et al., 1979
	73263	anorthosite clast	wr	Ar–Ar	4.23 ± 0.05	G. A. Schaeffer and Schaeffer, 1977
	76503	anorthosite clast	wr	Ar–Ar	4.23 ± 0.01	O. A. Schaeffer, Husain, and Schaeffer, 1976
	76535	troctolite	5	Sm–Nd	<u>4.26 ± 0.06</u>	Lugmair et al., 1976
			10	Rb–Sr	<u>4.51 ± 0.07</u>	Papanastassiou and Wasserburg, 1976
			wr	K–Ar	4.27 ± 0.08	Bogard et al., 1975
			wr	Ar–Ar	4.16 ± 0.04	Huneke and Wasserburg, 1975
			wr	Ar–Ar	<u>4.19 ± 0.02</u>	Husain and Schaeffer, 1975
			wr	Ar–Ar	4.20 ± 0.03	Husain and Schaeffer, 1975
			pl	Ar–Ar	<u>4.19 ± 0.02</u>	Husain and Schaeffer, 1975
	77215	norite breccia clast	5	Sm–Nd	<u>4.37 ± 0.07</u>	Nakamura et al., 1976
			8	Rb–Sr	<u>4.33 ± 0.04</u>	Nakamura et al., 1976
			pl	Ar–Ar	3.92 ± 0.03	Stettler et al., 1978
			wr	Ar–Ar	3.99 ± 0.03	Stettler et al., 1974
			wr	Ar–Ar	3.97 ± 0.03	Stettler et al., 1974
			wr	Ar–Ar	3.90 ± 0.03	Stettler et al., 1974
	78236	norite	9	Sm–Nd	<u>4.34 ± 0.05</u>	Carlson and Lugmair, 1981
			3	Sm–Nd	<u>4.43 ± 0.05</u>	Nyquist et al., 1981
			3	Rb–Sr	<u>4.29 ± 0.02</u>	Nyquist et al., 1981
			pl	Ar–Ar	<u>ca. 4.36</u>	Nyquist et al., 1981
Luna 20	L2015	anorthosite	wr	Ar–Ar	4.40 ± 0.10	Cadogan and Turner, 1977
	22013	anorthosite, dark	wr	Ar–Ar	4.36	Huneke and Wasserburg, 1979
		anorthosite, light	wr	Ar–Ar	4.51	Huneke and Wasserburg, 1979

[a] Table 5.3.

[b] Numbers represent number of points in isochron; wr, whole rock; pl, plagioclase; z, zircon. All of the dated samples were found as clasts in larger pieces of breccia or as fragments in the lunar soil (76535 and Luna 20 samples).

[c] All ages calculated with the decay constants in Table 3.1. Errors are at the 95% confidence level (two standard deviations). Ages underlined are shown as figures. Ages are based on either isochron (Rb–Sr, Sm–Nd) or $^{40}Ar/^{39}Ar$ age-spectrum (Ar–Ar) methods, except that one conventional K–Ar age has been included for sample 76535 and the U–Pb discordia age has been included for sample 73217.

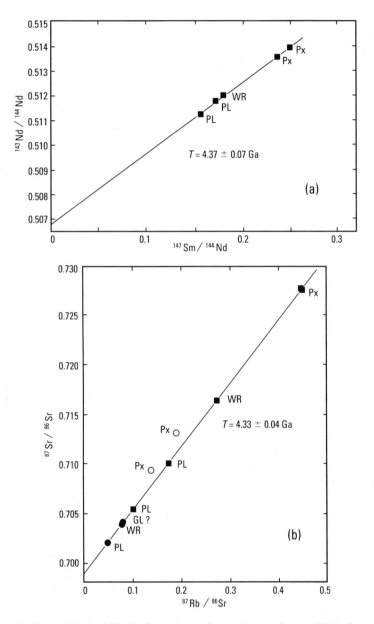

Fig. 5.15. Sm–Nd (a) and Rb–Sr (b) isochrons for noritic microbreccia 77215. Squares are data for chip 37, circles for chip 145. Isochrons are based on data represented by filled symbols. PL, plagioclase feldspar; PX, pyroxene; GL, glass; WR, whole rock. (After Nakamura et al., 1976.)

3.99 Ga, indicating that the norite breccia was affected by later heating sufficient to reset or partially reset the K–Ar system.

The simplest interpretation of the age results for 77215 is that the norite formed about 4.35 Ga and was incorporated into the breccia by an impact 3.92 Ga or later. The latter age is the result from the plagioclase, which has a simple high-temperature plateau age for 90% of the ^{39}Ar released, is the least likely to be contaminated by other material, and most likely represents the time of the last significant heating of the norite. ^{40}Ar/^{39}Ar ages of the various melt rocks that enclose and cut the norite breccia clast range from 3.73 to 3.93 Ga with a mode at 3.90–3.94 Ga. An alternative explanation for the age results is that the concordant Sm–Nd and Rb–Sr ages represent the time that the norite was incorporated into the breccia boulder and that the younger Ar–Ar ages reflect later impact events. Thus, the crystallization age of the norite cumulate is uncertain but must be equal to or greater than about 4.35 Ga.

Sample 73217 is a breccia that was recovered from landslide material at the South Massif of the Apollo 17 landing site. The breccia contains a clast of glassy material that is granitic in composition. Within this clast are zoned zircon crystals, four of which have been analyzed with the ion microprobe (Compston, Williams, and Meyer, 1984). Despite the obvious impact history of the breccia, these zircons are only slightly discordant, and the analyses of them generate a discordia line with an older intercept of 4.36 ± 0.02 Ga. The authors interpret this age as the crystallization age of the igneous rock from which the granitic clast was derived.

Sample 78236 was chipped by the Apollo 17 astronauts from a shattered and glass-coated boulder, about one-half meter in diameter, found at Station 8. It is a norite cumulate whose texture and mineralogy indicate that it was formed and cooled slowly at a depth within the lunar crust of between 8 and 30 km (Jackson, Sutton, and Wilshire, 1975, Nyquist et al., 1981). It is highly shocked and was probably excavated from depth and transported to the Apollo 17 site by a major basin-forming impact, but which impact is unknown.

The Sm–Nd, Rb–Sr, and ^{40}Ar/^{39}Ar relationships in this highland sample have been studied by R. W. Carlson and G. W. Lugmair (1981) of the University of California at San Diego, and by L. E. Nyquist of Lockheed Corporation and his associates (1981). All three isotopic systems in this rock show clear evidence of postcrystallization disturbance, which is not surprising in view of the degree of impact shock to which the sample has been subjected. Results of Sm–Nd analyses of plagioclase, maskelynite,[21] pyroxene, and the whole rock

21. Plagioclase glass, a product of impact melting.

on two different subsamples of 78236 fall near isochrons of about 4.3 to 4.4 Ga (Fig. 5.16a). The data, however, show somewhat more scatter about their respective isochrons than expected solely from analytical errors, indicating that the Sm–Nd system has been disturbed since the norite first crystallized. If only the most pristine mineral phases, i.e. those least likely to have been "contaminated" by phases formed or altered during impact, and the whole-rock point are considered,[22] the Sm–Nd isochron age for the sample is 4.43 ± 0.05 Ga (Fig. 5.16a). If all of the data are used the two studies give isochron ages of 4.34 ± 0.05 and 4.33 ± 0.09 Ga.

The Rb–Sr data for sample 78236 (Fig. 5.16b) indicate more disturbance than do the Sm–Nd data, which is in accord with the general view that the Sm–Nd isotopic system is more resistant to post-crystallization heating and metamorphism than is the Rb–Sr system. Calculated from the phases least likely to be contaminated or reequilibrated during impact, the Rb–Sr isochron age is 4.29 ± 0.02 Ga. The $^{40}Ar/^{39}Ar$ age spectrum for 78236 (Fig. 5.16c) shows evidence of severe disturbance and does not have a plateau. The only certain interpretation of the age spectrum is that the sample has undergone significant Ar loss since formation and has a minimum crystallization age of about 4.36 Ga.

All of the isotopic data indicate that 78236 crystallized at least 4.3 Ga. How much earlier is uncertain, but it is likely that the Sm–Nd isochron age of 4.43 ± 0.05 Ga for the most pristine phases is very close to the time of formation of the norite in the lunar crust.

Sample 76535 is a small (156 g) rock that was raked from the lunar regolith at Station 6 near the base of North Massif. It is a coarse-grained troctolite cumulate whose mineralogy, texture, and phase relationships indicate that it formed at depths of 10 to 30 km in the lunar crust and cooled at a rate of only a few tens of degrees Celsius per million years (Gooley, Brett, and Warner, 1974). The rock shows no mineralogical or textural evidence of impact processes, i.e. heating or shock, although logic dictates that it was probably excavated from its place of origin at depth by a major impact.

The chronology of 76535 has been investigated by Sm–Nd, Rb–Sr, conventional K–Ar, and $^{40}Ar/^{39}Ar$ techniques with some curious results (Table 5.6). The Sm–Nd isochron age, four $^{40}Ar/^{39}Ar$ age-spectrum plateau ages, and a conventional K–Ar age all agree within analytical uncertainty and have a mean of 4.21 ± 0.03 Ga. The $^{40}Ar/^{39}Ar$ age spectra on plagioclase and the whole rock (Fig. 5.17c), which

22. The reason for selecting these particular phases involves their rare-earth compositions. For detailed arguments, see Carlson and Lugmair (1981: 233–34) and Nyquist et al. (1981: 79–82).

show virtually no postcrystallization Ar loss,[23] combined with the concordant K–Ar and Sm–Nd ages indicate that the sample has been undisturbed for the past 4.2 Ga. The Rb–Sr isochron age, however, is 4.51 ± 0.07 Ga (Fig. 5.17b). The cause of this discrepancy is not understood, but D. A. Papanastassiou and G. J. Wasserburg of the California Institute of Technology have speculated that the Rb–Sr system in the mineral olivine, which, to a large degree, controls the slope of the Rb–Sr isochron, is highly resistant to disturbance. Their reasoning is that because olivine has no capacity to accept Rb or Sr into its crystal lattice those elements must reside in tiny inclusions within the olivine grains. Even if the rock is disturbed by heating, the olivine acts as an impenetrable barrier to Rb and Sr migration, preventing isotopic exchange with other mineral grains during metamorphism or reheating. Thus, once an inclusion is trapped within an olivine grain it remains a system closed to Rb and Sr unless the metamorphism is sufficient to reconstitute the olivine grain. In contrast, the plagioclase may exchange Rb and Sr freely during metamorphism, but as the bulk of the Rb and Sr in the troctolite resides in the plagioclase, isotopic exchange with other mineral phases has little or no effect on the Rb or Sr isotopic composition of the plagioclase. Finally, because isotopic exchange is between minerals, the isotopic composition of the whole rock is also unaffected by impact metamorphism. Accordingly, Papanastassiou and Wasserburg have interpreted the Rb–Sr result as the crystallization age of the troctolite and attributed the younger Sm–Nd, K–Ar, and ^{40}Ar/^{39}Ar ages to isotopic exchange and Ar loss due to elevated temperatures within the lunar crust between 4.5 and 4.2 Ga, after which the troctolite was excavated by impact and its residence at depth was terminated.

The proposed "immunity" of the Rb–Sr system within olivine to prolonged thermal stress is an interesting possibility and one for which troctolite 76535 seems to provide the only concrete example,[24] but in view of the known behavior of the Sm–Nd and Rb–Sr systems in mafic rocks, it is the most logical explanation proposed so far. In this regard, it is interesting to note that the result from olivine does not lie on the Sm–Nd isochron but above it (Fig. 5.17a), as it would if it was not reequilibrated with the other mineral phases and retained a "memory" of an older crystallization age.

23. Huneke and Wasserburg (1975) and Bogard et al. (1975) have shown that the olivine in 76535 contains trapped (excess) ^{40}Ar, but the age spectra of Husain and Schaeffer (1975) on plagioclase feldspar and on whole rock show no evidence of trapped ^{40}Ar (Fig. 5.17c).

24. Papanastassiou and Wasserburg (1976: 2043–44) also discussed the possibility that the Rb–Sr data represent an accidental mixing line and conclude that it is unlikely.

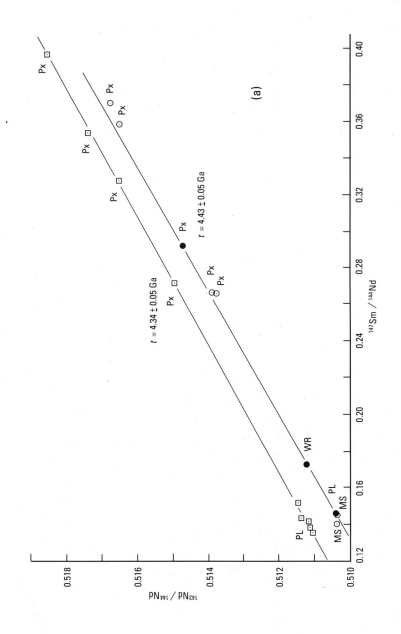

(a)

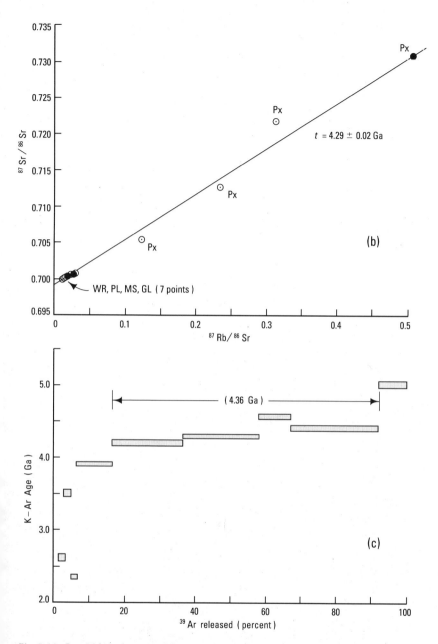

Fig. 5.16. Sm–Nd isochron (a), Rb–Sr isochron (b), and $^{40}Ar/^{39}Ar$ age spectrum (c) for shocked norite 78236. Squares in (a) are data from the study of Carlson and Lugmair, 1981, circles in (a) and (b) are data from Nyquist et al., 1981. Isochrons for the latter are based on the filled circles only, for the former on all nine data. WR, whole rock; PL, plagioclase feldspar; PX, pyroxene; MS, maskelynite; GL, glass. The Ar–Ar age spectrum, also from Nyquist et al., 1981, does not have a plateau and the age given is the mean of the four gas increments indicated. (After Carlson and Lugmair, 1981, and Nyquist et al., 1981.)

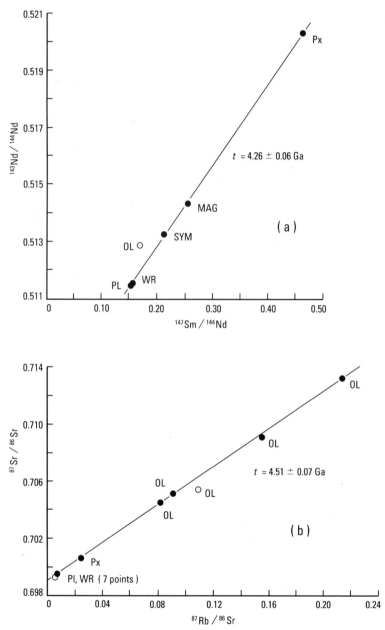

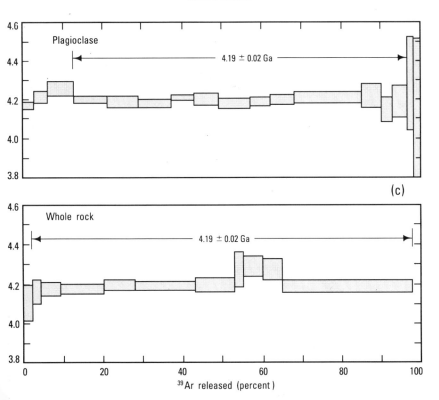

Fig. 5.17. Sm–Nd (a), Rb–Sr (b), and $^{40}Ar/^{39}Ar$ (c) data for troctolite 76535. For (a) and (b) the filled circles represent data used for the isochron fit; open circles represent data omitted in the calculation. OL, olivine; PX, pyroxene; PL, plagioclase feldspar; WR, whole rock; SYM, symplectite, an intergrowth of *groundmass* minerals; MAG, magnetic fraction. [After Lugmair et al., 1976 (a), Papanastassiou and Wasserburg, 1976 (b), and Husain and Schaeffer, 1975 (c).]

The Rb–Sr isochron age from troctolite 76535 remains somewhat enigmatic, although the best explanation is that it represents the time of formation of the troctolite at 4.51 Ga. The Rb–Sr age aside, the Sm–Nd, K–Ar, and $^{40}Ar/^{39}Ar$ results leave no doubt that 76535 is at least 4.2 Ga and has been virtually unaffected by impact heating since that time.

Sample 72417 is one of five chips from a clast of dunite in a *metaclastic* breccia boulder, about one-half meter in diameter, found at Station 2 at the base of South Massif, Apollo 17. The dunite has been deformed and crushed by one or more impacts and shows evidence of shock-induced melting and recrystallization. Its texture and mineralogy indicate that it formed as an igneous cumulate within the lunar crust, probably from a melt similar in composition to gabbro (Dymek, Albee, and Chodos, 1975). The ten-point Rb–Sr isochron for this sample, obtained by Papanastassiou and Wasserburg on whole rock chips, olivine, and symplectite, is 4.47 ± 0.10 Ga (Fig. 5.18), which the authors interpret as an original crystallization age. This ancient age is surprising in view of the extent to which the dunite has been deformed and recrystallized subsequent to formation, and suggests that the Rb–Sr isotopic system in olivine may, indeed, be highly resistant to metamorphism. Unfortunately, 72417 has not been studied by either Sm–Nd or $^{40}Ar/^{39}Ar$ techniques, so whether or not its isotopic systematics are similar to those of troctolite 76535 is unknown. The Rb–Sr result must stand on its own, but it would be difficult to find a more convincing isochron than the one shown in Figure 5.18.

The radiometric ages for the remaining samples listed in Table

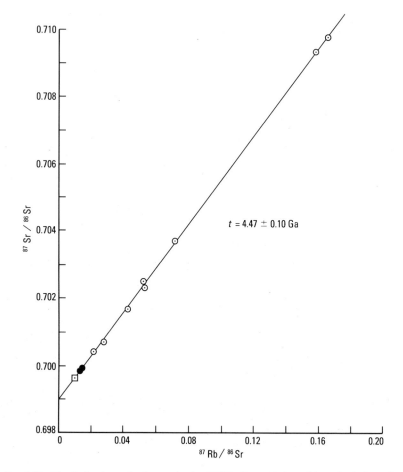

Fig. 5.18. Rb–Sr isochron for lunar dunite 72417. Open circles, chips of whole rock; filled circles, symplectites; square, olivine. All data were used in the isochron fit. (After Papanastassiou and Wasserburg, 1975.)

5.6 are based, by and large, on measurements by a single isotopic method. Sample 73215 has been discussed in some detail above. Several clasts from this breccia have $^{40}Ar/^{39}Ar$ plateau ages between 4.1 and 4.2 Ga. Only one age, for a feldspathic clast, exceeds 4.2 Ga, but a second measurement gave a somewhat lower result. Sample 73255, also a breccia, has two clasts with Sm–Nd and $^{40}Ar/^{39}Ar$ ages of more than 4.2 Ga (Table 5.6). Eight samples of the *aphanitic* matrix from the breccia give $^{40}Ar/^{39}Ar$ high-temperature plateau ages of 3.88 to 4.01 Ga, the lower of which may represent the Serenitatis impact (Jessberger et al., 1978, Staudacher et al., 1979). The isotopic ages of the older clasts may represent initial crystallization ages or a later event that reset the isotopic clocks.

Three small anorthosite fragments returned by Luna 20 have been studied by $^{40}Ar/^{39}Ar$ techniques. The data suggest that these fragments crystallized more than 4.3 Ga, but in all three cases the age spectra are complex. P. H. Cadogan and G. Turner (1977) of the University of Sheffield found that one of the fragments, L2015,3,6, contained at least two distinct components of trapped argon in addition to radiogenic ^{40}Ar. As the composition of the trapped Ar is unknown, it is not possible to find the precise crystallization age of the sample, but the most reasonable model results in a probable crystallization age of 4.40 ± 0.10 Ga and a minimum age of 4.30 ± 0.10 Ga.[25]

The $^{40}Ar/^{39}Ar$ age spectra for two fragments of anorthosite from depths of 32 to 41 cm in the Luna 20 core (20013) have multiple age plateaus. The darker fragment has a low-temperature plateau at 4.17 Ga and a high-temperature plateau, representing over more than 50% of the gas released, of 4.36 Ga. The light fragment has low- and intermediate-temperature plateaus of 4.17 and 4.36 Ga, and a high-temperature plateau of 4.51 Ga representing more than 50% of the gas released. J. C. Huneke and G. J. Wasserburg (1979) of the California Institute of Technology have interpreted these results as evidence that the anorthosite crystallized at 4.51 Ga or more and was partially degassed by events at 4.36 and 4.17 Ga.

Ages for the remaining four samples listed in Table 5.6 are straightforward and there is no reason to suspect that they do not represent either crystallization ages or the ages of major impact events that completely reset the isotopic clocks. They are, however, based on single measurements and are unconfirmed.

In summary, there is ample evidence that highland cumulate rocks crystallized in the lunar crust more than 4.2 Ga and were excavated from depth and transported to their recovery sites by major impacts. The best candidates for the oldest rocks include anorthosite

25. The composition of the trapped components can be estimated from an argon correlation diagram. For details see Cadogan and Turner (1977: 172–74).

15455 (4.42 Ga), dunite 72417 (4.47 Ga), troctolite 76535 (4.51 Ga), and the Luna 20 anorthosites L2015 (4.40 Ga) and 22013 (4.51 Ga). If these ages represent crystallization and cooling within the lunar crust, then differentiation of the proposed primeval magma sea and formation of the cumulates must have begun more than 4.5 Ga. Accordingly, the Moon must be older than 4.5 Ga. Even the most conservative interpretation of the data in Table 5.6 leads to the conclusion that the Moon's age must exceed 4.4 Ga.

Model Ages

Model ages are used by geochronologists to estimate the timing of precrystallization events, i.e. differentiation and formation of parent materials, under hypothetical conditions that may or may not obtain in nature. The interpretation of model-age results is commonly complex, and their precise meaning is uncertain. Here I summarize a few of the more common models for lunar rocks but do not discuss them in any detail because model ages do not provide conclusive or precise evidence about the ages of lunar rocks or of events in lunar history.

As mentioned earlier, the U–Pb method has proven of minimal value in determining the ages of individual lunar rocks but has contributed valuable information about lunar evolution. Most U–Pb isotopic data for lunar rocks, when plotted on the conventional $^{206}Pb/^{238}U$ vs $^{207}Pb/^{235}U$ concordia diagram, are highly discordant and do not fall on concordia. The few that do typically give corcordant U–Pb ages of 4.4–4.5 Ga, which is greater than the rocks' ages as determined by other radiometric methods. The reason for the high frequency and degree of discordance is that most lunar samples contain initial lead.

The U–Pb data are best understood by means of another type of concordia diagram in which $^{238}U/^{206}Pb$ is plotted against $^{207}Pb/^{206}Pb$ (Fig. 5.19). This diagram, called the "coupled U–Pb evolution diagram" by F. Tera and G. J. Wasserburg of the California Institute of Technology, who first introduced it to lunar studies (Tera, Papanastassiou, and Wasserburg, 1974, Tera and Wasserburg, 1972, 1974, Oberli et al., 1978, Oberli, Huneke, and Wasserburg, 1979),[26] is similar in one respect to the more familiar concordia diagram of Figure 3.13: a closed uranium-bearing system that is free of initial lead will plot on a concordia at an age, t_1, that represents the time the system formed. If this system is recrystallized at some later time, t_2, then the lead formed between times t_1 and t_2 becomes the initial lead of the new

26. Gale (1972) disputed the utility of this diagram on the grounds that it provides no information that cannot be determined from the conventional concordia diagram.

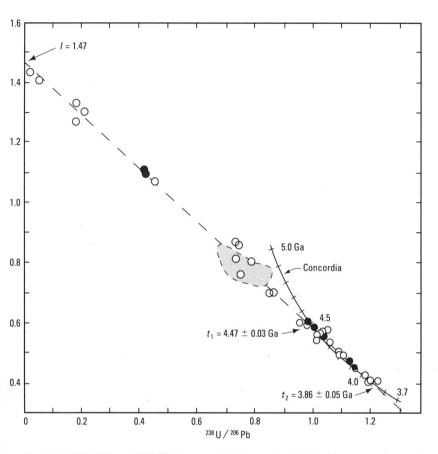

Fig. 5.19. $^{238}U/^{206}Pb$ vs $^{207}Pb/^{206}Pb$ concordia diagram for highland igneous rocks and breccias. Apollo 16 and Luna 20 soils plot within the shaded area. The isochron (dashed line) is fit to seven Apollo 14, 16, and 17 breccias (filled circles) and intersects concordia at 3.86 ± 0.05 Ga and 4.47 ± 0.03 Ga. *I*, composition of initial lead indicated by the model isochron. Open circles are data from other highland igneous rocks and breccias. All data have been corrected for primordial Pb, i.e., the Pb composition of U-free iron meteorites. (After Oberli, Huneke, and Wasserburg, 1979, with additional data from Tera and Wasserburg, 1974.)

system. The composition of this initial lead is indicated by the intersection, *I*, with the ordinate of a line passing through t_1 and t_2. Samples from this recrystallized system will plot on the line, or isochron. Those samples low in uranium relative to lead will plot nearer the ordinate, whereas an initially lead-free sample will plot on concordia at t_2. The isochron in the coupled U–Pb evolution diagram is, in effect, a mixing line of two components, and samples plotting along the isochron can be thought of as mixtures of varying proportions of (1) initial lead of composition *I*, and (2) radiogenic lead from the decay of uranium during the interval $t_1 - t_2$. In this diagram the effect of lead loss is to displace a point off of the line to the right.

Results of precise analysis of seven highland breccias plot on a line with a lower intersection of 3.86 ± 0.05 Ga and an upper intersection of 4.47 ± 0.03 Ga. The simplest interpretation of these data, following the two-stage model described above, is that the lower intersection represents the time when these breccias were last recrystallized or metamorphosed, the upper intersection represents the age of the parent materials from which the breccias were derived, i.e. the time of lunar crust formation, and I represents the composition of the initial lead in the current "system."

How realistic is this interpretation? Comparison of the lower intercept with the ages of the breccias as determined by other radiometric methods shows that it is, indeed, a reasonable value for the age of the breccia formation. The meaning of the upper intercept is less certain, but the most straightforward interpretation is that the value of 4.47 Ga is the time when the constituents of the breccia formed, which is consistent with the other radiometric evidence for a period of lunar crust formation and differentiation lasting from about 4.5 to 4.3 Ga.

One of the interesting things about the U–Pb data for highland rocks is that values derived from a wide variety of lithologic types, including igneous rocks, breccias, and soils, fall either on or very near this same isochron (Fig. 5.19). Why should this be so? The most likely explanation, and one that is consistent with the other radiometric age data for highland rocks, is that the lower and upper intercepts represent the average ages of recrystallization and impact metamorphism (3.8–3.9 Ga) and of lunar crust formation (4.4–4.5 Ga), respectively, and that I is an average value for lunar initial Pb about 3.8–3.9 Ga. Deviations of some data from the isochron are due to differences of a few hundred million years in the ages of source material, impact metamorphism, and initial lead compositions. It is also quite likely that some of the deviations are due to analytical errors, as the lead concentrations in these rocks are very low and precise analyses are difficult. Finally, some of the departures from exact linearity may be because some of the rocks have not had the simple, two-stage history that the model presumes.

Another interesting feature of U–Pb data is that the ages they specify for some rocks, including some mare basalts, are concordant, i.e. the values lie on concordia. Where this is so the indicated age is between 4.4 and 4.5 Ga even though the crystallization ages determined by Rb–Sr and $^{40}Ar/^{39}Ar$ are much younger. Where it has been possible to obtain internal U–Pb isochrons on individual samples, again including mare basalts, the line commonly intersects concordia at the crystallization age, as determined by Rb–Sr and ^{40}Ar/^{39}Ar methods, and at an age of about 4.4–4.5 Ga (Tera and Wasserburg, 1974, 1975). Why do mare basalts and highland rocks have the same upper intersection? The answer seems to be that highland rocks and the source regions (parent materials) of mare basalts were both formed at the same time, i.e. during differentiation and formation of the lunar crust. Thus, the U–Pb system, although of minimal value for measuring the crystallization ages of individual samples, seems to provide unique information about the time (or at least the average time) of formation of the lunar crust. Even though this crustal age is model-dependent, the consistency and coherence of the data lend considerable credibility to the validity of the model.

In Chapter 4 the basis for Sm–Nd model ages for the Watersmeet gneisses was discussed. Similar model ages can be calculated for lunar rocks once the crystallization age and initial ^{143}Nd/^{144}Nd ratio of a sample have been determined from either an internal isochron or from other age data (Fig. 5.20). On the basis of this information and

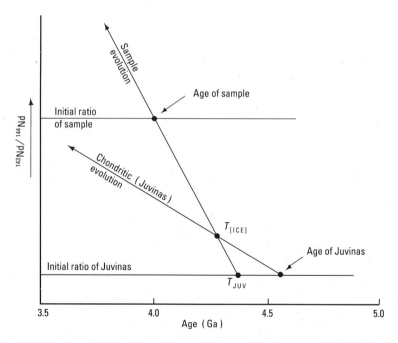

Fig. 5.20. Two types of Sm–Nd model ages. The crystallization age and initial ratio of the samples are determined from an internal isochron. From these values and the Sm/Nd ratio in the sample, the ^{143}Nd/^{144}Nd ratio can be extrapolated back in time. $T_{[ICE]}$ is the hypothetical time the sample parent material was separated from a reservoir of chondritic composition. $T_{[JUV]}$ is the hypothetical time the sample last had an initial ratio equal to that of the meteorite Juvinas. (After Papanastassiou, DePaolo, and Wasserburg, 1977.)

the Sm/Nd ratio of the sample the evolution of the ^{143}Nd/^{144}Nd ratio in the sample can be projected back in time. The intersection of this sample evolution line with the *chondritic* growth curve as determined from data on the meteorite Juvinas is defined as the model age $T_{[ICE]}$ (Intersection with Chondritic Sm–Nd Evolution) or $T_{[CUR]}$ (Chondritic Uniform Reservoir), which represents the hypothetical time in the past when the Sm/Nd ratio was changed (fractionated) from the chondritic growth value (Lugmair, Scheinin, and Marti, 1975, Lugmair et al., 1976, Papanastassiou, DePaolo, and Wasserburg, 1977, Lugmair and Marti, 1978, Lugmair and Carlson, 1978, Nyquist et al., 1979a, papers which also present data for various lunar rocks). Ideally, this time is the time of differentiation of the lunar crust, but this is true only if the samples under study evolved according to the two-stage closed-system model. $T_{[ICE]}$ values for both mare and highland lunar rocks typically fall within the range of 4.3 to 4.8 Ga. That the values, some of which exceed the age of the Moon, extend over this range is partly due to some, perhaps many, lunar rocks having multistage histories rather than the simple two-stage history of the $T_{[ICE]}$ model. In addition, $T_{[ICE]}$ ages are extremely sensitive to the choice of the model's parameters. Thus, $T_{[ICE]}$ model ages indicate a long period of differentiation early in lunar history but do not provide precise estimates of the ages of formation of the lunar source rocks.

A second type of Sm–Nd model age, $T_{[JUV]}$, is defined as the intersection of the sample evolution line with the initial ^{143}Nd/^{144}Nd ratio of the meteorite Juvinas. In other words, it is a Sm–Nd age calculated on the assumption that the sample had an initial ratio the same as Juvinas'. Because of the small amount of fractionation between Sm and Nd in the lunar rocks, however, all values of $T_{[JUV]}$ for lunar rocks are very near the age of Juvinas and $T_{[JUV]}$ model ages provide insufficient chronological resolution to be of significant value in lunar chronological studies.

The Rb–Sr model age known as $T_{[BABI]}$ (Basaltic Achondrite Best Initial) is exactly analogous to $T_{[JUV]}$ in the Sm–Nd system, but because variations in the Rb/Sr ratios of lunar rocks commonly exceed an order of magnitude,[27] $T_{[BABI]}$ is a much more sensitive model age than is $T_{[JUV]}$. $T_{[BABI]}$ ages for individual rocks are calculated by using BABI for the initial ^{87}Sr/^{86}Sr ratio of the sample and are invariably older than the crystallization ages of the samples.[28] Most values for

27. This is because Sm and Nd are chemically similar whereas Rb and Sr are not. Thus, processes such as differentiation and partial melting result in much more separation (fractionation) of Rb and Sr than of Sm and Nd.

28. BABI is the average initial ^{87}Sr/^{86}Sr ratio of the basaltic achondrite meteorites and appears to be the initial ratio of basaltic achondrite parent bodies at their time of formation. The $T_{[BABI]}$ model was devised by Papanastassiou and Wasserburg (1969).

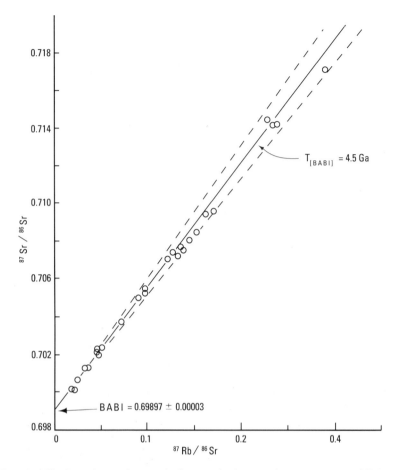

Fig. 5.21. Rb–Sr evolution diagram for lunar soils showing the narrow range of $T_{[BABI]}$ model ages. The samples are from Apollo missions 11, 12, 14, 15, and 16 and Luna 16 and 20. The dashed lines lie within 0.3 Ga of the reference isochron. This "isochron" diagram has no direct age significance but suggests an early period of lunar differentiation. (After Papanastassiou and Wasserburg, 1972a.)

$T_{[BABI]}$ fall in the range 4.2–4.6 Ga indicating, like the $T_{[ICE]}$ ages, an early period of crustal differentiation. Notable exceptions are the Apollo 11 high-K basalts, which have $T_{[BABI]}$ model ages of 3.7–3.8 Ga, presumably because the parent materials of these rocks were the products of multi-stage processes within the lunar crust.

Finally, it is interesting that lunar soils have $T_{[BABI]}$ ages of 4.5 ± 0.3 Ga, and that this seems to be a uniform characteristic of lunar soils

For discussions and examples of $T_{[BABI]}$ ages see Papanastassiou and Wasserburg (1971a,b, 1972a) and Oberli et al. (1978).

from all mission landing sites (Fig. 5.21). The scatter of the data about the reference isochron is probably due to multi-stage evolution of the soil parent materials, i.e. the lunar crustal rocks, but the overall consistency of the data suggests that the lunar crust formed early in the history of the moon, i.e. about 4.5 Ga.

Summary

Despite the intense early bombardment that brecciated and fractured the lunar crust to depths of perhaps 20 km, slightly more than a dozen lunar highlands samples with radiometric ages exceeding 4.2 Ga have been found. The oldest of these include a dunite (4.47 ± 0.10 Ga), an anorthosite (4.42 ± 0.10 Ga), and a troctolite (4.51 ± 0.07 Ga) whose ages are thought to represent crystallization associated with the formation of the lunar crust.

The timing of basin formation has been determined from the several hundred radiometric ages calculated for breccias recovered from the thick and extensive blankets of ejecta that cover much of the lunar highlands. The ages indicate that basin formation began about 4.4–4.5 Ga (Mare Tranquillitatiis, Oceanus Procellarum) and continued until about 3.8 Ga (maria Orientale and Imbrium).

The filling of the basins by mare lavas was not directly associated with the basin-forming impacts but, in general, occurred several hundred million years later. More than 100 radiometric ages on basalts from six mare sites show that mare volcanism at these sites occurred between about 3.9 and 3.1 Ga. Some radiometric data, crater density studies, and stratigraphic relations indicate, however, that mare volcanism began more than 4.1 Ga and may have continued, at a reduced rate, to as recently as 1 Ga.

The radiometric data, including both rock and model ages, show clearly that the Moon is at least 4.5 Ga in age. Between 4.5 and 3.8 Ga, the Moon was subjected to intense bombardment from material in its orbital path. This was followed by extensive infilling of the basins by mare volcanism from 3.9 to 3.1 Ga. From 3.1 to perhaps 1 Ga, lunar volcanism continually diminished as the Moon cooled, and for the past 1 Ga the Moon has been inactive, its features modified only by impacts of meteorites, meteoritic dust, and particles of the solar wind, and by gravity.

CHAPTER SIX

Meteorites:
Visitors from Space

It is easier to believe that Yankee Professors would lie, than that stones
would fall from the sky.   THOMAS JEFFERSON, 1807[1]

For a few brief seconds a thin white streak of light moves across
the clear night sky at impossible speeds, then vanishes. Most observ-
ers (the author included) call them "shooting" or "falling stars," but
they are *meteors* caused by frictional heating and ionization of air
molecules as a tiny bit of cosmic debris, a *meteoroid*, penetrates Earth's
atmosphere at hypersonic velocities. Nearly all meteoroids disintegrate in their brief flight through the upper atmosphere and reach the
Earth's surface only as *meteoritic dust*. A few of the larger meteoroids,
however, survive their fiery passage and strike the Earth, becoming
meteorites. These occasional visitors from space are more than curi-
osities, they are valuable samples of objects created when the planets
formed—bits of debris left over from the process. Many contain droplets of matter formed by condensation directly from the Solar Nebula
and represent some of the most primitive solid material in the Solar
System. Others are the fragments of chemically differentiated planetoids that were obliterated by one or more violent collisions with
similar bodies. Thus, meteorites can tell us much about the age and
early history of the Sun and planets.[2]

1. Upon learning of the report of Yale professors Benjamin Silliman and James
Kingsley on the 1807 fall at Weston, Connecticut. Quoted by J. A. Wood (1968: 4). Although widely cited, the quote is of dubious authenticity according to Marvin (1986:
146).
2. There are several good books on meteorites, of which Dodd (1981) is both current and comprehensive, especially on the subjects of composition, petrogenesis, and
classification. The books by McSween (1987), Hutchison (1983), and Sears (1978) are
excellent and easily readable introductory texts. Hartmann (1983) contains a chapter on
meteorites, which is an excellent short summary, as well as worthwhile discussions of
asteroids and other relevant topics. Marvin (1986) is a fine summary of the history of
meteorite research. Additional general works on meteorites include Nininger (1952), B.

The notion that stones occasionally fall from the sky is now generally accepted, and the scientific evidence for the phenomenon is conclusive, but it was not always so. For centuries, perhaps since man first noticed "falling stars," the origin of meteorites was equivocal and confused by myth and mysticism. On occasion, stones from the sky apparently were objects of religious significance. Several have been found carefully interred at Indian burial grounds in North America, and the Black Stone built into the Kaaba in Mecca, which was enshrined before Muhammad conquered the city in A.D. 624, reportedly fell from the sky and is probably a meteorite. The "image which fell down from Jupiter" mentioned in the Bible (Acts 19:35) may have been a meteorite. The stones that fell at Ogi, Hizen, Japan, in 1741 were thought to have fallen from the loom of the goddess Shokujo and were worshipped for 150 years. Buchwald (1975: 165–66) listed 26 venerated iron meteorites.

The ancient Greek, Roman, Egyptian, and Chinese literature contains numerous descriptions of meteors, meteorites, and meteorite falls. Perhaps the earliest are the descriptions of meteors contained in the Egyptian papyrus writings of about 2000 B.C. The philosophers of ancient Greece and Rome, including Anaxagoras (ca. 500–430 B.C.), Diogenes of Apollonia (ca. 412–323 B.C.), and Plutarch (ca. A.D. 45–120) described the fall of stones, sometimes attributing the phenomenon (not entirely inaccurately) to heavenly bodies that had slipped their moorings and fell to Earth. The Parian chronicle, whose marble tablets record fragments of pre-Homerian and classical Greek history, and other ancient sources describe stone and iron meteorite falls in Crete in 1478 and 1168 B.C., the fall of stones in Boeotia in about 1200 B.C., and falls near Rome that demolished several chariots and killed ten men in 616 B.C. Writings of the Han Dynasty contain numerous reports of meteorite falls in China within the period 89 B.C.–A.D. 29.

Meteorites were a source of metal for some ancient cultures. An iron-bladed axe of the Shang Dynasty (ca. 1400 B.C.) was forged from an iron meteorite, as were some Chinese coins minted from 400 B.C. to A.D. 300. Prior to the arrival of European explorers in the early nineteenth century, the Eskimos of northwest Greenland worked meteoritic iron for use as cutting edges in bone harpoons and knives. V. F. Buchwald (1975: 41) has estimated that as many as 18% of the known nonantarctic iron meteorites have been reheated and forged by our forefathers.

Mason (1962, 1967), J. A. Wood (1968), McCall (1973), and Wasson (1974). Volume I of the work by Buchwald (1975) is an excellent summary of information on iron meteorites. I have relied on these books for the content of much of this chapter.

The earliest observed fall for which the meteorite was recovered and is still preserved occurred near the village of Ensisheim, Alsace Province, France, on November 16, 1492. The 127-kg stone meteorite still resides in the church, where it was brought by the local residents nearly five centuries ago.

Several early philosopher-scientists, including Diogenes of Apollonia, Dominic Troili, who described a fall at Albareto, Italy, in July of 1766, and Jerome de la Lande, who described one at Bresse, France, in the mid-1700's concluded that meteorites were of cosmic origin and were associated with meteors. Most of their contemporaries as well as subsequent scholars, however, were skeptical and preferred to ascribe meteorites to a terrestrial origin. One common explanation was that stones had been carried into the clouds, perhaps by waterspouts, from where they fell again to Earth:

May it not be a reasonable conjecture, that all the various substances which have fallen from the atmosphere, in later as well as in former times, are nothing more than the sands, and other contents, found at the bottom of lakes and large rivers, and from the shores of the sea, naturally produced by the powerful influence or the attraction of the clouds? It is but a trite observation to say, that the clouds make frequent visits to the waters of the earth, from which they usually carry away large quantities of that element, and with it, no doubt, the substances (even with some of the fish) which form the beds, in proportion to the heat of the weather, and the depth of those waters which the clouds, when they fall, happen to attach upon. It is self-evident, that the streams which ascend with the clouds are sometimes clear as crystal, at other times thick and muddy. When the latter is the case, then it is that these substances may be concreted; and, by some extraordinary concussion in the atmosphere, return to the earth. W. Bingley (1796).

In 1769 Father Bachelay had presented the Academy of Sciences in Paris, perhaps the most prestigious scientific body of the time, with the stone meteorite and a detailed account of its fall the previous year near Luce. The Academy appointed a commission, which included the renowned chemist A. L. Lavoisier, to investigate, but their report discredited Bachelay's observations and attributed the fused crust on the meteorite to lightning; their conclusion was that the stone was of terrestrial origin.

E. E. F. Chladni, a German lawyer, physicist specializing in acoustics, and corresponding member of the Academy of Science of St. Petersburg, is generally credited with initiating the acceptance by the scientific community of the idea that meteorites are extraterrestrial. Chladni (1794) described the 160-kg Pallas iron meteorite from Krasnoyarsk in Siberia and the 13,500-kg Otumpa iron meteorite, found in the Gran Chaco region of Argentina. He concluded that these objects had been cooled slowly and, because they were unlike the rocks of the regions in which they were found, must have come from a distant source. By a careful process of elimination in which he considered and rejected works of man, electricity, volcanic action, and any mysterious and accidental "conflagration" as possible agents, Chladni concluded that iron meteorites were of cosmic origin, were the spent remains of meteors, had been heated by friction during their passage through the Earth's atmosphere, and possibly were the fragments of a disrupted planet. In a subsequent paper in 1799 he extended these conclusions to include stone meteorites.

Although Chladni's conclusions were not immediately and universally accepted, his careful investigation and flawless reasoning convinced much of the scientific community that meteorites might be extraterrestrial and that the idea was worthy of serious consideration and careful investigation. Most doubts were removed within the next decade by two important scientific events. One was the discovery of the first asteroid, Ceres, in 1801 by Giuseppe Piazzi, followed the succeeding year by the discovery of Pallas, a second, smaller body. The discovery of small bodies or planetoids showed that the Solar System was populated by more than a few planets and a star, and provided the first reasonable source for meteorites.

The second event was the fall of several thousand stone meteorites near L'Aigle, France, in April 26, 1803. Once again, the Academy of Sciences, in Paris, at the time called Section One of the Institute of France, appointed a commission, this time headed by the mathematician, physicist, and astronomer J. B. Biot. From the accounts of eyewitnesses, the composition and appearance of the stones, and the elliptical distribution of the fragments, the Commission concluded that the meteorites were undoubtedly fragments of a single body that was associated with an observed meteor and that the body had entered Earth's atmosphere from space. The evidence was undeniable and the scientific world was at last convinced that stones of extraterrestrial origin did, indeed, fall from the sky.

The fall of meteorites is no longer mysterious and it is now clear that the collision of cosmic debris with the Earth is an ongoing, daily process. Recent estimates of the mass of meteoritic influx based on satellite penetration studies, meteor surveys, and the meteoritic content of deep-sea sediments indicate that the Earth received between 9×10^6 and 3.2×10^7 kg of cosmic debris each year, mostly in the form of meteoritic dust, with the best estimate being about 2×10^7 kg/yr (Dohnanyi, 1972).[3] Although this may seem like a large mass of mate-

3. Nearly three-fourths of this material falls into the oceans.

rial, uniformly distributed over Earth's surface and left undisturbed in a century it would amount to a layer only 2×10^{-6} mm thick.

As might be expected, the number of meteoroids entering Earth's atmosphere each year is a function of their size, with the larger bodies being considerably less numerous than the smaller (Fig. 6.1). For example, objects with masses of 10^{12} kg or greater enter Earth's upper atmosphere only about once every million years. Such an object would have a diameter of a kilometer or so and if it struck land would create a crater more than 50 km in diameter. In contrast, more than 10^5 meteoroids with masses of 1 kg or more strike the upper atmosphere each year, and the flux rate for particles weighing 10^{-12} kg or more is about 10^{17}/yr (Dohnanyi, 1972).

Of the many tons of material that reaches the Earth's surface each year, less than one percent arrives in pieces large enough to be called rocks. The fate of a meteoroid entering the atmosphere is primarily a function of its mass and velocity. Objects with masses between about 10^3 and 10^{-7} g disintegrate in the atmosphere, whereas smaller particles are merely slowed and reach Earth's surface carried by air currents. Meteoroids with masses between about 1 and 9×10^5 kg also lose substantial velocity because of atmospheric drag but still penetrate the atmosphere and strike the surface. The atmosphere has no significant slowing effect on meteoroids with masses exceeding 9×10^5 kg. Meteoroid velocities range from 10 to more than 70 km/s but typically are 10–20 km/s. Those with high velocities are more likely to disintegrate in the atmosphere than those with lower velocities. Not surprisingly, composition also plays a role in meteoroid survival, with the iron and harder stone meteoroids being more durable than the friable stones.

Few of us ever see a meteorite except in a museum, but probably all have observed meteors, or shooting stars. This dazzling phenomenon is due primarily to ionization of air molecules by the rapid passage of the meteoroid through the atmosphere. Fireballs typically become visible at an altitude of about 130 km and disappear when the meteoroid reaches an altitude of about 20 km. Friction between the meteoroid and the air heats and may melt the surface of the meteoroid, but the passage of the meteoroid through the atmosphere is so rapid (lasting only a few seconds) that the body, as a whole, retains the chill of space. In fact, frost has been observed to form on meteorites immediately after their fall.

Although many sizable objects enter Earth's atmosphere each day, few survive intact. On the average, only about one object large enough to form a substantial crater actually reaches the Earth's surface each year, and large meteorites capable of generating giant explo-

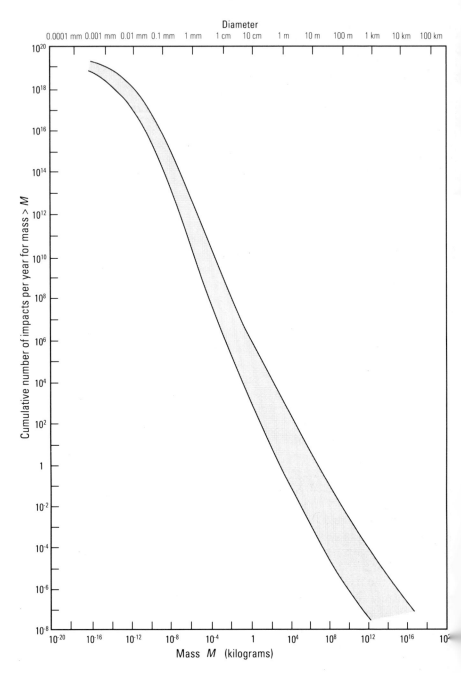

Fig. 6.1. Cumulative rate at which meteoroids of mass greater than M impact Earth's upper atmosphere. The diameters of the objects are for meteoroids with a density of $3.6\,g/cm^3$, the average density of chondritic meteorites. (Data from compilation of Dohnanyi, 1972: 3).

Fig. 6.2. Meteor, or Barringer, Crater, Arizona, is 1.2 km in diameter and 170 m deep. It was made by the prehistoric impact of an iron meteorite known as Canyon Diablo. More than 2.7×10^4 kg of meteorite fragments have been recovered from the surrounding desert.

sions strike the Earth only once or twice per century (Hartmann, 1983: 162). Meteorites that produce geologically durable craters are even rarer. Only 116 structures, ranging in age from Cenozoic to Precambrian, are known to be of meteoritic or probably meteoritic origin (Guest and Greeley, 1977: 89, Grieve, 1982, Hutchison, 1983: 18–24). One of the best known of these is Meteor (or Barringer) Crater near Flagstaff, Arizona, which was formed by the impact of the Canyon Diablo Meteorite (Fig. 6.2). A current hypothesis attributes the extinction of the dinosaurs and other species of plants and animals at the end of the Cretaceous Period to the impact of an extraterrestrial object perhaps 10 km or more in diameter (Chapter 5, note 11), but so far no associated crater has been found. Impacts of this magnitude are very rare and are estimated to occur only once every 100 Ma or so. In contrast, cratering on the scale of Meteor Crater occurs approximately once every 1,300 years, but most of these smaller craters do not long survive the attack of erosion (Hutchison, 1983: 24).

Finding a meteorite is a rare occasion, and meteorites are highly prized for their scientific value. As of 1986, slightly more than 10,000 meteorites had been found, more than 7,500 of these since 1969, the year it was discovered that meteorites are preserved and easily seen on the surface of the Antarctic ice fields (Lipschutz and Cassidy, 1986).[4] Graham, Bevan, and Hutchison (1985) catalogued authenticated meteorites. Only about 900 meteorites of the total collection were observed to fall; the others are "finds." The latter tend to be weathered because of (often) long exposure to Earth's atmosphere and weather and so are not as desirable for scientific study as falls, which are not only fresher but provide a more representative sample of the proportions of the various types of meteorites that reach Earth.

Types of Meteorites

There are many different types of meteorites, and they are generally classified by the degree to which they are *differentiated*, or chemically evolved, and by their relative proportions of silicate and metallic minerals (Table 6.1). *Stony meteorites*, or *stones*, are composed primarily of silicate minerals and, in a general way, resemble rocks found on the surfaces of the Earth, Moon, Mars, Venus, and, presumably, Mercury.[5] *Iron meteorites*, or *irons*, are metallic, composed primarily of an alloy of nickel and iron (Fig. 6.3). Rocks resembling irons are essentially unknown on planetary surfaces, but the core of the Earth probably consists of material very similar to that in the iron meteorites. In between the stones and irons are the *stony–iron meteorites*, which are, as their name implies, mixtures of both silicate and metallic phases. Even though this threefold classification is useful, it should be noted that there are intermediate varieties. Thus, many stones contain metallic phases, and many irons contain small amounts of silicates.

From the proportions of well-classified and observed meteorite falls, stone meteorites are more abundant than iron and stony–iron meteorites combined by about 20 to 1 (Table 6.1). This indicates that stones are more abundant than irons in our, i.e. the Earth's, region of the Solar System, but the ratio is probably even higher than indicated

4. The number of meteorites found in Antarctica is based on the total number of fragments, whereas the number of nonantarctic meteorites is the number of occurrences, each of which may contain more than one fragment. Thus the numbers are somewhat misleading.

5. Remote landers have provided photographs and analyses of rocks on the surface of Mars and Venus but not, as yet, on Mercury.

TABLE 6.1

Classification and Approximate Abundances of Meteorites

Meteorite type[a]	Approximate percentage of observed falls			
Undifferentiated stones (aerolites)	95			
Chondrites		86		
Carbonaceous			5	
CI (C1, CI)				0.7
CM (C2, CII)				2.0
CO (C3, CIII)				0.9
CV (C3, CIII)				1.1
Ordinary			79	
H (bronzite)				32.3
L (hypersthene)				39.3
LL (amphoterites)				7.2
Enstatite (E)			1	
Differentiated stones (aerolites)				
Achondrites		9		
Calcium-rich				
Basaltic (pyroxene–plagioclase)			5.3	
Eucrites				2.6
Howardites				2.4
Shergottites				0.3
Angrites (augite)			0.1	
Nakhlites (diopside–olivine)			0.1	
Calcium-poor				
Aubrites (enstatite)			1.1	
Diogenites (hypersthene)			1.1	
Chassignites (olivine)			0.1	
Ureilites (olivine–pigeonite)			0.4	
Stony irons (siderolites)	1			
Mesosiderites				
(pyroxene–plagioclase)			0.9	
Pallasites (olivine)			0.3	
Lodranites (bronzite–olivine)			0.1	
Irons (siderites)	4			
Hexahedrites			0.5	
Octahedrites			2.7	
Ataxites			1.3	

SOURCES: Dodd, 1981; Hartmann, 1983; Wasson, 1974; B. Mason, 1962, 1967; Van Schmus and Wood, 1967.
[a] Alternative names are in parentheses.

by the statistics of finds because stones tend to be more friable than irons and a higher proportion of stones probably disintegrates in the atmosphere. Of meteorite finds the combined frequency of irons and stony irons is much greater than that of stones. The inconsistency between the statistics of falls and finds has a very simple explanation: once on the ground stones are more susceptible to weathering and disintegration and are much more difficult to recognize than irons and stony irons because stones, to the untrained eye, closely resemble ordinary rocks, whereas irons appear unusual and are more apt to be "found."

Fig. 6.3. Cut and etched face of the Edmonton, Kentucky, iron meteorite (octahedrite or group IIIC). The distinctive Widmanstätten pattern is caused by the intergrowth of two polymorphs of nickel–iron alloy during slow cooling. The cut face is approximately 12 cm across. (Smithsonian Institution photograph M-375.)

Stone meteorites are divided into two main types: *chondrites*, which constitute about seven-eighths of all stone falls, and *achondrites* (Table 6.1). Chondrites are so named because most contain *chondrules*,[6] small spherules ranging in diameter from a few tenths to several millimeters (Fig. 6.4). These chondrules are composed primarily of high-temperature silicates, commonly olivine and pyroxene, with only minor amounts, if any, of metallic phases. The composition of individual chondrules varies widely, and chondrules may contain one or more of any number of silicate minerals or glass (Wasson, 1974: 63–68, Dodd, 1981: 30–36). Some chondrules are spherical and either glassy or very finely crystalline, indicating that they formed by rapid cooling of small liquid droplets, whereas others are irregular and coarsely crystalline, indicating slower cooling. The precise origin of chondrules is still not entirely clear and probably not limited to a single mechanism. Those that clearly formed from liquid droplets may have condensed directly from the Solar Nebula, the most commonly accepted hypothesis, or were created by fusion of silicate dust before accretion of the meteorites' parent bodies. Alternatively, they may have originated during accretion by impact melting on meter-sized planetesimals. Many of the coarsely crystalline chondrules with somewhat irregular shapes are clearly clastic and may be abraded fragments of once larger rocks. Dodd (1981: 61–68, 121–31) and Hartmann (1983: 166–68) discussed current thinking on the origin of chondrules. Whatever their precise origin, there is general agreement that chondrules probably represent some of the most primitive solid material in the Solar System that is available for scientific study. As we shall see, the chondrites are also among the oldest dated objects in the Solar System, with radiometric ages as great as 4.55 Ga.

Chondrites are undifferentiated and, like the chondrules most contain, are thought to be primitive objects that record very early events in the Solar System (McSween, 1979). They are agglomerates composed of chondrules, which may constitute anywhere from zero to as much as 70% of a chondrite's mass, in a fine-grained and often opaque matrix of mineral grains and fragments of broken chondrules. Aside from the presence, in most, of chondrules, the chondrites are distinctive in their chemical composition, which, except for highly volatile elements such as hydrogen and helium, closely resembles the composition of the Sun and, presumably, the Solar Nebula. The primitive, Solar composition of chondrites sets them apart from Earth rocks, Moon rocks, and other types of meteorites, all of which have been chemically differentiated by processing within the interiors of planets or asteroids.

6. From the Greek word for grain.

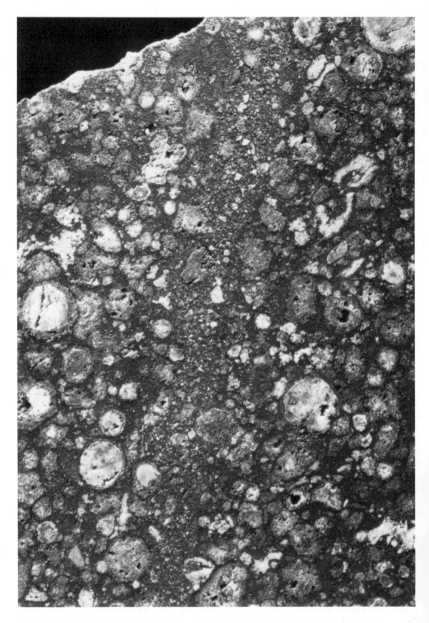

Fig. 6.4. The Allende carbonaceous chondrite (group CV), showing chondrules embedded in a fine-grained matrix composed primarily of olivine, hydrous silicates, and sulfide minerals. The largest chondrules in the photograph are approximately 2 mm in diameter. (Smithsonian Institution photograph M-1752H.)

Chondrites are divided into three classes primarily on the basis of their chemistry. *Carbonaceous chondrites* are both chemically and physically the most primitive of the three. Of all of the meteorites, the carbonaceous chondrites have compositions that most nearly resemble the Sun's. They are relatively high in volatiles and may contain more than 20% by weight of water bound within hydrous minerals. They contain no metallic iron or nickel. In addition, they contain carbon, typically 1–4% and primarily as graphite, and as much as 1% of organic molecules, including amino acids. Their high content of volatile and organic substances, their low densities, and their low to negligible degree of metamorphism and alteration indicate that carbonaceous chondrites have never been subjected to temperatures appreciably above 200°C nor to substantial pressures. The carbonaceous chondrites are subdivided into four gradational sub-groups (CI, CM, CO, and CV) on the basis of differences in chemistry, with the most primitive CI subgroup having the lowest density, the highest content of volatiles and carbon, the least postformation alteration, and compositions most nearly identical to the Sun's. Except for the CI subgroup, which is devoid of chondrules, carbonaceous chondrites consist of varying proportions of chondrules in a very fine-grained, black, and opaque matrix of graphite, magnetite, clay minerals, iron and nickel sulfides, complex organic molecules, and rare (perhaps) interstellar grains formed at high temperatures. Unlike the other types of chondrites, the carbonaceous chondrites contain only chondrules that formed from liquid droplets.

The mineralogy, texture, and chemistry of the carbonaceous chondrites suggest that, of all the meteorites, they are the most primitive, having features that, in large part, reflect complex and poorly understood processes that occurred early in the formation of the Solar System (Dodd, 1981: 68–69). The matrix and the chondrules of the carbonaceous chondrites appear to have formed separately by the vaporization and distillation of nebular gas and dust, followed by condensation. Matrix and chondrules then accreted to form the chondrite parent bodies. The amount of postformation metamorphism has been small, involving only mild thermal metamorphism, some alteration by water, and brecciation and slight shock metamorphism of some, but not all, individual meteorites. The CI carbonaceous chondrites are the least disturbed and were always within 300 m of the surface of their parent bodies. The other groups are progressively more altered and metamorphosed but not to any great degree and appear to have had similarly simple postformation histories at very shallow depths. The final stages in the histories of the carbonaceous chondrites were separation of the meteoroids from their parent bodies by collisions and, finally, their fall to Earth.

In the early days of modern meteorite study there was much debate about the origin of the organic material in carbonaceous chondrites. In particular, there was speculation that it was evidence of life elsewhere. The question was resolved in the early 1970's, however, when it was shown that the organic compounds could easily have originated by abiogenic processes (Wasson, 1974, Nagy, 1975).

Ordinary chondrites are the most numerous of the observed falls. They are similar in bulk composition to the carbonaceous chondrites but have been subjected to slightly more postformation "processing." The presence of chondrules shows that the ordinary chondrites have not been melted since they formed. The texture and mineralogy of these chondrites does, however, indicate mild metamorphism in a dry environment at low pressures (less than 1 kilobar) and at temperatures most commonly less than 400°C but in rare instances exceeding 900°C. One of the primary differences between the ordinary chondrites and the carbonaceous chondrites is that the former have a lower content of volatiles and are less oxidized than the latter, which is consistent with low-temperature, low-pressure metamorphism within their parent bodies at depths of from about 2 to 5 km beneath the surface. In addition to differences in content of volatiles, the ordinary chondrites also differ from carbonaceous chondrites in having lower ratios of magnesium, calcium, titanium, and aluminum to silicon (Table 6.2). Although some of the chondrules in ordinary chondrites are crystallized droplets, most are clasts that were probably formed by impact processes during accretion of the chondrites' parent bodies (Dodd, 1981: 121–30).

The ordinary chondrites are subdivided into three groups on the basis of their content of iron and related elements, such as nickel; the H group is high (25–30% total Fe), the L is low (20–25%), and the LL is very low (18–20%). Another difference between the groups is the ratio of oxidized to metallic iron, which is about 2 in the H group, 3 in the L group, and 10 in the LL group. The lower proportion of oxidized iron indicates that the H group has been subjected to somewhat higher temperatures than have the L and LL groups.

Unlike the carbonaceous chondrites, among which the chemical differences between groups are gradational, the three groups of ordinary chondrites are chemically distinct. In addition, a substantial fraction (> 20%) of all three of the groups are impact breccias, analogous to the lunar breccias, but clasts of one type do not occur in another. Their distinct chemistry and lack of clast "sharing" suggests that each of the three groups originated in separate parent bodies.

The histories of the ordinary chondrites are more complex than those of the carbonaceous chondrites in several ways. For example,

TABLE 6.2

Approximate Chemical Compositions of Several Types of Stony Meteorites

Meteorite type	Approximate content (weight %)								
	SiO_2	Al_2O_3	K_2O	CaO	TiO_2	FeO	MgO	H_2O	C
Carbonaceous chondrite									
CI	23	1.8	0.07	1.5	0.08	24	16	21	3.6
CM	27	2.3	0.05	2.0	0.10	28	19	13	2.4
CO, CV	34	2.6	0.05	2.3	0.12	32	24	1	0.5
Ordinary chondrite									
H	36	2.6	0.10	1.9	0.11	36	23	0.3	—
L, LL	40	2.8	0.10	1.9	0.11	28	25	0.2	—
Enstatite chondrite	39	1.9	0.11	1.0	0.06	36	21	0.6	0.4
Basaltic achondrite									
Eucrite	48	13	0.06	10	0.43	18	8	0.6	—
Howardite	49	10	0.36	8	0.10	18	12	0.3	—
Shergottite	50	6	1.8	10	—	22	10	—	—
Diogenite	52	1.2	0.001	1.4	0.19	18	26	—	—

SOURCES: McCall, 1973: 154; Dodd, 1981: 19, 243.

the nature of their chondrules suggests that impact and shock metamorphism played a more important role in their initial formation. In addition, postformation thermal and impact metamorphism were clearly important secondary and tertiary processes, whereas their effects on the carbonaceous chondrites was minimal. Still, the ordinary chondrites are not highly evolved objects and their features record relatively early, although probably postnebular, events in the formation of the Solar System.

The *enstatite chondrites* are the least abundant, most poorly studied, and, in general, least primitive of the three classes of chondrites. They are much less oxidized than the ordinary chondrites and contain lower rations of Mg, Ca, Ti, and Al to Si. They are mostly breccias composed primarily of the pyroxene mineral enstatite. Their texture and mineralogy indicate metamorphism at temperatures between about 600°C and 870°C. They are chemically similar to the *aubrites* (enstatite achondrites), and both may have come from the same parent body (Dodd, 1981: 133–53).

In addition to being classified into the chemical groups (CI, LL, E, etc.), chondrites are divided into petrologic types 1 through 7 on the basis of textural and mineralogical variations (Van Schmus and Wood, 1967, Dodd, 1981: 23–28). These petrologic types will not be discussed in any detail here but, in general, the variations are controlled by the degree of metamorphism and recrystallization, with the higher numbers designating meteorites that have been extensively heated and recrystallized. There is a fair degree of correspondence between the chemical groups and the petrologic types. Types 1 and 2, for example, are found only among the CI and CM carbonaceous chondrites, whereas the enstatite chondrites are invariably of types 4–7. The common descriptive notation for chondritic meteorites consists of the letter group designation followed by the type number. Thus, H3 describes a relatively unmetamorphosed ordinary chondrite of group H with primary feldspar, clear and isotropic chondrule glass, and well-defined chondrules, whereas LL6 designates a highly metamorphosed and recrystallized ordinary chondrite of group LL with coarse, secondary feldspar grains, recrystallized matrix, poorly defined chondrules, and no chondrule glass.

The achondrites differ from chondrites in several ways. They lack chondrules, have compositions that are distinctly nonsolar, and have igneous textures, which indicates that they formed by crystallization and cooling of a melt. Of all the types of meteorites the achondrites most closely resemble terrestrial rocks and, like their terrestrial counterparts, have complex histories consisting of differentiation in one or more episodes of partial melting and crystal fractionation within the interiors of their parent bodies, crystallization and cooling, metamorphism, brecciation by impact processes, separation from their parent bodies, and, finally, collision with Earth.

The achondrites range in composition from virtually monominerallic rocks, composed of olivine or pyroxene and resembling terrestrial and lunar cumulates, to rocks very much like terrestrial and lunar basaltic lavas. The latter group, the *basaltic achondrites*, are very interesting rocks, in part because they confirm the ubiquity and importance of basaltic volcanism in the inner Solar System. Their chemistry and mineralogy were addressed by the Basaltic Volcanism Study Project (1981: 214–35, 591–96). The basaltic achrondrites consist of three groups, two of which, the *eucrites* and the *howardites*, are closely related. The eucrites are essentially basalts that crystallized as lava flows (many are vesicular) and as shallow intrusive rocks after differentiation of their magma from a chondritelike parent rock. Their mineralogy, texture, and composition closely resemble those of terrestrial and lunar basalts (compare Table 6.2 with Table 5.2). Most of the eucrites have been brecciated, and some have been subjected to mild thermal and shock metamorphism.

Related to the eucrites are the *diogenites*, which are magmatic cumulates composed almost entirely of the pyroxene *bronzite*. The diogenites formed by partial melting of a chondritelike parent rock followed by crystal accumulation and slow cooling within the crust of their parent body. Most diogenites have undergone thermal metamorphism and brecciation. The howardites are surface breccias whose texture resembles that of the breccias found in the lunar regolith and were apparently formed on the surface of their parent body by impact processes. The howardites are mixed rocks composed of fragments of both eucrite and diogenite material with, occasionally, some chondritic fragments. The *mesosiderites* are also impact breccias that consist of fragments of eucrite and diogenite material mixed with a large proportion of metallic nickel-iron alloy. Their textures indicate that they have been subjected to repeated impacts, shock melting, and metamorphism at temperatures as high as 1,000°C. Dodd (1981: 236–74) discussed the eucrite association—the eucrites, diogenites, howardites, and mesosiderites—in detail.

The compositional similarities between eucrites, diogenites, howardites, and mesosiderites suggest that these four meteorite types may have come from the same parent body, which had an initial composition similar to that of the chondrites. The eucrites and diogenites were formed by differentiation and crystallization as shallow intrusive rocks (diogenites and some eucrites) and as surface lava flows (eucrites); many were brecciated by subsequent impacts. The howardites and mesosiderites represent surface breccias, more or less analogous to those found on the lunar surface, the mesosiderites containing a highly differentiated component of metallic nickel-iron alloy. Although these meteorites represent a complex series of postnebular events within and on the surface of a planetoid, radiometric dating of the meteorites shows that these events occurred within 100 Ma or less after the formation of the chondrites.

The remaining groups of stone meteorites are unassociated and relatively few in number, and each type may have originated in a different parent body. The *ureilites* consist of ultramafic cumulate fragments in a carbonaceous matrix that includes shock-produced diamond. Although their precise age, origin, and history is uncertain, the isotopic composition of their oxygen is similar to that of the carbonaceous chondrites, to which they may be related. The *shergottites, nakhlites, chassignites,* and *angrites* are all magmatic cumulates. The shergottites resemble terrestrial diabase except that the plagioclase is replaced by maskelynite, its highly shocked and glassy equivalent. They are similar to the eucrites but the details of their chemistry, mineralogy, and oxygen isotopic composition indicate that they are not directly related. The nakhlites are similar to terrestrial pyroxenites, whereas the chassignites resemble terrestrial and lunar dunites. An angrite, of which the meteorite Angra dos Reis is the only specimen, is composed primarily of pyroxene with some olivine, spinel, and a variety of minor minerals. Its age and oxygen isotope composition are similar to the eucrites'. In contrast, isotopic dating indicates that the nakhlites and perhaps the chassignites were formed at about 1.4 Ga, whereas the shergottites have a radiometric formation age of only about 650 Ma.

The *pallasites, lodranites,* and other stony–iron meteorites consist primarily of olivine or pyroxene mixed with metallic nickel-iron alloy. They apparently formed within the interiors of asteroid-sized bodies. Because of their mineralogy, these meteorites have not been dated.

The iron meteorites, or *siderites,* are subdivided into nine structural groups on the basis of the crystal structure of their nickel-iron alloy, which is controlled primarily by composition and partly by cooling history. These groups include the hexahedrites, six types of octahedrites, the ataxites, and the "anomalous" irons that do not fall into the eight primary groups. In addition to the structural classification, there are 16 chemical groups designated by roman numerals and letters (IA–IVB). As the details of structural and chemical classifications of the irons are not particularly germane to the radiometric ages discussed in this chapter, they will not be discussed further. For more on the stony iron and iron meteorites, see Dodd (1981: 192–235, 275–308) and Buchwald (1975). Some of the iron meteorites apparently sample the cores of asteroid-sized parent bodies, whereas others probably originated as metallic segregations at shallower depths. Because the nickel-iron alloy in iron meteorites does not contain sufficient quantities of the long-lived radioisotopes, most iron meteorites cannot be dated; however, some with silicate inclusions have radiometric ages that range from about 3.8 to nearly 4.6 Ga.

Where Do the Meteoroids Come From?

Although meteorites were once thought to be the remains of a shattered planet, their chemistry, mineralogy, and textures now clearly indicate that they are derived from perhaps as many as 70 to 80 separate parent bodies with diameters ranging from about 200 to 600 km (Dodd, 1981: 309–28, McSween, 1987: 67–99, 129–55, 185–200). Where are these bodies and how do fragments of them end up on Earth?

There is little doubt that meteorites originate in the Solar System. The high frequency of meteors and of meteorite falls requires a voluminous and nearby source. Other stars are ruled out because of their distance—the nearest star, Alpha Centauri, is 4.3 light-years away—and because there is insufficient large debris in interstellar space to account for the material that strikes Earth on a daily basis. Moreover, photographs of meteors and fireballs show that these objects entered Earth's atmosphere from orbits around the Sun. For example, two stations of the Prairie Network, operated from 1965 through 1975 by the Smithsonian Astrophysical Observatory, photo-

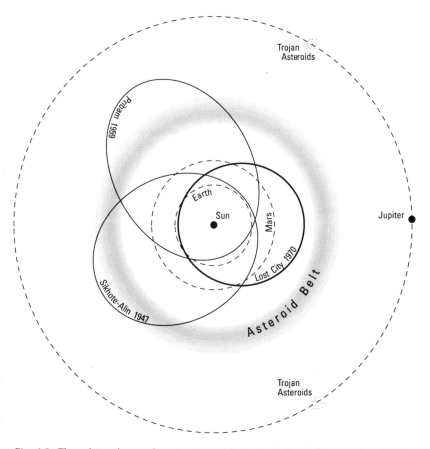

Fig. 6.5. The orbits of several meteorites as determined from photographs of their flight through the atmosphere, from orbital data in Buchwald (1975: 7, 18). Lost City (Oklahoma) and Pribam (Czechoslovakia) are H chondrites. Sikhote-Alin (USSR) is an iron meteorite.

graphically recorded the fiery flight of the Lost City chondrite before it struck the ground near Lost City, Oklahoma, on January 9, 1970. From the synchronized and timed photographic data the velocity and orbit of Lost City were calculated (Fig. 6.5). Similar data have been collected on several hundred meteors, including several recovered meteorites, by the Prairie Network and by similar meteor camera networks in western Canada and an area including Czechoslovakia and part of Germany. The data show that before their collisions with Earth, all of the objects orbited the Sun counterclockwise, like the planets, at low inclinations to the plane of the ecliptic, and with average velocities of about 17 km/s. In addition, the objects typically had *aphelia* in or beyond the asteroid belt.

Of the three possible sources of meteorites—comets, planets, and asteroids—the evidence indicates that asteroids are by far the largest suppliers. There are many tens of thousands of asteroids in the Solar System with an aggregate mass of some 2.7×10^{24} g, or about 4% of the mass of the Moon. The asteroids' relation to meteorites has been discussed by Dodd (1981: 309–27), Hartmann (1983: 189–213), McSween (1987: 67–99, 129–55, 185–211), Buchwald (1975: 5–13), and Chapman (1975). Of these, several thousand have been observed and cataloged. Asteroids range in size from minor planets like Ceres, the largest asteroid, with a diameter of 1000 km and half the mass of all asteroids put together, to numerous smaller bodies. Estimates of the size distribution of the asteroids vary, but there are at least a dozen with diameters exceeding 250 km, hundreds with diameters greater than 100 km, 10,000 or so whose diameters exceed 10 km, and perhaps as many as 10^{14} larger than one meter (J. A. Wood, 1968: 8, Hartmann, 1983).

Asteroids occur throughout the Solar System, but the majority are concentrated in the asteroid belt, a band of rocky debris and minor planets that occupies a gap between Mars and Jupiter. Kepler first suggested that a planet might be found in this position in the Solar System, and in 1772 J. E. Bode, a German astronomer, strengthened Kepler's prediction with the discovery of Bode's Rule, which predicts the distances of the planets from the Sun and includes a position between Mars and Jupiter.[7] The discovery of Ceres and of several other large asteroids orbiting in this position in the early 1800's confirmed the predictions of Kepler and Bode and led to speculation that the belt asteroids were the remains of a former planet. It is now known, however, that the asteroids are simply debris left over from the process of planet formation and were never part of a single, large planet. The bits and pieces of the asteroid belt were precluded from aggregating into a planet by their proximity to Jupiter, whose gravitational field would have torn apart a body of any significant size (McSween, 1987: 70).

The belt asteroids are irregularly distributed between about 2.1 and 3.5 *astronomical units* (A. U.) from the Sun and orbit the Sun in the same direction as the planets with periods of from three to nine years and orbital inclinations ranging from 0° to 35°. Many asteroids appear to be irregularly shaped, a fact that can be determined from changes in the brightness of reflected sunlight as these bodies tumble in space (Fig. 6.6).

Although a large majority of the asteroids occur within the as-

---

7. According to Bode's rule the distances in A. U. of the planets from the sun are obtained by adding 0.4 to the following sequence: 0 (Mercury), 0.3 (Venus), 0.6 (Earth), 1.2 (Mars), 2.4 (asteroid belt), 4.8 (Jupiter), etc.

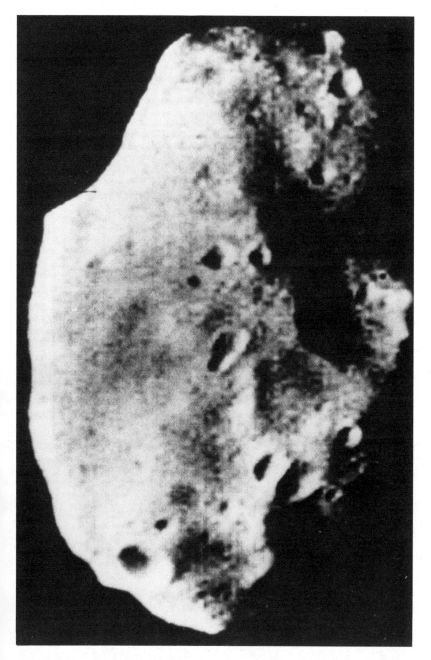

Fig. 6.6. Although there are no close-up photographs of asteroids, many probably look like Phobos, one of the irregularly shaped moons of Mars, here seen from a distance of 5540 km by the Mariner 9 spacecraft. Phobos is approximately 13.5 × 9.5 km and may be a captured asteroid. Note the large and small craters. (NASA photograph P12694.)

teroid belt, there are significant numbers elsewhere in the Solar System. The *Trojan asteroids* orbit the Sun at the stable Lagrangian points about 60° ahead and behind Jupiter (Fig. 6.5).[8] A few dozen Trojans, with mean diameters of 100–200 km, have been cataloged, although the probable population is several thousand. Asteroids also occur in orbits between the asteroid belt and Jupiter, and some, such as the Hidalgo asteroids, occupy orbits beyond Jupiter at distances of about 5.8 A. U. from the Sun.

Some of the most interesting are the planet-crossing asteroids (surveyed by Helin and Shoemaker, 1979), whose orbits intersect those of the inner planets. An estimated 10,000 ± 5,000 asteroids that occupy the innermost fringe of the asteroid belt travel paths that cross the orbit of Mars, and another 500, the Mars grazers, pass very near Mars' orbit. Dozens of these Mars crossers and Mars grazers have been observed.

Of some interest to Earthlings are the *Apollo asteroids*, which cross Earth's orbit. There are approximately 30 Apollo asteroids with diameters as great as 9 km that have been observed and cataloged, but the estimated population includes about 1,000 with diameters greater than 1 km. Because the Apollo asteroids cross Earth's orbit it is only a matter of time before one of them collides with our planet. Calculations show that, on the average, three such bodies with diameters of a kilometer or more strike Earth every million years and that a large asteroid of several kilometers diameter strikes the Earth every 40 Ma or so. For asteroids of 0.5 km or more in diameter and with sufficient energy to produce a crater at least 10 km in diameter the impact frequency is estimated at about one every 100,000 years (Wetherill and Shoemaker, 1982). These estimates of the frequency of asteroid impact are, in general, substantiated by the number of impact structures that have been identified on Earth. There are 116 known craters of Phanerozoic age (Grieve, 1982), but as three-fourths of the Earth is covered with water and as it is a virtual certainty that not all craters formed over the past 570 Ma have been preserved to the present, the number of impacts during Phanerozoic time must have been several times the number that have been identified.

Even though asteroids have yet to be visited and sampled by either man or machine, a great deal is known about their probable compositions. T. B. McCord and T. V. Johnson of the Massachusetts Institute of Technology and J. B. Adams of the Caribbean Research Institute applied the technique of the reflectance spectrophotometry, previously used on the Moon and Mars to determine the approximate

8. At these Lagrangian points the Trojan asteroids form equilateral triangles with the Sun and Jupiter. Because of the interaction of the gravitational fields of the Sun and Jupiter, the Lagrangian points are stable orbital positions.

mineralogic composition of those planetary surfaces, to Vesta, a belt asteroid of 515 km diameter that orbits the Sun at a mean distance of 2.36 A. U. (McCord, Johnson, and Adams, 1970). This method utilizes the spectra of reflected sunlight in the visible and *near infrared*. What these investigators found was that the spectrum of Vesta contains a strong infrared absorption band at about 0.9 μm that corresponds to the transition energies of electrons in the pyroxene mineral species pigeonite, a common constituent in basaltic achondrites, in particular the eucrites. Furthermore, the entire spectrum of Vesta closely matches the spectrum of eucrites as determined in the laboratory.

The surface compositions of more than 500 asteroids have now been determined by the reflectance spectrophotometric method, as summarized by Chapman, Morrison, and Zellner (1975), Chapman (1975), Gaffey and McCord (1978), Bowell et al. (1978), and McSween (1987). Some three dozen distinct types of spectra have been identified among the asteroids, and most of these can be matched with spectra from various meteorites (Fig. 6.7). Despite the variety of spectra, 90% of those that can be unambiguously classified fall into two major groups. The largest group, the C-type, comprises about 80% of all asteroids with diameters of more than 120 km. Their spectra match the laboratory spectra of carbonaceous chondrites. Most S-type asteroids have spectra resembling those of stony–iron meteorites and tend to have diameters of 100 to 200 km.

An interesting result of the compositional studies is that spectra matching those of the ordinary chondrites, the most common by far of meteorites, are rare among the asteroids—only three have been identified. In contrast, asteroids resembling the carbonaceous chondrites in composition appear to be very common even though such meteorites are rare. There are several explanations of why this is so. First, the carbonaceous chondrites are very fragile and most may not survive passage through the atmosphere, so in spite of their abundance in space few may reach Earth intact. Second, the frequency of meteorite types probably depends primarily on the composition of only a few asteroids rather than of the entire asteroid population, as discussed below. Interestingly, two of the three asteroids with spectra resembling those of ordinary chondrites are Apollo, or Earth-crossing, asteroids, whereas few Apollo objects have the spectra of carbonaceous chondrites.

Thus, spectral studies indicate that meteorites are very similar in mineralogic composition to asteroids, although they represent a statistically biased sample of compositional types. But how do they get here? Again, there is a relatively simple explanation. The orbits and velocities of individual asteroids differ greatly, and because of gravitational interactions with each other and with nearby planets, espe-

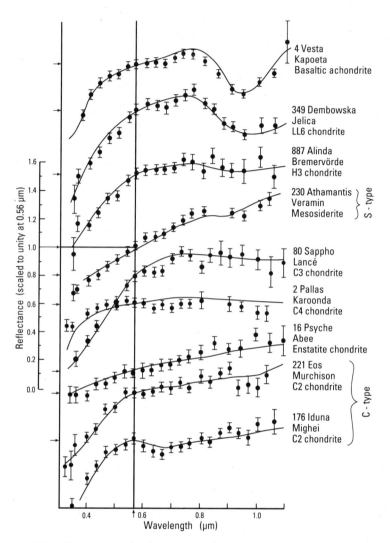

Fig. 6.7. The reflectance spectra of some asteroids (points with error bars, catalog number, and name) match closely the spectra of certain types of meteorites (lines, name of individual meteorite, and its type) as determined from laboratory measurements. These spectral similarities suggest that the mineralogical compositions may also be similar. Note the strong absorption band at about 0.95 μm in the spectrum of Vesta, which identifies the mineral pigeonite, a major constituent of the basaltic achondrites. Because of weathering some chondrites fall within the S-type spectra. (After Chapman, 1976.)

cially massive Jupiter, collisions are inevitable (Fig. 6.8). The resulting fragments are ejected into new orbits that differ from those of their parent bodies. Calculations (for example, Wetherill, 1985a) show that these initial new orbits would not be Earth-crossing, but resonance of the fragments with the gravitational field of Jupiter would result in

Fig. 6.8. This painting by W. K. Hartmann shows the collision of two asteroids. The larger one is heavily fractured while the small one disintegrates, ejecting fragments into different orbits. Resonance with the gravity field of Jupiter may inject such fragments into Earth-crossing orbits where they may collide with Earth, becoming meteors and meteorites. (From W. K. Hartmann, *Moons and Planets*, 2nd edition, © 1983 by Wadsworth, Inc. Reprinted by permission of publisher and author.)

some of the fragments being injected into Earth-crossing orbits, where they eventually would collide with our planet. Thus, the compositions of the contemporary meteorites may be largely governed by the compositions of the last few asteroids to collide and fragment, rather than by the asteroid population as a whole, and the fragments themselves must be relatively young even though the material of which they are made might be very old.

Collisions between asteroids also account for the dust in the inner solar system that is responsible for the phenomenon known as the zodiacal light. If the dust were not continuously resupplied it would disappear entirely within a few million years, because the very tiniest dust particles are blown out of the Solar System by radiation from the Sun and the larger dust particles spiral inward to the Sun because of the *Poynting–Robertson effect*.[9]

Cosmic ray exposure (CRE) *ages* confirm that meteorites are rather youthful fragments. CRE ages are determined by measuring the amounts of certain cosmogenic nuclides, such as ^{54}Mn, ^{3}He, ^{21}Ne, ^{38}Ar, and ^{41}K, that are produced by the interactions of cosmic-ray protons with target nuclides within the meteorites (Wasson, 1974: 122–32). Since *cosmic rays* penetrate rock to a depth of only a meter or less, CRE ages represent the length of time that the meteorite was exposed in space, either as a small, independent fragment or on the surface of its parent body.

CRE ages for iron meteorites range from less than 100 Ma to 2.3 Ga but show a pronounced mode near 650 Ma, an age that perhaps represents the breakup of a major iron meteorite parent body. In contrast, CRE ages for chondrites are invariably less than 60 Ma, and a large majority are less than 30 Ma. The CRE ages of the H chondrites, for example, cluster at about 4 Ma, and the majority of CRE ages for carbonaceous chondrites are less than 4 Ma. CRE ages for the achondrites are greater than those of the chondrites but still relatively young when compared to the CRE ages of iron meteorites. The majority of the aubrites, for example, have CRE ages near 40 Ma.

The tendency of the CRE ages of different meteorite types to cluster suggests that the CRE ages reflect discrete fragmentation events, i.e. asteroidal collisions. The difference in CRE ages between meteorite types may be due to differences in erosional or fragmentational susceptibility. Iron meteorites, for example, may have CRE ages older than do stone meteorites because the latter are more friable and easily fragmented than the former. If this is so, then stones should

9. Dust particles absorb energy from the Sun and reradiate it in all directions. This causes drag tangential to the orbit, which in turn decreases the particle's angular momentum and orbital radius. The outward pressure of radiation from the Sun is insufficient to overcome this effect except for the tiniest dust particles.

fragment more frequently and their CRE ages should be younger, which they are.

In summary, the evidence seems to suggest that meteorites are fragments of asteroids broken up by mutual collisions. The chondrites are chips from relatively primitive and small parent objects with diameters of only 100–200 km, whereas the differentiated meteorites probably come from larger parent bodies as much as 500–600 km in diameter. These larger bodies were heated early in their history, either by internal radioactivity from short-lived nuclides such as ^{26}Al or by a strong primitive solar wind, causing melting and differentiation into crust (basaltic achondrites), mantle (aubrites, diogenites), and metallic core (stony irons and irons). On some asteroids magma flowed onto the surface as basaltic lava flows (eucrites), and some of this surface material was converted into breccia (howardites) by subsequent bombardment over time. Successive collisions have now exposed asteroids to various depths. Thus, Vesta, with a surface composition like that of the eucrites, is probably largely intact and its surface covered with basaltic lava flows and related breccias. In contrast, Athamantis, with a surface composition resembling that of stony iron meteorites (Fig. 6.7) is probably the residual core of a once-larger body. Its surface, too, may be covered with impact breccia resembling the mesosiderites.

Although most meteorites probably sample asteroids, there is growing evidence that a few probably came from the Moon and Mars. At one time this was thought to be physically impossible because rocks would melt at the ejection energies required. Theoretical studies (Wetherill, 1984, Melosh, 1985), however, have shown that relatively unshocked pieces of rock can be ejected from planetary surfaces, and that some of this material can be injected into Earth-crossing orbits.

In January of 1982 an unusual 31-g meteorite designated ALHA-81005 was found in the Allan Hills region of Antarctica. It was described by Marvin (1983) and 17 companion papers in *Geophysical Research Letters* (see also Chaikin, 1983, Beatty, 1984, Marvin, 1984). Subsequent studies have identified this small specimen as a breccia from the lunar highlands. It contains clasts of anorthosite and of low-Ti mare basalt. Detailed petrographic, chemical, and isotope studies confirm a lunar origin. Since the discovery of ALHA81005, two additional small lunar meteorites have been identified, both collected in the Yamato Mountains of Antarctica.

It now seems probable that the shergottites, nakhlites, and chassignites (SNC) are pieces of the planet Mars. The evidence for such an exotic origin, though not entirely conclusive, is compelling (C. A.

Wood and Ashwal, 1981, R. H. Becker and Pepin, 1984, McSween, 1984, Basaltic Volcanism Study Project, 1981: 944). First, the crystallization ages of the SNC meteorites are much younger than those of other meteorites—about 1.3 Ga. This suggests that the parent body was a planet large enough and with sufficient internal heat to be actively volcanic long after the Moon and asteroids had cooled and become internally inactive. Second, most SNC meteorites are cumulate rocks whose mineralogies, textures, and rare-earth-element fractionation patterns indicate multi-stage magmatic histories that are too complex for small bodies like the asteroids but are consistent with an origin within a planet the size of the Moon or Mars. Third, the small grain size of the pyroxene phenocrysts in shergottites indicates segregation under the influence of a large, i.e. planetary, gravitational field. Finally, the abundances and isotopic compositions of rare gases in a severely shocked shergottite are nearly the same as the composition of the Martian atmosphere as analyzed by the Viking lander. Presumably, these gases were incorporated during the shock and melting that generated the meteorite and thus sample the ambient atmosphere of the parent body.

Another potential source of meteorites is comets, but for several reasons it is improbable that comets provide more than a small fraction of the meteorites, if any. One is that the orbits of meteorites and comets are greatly dissimilar. Except for a few short-period comets, the orbits of most comets are highly eccentric, with aphelia well beyond Jupiter. Another is that comets have diameters of less than about 20 km and are too small and of too low density to provide the chemical processing required to form the highly differentiated meteorites. Thus, if any meteorites did come from comets, they must be chondritic. Finally, the lifetime of an active (i.e. short-period) comet transiting the inner Solar System is on the order of only 10^4 years, whereas the most recent CRE ages found for meteorites are nearly twice as old, so it is improbable that any of the analyzed meteorites samples a presently active comet (Dodd, 1981: 315–16).

Although comets are unlikely sources of meteorites, on occasion one may strike the Earth. Probably the last such occurrence was the famed "Tunguska event," in which a fiery object struck the Earth near the Vanovara trading station on the Tunguska River, Siberia, on the morning of June 30, 1908. The resulting explosion devastated the forest for miles around, but no crater was formed and no trace of a meteorite has ever been found. It is likely that this event was caused by a small comet, though in their fascinating account of it, Baxter and Atkins (1976) concluded (not very credibly) that it may have been caused by an extraterrestrial spacecraft!

The Ages of Meteorites

Various types of radiometric ages have been obtained for meteorites, including (1) exposure ages, discussed in the previous section, (2) degassing ages, which indicate the time of last major impact heating, (3) metamorphic ages, which date the time of major reheating and metamorphism within the parent body, and (4) crystallization ages, which date the time of crystallization from a rock melt in or on the parent body or, for the more primitive meteorites, the time of condensation and aggregation from the Solar Nebula. For the purposes of this discussion the latter ages are the most interesting because they date the time that the meteorite first formed as a part of a homogeneous and distinct chemical system and represent an event near the time of formation of the Solar System. As mentioned above, however, many meteorites were heated and metamorphosed after crystallization in their parent body and have been subjected to subsequent brecciation as well as shock heating by collisions. In these meteorites, original crystallization ages have been partly or wholly obscured. As it turns out, however, the time separating crystallization and planetary metamorphism for most meteorites was very short—less than 100 Ma—so ages of metamorphism represent events very close to the time of formation of the Solar System. Thus, we will concentrate on the oldest ages of crystallization and metamorphism because they provide lower limits for the age of the Solar System. Such ages have been determined by the Rb–Sr, Sm–Nd, Lu–Hf, Re–Os, ^{40}Ar/^{39}Ar, and U–Pb methods. Discussion of the results of U–Pb measurements is deferred to Chapter 7.

As with rocks from the Earth and Moon, there are two basic approaches to the radiometric dating of meteorites. One involves analysis of different phases from the same meteorite and results in an internal isochron (Rb–Sr, Sm–Nd) or age spectrum (^{40}Ar/^{39}Ar). Such ages represent the time of the latest isotopic equilibration, which may be either initial crystallization or a later metamorphism, for the individual meteorite. The ^{40}Ar/^{39}Ar method is especially susceptible to thermal resetting, and many of the age spectra obtained by this technique probably date metamorphism or shocks. The Rb–Sr and, especially, the Sm–Nd systems are not so easily reset, and ages obtained by these methods probably reflect, in many instances, the ages of crystallization (achondrites) or metamorphism (ordinary chondrites).

The second approach involves analysis of whole-rock samples of a number of meteorites thought to be genetically related. Theoretically, these whole-rock isochrons may "see through" later metamorphisms or thermal events that redistribute isotopes within but not

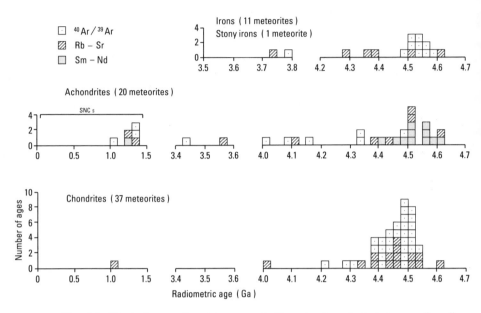

Fig. 6.9. Ninety-four radiometric ages of 69 meteorites. Included are all well-documented and valid isochron and age-spectrum ages with analytical uncertainties of 0.2 Ga or less at the 95% confidence level (two standard deviations) known to the author. The data were compiled from primary literature sources and the ages were converted, if necessary, to the decay constants of Table 3.1.

between the whole-rock samples. In both approaches the internal diagnostic characteristics of the isochron and age-spectrum methods provide information about the validity of the age results, although the methods do not identify whether the resulting age represents the time of original crystallization or of a later metamorphism. A rather substantial number of radiometric ages for meteorites and meteorite groups have been obtained by both approaches, and the results show unequivocally that the large majority of meteorites formed about 4.5–4.6 Ga.

Nearly 100 meteorites have been analyzed with the intent of determining internal radiometric ages. Of these, approximately 70 or so have yielded Rb–Sr or Sm–Nd isochrons or $^{40}Ar/^{39}Ar$ age spectra thought to indicate the time of initial crystallization or of a subsequent metamorphism (Fig. 6.9, Table 6.3). It is apparent from Figure 6.9 that a majority of the ages fall between 4.4 and 4.6 Ga. There are some meteorites with apparent radiometric ages between 3.4 and 4.4 Ga and a few with ages near 1 Ga (primarily the SNC meteorites), but they invariably show petrographic evidence of severe shock heating and metamorphism. Thus, these younger ages are thought to be a

TABLE 6.3

Meteorites Whose Radiometric Ages Exceed 4.4 Ga

Meteorite name	Type[a]	Material dated[b]	Method	Age (Ga)[c]	Source
CHONDRITES					
Allende	CV3	wr	Ar–Ar	4.52 ± 0.02	Dominik et al., 1978
		wr	Ar–Ar	4.53 ± 0.02	Dominik et al., 1978
		wr	Ar–Ar	4.48 ± 0.02	Dominik et al., 1978
		wr	Ar–Ar	4.55 ± 0.03	Jessberger et al., 1980
		wr	Ar–Ar	4.55 ± 0.03	Jessberger et al., 1980
		wr	Ar–Ar	4.57 ± 0.03	Jessberger et al., 1980
		wr	Ar–Ar	4.50 ± 0.02	Jessberger et al., 1980
		wr	Ar–Ar	4.56 ± 0.05	Jessberger et al., 1980
ALH-77288	H6	wr	Ar–Ar	4.50 ± 0.08	Kaneoka, 1983
Butsura	H6	wr	Ar–Ar	4.48 ± 0.06	Turner, Enright, and Cadogan, 1978
Guarena	H6	13	Rb–Sr	4.46 ± 0.08	Wasserburg, Papanastassiou, and Sanz, 1969
		wr	Ar–Ar	4.44 ± 0.06	Turner, Enright, and Cadogan, 1978
Kernouve	H6	wr	Ar–Ar	4.45 ± 0.06	Turner, Enright, and Cadogan, 1978
Menow	H4	wr	Ar–Ar	<u>4.48 ± 0.06</u>	Turner, Enright, and Cadogan, 1978
Mt. Browne	H6	wr	Ar–Ar	4.50 ± 0.06	Turner, Enright, and Cadogan, 1978
Ochansk	H4	wr	Ar–Ar	4.48 ± 0.06	Turner, Enright, and Cadogan, 1978
Queen's Mercy	H6	wr	Ar–Ar	4.49 ± 0.06	Turner, Enright, and Cadogan, 1978
Sutton	H6	wr	Ar–Ar	4.50 ± 0.04	Bogard, Husain, and Wright, 1976
Tieschitz	H3	7	Rb–Sr	<u>4.52 ± 0.05</u>	Minster and Allègre, 1979a
Wellington	H	wr	Ar–Ar	4.42 ± 0.02	Bogard, Husain, and Wright, 1976
ALH-761	L	wr	Ar–Ar	4.49 ± 0.10	Kaneoka, 1983
ALHA-77015	L3	wr	Ar–Ar	4.51 ± 0.10	Kaneoka, 1980
Barwell	L5	wr	Ar–Ar	4.45 ± 0.06	Turner, Enright, and Cadogan, 1978
Bjurbole	L4	wr	Ar–Ar	4.51 ± 0.08	Turner, 1969
Shaw	L6	wr	Ar–Ar	4.43 ± 0.06	Turner, Enright, and Cadogan, 1978
		wr	Ar–Ar	4.40 ± 0.06	Turner, Enright, and Cadogan, 1978
		wr	Ar–Ar	4.29 ± 0.06	Bogard and Hirsch, 1980
Jelica	LL6	5	Rb–Sr	4.42 ± 0.04	Minster and Allègre, 1981
Krahenberg	LL5	19	Rb–Sr	4.60 ± 0.03	Kempe and Muller, 1969
Olivenza	LL5	18	Rb–Sr	4.53 ± 0.16	Sanz and Wasserburg, 1969
		wr	Ar–Ar	4.49 ± 0.06	Turner, Enright, and Cadogan, 1978

(continued on following page)

TABLE 6.3 *(continued)*

Meteorite name	Type[a]	Material dated[b]	Method	Age (Ga)[c]	Source
Parnalee	LL3	5	Rb–Sr	4.53 ± 0.04	Hamilton, Evensen, and O'Nions, 1979
St. Severin	LL6	4	Sm–Nd	4.55 ± 0.33	Jacobsen and Wasserburg, 1984
		10	Rb–Sr	4.51 ± 0.15	Manhes, Minster, and Allègre, 1978
		wr	Ar–Ar	4.43 ± 0.04	Alexander and Davis, 1974
		wr	Ar–Ar	4.38 ± 0.04	Hohenberg et al., 1981
		wr	Ar–Ar	4.42 ± 0.04	Hohenberg et al., 1981
Soko Banja	LL4	10	Rb–Sr	4.45 ± 0.02	Minster and Allègre, 1981
Abee	E4	wr	Ar–Ar	4.52 ± 0.03	Bogard, Unruh, and Tatsumoto, 1983
Indarch	E4	9	Rb–Sr	4.46 ± 0.08	Gopolan and Wetherill, 1970
		12	Rb–Sr	4.39 ± 0.04	Minster, Ricard, and Allègre, 1979
St. Sauveur	E5	9	Rb–Sr	4.46 ± 0.05	Minster, Ricard, and Allègre, 1979
ACHONDRITES					
ALH-765	eucrite	4	Sm–Nd	4.52 ± 0.09	Nakamura, Tatsumoto, and Coffraut, 1983
Juvinas	eucrite	5	Sm–Nd	4.56 ± 0.08	Nakamura et al., 1976; Lugmair, 1974
		5	Rb–Sr	4.50 ± 0.07	Allègre et al., 1975
Moama	eucrite	3	Sm–Nd	4.46 ± 0.03	Jacobsen and Wasserburg, 1984
		4	Sm–Nd	4.52 ± 0.05	Hamet et al., 1978; Jacobsen and Wasserburg, 1984
Sierra de Mage	eucrite	3	Sm–Nd	4.41 ± 0.02	Lugmair, Scheinin, and Carlson, 1977
Stannern	eucrite	6	Sm–Nd	4.48 ± 0.07	Lugmair and Scheinin, 1975
Y-7308	howardite	wr	Ar–Ar	4.48 ± 0.06	Kaneoka, 1981
Y-75011	eucrite, basalt clast	9	Rb–Sr	4.50 ± 0.05	Nyquist et al., 1986
		7	Sm–Nd	4.52 ± 0.16	Nyquist et al., 1986
	eucrite, matrix	5	Rb–Sr	4.46 ± 0.06	Nyquist et al., 1986
		4	Sm–Nd	4.52 ± 0.33	Nyquist et al., 1986
Angra dos Reis	angrite	7	Sm–Nd	4.55 ± 0.04	Lugmair and Marti, 1977
		3	Sm–Nd	4.56 ± 0.04	Jacobsen and Wasserburg, 1984
IRONS					
Copiapo	IAB	si	Ar–Ar	4.50 ± 0.06	Niemeyer, 1979
Mundrabilla	IAB	si	Ar–Ar	4.57 ± 0.06	Niemeyer, 1979
		ol	Ar–Ar	4.54 ± 0.04	Kirsten, 1973
		pl	Ar–Ar	4.50 ± 0.04	Kirsten, 1973

TABLE 6.3 (*continued*)

Meteorite name	Type[a]	Material dated[b]	Method	Age (Ga)[c]	Source
Pitts	IAB	si	Ar–Ar	<u>4.54 ± 0.06</u>	Niemeyer, 1979
Pontlyfni	IAB	si	Ar–Ar	<u>4.52 ± 0.06</u>	Turner, Enright, and Cadogan, 1978
Woodbine	IAB	si	Ar–Ar	<u>4.57 ± 0.06</u>	Niemeyer, 1979
Colomera	IIE	10	Rb–Sr	<u>4.51 ± 0.04</u>	Sanz, Burnett, and Wasserburg, 1970
Weekeroo Station	IIE	4	Rb–Sr	4.39 ± 0.07	Evensen et al., 1979
		si	Ar–Ar	4.54 ± 0.03	Niemeyer, 1980
Enon	anomalous	si	Ar–Ar	4.59 ± 0.06	Niemeyer, 1983

[a]Table 6.1.

[b]Included are meteorites whose age uncertainties are 0.1 Ga or less, or that have been dated by more than one decay method. Numbers represent number of points in isochron; wr, whole rock; pl, plagioclase; ol, olivine; si, silicate inclusions.

[c]All ages calculated with the decay constants listed in Table 3.1. Errors are at the 95% confidence level (two standard deviations). Ages underlined are shown as figures. Ages are based on either internal isochron (Rb–Sr, Sm–Nd) or $^{40}Ar/^{39}Ar$ age-spectrum (Ar–Ar) methods.

consequence of impact phenomena that occurred long after initial crystallization or planetary metamorphism.

Only the SNC meteorites appear to be truly "young." The three dated nakhlites (Nakhla, Lafayette, and Governador Valadares) have Rb–Sr, Sm–Nd, and $^{40}Ar/^{39}Ar$ ages of 1.22 to 1.34 Ga, which are thought to record their time of crystallization as lava on the SNC parent planet, probably Mars. Numerous attempts to date Shergotty, which is highly shocked, have yielded only age spectra and isochron plots that show evidence of strong disturbance and resetting (Podosek, 1973, Gale, Arden, and Hutchison, 1975, Bogard and Husain, 1977, Papanastassiou and Wasserburg, 1974, Nyquist et al., 1979b, Bogard, Husain and Nyquist, 1979, Nakamura et al., 1982). Chassigny, the only chassignite, has not been dated because it is composed entirely of olivine and contains no datable minerals.

Because of their scarcity, small size, paucity of datable material, and fragile nature, individual carbonaceous chondrites have proved to be virtually undatable and their precise age of formation is unknown. Several attempts to determine internal Rb–Sr isochrons for these meteorites have not been successful because their Rb–Sr isotopic systems have been disturbed and the data scatter rather than fall on an isochron (Mittlefehldt and Wetherill, 1979, Nagasawa and Jahn, 1976, Gray, Papanastassiou, and Wasserburg, 1973). Some inclusions from Allende, a brecciated CV3 carbonaceous chondrite, however, have provided a number of $^{40}Ar/^{39}Ar$ age spectra indicating ages ranging from 4.48 to 4.57 Ga (Table 6.3). These ages, which cluster around

4.53 Ga, are thought to record a time of major resetting of the K–Ar system early in the history of Allende (Jessberger et al., 1980). The exact nature of this event is not known but the likely possibilities include the shock that formed Allende or metamorphism of the individual inclusions within Allende's parent body. Even though Allende's formation age is not precisely known, there is evidence that Allende chondrules are some of the earliest material to condense from the Solar Nebula (Gray, Papanastassiou, and Wasserburg, 1973). This conclusion is based on the observation that the $^{87}Sr/^{86}Sr$ ratios in these very low-rubidium phases are among the lowest known, indicating an early separation of the chondrules from the Solar Nebula.

Only one whole-rock isochron for a carbonaceous chondrite has been obtained (Table 6.4), but the resulting age of 4.37 ± 0.34 Ga has a large uncertainty. Results for many carbonaceous chondrites, however, do fall on or very near the whole-rock isochrons formed by analyses of other types of chondrites, and so there is little reason to doubt that they are ancient relics of the Solar System and are about 4.5–4.6 Ga in age (Kaushal and Wetherill, 1970, Jacobsen and Wasserburg, 1984).

In contrast to the carbonaceous chondrites, the ordinary and enstatite chondrites have provided a wealth of radiometric age data, in the form of both internal isochrons and age spectra for individual meteorites (Table 6.3) and whole-rock isochrons for the different chondrite groups (Table 6.4, Fig. 6.10). The majority of well-dated ordinary chondrites have Rb–Sr and $^{40}Ar/^{39}Ar$ ages between 4.40 and 4.55 Ga. Because of their low content of rare earth elements, chondritic meteorites are very difficult to date with the Sm–Nd method and so Sm–Nd ages for these meteorites are rare.

Most of the ordinary chondrites have been dated by only a single method, primarily $^{40}Ar/^{39}Ar$, but there are a few exceptions. Guarena and Olivenza give concordant Rb–Sr and $^{40}Ar/^{39}Ar$ ages of about 4.45 and 4.50 Ga, respectively (Table 6.3). St. Severin has been dated by Rb–Sr, Sm–Nd, and $^{40}Ar/^{39}Ar$ methods, and the ages obtained with these three decay schemes are not different within the limits of the analytical errors; they have a weighted mean of 4.41 Ga.

The data in Table 6.3 indicate that there are neither obvious age differences between the meteorites of the various chondrite groups (H, L, LL, and E) nor between ages measured by Rb–Sr and $^{40}Ar/^{39}Ar$ methods (Fig. 6.9). Both radiometric clocks appear to record an event or events about 4.45 to 4.50 Ga. But was this event initial condensation and aggregation or subsequent metamorphism? Dodd (1981: 166) has emphasized that most of the dated meteorites are of petrologic types 4–6 and that ages of meteorites of the less metamorphosed and

TABLE 6.4

Whole-Rock Isochron Ages of the Different Meteorite Groups

Meteorite class or group[a]	Number of meteorites	Method	Age (Ga)[b]	Source
Chondrites (CM, CV, H, L, LL, E)	13	Sm–Nd	4.21 ± 0.76	Jacobsen and Wasserburg, 1984
Carbonaceous chondrites	4	Rb–Sr	4.37 ± 0.34	Murthy and Compston, 1965
Chondrites (undisturbed H, LL, E)	38	Rb–Sr	<u>4.50 ± 0.02</u>	Minster, Birck, and Allègre, 1982
Chondrites (H, L, LL, E)	50	Rb–Sr	4.43 ± 0.04	Wetherill, 1975
H chondrites (undisturbed)	17	Rb–Sr	4.52 ± 0.04	Minster, Birck, and Allègre, 1982; Minster and Allègre, 1979a
H chondrites (undisturbed)	16	Rb–Sr	4.54 ± 0.11	Mittlefehldt and Wetherill, 1979
H chondrites	15	Rb–Sr	4.59 ± 0.06	Kaushal and Wetherill, 1969
L chondrites (relatively undisturbed)	6	Rb–Sr	4.44 ± 0.12	Gopolan and Wetherill, 1971
L chondrites	5	Rb–Sr	4.38 ± 0.12	Gopolan and Wetherill, 1968
LL chondrites (undisturbed)	13	Rb–Sr	4.49 ± 0.02	Minster, Birck, and Allègre, 1982; Minster and Allègre, 1981
LL chondrites	10	Rb–Sr	4.46 ± 0.06	Gopolan and Wetherill, 1969
E chondrites (undisturbed)	8	Rb–Sr	4.51 ± 0.04	Minster, Birck, and Allègre, 1982; Minster, Ricard, and Allègre, 1979
E chondrites	8	Rb–Sr	4.44 ± 0.13	Gopolan and Wetherill, 1970
Eucrites (polymict)	23	Rb–Sr	4.53 ± 0.19	Wooden et al., 1983
Eucrites	11	Rb–Sr	4.44 ± 0.30	Basaltic Volcanism Study Project, 1981
Eucrites	11	Rb–Sr	4.48 ± 0.14	Birck and Allègre, 1978
Eucrites	7	Rb–Sr	4.30 ± 0.25	Papanastassiou and Wasserburg, 1969
Eucrites	13	Lu–Hf	<u>4.57 ± 0.19</u>	Patchett and Tatsumoto, 1980; Tatsumoto et al., 1981
Diogenites	5	Rb–Sr	4.45 ± 0.18	Birck and Allègre, 1981
Irons (+ iron from St. Severin)	8	Re–Os	<u>4.57 ± 0.21</u>	Luck et al., 1980

[a]Table 6.1.
[b]All ages calculated with the decay constants listed in Table 3.1. Errors are at the 95% confidence level (two standard deviations). Ages underlined are shown as figures.

less disturbed type 3 are rare. As he pointed out, the change from un-equilibrated type 3 to highly metamorphosed types 6 and 7 involves virtually complete recrystallization of the major mineralogical con-stituents and it is difficult to see how such changes could not affect, i.e. reset, the Rb–Sr isotopic system. It is even more likely that the ^{40}Ar/^{39}Ar "clock" would be reset by the metamorphism required to produce meteorites of petrologic types 4–6. Thus, it appears probable that the ages for individual ordinary chondrites listed in Table 6.3

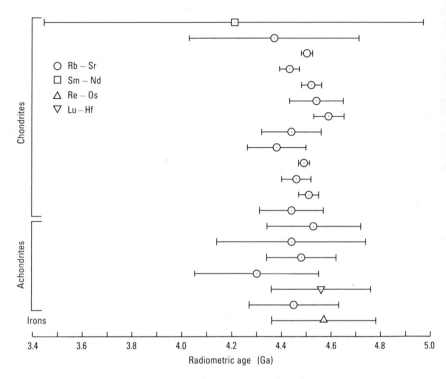

Fig. 6.10. Whole-rock isochron ages for meteorites. (Data from Table 6.4.)

record metamorphism and subsequent cooling, perhaps within their parent bodies, rather than the times of condensation and aggregation from the Solar Nebula. The consistency of the older internal meteorite ages indicates that metamorphism occurred within a relatively short period of time about 4.4 to 4.5 Ga ago. M. C. Enright and G. Turner (1977), of Sheffield University, have examined the $^{40}Ar/^{39}Ar$ data of 11 ordinary chondrites whose age spectra show patterns of minimal Ar loss and concluded that the results are consistent with a single cooling age (perhaps following metamorphism) of 4.465 ± 0.030 Ga.

As mentioned above, one possible way to determine formation ages for meteorites is to use whole-rock isochrons rather than internal isochrons. This method has the advantage that the whole rock may remain a closed system during shock and thermal metamorphism even though the individual mineral grains do not. The first comprehensive study of this type for the chondrite meteorites was by G. W. Wetherill and his students K. Gopolan and S. K. Kaushal of the University of California, Los Angeles. In a series of classic papers (Kaushal and Wetherill, 1969, Gopolan and Wetherill, 1968, 1969, 1970,

1971), these investigators systematically determined whole-rock iso-chrons for more than 35 H, L, LL, and E chondrites and found that the ages, which ranged from 4.38 ± 0.12 Ga for 5 L chondrites to 4.59 ± 0.06 Ga for 15 H chondrites (Table 6.4), were the same within the analytical errors and gave a combined chondrite isochron of 4.43 ± 0.04 Ga (Wetherill, 1975: 310–12).

More recently, a group-by-group Rb–Sr study of the chondrites has been repeated by J.-F. Minster and his colleagues at the University of Paris using modern, high-precision techniques (Minster and Allègre, 1979a, b, 1981, Minster, Ricard, and Allègre, 1979, Minster, Birck, and Allègre, 1982). They found that, with the exception of a few highly disturbed specimens, the H, LL, and E chondrites defined precise isochrons with indicated ages as follows:

17 H chondrites	4.518 ± 0.039 Ga
13 LL chondrites	4.486 ± 0.020 Ga
8 E chondrites	4.508 ± 0.037 Ga

Because these ages are not different within the analytical errors and because the initial ratios are also identical, Minster and his colleagues argued that these three groups of chondrites can be treated as a single isotopic system with a common age. The resulting isochron age for all 38 chondrites is 4.498 ± 0.015 Ga (Fig. 6.11), an age that probably represents the approximate time of condensation and aggregation of these objects from the Solar Nebula. The precision with which all of these specimens fit the same isochron is quite remarkable and indicates that the ordinary and enstatite chondrites formed either simultaneously or within a period of 10–20 Ma or less. Data from the L chondrites do not fall on a precise isochron because most are highly disturbed. The few specimens that are relatively undisturbed, however, fall very close to the 4.50-Ga isochron defined by the other chondrites and so are probably of similar age.

Fewer radiometric ages have been determined for individual achondrites than for chondrites (Fig. 6.9, Table 6.3), largely because far fewer achondrites have been found but partly because their low content of alkali metals, including rubidium, makes the application of Rb–Sr dating extremely difficult. Of the 20 or so meteorites for which high-quality radiometric data have been obtained, nearly half have Rb–Sr or $^{40}Ar/^{39}Ar$ ages of less than 4.4 Ga, and their K–Ar gas retention ages are also uniformly low (Basaltic Volcanism Study Project, 1981: 940).[10] Except for the SNC meteorites, discussed above, these relatively low ages for the individual achondrite meteorites probably

---

10. Section 7.2 of this volume is an excellent synthesis of the radiometric data for the achondrites.

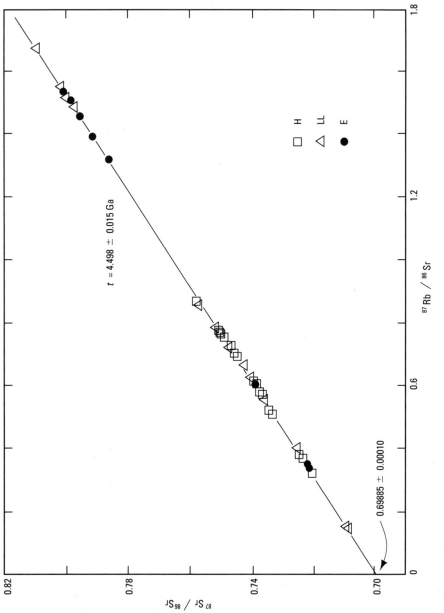

Fig. 6.11. Rb–Sr whole-rock isochron for 38 undisturbed H, LL, and E chondrites. (After Minster, Birck, and Allègre, 1982.)

represent postformation impact metamorphism (recall that most of the achondrites are brecciated). Fortunately, the major minerals in the achondrites, pyroxene and calcic plagioclase, though very low in rubidium are high in the rare earth elements and so the Sm–Nd method can be applied with considerable success. On the whole, the Sm–Nd isochron ages of individual achondrites are older than their Rb–Sr ages, reinforcing the conclusion that the younger of the Rb–Sr ages are the result of postcrystallization impact events (SNC meteorites excepted). The older values obtained by the Sm–Nd method are not surprising in view of the observation that the Sm–Nd isotopic system tends to be more resistant to postformational heating than either the Rb–Sr or ^{40}Ar/^{39}Ar systems.

Most of the older, well-dated achondrites are eucrites, and it is likely that the Sm–Nd ages represent the approximate times of eruption and cooling of the basaltic eucrite material on the parent body. This supposition is reinforced by the Rb–Sr and Lu–Hf isochrons derived from whole-rock samples of the eucrites (Table 6.4, Fig. 6.12), which are generally concordant with the Sm–Nd ages of individual eucrites. A Rb–Sr whole-rock isochron for five diogenites yields an age (4.45 ± 0.18 Ga), similar to those obtained for the eucrites. Only one eucrite, Juvinas, has been dated precisely by both Rb–Sr (4.50 ± 0.07 Ga) and Sm–Nd (4.56 ± 0.08 Ga) methods, and the ages agree within analytical uncertainty.

One of the oldest achondrites is the angrite Angra dos Reis, which has a Sm–Nd age of 4.55 ± 0.04 Ga. As we shall see in the next section, this meteorite has the lowest initial ^{87}Sr/^{86}Sr ratio of any of the achondrites and may, in fact, be one of the oldest meteorites yet studied.

The Sm–Nd ages for individual eucrites and the Rb–Sr and Lu–Hf whole-rock isochrons for eucrites and diogenites show that these meteorites formed about 4.5 Ga and that their radiometric ages are indistinguishable from those determined for the chondrites. As both the eucrites and diogenites are the products of magma formation and crystallization on the parent body, the results indicate that melting and eruption on the eucrite/diogenite parent body occurred very shortly after condensation and aggregation from the Solar Nebula, perhaps within a period of time as short as a few tens of millions of years.

Attempts in the late 1950's and 1960's to date iron meteorites by applying the K–Ar method to the metallic phases of these objects produced a number of conventional K–Ar ages that ranged from 6 to 13 Ga (Stoenner and Zähringer, 1958, Fisher, 1965, Müller and Zährin-

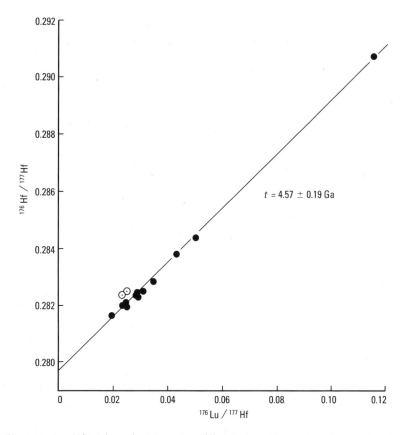

Fig. 6.12. Lu–Hf isochron for 13 eucrites (filled circles). The open circles are data for the antarctic eucrite ALHA77302, which do not fall on the isochron defined by the other eucrites. The age was calculated by using the experimentally determined value for the half-life of ^{176}Lu (Table 3.1). (After Patchett and Tatsumoto, 1980, with additional data from Tatsumoto, Unruh, and Patchett, 1981.)

ger, 1966, Rancitelli et al., 1967). These early K–Ar results were quite puzzling because even then the evidence from Rb–Sr and Pb–Pb methods strongly indicated that the oldest of the meteorites, including the irons, were about 4.5 Ga. These inconsistent and anomalously high K–Ar ages were the result of the loss of potassium due to terrestrial weathering as well as to the exceedingly low abundances and nonuniform distribution of K and Ar in the metallic minerals, inadequate analytical techniques, and large interferences from cosmic rays.

The key to the measurement of reliable ages for iron meteorites was to analyze the small silicate inclusions found in many of them rather than the metal phases, which contain only exceedingly small quantities of the alkali metals potassium and rubidium. This approach, using the Rb–Sr method, was first applied to iron meteorites by D. S. Burnett and G. J. Wasserburg of the California Institute of Technology in the late 1960's (Wasserburg, Burnett, and Frondel, 1965, Burnett and Wasserburg, 1967a,b). While their methods then lacked the precision and sensitivity of today's, Wasserburg and Burnett were able to show that several iron meteorites were approximately the same age as the chondrites and achondrites and not several billion years older as the early K–Ar results led some to suggest.

To date there are only a few iron meteorites for which precise Rb–Sr ages have been measured. These include Colomera and Weekeroo Station, which have calculated ages of 4.51 and 4.39 Ga, respectively (Table 6.3). Kodaikanal has a Rb–Sr age of only 3.73 ± 0.10 Ga and a conventional K–Ar age of 3.49 ± 0.10 Ga (Burnett and Wasserburg, 1967b, Bogard et al., 1968).

There are a number of precise ^{40}Ar/^{39}Ar ages for iron meteorites, mostly as a result of a careful and systematic study by S. Niemeyer (1979, 1980, 1983) of the University of California at Berkeley, who obtained age spectra for silicate inclusions in eight type I and type II irons. Seven of these meteorites gave excellent high-temperature plateaus (two of which are illustrated in Fig. 6.13), six with indicated ages from 4.50 to 4.59 Ga (Table 6.3). The seventh, Netschaevo, gave a high-temperature plateau age of 3.79 ± 0.06 Ga, in agreement with the Rb–Sr age of Kodaikanal. This suggests that there may have been at least two distinct formation times for iron meteorite parent bodies, one very early in the history of the solar system and another about 3.7–3.8 Ga. Alternatively, the isotopic systems in Kodaikanal and Netschaevo may have been reset by some postformational event, perhaps collision and destruction of their parent body, about 3.8 Ga.

Because of the paucity of alkali metals and rare earth elements in the metal phases of iron meteorites and because those with silicate inclusions are heterogenous, the iron meteorites do not lend themselves to analysis by Rb–Sr or Sm–Nd whole-rock techniques. The Re–Os method, however, has been applied to the metallic phases of seven irons (Fig. 6.14, Table 6.4). The result, derived with the experimentally determined value for the half-life of ^{187}Re, is an isochron age of 4.57 ± 0.21 Ga, which is in agreement with the Rb–Sr and ^{40}Ar/^{39}Ar ages of individual iron meteorites. Metal inclusions in the St. Severin LL chondrite also fall on the Re–Os isochron for iron meteorites, indicating that the irons and the chondrites (at least St. Severin) formed at the same time and from parent material with the same osmium isotopic composition.

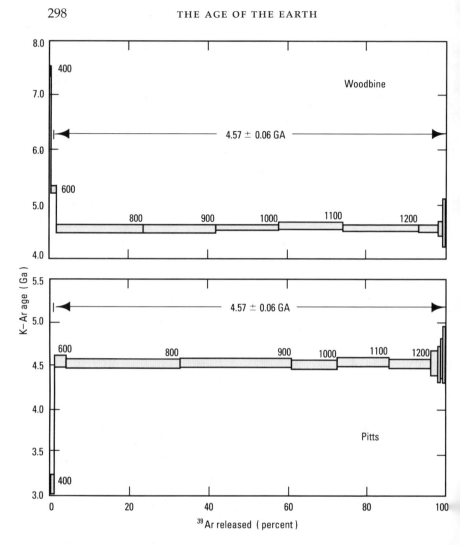

Fig. 6.13. ^{40}Ar/^{39}Ar age spectra for silicate inclusions in two iron meteorites. (After Niemeyer, 1979.)

In summary, the radiometric data show that the majority of iron meteorites formed about 4.5–4.6 Ga. Presumably, the ages represent the time of differentiation and the formation of the metal cores in the iron meteorites' parent bodies. The ages of the oldest iron meteorites are indistinguishable from the ages of the oldest chondrites and achondrites, suggesting that the material of most meteorites formed within a period of less than 100 Ma at about 4.5–4.6 Ga.

But is there a "best" age for meteorite formation and is there some method of distinguishing between the formation times of mete-

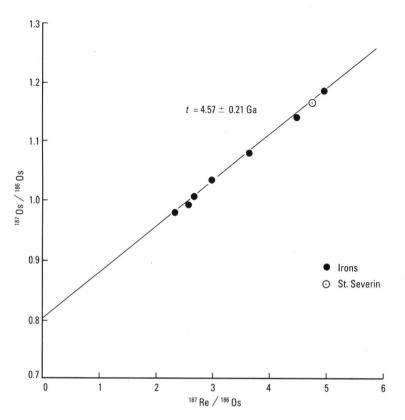

Fig. 6.14. Re–Os isochron for the metal phases of iron meteorites and the St. Severin LL chondrite. The age was calculated by using the experimentally determined value for the half-life of ^{187}Re. (After Luck, Birck, and Allègre, 1980.)

orites? The answer to the first question will have to wait until Chapter 7 but the second will be briefly discussed in the following section.

Formation Intervals

In chapters 4 and 5 we discussed Sm–Nd and Rb–Sr model ages for some terrestrial and lunar samples and the use of such ages in estimating the time of separation of magma from a mantle reservoir or the time of crustal differentiation. Similar calculations can be done for meteorites, but the resulting ages are usually quite close to the ages determined by isochron methods, indicating that the formation of the various meteorite types occurred very close to the time of aggregation of the parent bodies from the Solar Nebula about 4.5–4.6 Ga. Because so much of the formational history of meteorites seems to have occurred within a very short interval of time at about 4.5 Ga, neither the usual model nor isochron ages have the resolution to determine clearly the timing of these early events. But are there ways to estimate the order of origin of objects in the early Solar System and the elapsed time between the onset of aggregation and the formation of meteorites? The answer is a qualified "yes."

Calculations analogous to the usual Rb–Sr model ages can be used to estimate what are called *formation intervals*, i.e. the differences between the formation times of individual meteorites and meteorite types, with a resolution of a few Ma. The method was first suggested and demonstrated by D. A. Papanastassiou and G. J. Wasserburg (1969) of the California Institute of Technology following a significant improvement in the precision of analytical methods. The idea behind the calculation of meteorite formation intervals is illustrated in Figure 6.15. The strontium isotopic composition ($^{87}Sr/^{86}Sr$) of an initial reser-

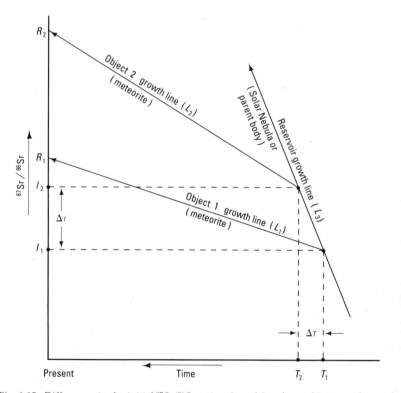

Fig. 6.15. Differences in the initial $^{87}Sr/^{86}Sr$ ratios, I_1 and I_2, of two objects can be used to determine the formation interval, ΔT, which is the time between their times of separation (T_1 and T_2) from a common reservoir, with a resolution of a few million years. The slope of a growth line (L_1–L_3) is a function of the Rb/Sr ratio in the reservoir or object. (After Minster, Birck, and Allègre, 1982.)

voir, such as the Solar Nebula or a meteorite's parent body, will change with time along a growth line, L_3, whose slope is a function of the Rb/Sr ratio of the reservoir. If an object, say a meteorite, with a lower Rb/Sr ratio is formed from this reservoir at time T_1, then its initial Sr isotopic composition, I_1, will be that of the reservoir at T_1 and subsequently will change with time along a new growth line (L_1) of lesser slope until at the present it will have a $^{87}Sr/^{86}Sr$ ratio of R_1. A second object (meteorite) formed at T_2 will have an initial ratio of I_2, growth line L_2, and a present-day ratio of R_2. The times of separation of the objects from the reservoir, i.e. their formation times, T_1 and T_2, can be determined by the internal isochron method to a precision of at best perhaps 20 to 60 Ma (see Table 6.3), but this precision is inadequate to distinguish formation intervals, $\Delta T = T_1 - T_2$, of only a few million years.

The approach suggested by Papanastassiou and Wasserburg was to use the difference in the initial $^{87}Sr/^{86}Sr$ ratios between meteorites (ΔI in Fig. 6.15) to calculate ΔT rather than the difference in isochron ages, $T_1 - T_2$. Their reasoning was that for meteorites with phases that have very low Rb/Sr ratios the initial $^{87}Sr/^{86}Sr$ ratio is relatively insensitive to the slope of the isochron and thus the error in the initial ratio is several times smaller than the error in the isochron age. Accordingly, ΔT calculated from ΔI should be more precise and have better resolution than ΔT calculated from $T_1 - T_2$. To calculate ΔT from ΔI, however, the Rb/Sr ratio of the reservoir must be estimated or assumed because there is no direct way to measure the Rb/Sr ratio of the Solar Nebula or of a meteorite parent body.

The key to the use of initial ratios to determine formation intervals was the development, also pioneered by the Cal Tech group, of high-precision techniques for measuring strontium isotopes. With these methods differences in $^{87}Sr/^{86}Sr$ of less than 0.01% can be resolved. For a reservoir with Rb/Sr equal to the solar value of 0.65, a value thought to be a reasonable estimate of Rb/Sr in the Solar Nebula, the change in $^{87}Sr/^{86}Sr$ over time is about 0.004%/Ma, and the method thus has a potential resolution of about 2–3 Ma.

Papanastassiou and Wasserburg applied the initial Sr isotopic ratio method first to seven basaltic achondrites, chosen because these meteorites are very depleted in Rb, having Rb/Sr ratios typically 200–300 times lower than the solar value. They found that the data for these meteorites fit an isochron with an age of 4.30 ± 0.25 Ga and an initial $^{87}Sr/^{86}Sr$ ratio of 0.698990 ± 0.000047, which they named BABI for Basaltic Achondrite Best Initial. Assuming an age of 4.5 Ga for the basaltic achondrites, the Cal Tech investigators calculated model initial strontium ratios for each of the meteorites analyzed. They found that the highest initial ratio of 0.699047 ± 0.000118 (Nuevo Laredo) was within 0.000124 of the lowest of 0.0698928 ± 0.000137 (Stannern), where the errors include the uncertainty in the assumed formation age. If these meteorites were derived from material with Rb/Sr equal to the solar value of 0.65, then all of the basaltic achondrites analyzed must have been formed within a time period ΔT of only 1.6 Ma. Because the basaltic achondrites are differentiated, however, it is virtually certain that they did not form directly from the Solar Nebula but from a parent body. If the parent body had a Rb/Sr ratio equal to the value for chondrites of about 0.25, then the calculated formation interval ΔT is 4.0 Ma. Thus despite the value chosen for the Rb/Sr ratio of the reservoir, the data indicate that the basaltic achondrites formed within a very short period of time.

This type of formation-interval analysis has been applied to several meteorites and meteorite groups including chondrules from the carbonaceous chondrite Allende, which give an initial $^{87}Sr/^{86}Sr$ ratio (known as ALL) of 0.69877 ± 0.00002 and is the lowest value yet found (Gray, Papanastassiou, and Wasserburg, 1973), refractory white inclusions from the angrite Angra dos Reis, which give an initial ratio (known as ADOR) of 0.69889 ± 0.00002 (Wasserburg et al., 1977), the chondrites as a group, which give an initial ratio of 0.69885 ± 0.00010 (Minster, Birck, and Allègre, 1982), and the Moon, which has an estimated initial ratio of about 0.69870 (Gray, Papanastassiou, and Wasserburg, 1973). These important results indicate that the events leading to the formation of the meteorites and planets, including aggregation from the Solar Nebula, formation of planetoids as discrete bodies, and differentiation on the basaltic achondrite parent body occurred within a period of only about 10 Ma.

Another method for estimating formation intervals involves the use of a radioisotope whose half-life is so short that it no longer exists in nature. This isotope is ^{129}I (iodine), which decays to ^{129}Xe (xenon) with a half-life of 17 Ma. The method was pioneered by J. H. Reynolds and his students at the University of California at Berkeley (see Podosek, 1970, Jordan, Kirsten, and Richter, 1980, Drozd and Podosek, 1977, and Dodd, 1981: 173–77, for more detailed discussions of methods and results). The basic idea is relatively simple. The creation of new ^{129}I ceased at the end of galactic nucleosynthesis, after which its abundance decreased systematically as a result of decay to ^{129}Xe. Meteorites will incorporate iodine at formation but tend to exclude xenon because it is an inert gas and does not react with other elements to form compounds, including minerals. Any xenon formed within the meteorite by the decay of ^{129}I, however, will be retained much as ^{40}Ar from the decay of ^{40}K is retained. The longer the elapsed time between meteorite formation and the cessation of nucleosynthesis, the less ^{129}I was available for incorporation into the meteorite and the less radiogenic ^{129}Xe it will contain today. As in the other radiometric methods it is convenient to normalize the quantities of the parent and daughter isotopes by dividing by the quantity of a stable isotope of the parent element, in this case ^{127}I. The elapsed time between nucleo-synthesis and the formation of a meteorite could be calculated if the ratio of ^{129}I to ^{127}I at the end of nucleosynthesis was known. This ratio is not known well, but calculations using estimates of the ratio yield an elapsed time between nucleosynthesis and meteorite formation of a few hundred million years or less (Podosek, 1970: 360–61).

Because the ^{129}I/^{127}I ratio at the end of nucleosynthesis is not well known, the I–Xe method is usually used to determine differences in formation times (formation intervals) between meteorites, as such cal-culations do not depend on knowledge or assumption of an initial ratio. The results of I–Xe analyses on several dozen meteorites (Dodd, 1981: 175) generally confirm that all of the major meteorite groups formed within a few tens of millions of years. It is of particular inter-est to note that the I–Xe formation intervals for the carbonaceous chondrites, thought to have formed during the early stages of aggrega-tion from the Solar Nebula, fall within a time span of only 12 Ma.

There are numerous uncertainties and difficulties in interpreting results from both the Sr initial ratio and the I–Xe methods. For ex-ample, ^{87}Sr/^{86}Sr and ^{129}I/^{129}Xe results on the same meteorites and ^{129}I/^{129}Xe formation intervals on different phases within the same me-teorite do not always agree, so the precise chronology of the earliest events in the Solar System cannot yet be determined with any degree of certainty. Also it is not entirely clear what sorts of events the meth-ods record. These are complicated subjects that will not be further discussed here. The results from these methods, however, generally indicate that the early history of the Solar System, from initial ag-gregation to the formation of the basaltic achondrites on their parent bodies, occurred primarily within a period of a few tens of millions of years. Dating of individual meteorites and of meteorite groups by Rb–Sr, Sm–Nd, and ^{40}Ar/^{39}Ar methods indicates that these events occurred about 4.5 Ga. This "age" for the Solar System, or more pre-cisely for the aggregation of solid entities from the Solar Nebula, has been determined more precisely and related to the "age of the Earth" by the study of lead isotopes, a topic that has been studiously avoided in this chapter. Lead isotopes and their bearing on the age of the Earth and Solar System are the subject of Chapter 7.

Summary

Meteorites are among the most primitive and oldest rocks in the Solar System. Their abundance, compositions, and orbits indicate that nearly all are fragments of asteroids, although a few probably came from Mars and the Moon. Meteorites collide with the Earth because impacts and the gravitational effects of Jupiter inject them into Earth-crossing orbits.

The chondrites are the most primitive of the meteorites. They are aggregates of silicate minerals and were formed by accretion of material from the Solar Nebula. The achondrites, stony–iron, and iron meteorites are secondary and were formed from (probably chondritic) parent material by processing in the cores and mantles of the larger asteroids and, in the case of the achondrites, eruption as lavas onto their surfaces.

The majority of meteorites are old, with individual Rb–Sr, Sm–Nd, and $^{40}Ar/^{39}Ar$ ages of 4.4–4.6 Ga. Internal and whole-rock isochrons for chondrites, achondrites, and iron meteorites show that the major meteorite types were formed about 4.5–4.6 Ga. The similarity of their ages, as well as formation-interval ages determined from I–Xe systematics and Sr isotope initial ratios, indicates that at most a few tens of millions of years passed between the formation of the primitive chondrites and the differentiated meteorites.

CHAPTER SEVEN

Isotopes of Lead: The Hourglass of the Solar System

Liquids and gases forget, but rocks remember.   J. A. O'KEEFE[1]

Up to this point any substantial discussion of lead isotopes and their bearing on the ages of meteorites and the Earth has been largely avoided. This is not because lead isotopes are unimportant but, on the contrary, because they are so important that they deserve a chapter of their own.

Ask any geologist how old the Earth is and the odds are very good that he or she will provide an answer very close to 4.54 Ga. This number comes not from direct radiometric dating of lunar rocks or meteorites but from a model that describes the evolution of lead isotopes in isolated systems, a model that can be applied quite successfully to meteorites, the Earth, and the Solar System. The basis for the model was formulated by E. K. Gerling of the Radium Institute of the Academy of Sciences of the U.S.S.R. in 1942, by Arthur Holmes of the University of Edinburgh in 1946, and by F. G. Houtermans of the University of Göttingen in 1946. Although neither Gerling, Holmes, nor Houtermans was successful in producing a valid age for the Earth, their general approach, known today as the *Holmes–Houtermans model*, was refined by later workers and provides a value today that geologists generally accept as the "age of the Earth." It is this Holmes–Houtermans model and its application to the problem of the Earth's age that is the subject of this chapter.

Before we inquire into the reasoning and results of Gerling, Holmes, and Houtermans and how their model was refined by suc-

---

1. Personal communication, 1987.

ceeding workers, it will be useful to consider how the proportions of
lead isotopes change with time in a Holmes–Houtermans uranium–
lead system.

The Growth of Lead Isotopes in the
Holmes–Houtermans System

Consider a system that contains uranium and originated at some
time T years ago with a uniform initial lead isotopic composition represented by

$$a_0 = (^{206}Pb/^{204}Pb)$$

$$b_0 = (^{207}Pb/^{204}Pb)$$

Over time, the decay of ^{238}U and ^{235}U will increase the proportions of
^{206}Pb and ^{207}Pb relative to ^{204}Pb so that at some later time t years ago the
lead-isotopic composition of the system will have been

$$\frac{^{206}Pb}{^{204}Pb} = a_0 + \frac{^{238}U}{^{204}Pb}(e^{\lambda_1 T} - e^{\lambda_1 t}) \tag{7.1}$$

$$\frac{^{207}Pb}{^{204}Pb} = b_0 + \frac{^{235}U}{^{204}Pb}(e^{\lambda_2 T} - e^{\lambda_2 t}) \tag{7.2}$$

where λ_1 and λ_2 are the decay constants of ^{238}U and ^{235}U (Table 3.1),
respectively, and $^{238}U/^{204}Pb$ and $^{235}U/^{204}Pb$ are the ratios in the system
today (for a derivation of these equations, see Faure, 1986: 310–11).[2]
Note that the lead-isotope ratios described by each of these equations
are a function of five quantities:

1, 2. The initial lead compositions, a_0 and b_0;
   3. The ratio of uranium to lead;
   4. The time of origin of the system, T; and
   5. The time at which the system is later sampled, t.

Given values of a_0 and T, we can use Equation 7.1 to construct a
family of growth curves that show the change of $^{206}Pb/^{204}Pb$ with time
(Fig. 7.1), where each curve represents a system with a different
present-day value of $^{238}U/^{204}Pb$, a quantity customarily known as μ.
Similarly, a family of curves can be drawn for $^{207}Pb/^{204}Pb$ if values of b_0

2. A similar equation can be written to describe the change in ^{208}Pb resulting
from the decay of ^{232}Th, but since Th is geochemically dissimilar to U, it is more useful
for the purpose of this book to restrict the discussion to the U–Pb system.

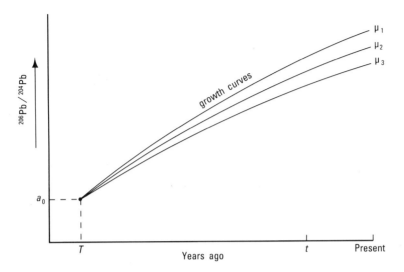

Fig. 7.1. ^{206}Pb/^{204}Pb growth curve for a system with an initial composition of a_0 at a time T years ago. Each curve represents the change in Pb-isotopic composition within a subsystem with a distinct value of μ, the ratio of ^{238}U to ^{204}Pb.

and T are known. The second set of curves can also be directly related to values of μ because

$$^{235}U = \frac{^{238}U}{137.88} \tag{7.3}$$

and so

$$\mu = \frac{^{238}U}{^{204}Pb} = \frac{137.88 \times ^{235}U}{^{204}Pb} \tag{7.4}$$

and Equation 7.2 becomes

$$\frac{^{207}Pb}{^{204}Pb} = b_0 + \frac{^{238}U}{^{204}Pb \times 137.88}\left(e^{\lambda_2 T} - e^{\lambda_2 t}\right) \tag{7.5}$$

Growth curves for the ratios ^{206}Pb/^{204}Pb and ^{207}Pb/^{204}Pb as determined from equations 7.1 and 7.5 can be plotted as a function of time for various values of μ as in Figure 7.1, but a more useful and common representation is to plot ^{206}Pb/^{204}Pb vs ^{207}Pb/^{204}Pb. This type of graph (Fig. 7.2) also contains growth curves that intersect at the initial lead-isotope composition a_0, b_0. Each growth curve represents the change over time in the composition of Pb isotopes in an isolated system with a discrete value of μ. In this graph, time may be represented

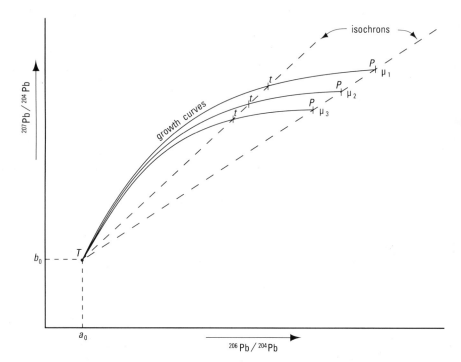

Fig. 7.2. Pb-isotope growth curves for a system with initial composition a_0, b_0 that originated at time T years ago. Each curve represents the change in isotopic composition within a subsystem with a distinct value of μ, the ratio of ^{238}U to ^{204}Pb. Provided that T is constant, the compositions of the subsystems at a later time t years ago will lie along an isochron that passes through a_0, b_0. P is the present time, i.e. zero years ago; $\mu_1 > \mu_2 > \mu_3$.

by a series of points along a growth curve: T is at the point of the initial Pb composition, and succeeding values of t are progressively farther along the growth curve.

To picture the physical situation represented by the growth curves of Figure 7.2, presume that the point a_0, b_0 represents the initial isotopic composition of Pb (commonly called *primeval* or *primordial* Pb) in the Solar Nebula at time T years ago. At that time, discrete bodies (meteorites, planets) were formed, each with the same initial Pb composition (a_0, b_0) but with different ratios of U to Pb, i.e. different values of μ. Each growth curve, then, represents the change in Pb-isotope composition within one of these bodies over time if, of course, the bodies neither lose nor gain U or Pb between times T and t. The composition of any of the bodies may be determined at any later time t years ago from its growth curve provided that a_0, b_0, and T

are known (it will be shown in Equation 7.6 that knowledge of μ is not necessary in this instance).

A system or body that begins at some initial point like a_0, b_0 and remains on the same growth curve is called a *single-stage system*, and the Pb composition that results from that system is called a *single-stage lead*. If growth continues to the present, then the composition of the system will today plot at point P in Figure 7.2. Alternatively, the system may have been disrupted at some time t years ago by geologic processes and its Pb extracted to form Pb ore devoid of U. In this instance the isotopic composition of the Pb will cease to change from t onward. This type of single-stage Pb will plot on the growth curve at t and represents a "frozen" or "fossil" record of the Pb-isotopic composition of a system at a specific time in the past t years ago. As we shall see, such leads are important in arriving at a value for the age of the Earth.

An alternative to single-stage growth is *multi-stage growth*. If at any time the U/Pb ratio of the system is changed, i.e. if the system becomes momentarily open to Pb or U, then a new system is created with a new initial Pb composition, and its Pb composition will depart from the original growth curve and follow another growth curve of different μ. Such a change may happen any number of times, and the leads produced by such systems are called *multi-stage leads*. As you might suspect, multi-stage leads are more difficult to interpret than single-stage leads. They have been discussed thoroughly by Kanasewich (1968), Doe (1970), and Gale and Mussett (1973). Multi-stage leads contain much information about the evolution of the crust and mantle but do not contribute much useful evidence concerning the age of the Earth, so this chapter will be concerned mostly with single-stage systems, or at least those that approximate single-stage behavior.

In addition to growth curves, the Pb–Pb graph, as it is sometimes called, has another interesting feature. Equations 7.1 and 7.5 can be combined by dividing the latter into the former to give

$$\frac{(^{207}\text{Pb}/^{204}\text{Pb}) - b_0}{(^{206}\text{Pb}/^{204}\text{Pb}) - a_0} = \frac{1}{137.88} \times \frac{(e^{\lambda_2 T} - e^{\lambda_2 t})}{(e^{\lambda_1 T} - e^{\lambda_1 t})} \tag{7.6}$$

Combining the equations in this way not only eliminates μ but results in the equation of a straight line whose slope is represented by the right-hand side of the equation. If T is a constant, i.e. if there is a single time of origin for the system, then the slope of the line is solely a function of t. Thus the lines defined by Equation 7.6 are isochrons that connect all points of equal t on the growth curves and pass through the initial Pb composition of the system a_0, b_0. Equation 7.6 is transcendental and cannot be solved directly for T or t but can be solved either graphically, by successive approximations, or from pre-computed tables.

Equations 7.1, 7.5, and 7.6 and Figure 7.2 constitute what today is known as the Holmes–Houtermans Pb isotope model, which is the basis for nearly all "age of the Earth" calculations. The major problem with the model is that it contains many unknowns, and no solution is possible unless all variables but one can be determined in some independent way. Suppose, for example, that we analyze a sample of lead ore and determine its Pb-isotopic ratios $^{207}Pb/^{204}Pb$ and $^{206}Pb/^{204}Pb$. Inserting these values into Equation 7.6 still leaves us with four unknowns: T, t, a_0, and b_0. Even if we know the age of the Pb, t, the equation still cannot be solved for T without values of the primeval ratios a_0 and b_0. As it turns out, a_0 and b_0 cannot be determined from any materials now accessible on the Earth, but the values are preserved in meteorites and so an ultimate solution for T through Equation 7.6 is possible provided that we are willing to accept some assumptions about the distribution of Pb isotopes in the early Solar System. We will return to this point later in the chapter, for it is an important one.

With this as background, let us return to Gerling, Holmes, and Houtermans and see how their early attempts to use the systematics of Pb isotopes to determine the age of the Earth's crust eventually led their successors to use the Holmes–Houtermans model to determine a value of 4.54 Ga for the age of the meteorites, the Earth, and the Solar System.

Gerling, Holmes, Houtermans, and the Age of the Crust

The explosive growth of physics during the early part of the twentieth century resulted in the development of many new instruments to explore the nature of matter and its constituents. One of these instruments was the mass spectrograph, a forerunner of the modern mass spectrometer now used to measure isotopic ratios.

In 1914 J. J. Thomson, of the Cavendish Laboratories at Cambridge, developed an instrument for detecting isotopes and was able to show that Ne had two isotopes of masses 20 and 22 (J. J. Thomson, 1914). Within a few short years, F. W. Aston, working in Thomson's laboratory, redesigned the apparatus, built the first quantitative "positive ray spectrograph," and set about to determine the isotopes of a variety of elements. In 1927 he turned his attentions to Pb and made the first successful measurements of the isotopic composition of *com-*

mon lead, i.e. Pb in minerals whose U content is negligible and so represent "frozen" Pb compositions. Aston showed that Pb had three isotopes, ^{206}Pb, ^{207}Pb, and ^{208}Pb, in approximately the right proportions to account for the then-accepted atomic weight of Pb. Two years later, Aston (1919, 1927, 1929, 1933) measured the lead-isotopic composition of a sample of uranium ore and found it greatly enriched in ^{206}Pb relative to ^{207}Pb. From Aston's data, C. N. Fenner and C. S. Piggot (1929) of the Geophysical Laboratory in Washington, D.C., calculated the first isotopic age based on the decay of ^{238}U to ^{206}Pb, and Lord Rutherford (1929) was able to show that ^{207}Pb was probably the product of another isotope of U of mass 235, estimated the half-life of ^{235}U, and estimated the age of the Earth to be 3.4 Ga, presuming that ^{238}U and ^{235}U were equal in abundance when the Earth formed out of matter from the Sun.

The first isotopic ages based on the ratio of ^{207}Pb to ^{206}Pb were published by J. L. Rose and R. K. Stranathan (1936) of New York University, who pointed out that this ratio must vary systematically with time as a function of the different decay rates of their parents ^{235}U and ^{238}U. In the years 1938 through 1941, A. O. Nier, who further developed Aston's positive ray spectrograph into a precision instrument with an amplifier instead of a photographic plate to measure the strengths of the *ion* beams, and his colleagues at the University of Minnesota published several papers in which they reported systematic variations in the proportions of ^{206}Pb and ^{207}Pb relative to ^{204}Pb in U and Pb ores (Nier, 1938, 1939, Nier, Thompson, and Murphey, 1941). They proposed that these variations were due to the admixture of "primeval" and radiogenic Pb, the latter of which was a function of geologic time. By 1941 the way to estimate the age of the Earth from new principles and reliable isotopic data had been prepared—and E. K. Gerling quickly seized the opportunity.

Nier and his co-workers (1941: 116) had found one sample of lead ore, a galena from Ivigtut, Greenland, whose ^{206}Pb/^{204}Pb and ^{207}Pb/^{204}Pb ratios were extremely low and speculated that the amount of radiogenic Pb in this sample was small or negligible. Gerling (1942) used the ratios in the Greenland sample as "primeval" for the purposes of his calculations and subtracted these values from the ratios of seven other lead ores whose geological ages were known to find the ratios of the radiogenic fraction in each sample. Dividing the average radiogenic ^{207}Pb/^{204}Pb by the average radiogenic ^{206}Pb/^{204}Pb gave him an average ^{207}Pb/^{206}Pb for the seven ore leads of 0.253. The calculation thus far is equivalent to evaluating the left-hand side of Equation 7.6 by using the ratios in the Greenland ore for a_0 and b_0 and the mean values for the seven additional ores for the ratios ^{207}Pb/^{204}Pb and ^{206}Pb/^{204}Pb.

For the age calculation itself, Gerling used a form of Equation 7.6. As the average age of the seven ores was about 130 Ma he set $t = 0$, which reduced $e^{\lambda_1 t}$ to 1, and used the value for $^{238}U/^{235}U$ as it would have been at 130 Ma:

$$0.253 = \frac{1}{126} \times \frac{(e^{\lambda_2 T} - 1)}{(e^{\lambda_1 T} - 1)}$$

Solving for T by using the decay constants as they were then known, Gerling obtained 3.1 Ga. To this he added the average age of the ores (130 Ma) to get an age for the Earth of 3.23 Ga, which he regarded as a minimum estimate. What Gerling did by setting $t = 0$ and $^{238}U/^{235}U = 126$ was, in effect, to solve Equation 7.6 from the point of view of an observer at 130 Ma and determine the time required for the ore leads to evolve from the "primeval" Greenland values to the observed values, then add the 130 Ma to T to bring the calculation back to the present.

Gerling presented no graphics but his calculation is shown graphically in Figure 7.3. He had, in effect, determined a two-point isochron as it would have appeared at 130 Ma by using the Greenland analysis for a_0, b_0 and the average of the seven lead ores for the upper point. The slope of 0.253 was equivalent to an age of 3.1 Ga, which was the age of the ore at 130 Ma.

Gerling went through the same exercise in comparing a "primitive" galena from Great Bear Lake, whose age was 1.25 Ga, to the Greenland ore. This calculation yielded $T = 2.7$ Ga and a minimum age for the Earth of 3.95 Ga. Gerling (1942: 261) concluded that "from these computations the age of the earth is not under $3 \times 10^9 - 4 \times 10^9$ years. This is certainly not too much, since the age of certain minerals, calculated with reference to AcD/RaG [$^{207}Pb/^{206}Pb$], was put at 2.2×10^9 years and even 2.5×10^9 years."[3]

Several assumptions are implicit in Gerling's calculation. Foremost of these is the presumption that the seven lead ores originated from the same homogeneous source, whose initial Pb-isotopic composition was identical to that of the Ivigtut galena, and that all of the leads are single-stage leads. Gerling's results, as he fully realized, are minimum values for the age of the Earth and, while of the correct order of magnitude, are too low primarily because the Pb isotopes in the Ivigtut galena are not of truly primeval composition. Nonetheless, he had devised a fruitful approach, the blossoming of which is described in the remainder of this chapter.[4]

3. The $^{207}Pb/^{206}Pb$ ages are referenced to Nier's paper.
4. It is with some injustice that the lead–lead isotope model is now known as the Holmes–Houtermans model, for Gerling clearly has priority in the literature. Gerling–Holmes–Houtermans would be more appropriate.

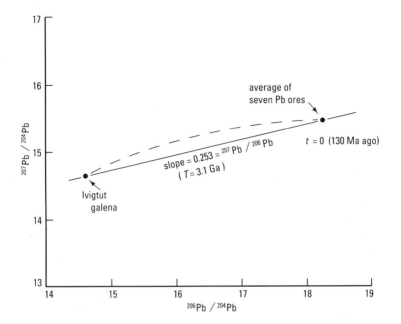

Fig. 7.3. Graphic representation of Gerling's (1942) calculation of a minimum age for the Earth using the Ivigtut galena to represent "primordial" lead. His result was 3.1 + 0.13 = 3.23 Ga.

Arthur Holmes, whom we met in Chapter 2, developed his ideas independently of Gerling and presented them in a series of papers published between 1946 and 1950 (Holmes, 1946, 1947a,b, 1948).[5] Like Gerling's, the basis of Holmes' calculations was Equation 7.6, and the data he used were those of Nier and his colleagues on 25 lead ores. Holmes noted that the ratios $^{206}Pb/^{204}Pb$ and $^{207}Pb/^{204}Pb$ in the leads varied sympathetically and systematically, i.e. as one increased so did the other. This meant to Holmes that the lead ores must be a mixture of some common primordial Pb and radiogenic Pb, and furthermore that the systems in which they evolved all originated at the time of formation of the crust, which he equated with the origin of the Earth.

Holmes broke the problem into several parts and arrived at a solution in a somewhat different way than Gerling had. To illustrate Holmes' approach, let us, as did Holmes, set

$$r = \frac{^{235}U}{^{238}U} \times \frac{e^{\lambda_2 T} - e^{\lambda_2 t}}{e^{\lambda_1 T} - e^{\lambda_1 t}} \qquad (7.7)$$

5. The first two papers are the most important of the four; the latter two are reviews that include clear explanations of the methods and results. Holmes did not cite Gerling in any of these papers and was undoubtedly unaware of his 1942 paper, which was in Russian and not then available in translation.

so that r represents the change in the lead-isotope ratio between times T and t. Then by substitution Equation 7.6 becomes

$$\frac{(^{207}\text{Pb}/^{204}\text{Pb}) - b_0}{(^{206}\text{Pb}/^{204}\text{Pb}) - a_0} = r \qquad (7.8)$$

For any particular sample Holmes knew the age of the lead ore, t, at least to a reasonable approximation, and the values of the lead ratios in the sample. What he did not know but wanted to find was the primeval composition of lead, a_0, b_0, and the age of the Earth, T. Holmes simply chose a value for T and used it in Equation 7.7 to find r for one lead ore (r_1). He then inserted this value of r in Equation 7.8 along with the lead ratios for the ore, leaving a_0 and b_0 as the only two unknown quantities. Holmes then did the same thing for a second lead ore by using the same value for T. Now he had two equations

$$\frac{(^{207}\text{Pb}/^{204}\text{Pb})_1 - b_0}{(^{206}\text{Pb}/^{204}\text{Pb})_1 - a_0} = r_1$$

and

$$\frac{(^{207}\text{Pb}/^{204}\text{Pb})_2 - b_0}{(^{206}\text{Pb}/^{204}\text{Pb})_2 - a_0} = r_2$$

which can be solved simultaneously for a_0 and b_0:

$$a_0 = \frac{(^{207}\text{Pb}/^{204}\text{Pb})_1 - (^{207}\text{Pb}/^{204}\text{Pb})_2 + r_2(^{206}\text{Pb}/^{204}\text{Pb})_2 - r_1(^{206}\text{Pb}/^{204}\text{Pb})_1}{r_2 - r_1}$$

$$b_0 = (^{207}\text{Pb}/^{204}\text{Pb})_1 + r_1 a_0 - r_1(^{206}\text{Pb}/^{204}\text{Pb})_1$$

Holmes now had a single composition for "primordial" lead for an arbitrarily chosen value of T for one pair of lead-ore samples.

Holmes' next step was to choose other values of T and find the corresponding values of a_0 and b_0, after which he could plot a_0 as a function of T for the pair of leads on a graph of $^{206}\text{Pb}/^{204}\text{Pb}$ (and $^{207}\text{Pb}/^{204}\text{Pb}$) versus T. The line so obtained represented the locus of possible solutions for that pair of samples (Fig. 7.4). A second pair of leads provided a second line, and the intersection of the two lines provided a unique solution for a_0 and T for the two pairs of leads. A similar plot of the same data on a graph of $^{207}\text{Pb}/^{204}\text{Pb}$ vs T yielded another unique solution for b_0 and T.

Holmes calculated solutions for a number of pairs of lead ores. Some of the pairs failed to give solutions because their lines did not intersect, while some of the leads had abnormal compositions, which Holmes recognized as being due to multi-stage processes, and gave

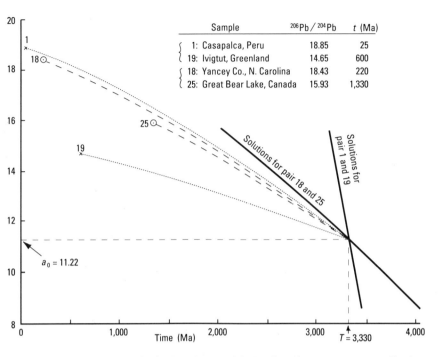

Fig. 7.4. Holmes' method for finding the age of the Earth, T, by using two pairs of lead-ore samples. Each pair of samples yields an infinite number of solutions for T as a function of primordial Pb composition, a_0. The solutions for each pair fall on a line and the intersection of the two lines defines unique values for a_0 and T. Growth lines are shown for each sample from T to t. (After Holmes, 1947b.)

solutions that were suspect.[6] In his 1946 paper (p. 683), Holmes presented 72 solutions that gave values of T ranging from 2.7 to 3.15 Ga, from which he concluded that "it may be concluded with a high degree of probability that the age of the earth is not far from 3,000 m.y. Adopting this estimate, the corresponding values for x and y [a_0 and b_0 in the notation used here] are 12.50 and 14.28."

The following year Holmes (1947a) increased the number of solutions for T to 1,419 by using all of the possible pairs from 19 of the 25 lead-ore analyses.[7] The six he excluded were, according to Holmes,

6. Few leads actually conform precisely to the single-stage Holmes–Houtermans model because most have had complex histories. The Joplin leads contain "excess" radiogenic lead and give negative (i.e. "future") model ages. For discussions of anomalous leads see Kanasewich (1968), Doe (1970), or Faure (1986).

7. The calculations were quite laborious. Holmes opens his 1946 paper as follows: "Ever since the publication by Nier and his co-workers of the relative abundances of the isotopes in twenty-five samples of lead from common lead minerals of various geologic ages, I have entertained the hope that from these precise data it might be possible to fathom the depths of geological time. The calculations involved are, however,

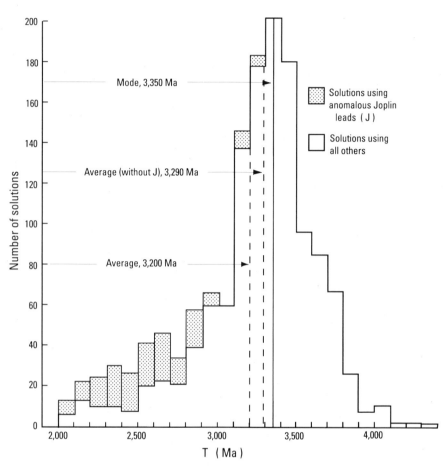

Fig. 7.5. Histogram of Holmes' 1,419 solutions for the age of the Earth using the isotope data for 25 lead ores. (After Holmes, 1947a.)

either highly abnormal in composition or of unknown age. The resulting values for T ranged from 2 Ga to more than 4 Ga with a pronounced mode at 3.35 Ga (Fig. 7.5). Holmes (1947a: 128) reasoned "it
is, therefore, concluded that on the evidence at present available, the
most probable age of the earth is about 3,350 million years."

Following Holmes' methods and using Nier's data, several sub-

somewhat formidable, and a systematic investigation became possible only recently,
with the acquisition of a calculating machine, for which grateful acknowledgment is
made to the Moray Endowment Research Fund for the University of Edinburgh." Today, such calculations can be done quickly and easily with an inexpensive pocket cal-
culator. One can only imagine what a man of Holmes' genius might have done with a
home computer.

sequent investigators found values for the age of the Earth similar to that of Holmes. Bullard and Stanley (1949) determined an age of 3.29 ± 0.2 Ga by using a refined statistical approach, while Collins, Russell, and Farquhar (1953), using additional analyses of lead ores and more rigorous criteria for selection of data, found an age of 3.5 Ga. Alpher and Herman (1951) found a maximum age for the Earth of 5.3 Ga by calculating the time when a_0 equaled zero while b_0 remained positive.

None of these ages for the Earth are correct because the samples do not conform to the assumptions of the methods. Holmes assumed that the lead ores evolved as separate systems within the crust, that each system had a distinct U/Pb ratio, that all of the systems originated at the same time and with the same primeval Pb composition, and that all of the systems remained closed until the Pb was separated in a single, short event (i.e. single-stage growth). If these assumptions were true then Holmes' method, which consists, in effect, of extrapolating isochrons back to a common intersection (although Holmes did not explain it in this way), would have yielded a single, unique solution for T, a_0, and b_0. Unfortunately for Holmes' method, leads on Earth very closely approximate single-stage growth, and the data used by Holmes represent primarily scatter about a single growth curve rather than evolution along separate growth curves. Holmes apparently realized this but not the consequences for his method, for in his 1946 publication (pp. 681–82) he plotted the Pb ratios as a function of total Pb content, noting that the values fall close to calculated growth lines, and remarked "this coincidence demonstrates that the excess isotopes are essentially of radioactive origin; and that in the source-materials the average values of the ratio lead/uranium have been everywhere nearly the same at any given time." The wide range in solutions, the fact that Holmes' values for a_0 and b_0 fall very close to the now-accepted terrestrial growth curve (discussed below), and the parallelism of some of the solution lines are additional evidence that Holmes' assumptions are invalid (Mussett, 1970: 67–68).

Houtermans (1946, 1947) used an approach slightly different from either Holmes' or Gerling's, although the basic method, which he too developed independently, was similar, as was his use of the Pb data of Nier and his colleagues.[8] Like Holmes, Houtermans noted that the isotopic ratios in the lead ores were interdependent and concluded that the leads were mixtures of radiogenic and primordial Pb. He pointed out that on a plot of $^{207}Pb/^{204}Pb$ vs $^{206}Pb/^{204}Pb$ lead minerals

8. Like Holmes and probably for the same reason (note 4) Houtermans was unaware of Gerling's paper. He referred to Holmes' 1946 paper in an addendum written after his 1946 paper had been set in type.

of the same age, t, must lie on straight lines, which he called "iso-chrones" (now isochrons), whose slope is expressed by the right-hand side of Equation 7.6 and is thus a function of T and t. Moreover, said Houtermans, two or more isochrons must intersect at a_0 and b_0 and give a unique solution for T. Houtermans found $T = 2.9 \pm 0.3$ Ga, $a_0 = 11.52 \pm 0.60$, and $b_0 = 14.03 \pm 0.20$, but he did not specify which of the 25 leads he used in the calculations. This age, T, he called "the age of uranium" and thought that it represented the age of the Solar System, i.e. the time of formation of terrestrial uranium. This was provided there was not enrichment of uranium relative to lead during formation of the Earth's crust. If such enrichment had oc-curred, and he cited some evidence that it had, then T was the age of the crust.

Houtermans' method was based on the same assumptions as Holmes' and suffered from the same flaws, but Houtermans ad-vanced the final concept in the Holmes–Houtermans model—the concept of lines of equal time—and gave isochrons their name. In ad-dition, he (1947) pointed the way for future work by suggesting that a better value for the composition of primeval Pb might be found by analyzing iron meteorites.[9]

In 1953, Patterson et al. determined the Pb-isotopic composition and the U and Pb concentrations in both the iron–nickel phase and in troilite (FeS) from the iron meteorite Canyon Diablo, which excavated Meteor Crater some 50 ka ago (Figure 6.2). The troilite contained the lowest Pb ratios ever measured and was also exceedingly low in U relative to Pb: $^{206}Pb/^{204}Pb = 9.41$, $^{207}Pb/^{204}Pb = 10.27$, $Pb = 18 \pm 1$ ppm, $U = 0.009 \pm 0.003$ ppm, and $U/Pb = 0.0005$. The low U/Pb ratio meant that the Pb-isotopic composition could not have changed sig-nificantly since the meteorite, even then known to be an ancient ob-ject, was formed. Thus, suggested Patterson and his colleagues, the Pb ratios in the troilite might be a record of the composition of pri-mordial Pb (Patterson et al., 1953).

Houtermans was quick to take advantage of the new Pb data for Canyon Diablo and in the same year (1953) published a paper in which he calculated an age for the Earth that is very close to the cur-rently accepted value. Houtermans made two principal assumptions: (1) that the isotopic composition of Pb at the time of formation of Earth's *lithosphere* was represented by the values found in the troilite of Canyon Diablo, and (2) that the majority of Tertiary leads whose Pb-isotopic compositions had been measured formed by single-stage growth from a common time of origin up to the time of formation of the lead ores in which they occur. He chose lead ores of Tertiary age

9. A similar suggestion was made by Brown (1947).

TABLE 7.1

Data Used by Houtermans (1953) to Determine an Age for the Earth

Mineral[a]	Locality	r[b]
Wulfenite, vanadinite	Tucson Mtns., Arizona	0.5851
Bournonite	Casapalca Mine, Peru	0.5649
		0.5528
Galena	Casapalca Mine, Peru	0.5757
		0.5669
Galena	Durango, Mexico	0.5839
Galena	Sonora Mine, Arizona	0.6014
Galena	Freiberg, Saxony	0.5924
Galena	Bolivia	0.5866
Galena	Montenegro, Bolivia	0.5911
Galena, 1	Wiesloch, Germany	0.5755
Galena, 2	Wiesloch, Germany	0.5755
Mean		0.5793 ± 0.004

[a] All of Tertiary age.
[b] The quantity r, the slope of a two-point isochron joining the value from the ore lead and that from the lead in troilite from the Canyon Diablo meteorite, was calculated by Houtermans with Equation 7.8.

because for these young leads the calculated age of the Earth is relatively insensitive to errors in the geological age of the samples. At that time the literature contained lead-isotopic compositions for 22 ores, from which Houtermans selected 10 (Table 7.1). For each ore he calculated a value for Equation 7.8 (the left-hand side of Equation 7.6) of

$$r = \frac{(^{207}\text{Pb}/^{204}\text{Pb}) - 10.27}{(^{206}\text{Pb}/^{204}\text{Pb}) - 9.41}$$

This value is the slope of a two-point isochron through the lead ore and a_0, b_0 as measured in Canyon Diablo. He took the mean of the ten lead ores and used the result in Equation 7.7 (the right-hand side of Equation 7.6) to find the age of the Earth, $T = 4.5 \pm 0.3$ Ga:

$$0.5793 = \frac{^{235}\text{U}}{^{238}\text{U}} \times \frac{e^{\lambda_2 T} - e^{\lambda_2 t}}{e^{\lambda_1 T} - e^{\lambda_1 t}}$$

by using $^{235}\text{U}/^{238}\text{U} = 1/139$, $\lambda_1 = 1.535 \times 10^{-10}$/yr, $\lambda_2 = 9.815 \times 10^{-10}$/yr, and $t = 30 \times 10^6$ yr (Tertiary). Graphically, Houtermans' solution is nearly identical to Gerling's (Fig. 7.3) except that he used the Canyon Diablo values instead of the Ivigtut galena for a_0 and b_0 and the mean ratios of ten young leads of similar age. The fact that the ten leads have similar slopes suggests that Houterman's assumption of a common time of origin, i.e. system closure, was not unreasonable.

In selecting the 10 leads Houtermans had rejected 12 others because their compositions were anomalous and their model $^{207}\text{Pb}/^{206}\text{Pb}$

ages were negative. He noted, however, that even if such data were included the results were not changed appreciably ($T = 4.05$ Ga). Therefore, he concluded, as long as a large number of carefully selected samples was used, any multi-stage leads inadvertently included in the calculations resulted in only small errors in T.

At a conference held three months before Houtermans' 1953 paper was published, Patterson (1953) presented the results of calculations that were virtually identical to Houtermans'. Patterson used the meteoritic lead composition as the composition of primordial lead, as did Houtermans, and two different types of materials to represent the composition of present-day lead. In one calculation he used the average lead composition of recent oceanic sediment and a manganese nodule. In the other he used the composition of lead in Columbia River Basalt, of Miocene age. His results for the two calculations were 4.51 and 4.56 Ga, respectively. The precise publication date of the conference proceedings is obscure, but according to the volume editor it was probably published in late 1953 or early 1954 (L. T. Aldrich, personal communication, 1990).

Houtermans' and Patterson's 1953 results for the age of the Earth are notable not only because they are near the present-day value but also because they were the first calculations to link the age of the Earth to the age of meteorites, thereby implying a genetic connection. Neither Houtermans nor Patterson made any attempt to justify this assumption of cogenesis, but as we shall see a reasonable case can be made for its validity.

Thus, in 1953 there were two somewhat different ways of applying the Holmes–Houtermans Pb–Pb model to the problem of the age of the Earth. One, the "ore method," used the growth in lead isotopes between the times of formation of an ancient lead ore and a recent one. This method gave ages of about 3.3 Ga that, in hindsight, were really the ages of the ancient lead ores and only minimum ages for the Earth. These ages, however, were within 30% of the modern value for the age of the Earth and were, at the time, important contributions to knowledge of geologic time. The second method, the "meteorite method," consisted of using the Pb-isotopic composition in the troilite phase of iron meteorites for the primeval Pb composition and calculating the time required to form Pb with the composition of young lead ores. This method required the assumption that the Pb in iron meteorites and the Pb of the Earth shared a common time and mechanism of origin, an assumption that had yet to be proved. Evaluating the situation in 1955, Patterson, Tilton, and Inghram concluded (p.74) that while there was some reason to think that there was a genetic connection between meteorites and the Earth, results based on such an assumption "should be viewed with considerable skepticism until the assumptions involved in the method are verified."

Patterson and the Meteoritic Lead Isochron

Houtermans had assumed that there was a genetic connection between meteoritic lead and young terrestrial lead but had not provided any arguments for the validity of this assumption. Moreover, Houtermans' isochron was based on only two points—troilite lead in the Canyon Diablo meteorite and the average lead in Tertiary ores. Both deficiencies were corrected by C. C. Patterson of the California Institute of Technology in a classic 1956 paper. Patterson (1956) used the lead-isotope analyses from three stone meteorites and the troilite phase of two iron meteorites and showed that these data fell precisely on an isochron (Fig. 7.6) whose slope indicated an age of 4.55 ± 0.07 Ga by using the decay and abundance constants then in use (the calculated age is 4.48 Ga if the constants of Table 3.1 are used). Such collinearity from a set of data representing a wide range of isotopic compositions, Patterson argued, strongly indicated that these five meteorites fulfilled the assumptions of the Pb–Pb age method. Data from any meteorite in whose history U had been fractionated from Pb would

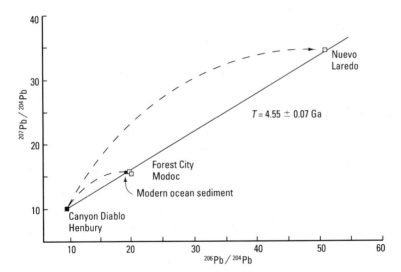

Fig. 7.6. Patterson's meteoritic lead isochron using three stone and two iron meteorites (squares). The lead composition of modern ocean sediment (filled circle) falls on the isochron, suggesting that meteorites and the Earth are genetically related and of the same age. The dashed lines are growth curves. (After Patterson, 1956.)

not fall on the isochron. Therefore, Patterson concluded, the isochron age represents the time of initial formation and differentiation of meteorites. He also noted that the Pb–Pb age was in agreement with the existing K–Ar and Rb–Sr ages for meteorites (which were then few and poor) but considerably more precise.

Patterson's next step was to make the genetic connection between meteorites and the Earth. By substituting T = 4.55 Ga, t = 0 years, and the troilite lead ratios into equations 7.1 and 7.5, he derived expressions for predicting the isotope ratios for any modern lead that belonged to the meteoritic system:

$$(^{206}Pb/^{207}Pb) = 9.50 + 1.014 \, (^{238}U/^{204}Pb)$$

$$(^{207}Pb/^{204}Pb) = 10.41 + 0.601 \, (^{238}U/^{204}Pb)$$

Measurement of any two of the remaining unknown quantities in a modern lead sample should satisfy these expressions if the sample belonged to the meteoritic system. This is simply another way of saying that if modern Earth lead falls on the meteoritic isochron then it must have evolved in a closed system from an initial composition of a_0, b_0, as measured in meteoritic troilite, from T = 4.55 Ga to the present (t = 0). Patterson realized that there were other ways in which an Earth lead could have developed a composition on the isochron, but these required complicated and improbable mechanisms.

The problem of choosing a representative sample of modern Earth lead is not simple because the crust of the Earth has had a complex history. Patterson proposed that modern sediment from the deep ocean might provide a reasonable sample of modern Earth lead because such sediment samples a wide volume of material from the present continents and thus represents average crustal lead. The lead-isotopic composition of Pacific deep-sea sediment had been measured in 1953 (Patterson, Goldberg, and Inghram, 1953), and the results, $^{206}Pb/^{204}Pb$ = 19.0 and $^{207}Pb/^{204}Pb$ = 15.8, satisfied the expressions very well (Fig. 7.6). Furthermore, he could independently predict the $^{238}U/^{204}Pb$ ratio of the subsystem (Earth's crust) in which the sediment lead evolved from the growth curve defined by the dated ore leads (more on this subject below). This ratio also satisfied the expressions for the meteoritic system and successfully predicted where on the isochron the sediment lead should and did plot. Patterson (1956: 138) concluded that "independently measured values for all three ratios adequately satisfy expressions (2) and (3), and therefore the time since the earth attained its present mass is 4.55 ± 0.07 × 10⁹ years." Patterson had determined a precise age for meteorites and had also shown it highly probable that the Earth was part of the meteoritic lead system and of the same age.

TABLE 7.2

Isotopic Composition of Primeval Lead as Measured in Troilite (FeS) from Iron Meteorites and One Chondrite

Meteorite	$^{206}Pb/^{204}Pb$	$^{207}Pb/^{204}Pb$	Source
Canyon Diablo	9.41	10.27	Patterson et al., 1953
Canyon Diablo, Henbury	9.50	10.36	Patterson, 1955
Canyon Diablo	9.61	10.38	Chow and Patterson, 1961
Mean of five irons	9.56 ± 0.28	10.42 ± 0.32	Murthy and Patterson, 1962
Mean of five irons	9.460 ± 0.122	10.327 ± 0.131	Cumming, 1969
Canyon Diablo	9.346 ± 0.018	10.218 ± 0.030	Oversby, 1970
Mezö-Madras (L3 chondrite)	9.310 ± 0.019	10.296 ± 0.021	Tilton, 1973
Canyon Diablo	<u>9.307 ± 0.006</u>	<u>10.294 ± 0.006</u>	Tatsumoto, Knight, and Allègre, 1973a

NOTE: Errors are at the 95% confidence level (two standard deviations). The composition measured by Tatsumoto, Knight, and Allègre (underlined) is the generally accepted value for primeval lead in the Solar System.

Six years later Patterson teamed with V. R. Murthy of the University of California at San Diego to refine Patterson's age of meteorites and to strengthen the hypothesis that the Earth was part of the meteoritic lead system. Murthy and Patterson (1962) selected lead analyses of five stone meteorites thought, on the basis of other isotopic age data, most likely to have been closed systems since formation. To this array of data they added the composition of primeval lead, which they took to be the average composition of lead in troilite from five iron meteorites as measured by three different laboratories (Table 7.2). The six meteorite lead compositions formed a linear array (Fig. 7.7). To determine the slope of the isochron, Murthy and Patterson calculated the mean of the five slopes defined by each of the stone meteorites and the primeval composition. This average slope of 0.59 ± 0.01 represented an age for meteorites of 4.55 Ga (4.48 with new constants). This isochron of "zero age" they named the *meteoritic geochron.*[10]

To show the relationship between the meteoritic geochron and the terrestrial geochron, Murthy and Patterson used two samples of terrestrial lead (Fig. 7.7, inset). One, the average composition of lead in more than 100 samples of recent North Pacific sediments, should, they reasoned, lie to the right of the terrestrial geochron because the marine sediments are eroded from rocks of the upper layers of the crust. At the time of their formation, these source rocks contained lead of a composition representative of that of the entire crust but were enriched in ^{238}U relative to ^{204}Pb by the crustal formation processes.

10. The "geochron" is a zero-age isochron because it represents the composition of modern, single-stage leads and $t = 0$ years. T, however, is 4.55 Ga, and the isochron so dates the meteorites. This dual meaning can be confusing at times.

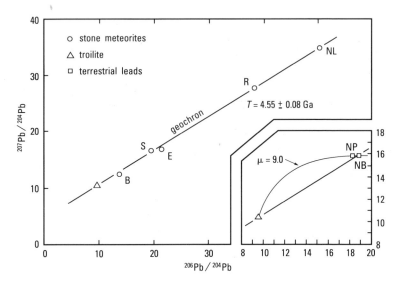

Fig. 7.7. Murthy and Patterson's meteoritic geochron, which is bracketed by terrestrial leads from North Pacific sediments (NP) and Bathurst, New Brunswick, lead ores (NB). The terrestrial leads lie on a growth curve of $\mu = 9.0$ that passes through the value for meteoritic troilite. B, Beardsley; E, Elenovka; NL, Nuevo Laredo; R, Richardton; S, Saratov stone meteorites. (After Murthy and Patterson, 1962.)

Thus, the marine leads should be displaced to the ^{206}Pb-enriched side of the average crustal lead composition. In other words, the marine sediments' lead composition should lie slightly to the right of the terrestrial (= crustal) geochron. On the left, the terrestrial geochron should be bracketed by single-stage ore leads that define the crustal growth curve. Murthy and Patterson chose the mean composition of ore leads from Bathurst, New Brunswick, which have a geologic age of about 350 Ma and which were, at the time (but no longer), thought to be single-stage leads. These two points, they reasoned, should limit the position of the terrestrial geochron, and because they also bracketed the meteoritic geochron, the two geochrons must be very nearly the same if not identical. Moreover, both the North Pacific and Bathurst leads lie on a primary (single-stage) growth curve that passes through the primordial (troilite) composition and satisfies what was then known about the average ^{238}U/^{204}Pb ($\mu = 9.0$) of the crust. Murthy and Patterson concluded, therefore, that meteorites and the Earth's crust are parts of the same Pb-isotopic system and that the age of meteorites and the age of the Earth are the same.

In addition to refining the Pb-isotopic age of meteorites, Murthy and Patterson provided a sound basis for connecting Pb growth in the Earth, a body whose time of origin cannot be determined directly, with Pb growth in meteorites, whose ages can be precisely measured.

Houtermans had assumed that meteorites and the Earth were *cogenetic*. Murthy and Patterson had shown that such an assumption was not only reasonable but probable. Subsequent workers would improve on this model and leave little doubt that the age of meteorites represents, to a very close approximation, the age of the Earth. But before continuing with Pb growth in the Earth, let us consider in more detail the current status of the Pb–Pb ages of meteorites.

Pb–Pb Ages of Meteorites

Not nearly so many Pb–Pb ages of meteorites have been determined as have K–Ar and Rb–Sr ages, primarily because precise Pb measurements are rather difficult. One reason is that most meteorites contain only a very small amount of Pb, typically a small fraction of one part per million. Another is the ease with which the samples may become contaminated. Lead is not only a volatile element but is ubiquitous in our environment. It occurs, for example, in most paints, and there are relatively high concentrations of lead in the atmosphere due to exhaust from vehicles that burn fuel containing lead tetraethyl. Because of the problem with contamination, precise lead-isotopic measurements must be made in special lead-free laboratories, and there are only a few institutions in the world where such facilities exist.

Despite the difficulties, Pb–Pb ages commonly are more precise than K–Ar or Rb–Sr ages. This is because their calculation is based on two daughter isotopes of the same element (Equation 7.6) rather than on a parent–daughter pair consisting of two geochemically different elements. This has two advantages. First, the critical measurement is of an isotopic ratio of the same element rather than of a ratio based on the concentrations of isotopes of two different elements, and the former can be made much more precisely than the latter. Second, the ^{207}Pb–^{206}Pb daughter–daughter isotope pair is less susceptible to fractionation during a postcrystallization disturbance than is a geochemically dissimilar parent–daughter pair.

As with the other types of radiometric methods, Pb–Pb ages can be grouped into three general categories: (1) internal isochron ages based on analyses of several phases from an individual meteorite, (2) whole-rock isochron ages based on the analysis of three or more different meteorites, and (3) model ages for individual meteorites calculated on the assumption of a value for the primordial lead composition (e.g. Canyon Diablo troilite). Neither (1) nor (2) requires a value for the composition of initial lead whereas (3) is, in effect, a two-point isochron defined by primordial lead as one datum and the meteorite lead as the other.

Precise internal isochron ages have yet been measured for only a

Pb–Pb Internal Isochron Ages for Some Individual Meteorites

Meteorite name	Type[a]	Number of data[b]	Age (Ga)[c]	Source
CHONDRITES				
Allende	CV3	27	4.553 ± 0.004	Tatsumoto, Unruh, and Desborough, 1976
		19	4.565 ± 0.004	Chen and Tilton, 1976
Mezö-Madras	L3	12	4.480 ± 0.002	Hanan and Tilton, 1985
Sharps	H3	16	4.472 ± 0.005	Hanan and Tilton, 1985
Barwell	L5–6	11	4.559 ± 0.005	Unruh, Hutchison, and Tatsumoto, 1979
Bruderheim	L6	11	4.482 ± 0.017	Gale, Arden, and Abranches, 1980
Appley Bridge	LL6	5 + P	4.569 ± 0.018	Gale, Arden, and Hutchison, 1979
St. Severin	LL6	5	4.543 ± 0.019	Manhes, Minster, and Allègre, 1978
ACHONDRITES				
Juvinas	eucrite	8	4.556 ± 0.012	Tatsumoto and Unruh, 1975
		9	4.5400 ± 0.0007	Manhes, Allègre, and Provost, 1984
Pasamonte	eucrite	15	4.53 ± 0.03	Unruh, Nakamura, and Tatsumoto, 1977
Angra dos Reis	angrite	3 + P	4.544 ± 0.002	Wasserburg et al., 1977

[a] Table 6.1.
[b] P indicates that the primordial Pb ratio as determined from iron meteorites was included as a datum in the isochron.
[c] Ages calculated with the current decay and abundance constants (Table 3.1). Errors are at the 95% confidence level (two standard deviations). Ages underlined are shown as figures (see Figure 3.14 for St. Severin).

handful of individual meteorites (Table 7.3), but all but three fall within the narrow range of 4.53 to 4.57 Ga, a scatter of less than 1%. In addition, there is no discernible difference in the Pb–Pb isochron ages of chondrites and achondrites.

Some of the internal isochrons are quite remarkable. The one obtained by M. Tatsumoto and his colleagues of the U.S. Geological Survey for the carbonaceous chondrite Allende, for example, is based on 27 analyses of chondrules, aggregate, and matrix fractions. These data form an extraordinary linear array that passes through the composition of Canyon Diablo troilite and has an error in the line of best fit (and the resulting age) of less than 0.1% (Fig. 7.8).

Although the internal isochron ages of both chondrites and achondrites are remarkably concordant, the ages of Bruderheim, Mezö-Madras, and Sharpes are anomalously low. Bruderheim has a Pb–Pb isochron age of 4.482 ± 0.017 Ga and a line of best fit that passes above, rather than through, the composition of troilite from Canyon Diablo. Calculations show, however, that Bruderheim's age

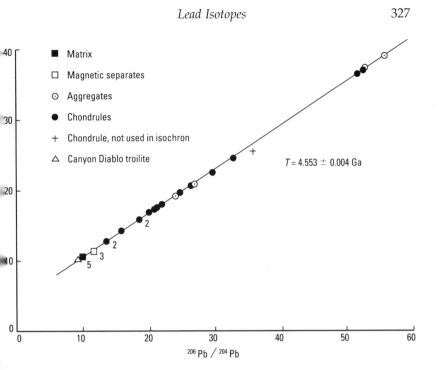

Fig. 7.8. Pb–Pb internal isochron for the carbonaceous chondrite Allende. The regression line through the data passes, within the errors, through the composition of Canyon Diablo troilite. Where data are too tightly clustered to be shown individually, the number of data represented by a single symbol is indicated. (After Tatsumoto, Unruh, and Desborough, 1976.)

can be explained by a two-stage history: formation at 4.536 Ga followed by a shock, perhaps while Bruderheim was still on its parent body, at about 500 Ma that disturbed the U–Pb system (Gale, Arden, and Abranches, 1980). Unlike Bruderheim, the isochrons for Mezö-Madras (4.480 ± 0.011 Ga) and Sharpes (4.472 ± 0.005 Ga) pass, within error limits, through the composition of Canyon Diablo troilite. It is thought that these chondrites record metamorphism at about 4.48 Ga, a conclusion that is supported by Rb–Sr, Sm–Nd, and K–Ar data (Hanan and Tilton, 1985, Tilton, 1988).

Whole-rock Pb–Pb isochrons for meteorite groups tend to fall within a somewhat wider range than do the internal isochrons for individual meteorites, although most of the ages exceed 4.5 Ga (Table 7.4). Many of these measurements, however, were made before the techniques for determining Pb-isotopic compositions were dramatically improved in the mid-1970's (note that all of the internal isochrons listed in Table 7.3 were measured after 1976). If we examine only the most recent isochrons, of which there are three, they fall within the range 4.54–4.58 Ga, which coincides very closely with the range of

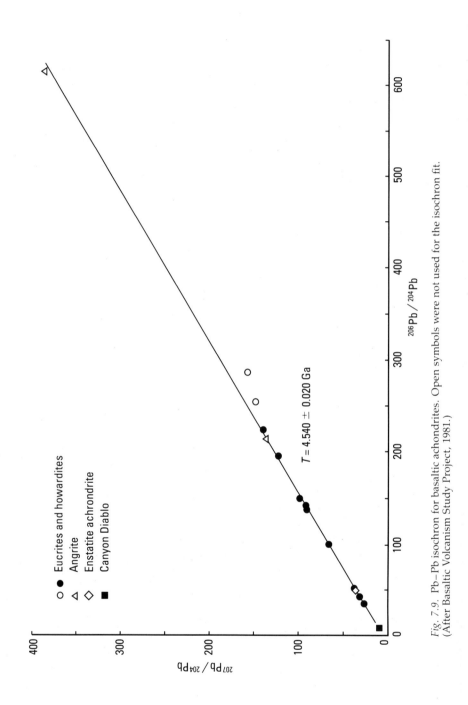

Fig. 7.9. Pb–Pb isochron for basaltic achondrites. Open symbols were not used for the isochron fit. (After Basaltic Volcanism Study Project, 1981.)

TABLE 7.4

Pb–Pb Whole-Rock Isochron Ages of Different Meteorite Groups

Meteorite class or group[a]	Number of meteorites[b]	Age (Ga)[c]	Source
Chondrites (C, H, L, E)	15	4.505 ± 0.016	Huey and Kohman, 1973
Ordinary chondrites (L)	5	4.551 ± 0.007	Unruh, 1982
Ordinary chondrites (L, H)	3 + P	4.562	Tilton, 1973
Ordinary chondrites (E)	4	4.577 ± 0.004	Manhes and Allègre, 1978
Ordinary chondrites (H, L) + eucrite	5 + P	4.48 ± 0.08	Murthy and Patterson, 1962
Ordinary chondrites (H, L) + eucrite	12 + P	4.49 ± 0.05	Cumming, 1969
Chondrites + achondrites	53	4.50 ± 0.03	Kanasewich, 1968
Eucrites and howardites	9 + P	<u>4.540 ± 0.020</u>	Basaltic Volcanism Study Project, 1981
Eucrites	4 + P	4.52 ± 0.02	Silver and Duke, 1971

[a] Table 6.1.
[b] P indicates that the primordial Pb ratio as determined from iron meteorites was included as a datum in the ,ochron.
[c] All ages calculated with the current decay and abundance constants (Table 3.1). Errors are at the 95% confi-ence level (two standard deviations). Age underlined is shown as figure.

values for individual meteorites. Like the data for the individual mete-orites, the whole-rock data form precise linear arrays that pass through the composition of Canyon Diablo troilite (Fig. 7.9), and the ages of chondrites and achondrites are not significantly different.

Model Pb–Pb ages for meteorites are calculated from Equation 7.6 with the assumption of a single-stage history for the meteorite and the composition of Canyon Diablo troilite for the composition of pri-meval lead. Each model age is, in effect, a "two-point" isochron, the points being the lead-isotopic compositions of Canyon Diablo troilite and of the meteorite. As these measurements can be made on nearly any stone meteorite and do not require nearly as much effort as either whole-rock or internal multi-point isochrons, model ages have been published for many meteorites (Table 7.5).

Like the internal and whole-rock isochrons, most of the Pb–Pb model ages fall within the relatively small range of 4.53 to 4.57 Ga, although there are some with values as high as 4.60 Ga (Karoonda) and as low as 4.50 Ga (Orgueil). Still, the total range of the model ages is only slightly more than 2% of the mean of the model ages of 4.550 ± 0.025 Ga.

There are several peculiar features of the Pb–Pb isochron and model ages. The first is that they tend to be somewhat higher, one percent or so, than the Rb–Sr and K–Ar ages for meteorites. This may be, at least in part, because the different radiometric systems are measuring slightly different events. This is quite likely so for the

TABLE 7.5

$^{207}Pb/^{206}Pb$ Model Ages of Meteorites

Meteorite name	Type[a]	Material[b]	Age (Ga)[c]	Source
CHONDRITES				
Orgueil	CI1	wr	4.496 ± 0.010	Tatsumoto, Unruh, and Desborough, 1976
Karoonda	CO4	wr	4.600 ± 0.028	Manhes and Allègre, 19?
Murray	CM2	wr	4.511 ± 0.042	Tatsumoto, Knight, and Allègre, 1973
Allende	CV3	wr	4.496 ± 0.010	Tatsumoto, Unruh, and Desborough, 1976
		6cl	4.559 ± 0.015	Chen and Wasserburg, 1981
Beardsley	H5	wr	4.574 ± 0.012	Tatsumoto, Knight, and Allègre, 1978
Beaver Creek	H4	wr	4.551 ± 0.016	Manhes and Allègre, 19?
Kirin	H4	wr	4.583 ± 0.010	Manhes and Allègre, 19
Plainview	H	wr	4.529 ± 0.010	Tatsumoto, Knight, and Allègre, 1973
Richardton	H5	wr	4.519 ± 0.015	Tatsumoto, Knight, and Allègre, 1973
Barwell	L5–6	wr	4.560 ± 0.004	Unruh, 1982
Bjurbole	L4	wr	4.590 ± 0.006	Unruh, 1982
Bruderheim	L6	wr	4.535 ± 0.004	Unruh, 1982
Harleton	L6	wr	4.550 ± 0.014	Unruh, 1982
Homestead	L5	wr	4.553 ± 0.002	Manhes and Allègre, 19
Knyahinya	L5	wr	4.552 ± 0.004	Unruh, 1982
		wr	4.552 ± 0.002	Manhes and Allègre, 19
Modoc	L6	wr	4.530 ± 0.015	Tatsumoto, Knight, and Allègre, 1973
		wr	4.546 ± 0.012	Unruh, 1982
Tennasilim	L4	wr	4.552 ± 0.013	Unruh, 1982
St. Severin	LL6	wr	4.549 ± 0.012	Manhes, Minster, and Allègre, 1978
		wr	4.555 ± 0.005	Tatsumoto, Knight, and Allègre, 1973
		wh	4.551 ± 0.004	Chen and Wasserburg, 1981
Hvittis	E6	wr	4.588 ± 0.008	Manhes and Allègre, 1?
St. Marks	E5	wr	4.582 ± 0.006	Manhes and Allègre, 1?
ACHONDRITES				
Ibitira	eucrite	wh	4.55 ± 0.02	Wasserburg et al., 1977
Juvinas	eucrite	cl	4.549 ± 0.006	Manhes and Allègre, 1
Nuevo Laredo	eucrite	wr	4.529 ± 0.005	Tatsumoto, Knight, an Allègre, 1973
Pasamonte	eucrite	wr	4.573 ± 0.011	Unruh, Nakamura, an? Tatsumoto, 1977
Sioux County	eucrite	wr	4.526 ± 0.010	Tatsumoto, Knight, an Allègre, 1973
Angra dos Reis	angrite	wr	4.555 ± 0.005	Tatsumoto, Knight, an Allègre, 1973
		wr	4.544 ± 0.002	Wasserburg et al., 197?
		wr + 2 wh	4.551 ± 0.004	Chen and Wasserburg, 1981

[a] Table 6.1.   [b] wr, whole rock; wh, whitlockite; cl, clast.
[c] All ages calculated with the current decay and abundance constants (Table 3.1). Errors are at the confidence level (two standard deviations). The composition of Pb in Canyon Diablo troilite was used a? composition of primordial Pb in the calculation of the single-stage model ages.

K–Ar ages, which measure the time of cooling rather than of formation or aggregation. Perhaps the Pb–Pb ages record the time of last homogenization of Pb isotopes throughout the Solar Nebula, the Rb–Sr ages the formation or aggregation times of the parent bodies, and the K–Ar ages the last major cooling. This plus metamorphic and shock disturbances of the parent bodies, which would disturb the Pb–Pb ages somewhat less than the Rb–Sr or K–Ar ages, might account for the slight differences as well as the overlap in the ages from the various methods.

Another possible explanation for the discrepancy between Rb–Sr and Pb–Pb ages is that the Rb–Sr decay constant, which is the least well known of the decay constants used in radiometric dating, is incorrectly estimated (Minster, Birck, and Allègre, 1982). A change in this constant from 1.42×10^{-11}/yr to 1.40×10^{-11}/yr would bring the Rb–Sr and Pb–Pb ages into general concordance.

Another interesting feature of the Pb–Pb ages is that even though they are remarkably consistent and tightly grouped, they scatter over a wider range of values than expected from their individual analytical errors. This is especially apparent in the model ages. Compare, for example, the differences in the model ages of the L chondrites Bjürbole and Bruderheim or the eucrites Juvinas and Nuevo Laredo with the errors, which are based on analytical considerations. In addition to the "excess scatter," many chondritic meteorites contain "excess lead," i.e. more radiogenic lead than can be accounted for simply by the decay over the past 4.55 Ga of the uranium (and thorium) that they contain. Three hypotheses have been advanced to explain these peculiarities in the lead data: (1) multi-stage histories, (2) variable primeval lead compositions, and (3) contamination by terrestrial lead (for detailed discussions and varying opinions see Tatsumoto, Unruh, and Desborough, 1976, Manhes and Allègre, 1978, Unruh, Hutchison, and Tatsumoto, 1979, Gale, Arden, and Abranches, 1980, and Unruh, 1982).

Multi-stage histories can account for some of the scatter in the ages but not for the excess radiogenic lead. A variable primeval lead composition can also account for some of the scatter in the results, but nearly all of the internal isochrons from the major meteorite types pass through the composition of Canyon Diablo troilite, so variable composition of primordial lead cannot be a widespread or prevalent phenomenon. Terrestrial contamination is a known factor in at least some meteorite analyses and has been proposed as the sole cause of excess lead (Manhes and Allègre, 1978: 545, but other workers—see Kanasewich, 1968—do not agree). Contamination of a meteorite sample by terrestrial lead can account for much of the excess radiogenic lead observed in meteorites. But such contamination, while causing some small degree of scatter in the meteorite ages, would not cause them to deviate very much from 4.55 Ga. The reason is that the contamination is with modern lead that also falls on or very close to the "zero-age" geochron. In other words, the contaminating lead also has an age of 4.5–4.6 Ga.

It seems likely that all three of the proposed mechanisms may be responsible, perhaps to varying degrees in different meteorites, for the excess scatter and excess radiogenic lead revealed in Pb–Pb analyses of meteorites. Nevertheless, the Pb–Pb ages (tables 7.3, 7.4, 7.5) form a remarkably consistent array of data that leads inevitably to two important conclusions. First, the isochron and model Pb–Pb ages of meteorites are concordant and have a mean of 4.55 Ga. This age represents, within a small fraction of one percent, the time when the composition of Pb isotopes was last uniform throughout the Solar System. This condition changed immediately upon condensation of solid matter and aggregation of planetoids. Second, there are no detectable differences between the Pb–Pb ages of chondrites and achondrites, which indicates that condensation from the Solar Nebula, aggregation of the meteorites' parent bodies, and differentiation of the achondrites on and in those parent bodies occurred within a period of only a few tens of millions of years. These two conclusions are the same as those reached on the basis of results from the other dating methods (Chapter 6), except that the average Rb–Sr formation ages are approximately 50 Ma or so younger than the Pb–Pb ages.

The Terrestrial Connection

There is incontrovertible evidence from lead-isotopic data that meteorites are approximately 4.55 ± 0.02 Ga. We can presume, as the evidence indicates, that the solid bodies of the Solar System formed nearly simultaneously, and conclude that the Pb–Pb age of meteorites also represents the age of the Earth. This is a perfectly reasonable conclusion and probably correct, but as Patterson recognized more than three decades ago it is much more convincing if we can demonstrate that the Earth is part of the same lead-isotopic system as are meteorites.

In the earlier sections of this chapter we saw that the present Pb composition of meteorites can be explained quite simply if meteorites were all formed at 4.55 Ga with the same Pb-isotopic composition but different ratios of U to Pb. Through time, the Pb-isotopic composition of each meteorite (or perhaps each parent body) has followed its own single-stage growth curve to the zero-age isochron, or geochron (Figs. 7.2 and 7.6).

One simple approach is to treat the Earth like a meteorite and presume that it formed at the same time as the meteorites with its own ratio of U to Pb and has remained a system closed to uranium and lead since formation. It then follows that the Pb-isotopic composition of the Earth, too, must have evolved along a similar single-stage growth curve that originated, as do the meteoritic growth curves, at the primeval composition of Pb, i.e. a_0, b_0 of Canyon Diablo. Such a growth curve need not involve the entire Earth but could apply to any U–Pb reservoir within the Earth, such as the mantle or crust, so long as that reservoir formed at the same time or very shortly after the Earth and has remained a system closed to U and Pb.

The hypothesis of a single growth curve for the Earth, or some substantial portion of it, was implicit in the early work of Gerling, Holmes, and Houtermans and was more fully developed in the 1950's by R. D. Russell of the University of British Columbia and his colleagues (for example, Collins, Russell, and Farquhar, 1953, R. D. Russell, 1956, R. D. Russell and Farquhar, 1960). The hypothesis, while somewhat of an oversimplification (see below), appears to be a reasonable approximation because, as discussed in the earlier sections of this chapter, modern terrestrial sediments and young lead ores plot very close to the meteoritic geochron (Fig. 7.7). But a more convincing case could be made if the terrestrial growth curve actually could be "traced" backward in time, i.e. reconstructed, and if it could be shown that the reconstructed curve passes through the Pb-isotopic composition of Canyon Diablo troilite at $T = 4.55$ Ga.

As was recognized by Gerling more than four decades ago, lead ores represent the isotopic composition of lead in their parent rocks at the time the ores formed and thus represent the "fossil" Pb-isotopic record of some large U–Pb reservoir within the Earth, probably the lower crust or upper mantle. Strictly speaking, the periodic withdrawal of lead ores from this reservoir constitutes a violation of the closed-system assumption, but as long as the reservoir is very large relative to the amount of lead withdrawn there is no significant change in the U–Pb composition of the reservoir except by radioactive decay of uranium. Thus the reservoir, even though technically "open," behaves and may be treated as an essentially closed system. So isotopic analyses of single-stage or ordinary lead ores should, theoretically, allow us to trace the evolution of lead isotopes in the Earth through time, i.e. reconstruct the *primary terrestrial growth curve*.

The method of fitting a single-stage growth curve to ordinary leads does not require any knowledge of the age of the lead ores (although, as discussed below, such knowledge is very useful) nor does it require that the Earth be the same age as meteorites. The method does require the assumption that the Earth, or more precisely the reservoir sampled by the lead ores, evolved from an initial lead composition similar to that of meteoritic troilite.

One difficulty with the method involves the selection of samples, because in order to define a single-stage growth curve for the Earth it is first necessary to identify which ores contain single-stage leads. Since it is known that most lead ores are the products of multi-stage processes and are anomalous in composition, how can single-stage ores be identified? A potential solution to this problem was suggested in 1959 by two geologists, R. L. Stanton of Australia and R. D. Russell of Canada, who proposed that *conformable* lead-ore deposits were probably composed of single-stage leads (Stanton and Russell, 1959).[11] Conformable ores, also called *stratiform* ores, are so named because they conform to the geometry of the sedimentary beds within which they are found, i.e. they do not intrude or cross-cut the sedimentary rocks but occur as lenticular, bedded deposits within the largely volcanogenic strata that invariably enclose them. They are thought to form offshore by deposition of sulfides from seawater during sedimentation and subsequent *diagenesis*, with the metal originating near the time of deposition from volcanic eruptions that brought metal sulfides directly from the mantle or lower crust. Accordingly, these conformable ores are the same age as the enclosing sedimentary rocks and have not been modified or contaminated by multi-stage processes within the crust as have the anomalous lead deposits. Stanton and Russell observed that the isotopic composition of leads from conformable deposits were ordinary and quite uniform, whereas those from other types of deposits commonly were anomalous and highly variable. They found that the leads from the nine then-analyzed conformable deposits fit a single-stage growth curve to within a few tenths of one percent. Subsequent work has shown that Stanton and Russell's hypothesis for the origin of conformable lead ore deposits and the identification of single-stage lead ores is generally correct, although some lead ores that appear to be stratiform deposits are the result of more complicated processes and contain leads of anomalous composition.

During the 1960's and early 1970's, the definition and refinement of the primary terrestrial growth curve was a major goal of lead-isotope studies. Conformable ores are not numerous but results from the dozen or so then-analyzed deposits appeared to fit rather precisely a single-stage growth curve that passed through the (then known) composition of Canyon Diablo troilite and substantiated an age of 4.55 Ga for the Earth. Figure 7.10 shows an example of one such single-stage

11. In this paper Stanton and Russell were not specifically concerned with the age of the Earth but with the origin of the various types of Pb-ore deposits.

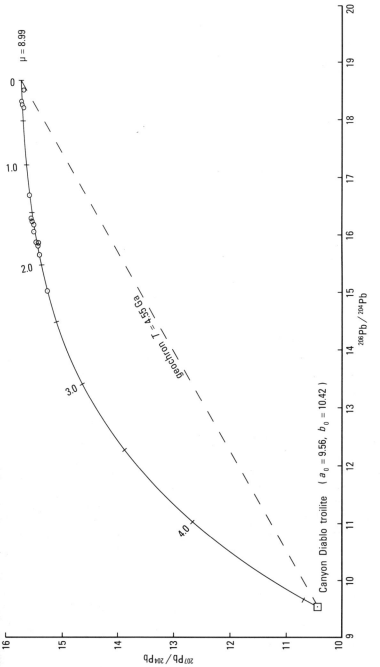

Fig. 7.10. The isotopic compositions of 14 conformable leads fit a single-stage growth curve if the old uranium decay constants (λ_{235} = 9.722 × 10^{-10}/yr, λ_{238} = 1.537 × 10^{-10}/yr), the value for Canyon Diablo determined by Murthy and Patterson (1962) (Table 7.2), and an age of 4.55 Ga for the Earth are used. (Data from compilation by Kanasewich, 1968.)

growth curve based on 14 conformable lead ores ranging in age from 0.1 to 2.2 Ga. Numerous authors of the period (for example, R. D. Russell and Farquhar, 1960, Ostic, Russell, and Reynolds, 1963, Ostic, Russell, and Stanton, 1967, R. D. Russell and Reynolds, 1965, Kanasewich, 1968, Cooper, Reynolds, and Richards, 1969, J. S. Stacey, Delevaux, and Ulrych, 1969, Doe, 1970, R. D. Russell, 1972) presented similar primary terrestrial growth curves, but all used the same basic data set, differing in only a few details, and came to the same basic conclusion that, to a first approximation, the source of ordinary leads has behaved as a single-stage system since formation of the Earth at approximately 4.55 Ga.

At the end of the 1960's, the data from lead ores seemed to be in excellent accord with the independently measured radiometric ages of meteorites. But in detail the beautiful concordance was partly fortuitous and was destined to degenerate somewhat with the more accurate measurement of the uranium decay constants and of the composition of lead in Canyon Diablo troilite. The simple and elegant single-stage model for lead evolution in the Earth was simply too good to be precisely true.

Virginia Oversby (1974), then of the Australian National University, was the first to publish a detailed analysis of the problems with the single-stage hypothesis for the evolution of lead ores. One problem arose when the new and highly precise values for the uranium decay constants (Table 3.1) and for Canyon Diablo troilite (Table 7.2) became available in 1972 and 1973, respectively. With these better values in hand, it was no longer possible to construct a single-stage growth curve that passed through the values from both conformable ores and Canyon Diablo troilite with $T = 4.55$ Ga (Fig. 7.11a). The deviation of the data from a single-stage growth curve was especially pronounced for ores younger than about 2.5 Ga.

A second problem was that the model Pb–Pb ages for the conformable ores, i.e. the age calculated from an "isochron" drawn through Canyon Diablo and through an individual ore datum, were 300–450 Ma younger than the known geological ages of the ore leads as measured by other means, both radiometric and stratigraphic. Take, for example, the galena from the Balmat deposit in New York. The known age of this deposit, determined by other radiometric methods, is about 1.06 Ga, yet the Pb–Pb model age is only about 0.7 Ga (Fig. 7.11a). This discordance is especially noticeable in the younger ore leads, which give negative model ages, i.e. plot to the right of the terrestrial geochron.

One solution to the discrepancy is to change the age of the Earth. B. R. Doe and J. S. Stacey (1974: 759–63) of the U.S. Geological

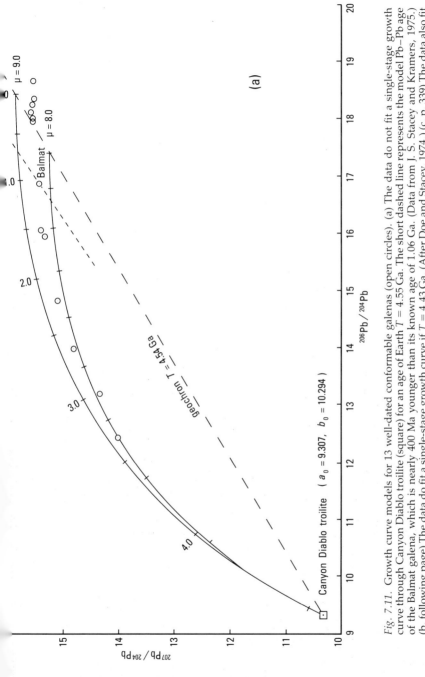

Fig. 7.11. Growth curve models for 13 well-dated conformable galenas (open circles). (a) The data do not fit a single-stage growth curve through Canyon Diablo troilite (square) for an age of Earth $T = 4.55$ Ga. The short dashed line represents the model Pb–Pb age of the Balmat galena, which is nearly 400 Ma younger than its known age of 1.06 Ga. (Data from J. S. Stacey and Kramers, 1975.) (b, following page) The data do fit a single-stage growth curve if $T = 4.43$ Ga. (After Doe and Stacey, 1974.) (c, p. 339) The data also fit a two-stage model for single-stage Pb evolution from 4.57 to 3.70 Ga followed by a second stage from 3.70 Ga to the present. (After J. S. Stacey and Kramers, 1975.) The three figures use the new U decay constants (Table 3.1) and the value for Canyon Diablo determined by Tatsumoto, Knight, and Allègre (1973) (Table 7.2).

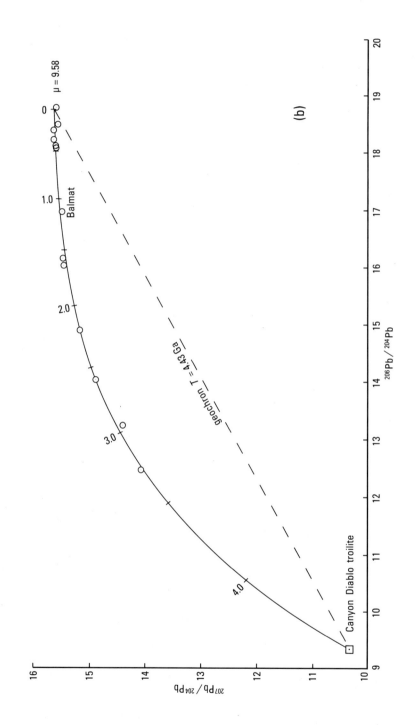

(b)

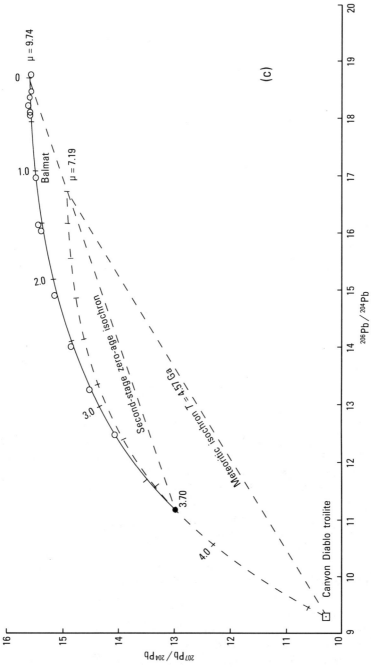

μ = 9.74

0

Balmat

1.0

μ = 7.19

2.0

Second-stage zero-age isochron

3.0

Meteoritic isochron T = 4.57 Ga

3.70

4.0

Canyon Diablo troilite

$^{206}Pb/^{204}Pb$

$^{207}Pb/^{204}Pb$

9   10   11   12   13   14   15   16   17   18   19   20

10   11   12   13   14   15   16

(c)

Survey found that the data from conformable leads fit a single-stage growth curve derived from the new decay and troilite values (Fig. 7.11b) provided that the age of the Earth, T, was adjusted downward to 4.43 Ga. This new value for the age of the Earth also brought the model ages of the ores into more reasonable agreement with their known ages. (Note, for example, the position, or model age, of the Balmat datum on the concordia curve of Figure 7.11b and compare with Figure 7.11a.) But this solution was unsatisfactory because the new T did not agree with the independently measured radiometric ages of meteorites. Doe and Stacey pointed out, as did numerous previous authors, that a more likely explanation for the imperfect fit of conformable leads to a single-stage model was that the reservoir for conformable leads had, in fact, not behaved as a single-stage system since the Earth formed. Instead, these leads, like the anomalous ones, were products of an environment in which the ratio of uranium to lead changed over time, albeit only slightly.

The demonstration that conformable leads did not precisely fit a single-stage growth curve led to the consideration of more complex models. J. S. Stacey and J. D. Kramers (1975), of the University of the Witwatersrand in Johannesburg, South Africa, proposed that the data for conformable leads could be fit quite precisely to a two-stage model consisting of single-stage growth from 4.57 to 3.70 Ga followed by a second stage of growth in a reservoir with a higher $^{238}U/^{204}Pb$ ratio (μ) from 3.70 Ga to the present (Fig. 7.11c). This change in μ at 3.70 Ga might be due, they speculated, to an episode of large-scale differentiation within the Earth that increased the value of μ in the upper mantle and lower crust. Although this model fits the data quite nicely and preserves the agreement between the age of the Earth and the age of meteorites, there is an unsatisfying lack of data to substantiate the first-stage growth and the proposed abrupt change in μ at 3.70 Ga ago.

Taking a slightly different approach, G. L. Cumming and J. R. Richards (1975) of the Australian National University developed a model, previously proposed by A. K. Sinha and G. R. Tilton (1975) of the University of California at Santa Barbara, in which the value of μ increased linearly from the time of formation of the Earth to the present. Like the Stacey and Kramers model, this model removes the large discrepancies between the Pb–Pb model ages and the known ages of the conformable ores and provides an acceptably precise fit to the conformable ore data with $T = 4.509$ Ga. There is no independent evidence that μ has increased linearly over time, but Cumming and Richards argued that their model might be a reasonable approximation of multiple episodes of differentiation and partial melting processes by which the incompatible elements, including uranium, are continually transferred upward and thus increasingly concentrated in the crust and upper mantle.

The precise history of the U–Pb "reservoir" that has acted as the source of conformable lead ores for the past 3.5 Ga or so is unknown. The Stacey and Kramers and Cumming and Richards models are probably reasonable but oversimplified approximations to a somewhat more complex situation. There is no doubt that differentiation and partial melting have transferred the incompatible elements like U and K upward, from mantle to crust, over geologic time. Probably this transfer has been a many-stage process consisting of multiple and frequent episodes irregular in both space and time. Indeed, the Earth is such a complicated and active body that it comes as no surprise that, in the strict sense, single-stage leads probably do not exist.

The abandonment of the concept of single-stage leads has resulted in the loss of some of the uniqueness of solutions for the age of the Earth. This is because models involving two or more stages have considerably more flexibility than a single-stage model and thus values for T determined from the former are not as tightly constrained by data as are solutions found from the latter. Nevertheless, the precise fit of the conformable-lead data to the two-stage and linear-increase models gives us some confidence that these models are not too far from providing a reasonable, if simplified, approximation of lead evolution in the Earth. In addition, we can have some confidence that the Earth is, indeed, part of the same lead-isotopic system as meteorites and is of the same age. Thus, the data from conformable leads provide a reasonable if qualitative connection between the age of meteorites and the age of the Earth. But there are other calculations that involve lead isotopes and result in more quantitative values for T. These will be examined in the next section.

Other (but Related) Calculations

In addition to reconstruction of the primary terrestrial growth curve, as discussed in the previous section, there are several other ways of calculating an age for the Earth from Pb–Pb data. All of these, however, rely in one way or another on the assumption that the isotopic composition of Pb in the Earth "reservoir" evolved in one or more stages from, or at least through, a primitive source now represented by the Pb in Canyon Diablo troilite. Thus, these calculations are closely related to those that involve fitting Pb data to a one- or two-stage growth curve.

One type of solution is similar to the one developed by Gerling in 1942, i.e. calculation from Equation 7.6 of the length of time re-

TABLE 7.6

Some Estimates of the "Age of the Earth" Based on Lead-Isotope Data

Basis[a]	Age[b] (Ga)	Source
Calculated with constants in use at the time of publication		
Terrestrial Pb ores	3.23, 3.95	Gerling, 1942
Terrestrial Pb ores	3.00	Holmes, 1946
Terrestrial Pb ores	2.9 ± 0.3	Houtermans, 1946
Terrestrial Pb ores	3.35	Holmes, 1947a
Terrestrial Pb ores	3.29 ± 0.20	Bullard and Stanley, 1949
Terrestrial Pb ores	5.3	Alpher and Herman, 1951
Terrestrial Pb ores	3.50	Collins, Russell, and Farquhar, 1953
Meteorite, young Pb ore	4.5 ± 0.3	Houtermans, 1953
Meteorites, modern ocean sediment	4.55 ± 0.07	Patterson, 1956
CDT, conformable Pb ores	4.56	R. D. Russell and Farquhar, 1960
Meteorites, modern ocean sediment, young Pb ores	4.55 ± 0.08	Murthy and Patterson, 1962
CDT, conformable Pb ores	4.53 ± 0.03	Ostic, Russell, and Reynolds, 1963
CDT, conformable Pb ores	4.54 ± 0.02	Ostic, Russell, and Reynolds, 1963
CDT, conformable Pb ores	4.56 ± 0.02	R. D. Russell and Reynolds, 1965
Oceanic basalts, U–Pb concordia, two-stage model	4.53 ± 0.04	Ulrych, 1967
CDT, conformable Pb ores	4.55	Kanasewich, 1968
CDT, conformable Pb ores	4.58	Cooper, Reynolds, and Richards, 1969
CDT, Manitouwadge Pb ore (2.7 Ga)	4.75[c]	Tilton and Steiger, 1969
CDT, conformable Pb ores, linear increase in μ	4.66	Sinha and Tilton, 1973
Calculated with constants now in use (Table 3.1)		
CDT, conformable Pb ores	4.43	Doe and Stacey, 1974
CDT, conformable Pb ores, two-stage model with increase in μ at 3.70 Ga	4.57	J. S. Stacey and Kramers, 1975
CDT, conformable Pb ores, linear increase in μ	4.509	Cumming and Richards, 1975
CDT, Amîtsoq gneiss feldspar (3.6 Ga)	4.47 ± 0.05	Gancarz and Wasserburg, 1977
As above, correcting for effect of metamorphism at 2.7 Ga	4.53	Gancarz and Wasserburg, 1977
CDT, Big Stubby Pb ore (3.5 Ga)	4.52	Pidgeon, 1978c
CDT, galenas and sulfides from Timmons, Ontario (2.6 Ga)	4.56 ± 0.03	Bugnon, Tera, and Brown, 1979
CDT, 20 rock units, two-stage model	4.49 ± 0.17	Manhes et al., 1979
CDT, congruency point of four oldest conformable galenas (2.7–3.4 Ga)	4.53	Tera, 1980

TABLE 7.6 (*continued*)

Basis[a]	Age[b] (Ga)	Source
CDT, congruency point of four younger conformable galenas (0.42–2.6 Ga)	4.46	Tera, 1980
CDT, congruency point of four oldest conformable galenas (2.7–3.4 Ga)	4.54	Tera, 1981

[a] CDT, Canyon Diablo troilite used for a_0, b_0.
[b] Based on single-stage Pb–Pb growth models unless noted otherwise. The values derived from the least objectionable methods are underlined (see text).
[c] 4.55 when calculated with the current constants.

quired for the composition of a Pb ore, whose age (t) is known from other data, to evolve from some primordial value (a_0, b_0). The value of T found from the equation then represents the age of the Earth, or at least the age of the Pb reservoir from which the ore was derived. Such calculations generally utilize old, primitive, conformable leads on the assumption that because old leads have spent less time evolving in the Pb reservoir than have young leads, and did so early in Earth's history, they are more likely to be single-stage leads. Also, any change in μ, such as the one proposed by Stacey and Kramers for 3.7 Ga (see previous section), will have a smaller effect on calculations that utilize an old lead than on those that utilize a young one. Not surprisingly, only a few calculations of this type are possible because there are so few really old well-dated conformable leads (Table 7.6).

One of the earliest calculations was by G. R. Tilton and R. H. Steiger (1965, 1969), then of the Carnegie Institution of Washington. They measured the Pb-isotopic composition of galena from Manitou-wadge, Ontario, Canada. This ore is associated with granitoids whose U–Pb and Rb–Sr ages are 2.72 ± 0.05 Ga (= 2.70 Ga by newer constants). Using a single-stage model (Equation 7.6), Murthy and Patterson's 1962 values for Canyon Diablo troilite (a_0, b_0), the age of the granitoids as the age of the galena (t), and the decay constants then in use they found an age for the Earth of 4.75 Ga, which, both then and now, seemed excessive in view of the age of 4.55 Ga for meteorites. Tilton and Steiger suggested three hypotheses to explain their results, none of which could then be eliminated: (1) the age of the Earth is 4.75 Ga, (2) the terrestrial primordial Pb ratio is different from that of the Canyon Diablo troilite, or (3) the Manitouwadge galena is not of single-stage origin. The discrepancy, however, has since disappeared, for when the calculation is done with the current decay constants and values for Canyon Diablo troilite, the age of the Earth found from the Manitouwadge ore is reduced to 4.53 ± 0.05 Ga.

M.-F. Bugnon, F. Tera, and L. Brown (1979) of the Carnegie Institution of Washington have made a similar calculation using galena from the Kidd Creek Mine near Timmins, an Ontario town about 550 km southeast of Manitouwadge. There five sulfides from the ore deposit, including galena, pyrite, sphalerite, pyrrhotite, and chalcopyrite, and two samples of the surrounding volcanic rock give a Pb–Pb isochron age of 2.64 ± 0.06 Ga. From Equation 7.6 Bugnon and her colleagues found a single-stage age of the source (Earth) of 4.56 ± 0.03 Ga by using values from Canyon Diablo troilite for a_0, b_0. They noted that lead deposits with single-stage model Pb ages of 2.7 Ga are widespread, occurring in Canada, Finland, India, Australia, and Rhodesia, and that their Pb-isotopic compositions are remarkably uniform even though their μ values vary. They attributed these occurrences to a global episode of volcanogenic lead deposition from a source with $\mu = 7.49$.

Another calculation of some interest is the one published by R. T. Pidgeon (1978c) of the Australian National University using lead from Big Stubby, the oldest conformable lead deposit known (Sangster and Brook, 1977). Big Stubby occurs within the Duffer Formation of the Warrawoona Group, Pilbara Block, Western Australia (Chapter 4) and contains galena with the lowest fraction of radiogenic lead found so far in any terrestrial lead deposit. Rb–Sr and U–Pb ages of a volcanic unit within the Duffer Formation show that Big Stubby is 3.5 Ga old (Table 4.4). Pidgeon used the U–Pb zircon age of the Duffer Formation (3.452 ± 0.016 Ga), the Pb-isotopic composition of the Big Stubby galenas, and the Pb-isotopic composition of Canyon Diablo troilite to calculate an age for the Earth of 4.52 Ga based on the assumption of single-stage evolution. According to Pidgeon, this age represents a maximum age for the formation of the Earth's core and the establishment of the primitive Pb-isotopic reservoir that was the source for the Big Stubby galenas. Comparing this age with the age of 4.55 Ga for chondritic meteorites, Pidgeon concluded that the Earth's core formed within 100 Ma after accretion.

A. J. Gancarz and G. J. Wasserburg (1977) of the California Institute of Technology have made a similar calculation using feldspar crystals from the 3.6-Ga Amîtsoq gneisses of western Greenland. Feldspar largely excludes U from its crystal lattice and so U is present in it only in trace amounts. The composition of Pb in feldspar, therefore, is very close to that of the source rocks from which the magma was derived and only a small correction need be made for the decay of U to Pb that has occurred within the feldspar from the time the feldspar formed, t, to the present. Gancarz and Wasserburg measured the Pb-isotopic composition of both potassium feldspar and plagioclase feldspar from four samples of the Amîtsoq gneisses and corrected the values for the Pb generated by decay of U in the feldspar. These corrections were small—only 0.5% for the $^{206}Pb/^{204}Pb$ ratio and 0.2% for the $^{207}Pb/^{204}Pb$ ratio—and resulted in mean corrected (initial) values of $^{206}Pb/^{204}Pb = 11.468$ and $^{207}Pb/^{204}Pb = 13.203$. Using the composition of Canyon Diablo troilite for a_0, b_0 and a U–Pb zircon age of 3.59 $\pm$ 0.05 Ga for the Amîtsoq gneisses, they found a single-stage model age for the Earth (Equation 7.6) of 4.47 $\pm$ 0.05 Ga and a value of $\mu = 8.5$ for the interval 4.47–3.59 Ga (Equation 7.5).

Because of the probability that the lead in the feldspars of the Amîtsoq gneisses did not evolve as a single-stage system, Gancarz and Wasserburg also considered a two-stage model in which the lead in the Amîtsoq feldspars was reequilibrated during the well-documented metamorphism that occurred at 2.7 Ga in western Greenland. They found that the age of the Earth was 4.53 Ga for a two-stage model in which μ in the source reservoir was 7.5 from 4.53 to 3.59 Ga ago and μ in the gneisses was 0.5 from 3.59 to 2.7 Ga ago.

F. Tera (1980, 1981) developed a slightly different and rather elegant method of determining the age of the Earth from ancient conformable lead deposits. Unlike the previous methods, this one does not require that the ages of the lead deposits be known but instead is based on the assumption that age-composition profiles of galenas of different ages but from a common source must have a single point of congruency that defines unique values for both μ and T.

Tera used lead-isotope data for the four oldest conformable galenas known (Table 7.7).[12] For each galena he assumed various values of T (age of source) and calculated the corresponding values of t (age of ore) using the measured $^{207}Pb/^{206}Pb$ ratio of the galena, values from the Canyon Diablo troilite for primordial lead (a_0, b_0), and Equation 7.6. For each pair of T and t values he then calculated the corresponding μ_s, which is the present-day value of μ for the source, using Equation 7.1. The calculated values of T and μ_s provide the data from which a "source profile" of T vs μ_s for the galena (Fig. 7.12) can be constructed. The profile is a unique function of the lead composition of the analyzed galena, but only one point on the curve can be correct, i.e. only one point represents the true composition (μ_s) and age (T) of the source. From the data of a single galena it is not possible to

12. Manitouwadge galenas have been analyzed by both Tilton and Steiger (1969) and J. S. Stacey and Kramers (1975), but the results of the latter workers are inconsistent with the otherwise uniform values found for all lead deposits in the Abitibi Belt of Ontario, Canada. On this basis Tera (1980) omitted the data of Stacey and Kramers from the profile analysis, and they are likewise omitted from Table 7.7 and Figure 7.12.

TABLE 7.7

Lead-Isotopic Data and Ages for Galenas Whose Source Profiles Are Shown in Figure 7.12

Locality	Age (Ga)	$\frac{^{206}Pb}{^{204}Pb}$	$\frac{^{207}Pb}{^{204}Pb}$	Model age[a] (Ga)	Source
Timmons, Ontario	2.64	13.27	14.48	2.70	Bugnon, Tera, and Brown, 1979
Manitouwadge, Ontario	2.68	13.30	14.52	2.71	Tilton and Steiger, 1969
Barberton, South Africa	3.23	12.46	14.08	3.13	J. S. Stacey and Kramers, 1975
Big Stubby, Western Australia	3.50	11.95	13.72	3.38	Sangster and Brook, 1977

[a]Calculated under the assumption that $T = 4.53$ Ga (Tera, 1980: 526).

determine this point. If we presume, however, that the four ancient galenas of Table 7.7 originated from the same source, then the individual profiles from these galenas should intersect at a point that represents the true age and composition of the source. For the four analyzed galenas, this congruency point is $T = 4.53$ Ga and $\mu = 8.04$ (Fig. 7.12).

The type of source profile shown in Figure 7.12 is only one of several that can be constructed from the same basic data by using quantities calculated from Equations 7.1 through 7.6. For example, Tera (1981) has also constructed source profiles for the same four ancient galenas by plotting the radiogenic $^{207}Pb/^{206}Pb$ against the source composition expressed as $^{204}Pb/^{206}Pb$. In this analysis he obtained a congruency point that translates to $T = 4.54$ Ga and $\mu = 8.0$.

As Tera observed, it is probably significant that ancient galenas from three continents seem to define a common source with a common age and lead composition, and also that the age obtained is similar to the age determined for meteorites. It is also satisfying that the model ages calculated for the galenas by the congruency result are within a few percent of the ages measured by other dating methods (Table 7.7). The precise nature of the event of 4.53–4.54 Ga is not entirely clear, but it is Tera's opinion that the age represents the time of U–Pb fractionation in the primary materials from which the Earth was formed. If this fractionation occurred at the time the Earth accreted, then the age is the age of the Earth; if not, then it represents the age of the debris from which the Earth was formed. Alternatively, the fractionation might be the result of separation of the Earth's materials into core and mantle. Regardless of the precise interpretation, the near equivalency of this "age of the Earth" and the ages of the

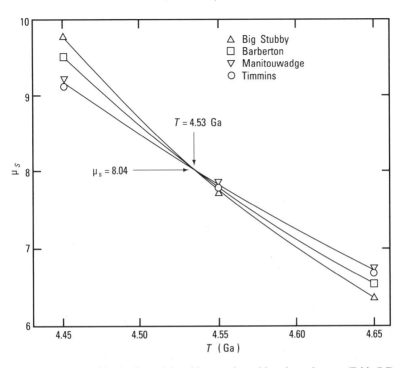

Fig. 7.12. Source profiles for four of the oldest conformable galenas known (Table 7.7). The profiles intersect at a congruency point corresponding to a parental source that is 4.53 Ga with $\mu_s = 8.04$. Because the galenas of the Abitibi belt, to which both Timmons and Manitouwadge belong, seem to be homogeneous with regard to Pb-isotopic composition, a single profile was constructed for these two localities. (After Tera, 1980.)

primitive meteorites indicates that the condensation of solid matter from the Solar Nebula, the formation of the meteorites' parent bodies, and the formation of the Earth as a planet occurred within a period of only about 20 Ma or so.

Tera (1980) has also applied his method to younger conformable ores, but the results are somewhat different. Galenas from Geneva Lake, Ontario (2.60 Ga), Cobalt, Ontario (2.17 Ga), Broken Hill, Australia (1.66 Ga), and Captain's Flat, Australia (0.42 Ga), have a congruency point at $T = 4.46$ Ga and $\mu = 9.3$. This result seems to indicate that the younger conformable deposits were derived from a younger source more enriched in uranium relative to lead than the source of the more ancient galenas shown in Figure 7.12. Tera concluded that "age of the Earth" calculations based on conformable leads younger than about 2.65 Ga may give results that are closer to the age of this second source than to the formation of the Earth. This conclusion is consistent with other evidence (for example, J. S. Stacey and Kramers, 1975), discussed above, that the younger conformable lead deposits are the products of multi-stage processes.

On the basis of his method and an evaluation of the calculations of previous workers Tera concluded that the least objectionable methods (Table 7.6)[13] fall within the range 4.53 to 4.56 Ga and that the best "age of the Earth" (age of source) is the median of this range, or 4.54 Ga.

In addition to those based on lead minerals from conformable lead-ore deposits and feldspar crystals, attempts have been made to use rock bodies to calculate an age for the Earth. An early attempt was by T. J. Ulrych (1967), of the University of British Columbia, who used U–Pb isotopic data of Zartman (1965) and Tatsumoto (1966a,b) on basaltic lava flows from Hawaii, Texas, Japan, Easter and Guadalupe islands, and mid-ocean ridges in both the Atlantic and Pacific oceans. From the lead-isotopic composition of each lava, Ulrych subtracted Murthy and Patterson's 1962 value for a_0, b_0 (Table 7.2), then plotted these "corrected" data from each area on a U–Pb concordia diagram. Ulrych found that the data from each area formed a linear array. His plot for eight basalts from Hawaii is shown as Figure 7.13a. According to Ulrych's interpretation, the upper intercept of the line with the concordia represents the age of the Earth and the lower intercept represents the age of the source reservoir from which the rock was extracted at the time of eruption. His results for the seven rock suites are summarized in Table 7.8. Ulrych's ages for the Earth range from 4.27 to 4.58 Ga and the source ages range from 1.23 to −1.8 Ga, i.e. 1.8 Ga in the future.

Implicit in Ulrych's model are the following assumptions:

1. The $^{238}U/^{206}Pb$ ratio, μ, of each rock is representative of the source reservoir from which it was derived.

2. The source reservoir of the basalt was formed at a time, t, represented by the lower intercept of the line with the concordia. This source reservoir was formed, presumably in the upper mantle, from some initial "parent" reservoir without fractionation or addition of lead of some different isotopic composition.

3. The parent reservoirs were formed in the mantle at the time of formation of the Earth, T, remained closed systems until time t, and had an initial Pb-isotopic composition represented by the Pb in the troilite phase of iron meteorites.

This model would, indeed, produce linear arrays on the concor-

13. In particular, Tera cited the calculations of Tilton and Steiger (1969), Gancarz and Wasserburg (1977), and Bugnon, Tera, and Brown (1979), as well as his own. Although not evaluated by Tera, the calculation of Pidgeon (1978c) also meets Tera's criteria as one of the "least objectionable" methods.

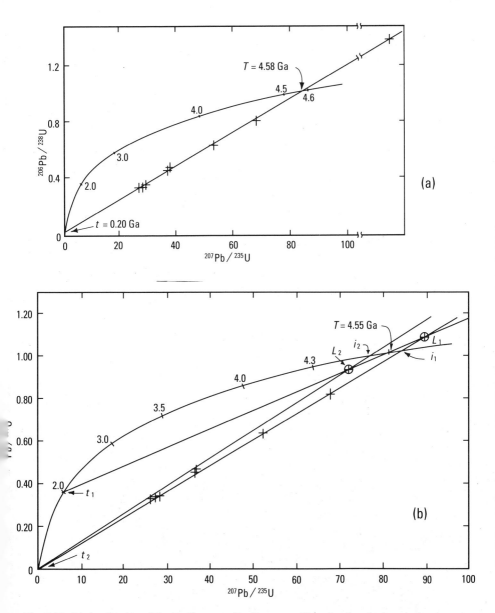

Fig. 7.13. (a) Application of the U–Pb concordia diagram to Hawaiian basalts (crosses) to determine the age of the Earth, *T*. (After Ulrych, 1967, copyright by the AAAS.) (b) Concordia plot showing the effect of hypothetical differentiation of the mantle into two independent reservoirs, L_1 and L_2, at 2.0 Ga (Hawaiian data from Tatsumoto, 1966a). In this case the upper intercept, *i*, has no physical meaning. (After Oversby and Gast, 1968, copyright by the AAAS.)

TABLE 7.8

Results Obtained by Ulrych (1967) from U–Pb Concordia Treatment
of Data on Basaltic Lava Flows

Location of samples	Age of Earth[a] (upper intercept) (Ga)	Age of source reservoir	
		Lower intercept (Ga)	Slope (Ga)
Mid-Atlantic ridge	4.51	1.15	1.23
East Pacific rise	4.54	0.20	0.15
Hawaii	4.58	0.32	0.09
Japan	4.56	0	−0.18
Easter Island	4.46	−0.10	0.12
Llano, Texas	4.53	1.00	1.03
Guadalupe Island	4.27	−1.80	−0.10
Mean	4.53 ± 0.04		

[a]Calculated with old decay constants.

dia diagram whose upper and lower intercepts had the meaning attributed to them by Ulrych, but V. M. Oversby and P. W. Gast (1968), then of Columbia University's Lamont Geological Observatory, have pointed out some serious objections to Ulrych's analysis. First, it is highly unlikely that basalt would form from a source rock (reservoir) without some change in μ. Such a scenario would require fractionation of U and Pb by processes unrelated to the formation of the rock, and from what is known about the processes of basalt formation this is geochemically unreasonable. In addition, several of the Hawaiian samples (as well as many others) have high U/Pb ratios but low ^{206}Pb/^{204}Pb ratios, and such samples could not possibly have formed from a source with the same U/Pb ratio.

Second, for samples derived from a homogeneous source by a single episode of fractionation, the age found from the lower concordia intercept should agree with the age found from the slope of the discordia line. For many of the sample suites studied by Ulrych, this is clearly not the case, as shown by the data from Hawaii, Japan, Easter Island, and Guadalupe Island (Table 7.8).

Finally, Oversby and Gast pointed out that the concordia method is relatively insensitive to multiple episodes of fractionation and may result in an apparent upper intercept that has no physical meaning. Their hypothetical two-stage example is shown as Figure 7.13b. Suppose that a reservoir formed at $T = 4.55$ Ga and that at some later time, $t_1 = 2.0$ Ga, the reservoir was fractionated into two secondary reservoirs, L_1 and L_2. Note that points L_1 and L_2 fall on a line whose upper intercept is the age of the initial reservoir (Earth) and whose lower intercept is the time of fractionation, in accord with Ulrych's interpretation. Now further suppose that reservoir L_1 was fractionated at time $t_2 = 0$ (today) and that samples were withdrawn from this system in the form of basalts derived by partial melting of the new reservoir. These basalts would form a new linear array whose lower intercept would be t_2 but whose upper intercept would be meaningless; the slope of the discordia would likewise have no physical meaning. This is because the upper intercept and the slope of the line are controlled by the extent to which uranium and lead were fractionated during the first fractionation episode. The same principle applies to basalts generated from secondary reservoir L_2. To complicate matters further, if the basalts were generated from several subreservoirs formed from system L_1, then the lower intercept would not necessarily have any time significance. Oversby and Gast pointed out that the disagreement between the lower intercepts and the slopes of the linear arrays (Table 7.8) is indicative of one or more episodes of intermediate fractionation.

If the concordia method is invalid for the oceanic basalts, why are the upper intercepts near 4.5 Ga? The method of Ulrych gives an approximate value for the age of the Earth simply because fractionation in natural uranium–lead systems does not cause large changes in composition and because the concordia method is relatively insensitive to multiple episodes of fractionation. Thus, the upper intercepts are usually near the age of the Earth. But the interpretation of the linear arrays of data from basalts is ambiguous, and the intercepts and slopes do not necessarily specify certain times. It is now known from extensive geochemical studies (not discussed here) that oceanic basalts are the products of complex processes involving multiple episodes of fractionation in the mantle and that Ulrych's interpretation is not physically realistic.

A more recent attempt to obtain the age of the Earth from lead-isotopic data on rocks is that of G. Manhes et al. (1979) of the University of Paris. They used results from 20 crystalline rock units of assorted compositions and ages (0.05 to 3.6 Ga) for which Pb–Pb data from each unit are sufficient to define an isochron. Instead of treating the data with conventional growth-curve systematics, however, they devised a new type of diagram, called the I vs S diagram, which is derived directly from the Pb–Pb growth curve and isochron diagram and its related equations.

Consider a rock that began with some initial Pb ratio, a_1, b_1, and evolved as a closed system over time t. Pb-isotope ratios from this unit would plot on an isochron with slope

$$S = \frac{1}{137.88} \frac{(e^{\lambda_1 t} - 1)}{(e^{\lambda_2 t} - 1)}$$

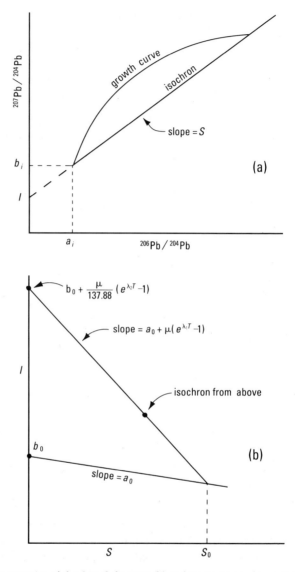

Fig. 7.14. Systematics of the I vs S diagram (b) and its correspondence to the Pb–Pb isochron/growth curve diagram (a). Note that the isochron for a rock body in (a) is represented by a point in (b). S_0 is a direct function of the age of the reservoir, T. (After Manhes et al., 1979.)

proportional to its age (Fig. 7.14a). This isochron passes through a_1, b_1, although that point cannot be directly determined. This isochron can be extrapolated, however, to find the hypothetical value $I = {}^{207}Pb/{}^{204}Pb$ at ${}^{206}Pb/{}^{204}Pb = 0$. Thus the isochron for the rock body can be characterized by the two values I and S, which plot as a point on the I vs S diagram (Fig. 7.14b). The isochron can also be described by the equation for a straight line

$$I = ({}^{207}Pb/{}^{204}Pb) - ({}^{206}Pb/{}^{204}Pb)S \qquad (7.9)$$

Note that all points on the isochron obey this equation, including the initial Pb ratio a_1, b_1, so that

$$I = b_1 - a_1 S \qquad (7.10)$$

Now consider several rock bodies (or suites of cogenetic rock units) of different ages that were derived from the same Pb-isotopic reservoir, which evolved as a closed system since the Earth formed. The composition of the initial leads in each of these bodies can be described as follows:

$$a_1 = a_0 + \mu(e^{\lambda_1 T} - e^{\lambda_1 t}) \qquad (7.11)$$

$$b_1 = b_0 + \frac{\mu}{137.88}(e^{\lambda_2 T} - e^{\lambda_2 t}) \qquad (7.12)$$

where T, a_0, and b_0 have their usual meanings and t is the age of the individual rock body. To describe this system of two-stage leads in terms of I and S, we substitute the values of a_1 and b_1 from equations 7.11 and 7.12 into Equation 7.10 and simplify to obtain

$$I = \left[b_0 + \frac{\mu}{137.88}(e^{\lambda_2 T} - 1) \right] - \left[a_0 + \mu(e^{\lambda_1 T} - 1) \right] S \quad (7.13)$$

This equation describes a straight line on the I vs S diagram whose intercept (first term in the equation) and slope (coefficient of the second term in the equation—see Fig. 7.14b) are a function of the age of the Earth, T, and the initial ratio of the Earth reservoir, a_0, b_0. Thus, the Pb-isotopic data from these individual two-stage rock bodies of different ages but from a common reservoir are related in such a way that they will define a straight line on the I vs S diagram. This line describes the closed-system Pb "growth curve" (although it is a line and not a curve on the I vs S diagram) for the Earth reservoir.

Manhes and his colleagues took high-quality Pb–Pb isochron data from 20 crystalline rock bodies ranging in age from 54 Ma (the Skaergaard layered ultramafic complex of Greenland) to 3.5 Ga (the

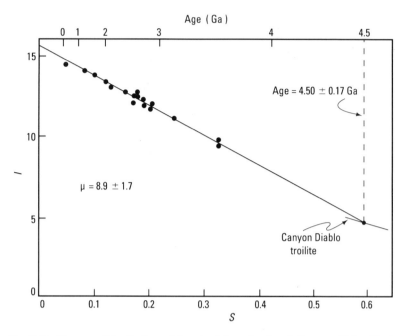

Fig. 7.15. The age of the Earth as determined from the *I* vs *S* diagram. (After Manhes et al., 1979.)

Isua supracrustal rocks described in Chapter 4) and found the straight line that best fit the data for the *I* vs *S* representation (Fig. 7.15). The intersection of this line with the straight line representing a_0, b_0 as determined from Canyon Diablo troilite yields an age for the reservoir (Earth) of 4.49 ± 0.17 Ga, which Manhes and his co-workers characterized as the "statistical best age of the Earth."

The principal advantage of the *I* vs *S* diagram seems to be the ease with which linear statistical methods can be applied to determine a "best-fit" straight line to the data as opposed to the methods required to fit two-stage data to growth curves on the conventional Pb–Pb diagram. Note, however, that the *I* vs *S* treatment differs from the growth curve method only in its mathematical approach and graphical representation and that there is no fundamental *physical* difference between the two techniques. The *I* vs *S* method is still fundamentally a two-stage calculation; it requires some value for a_0, b_0 (Canyon Diablo troilite) and the assumption of two (and only two) stages of Pb-isotopic evolution. It is this latter assumption that is open to serious question. As Tera (1981) has observed, it is highly unlikely that all of the leads in the rock bodies chosen by Manhes and his colleagues are the products of simple two-stage evolution. Many, especially the younger bodies, are probably the products of multi-stage processes. In addition, the method lacks the precision of several of the other methods reviewed above. This too is probably a result of incorporating data from multi-stage leads in the calculation. Thus, the "age of the Earth" of 4.49 ± 0.17 Ga calculated by Manhes and his colleagues is probably at best a first-order approximation.

Summary

The best value for the age of the Earth, 4.54 Ga, is based on a single-stage model for the evolution of lead isotopes in the Earth using data from a few ancient lead ores and from one special iron meteorite, Canyon Diablo. Thus, the "age of the Earth" is really the age of the meteorite–Earth system and probably represents the last time that the isotopic composition of lead was uniform throughout the Solar System, i.e. the time when solid bodies first formed from the Solar Nebula.

Early attempts by Gerling, Holmes, Houtermans, and others to determine the age of the Earth by using lead isotopes yielded values that were too low, primarily because there was no good way to determine the isotopic composition of primordial lead, but also because the samples used were not of single-stage origin. The problem of primordial lead composition was solved in 1953 by Patterson, who discovered that the composition of primordial lead was preserved in the troilite phase of certain iron meteorites, especially Canyon Diablo. In the same year Houtermans, using Patterson's value for a_0, b_0, calculated the first reasonable estimate of the age of the Earth: 4.5 ± 0.3 Ga.

Internal Pb–Pb isochrons for individual meteorites and whole-rock isochrons for several compositional groups of meteorites all fall within the range 4.53–4.58 Ga. These isochrons also pass through the composition of Canyon Diablo troilite, as they should if all types of meteorites originated from the same U–Pb system. Whole-rock Pb–Pb model ages for a large number of individual meteorites also fall within the range 4.53–4.57 Ga and have a mean of 4.550 ± 0.025 Ga.

The lead-isotopic composition of modern sediments and of modern single-stage (or near single-stage) lead ores fall very close to and on either side of the meteoritic geochron. In addition, the growth curve formed by the dozen or so well-documented conformable lead ores passes close to the composition of Canyon Diablo troilite. Thus, there is good reason to conclude that, to a first approximation, the meteorites and the Earth are the same age and can be treated as a single system for the evolution of lead isotopes.

The best values for the age of the Earth are based on the time required for the most ancient leads, which approximate single-stage systems, to evolve in isotopic composition from the value of Canyon Diablo troilite to their values at the time of separation from their mantle reservoirs. These calculations result in ages for the Earth of 4.52 to 4.56 Ga. Probably the best value is 4.54 Ga, found by Tera using the congruency point of the four oldest conformable lead ores. This value, which is known to within 1% or better, is consistent with the ages of meteorites, the ages of the oldest lunar samples, and the ages of the oldest Earth rocks.

CHAPTER EIGHT

The Universe and the Elements: Indirect Evidence

Ice is the silent language of the peak; and fire is the silent language of the star. CONRAD AIKEN [1]

In the preceding chapters, we have discussed the various radiometric methods of determining the ages of the Earth, the Moon, and the meteorites. These methods, based on the changes over time of the naturally occurring long-lived radioactive isotopes, lead to the incontrovertible conclusion that the solid bodies of the Solar System formed 4.54 Ga. Suppose, for the moment, that these radiometric methods did not exist or had not yet been discovered. Is there other physical evidence concerning the age of our universe? Are there other methods that substantiate the great antiquity of the Solar System? The answer to this question is yes, and the evidence is found in the properties of the very small, elements, and in the very large, the stars and galaxies.

There are several qualitative and quantitative ways to estimate the ages of the Solar System, the Galaxy, the elements, and the universe. Currently most of these methods do not lead to precise age values, but they do provide substantial evidence that our cosmic surroundings are billions, rather than thousands or even millions, of years old. These indirect methods are relevant to the theme of this book, because they provide consistency and support for the more direct radiometric techniques discussed in the preceding chapters. Thus, it will be worthwhile to review briefly the more important of these methods and the insight that they provide concerning the age of the Earth and its environment.

1. Sonnet 10, 1940.

Messages from the Stars

Expansion of the Universe and the Hubble Constant

As discussed in Chapter 1, the universe is expanding, or inflating. A result of this expansion is that each of the galaxies in the universe is moving away from all of the other galaxies, provided that local motions are excluded. As observed from the Earth, therefore, all of the distant galaxies are receding.

Astronomers can measure the velocity of distant objects either toward or away from the Earth because of the Doppler effect, or "red shift." As any source of electromagnetic radiation moves toward or away from an observer there is a velocity-dependent shift in both the wavelength and the frequency of the radiation detected by the observer. For non*relativistic speeds*, i.e. velocities that do not approach the speed of light, this shift is expressed by the formula

$$\lambda' = \lambda_0(1 + v/c) \tag{8.1}$$

where λ' is the observed wavelength of a given spectral line, λ_0 is the rest wavelength (the wavelength if the relative velocity is zero), v is the relative velocity of the source with respect to the observer, and c is the speed of light. Because λ' can be measured directly, Equation 8.1 can be solved for the velocity

$$v = c(\lambda'/\lambda_0 - 1)$$

provided that the rest wavelength is known, which it usually is, because the characteristic spectra of most types of stars have been determined. Note that when an object is moving toward the observer, v is negative and λ' is smaller than λ_0 (blue shift). Conversely, when an object is moving away from an observer, v is positive and λ' is larger than λ_0 (red shift).

In 1912 V. M. Slipher of the Lowell Observatory near Flagstaff, Arizona, observed the red shift and interpreted it as relative motion of the galaxies away from the Milky Way. Seventeen years later, E. P. Hubble (1929), of the Mt. Wilson Observatory in California, made the important observations that the motion of galaxies away from the Milky Way was isotropic, i.e. independent of the direction of observation, and that the relative velocity was proportional to distance. This relationship has been repeatedly verified and is now known as *Hubble's law*

$$v = H_0 d \tag{8.2}$$

where v is velocity, or *Hubble flow*, in kilometers per second, d is distance in *megaparsecs* (Mpc),[2] and H_0, the constant of proportionality, is the *Hubble constant*.

Equation 8.2 is an interesting equation. Not only does it relate velocity to distance but it also can be used to estimate the age of the universe. This is so because H_0 is in units of kilometers per second per megaparsec, and since both kilometers and megaparsecs are measures of distance, $1/H_0$ gives the time, known as the *Hubble time* or *Hubble age*, since the inflation of the universe began, provided that the inflation has been linear. The assumption about the linearity of inflation is probably incorrect, but that point will be discussed later. As it turns out, the units of H_0 are rather convenient: when megaparsecs are converted to kilometers and seconds are converted to years the result is that $1000/H_0$ is in units of Ga to within a few percent. As we shall see, the errors in the various estimates of H_0 are considerable, and so it is customary to use $1000/H_0$ (commonly if erroneously noted simply as $1/H_0$) as the Hubble time in Ga.

Although Equation 8.2 is straightforward and uncontroversial, its application to the universe is not without difficulty because of the problems of accurately measuring both velocities and distances. The measurement of relative velocities by using the Doppler shift is straightforward and accurate, but superimposed on the velocities due purely to universal expansion are the local motions of stars within galaxies, of galaxies within groups of galaxies, and of groups of galaxies relative to other groups. For close objects these local motions are very large in comparison to the Hubble flow, and so the v in Equation 8.2 is virtually impossible to measure. As an extreme example, the Hubble flow cannot be measured by observing the relative motion between the Sun and its nearest neighbor star Alpha Centauri, which is only 4×10^{-7} Mpc away, because the observed motion is entirely due to movements of the two stars within the Milky Way galaxy. For more distant objects, from a few to a few tens of megaparsecs away, corrections can be made for the local motions. For objects with distances that exceed 50 Mpc the local motions become negligible in comparison to the magnitude of the Hubble flow, and so the most accurate Hubble-flow velocities are commonly obtained on objects at some distance from the Milky Way.

Although there are some problems in measuring Hubble velocities for nearby objects, the main difficulty in finding the Hubble constant lies with the measurement of distances to extragalactic ob-

---

2. A parsec is the distance at which a star has a parallax of $1''$ of arc as viewed from opposite sides of Earth's orbit about the Sun. It is equivalent to 3.09×10^{13} km.

jects. Hodge (1981) listed 27 extragalactic distance indicators of vary-ing degrees of usefulness and with ranges of from a few megaparsecs to several thousand megaparsecs.[3] The most common distance indica-tors are the *standard candles*. These are based on the concept that cer-tain types of objects, regardless of their location, give off the same amount of radiant energy and so have the same *intrinsic brightness*, or *luminosity* (L). If this is so, then it is possible to calculate the distance to a standard candle from its apparent brightness (f) because the light diminishes as an inverse function of its distance (d) squared

$$f = L\sqrt{4\pi d^2}$$

Certain individual stars are often used as standard candles, and one of the most widely used is the period–luminosity relationship for *Cepheids*, a class of variable or pulsating stars whose mean luminosi-ties are a function of their periods. For reasons too complicated to ex-plain here, Cepheids expand and contract with periods of from 1 to 50 days.[4] As a Cepheid expands its surface temperature and luminosity increase, and as it contracts they decrease. The mean luminosity of any given Cepheid is a regular and predictable function of its period of expansion and contraction, with the longer-period Cepheids hav-ing the higher mean luminosities.

An example of the use of Cepheids to determine distances is a study by Allan Sandage (1986b) of the Mt. Wilson Observatory. San-dage determined the distances to 18 nearby galaxies by using data pri-marily from Cepheids. When the distances are plotted as a function of velocity, as determined from red shifts and corrected for the gravi-tational effects of the Local Group of galaxies, the data fall about a straight line whose slope gives a value of H_0 of 55 km/s/Mpc (Fig. 8.1) and a Hubble age of 18 Ga.

In general, standard candles based on individual stars are useful only at distances of a few megaparsecs. Beyond that ordinary stars cannot be adequately resolved and astronomers must rely on the properties of exceptional or larger objects, like novae, supernovae, and galaxies. For example, Arnett, Branch, and Wheeler (1985) calculated the expected luminosity of Type I supernovae, each of which is the product of an exploding white dwarf star. Their calculations are based on the theoretical behavior of gamma rays and positrons emitted dur-

3. A megaparsec is thus equivalent to 3.09×10^{19} km. For reviews of the extra-galactic distance scale and determination of the Hubble constant see Hodge (1981), van den Bergh (1981, 1984), or de Vaucouleurs (1986). Many modern astronomy texts, such as Shu (1982) and Rowan-Robinson (1985), also discuss these subjects.
4. Variable stars with periods of less than one day are called RR Lyrae variables, while those with periods greater than 50 days are known as long-period variables. The classification is somewhat arbitrary.

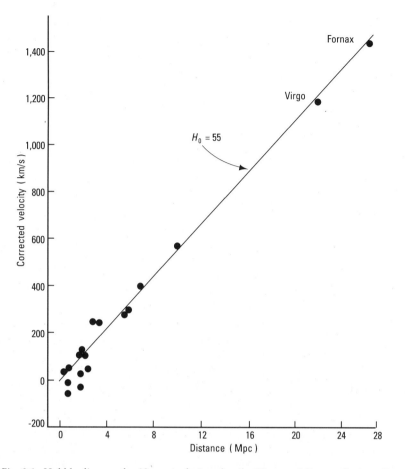

Fig. 8.1. Hubble diagram for 18 near galaxies plus the Virgo and Fornax clusters. Velocities are corrected for gravitational effects of the Local Group. Distances are based on the luminosities of Cepheid variables, bright stars, and type I supernovae. (After Sandage, 1986b, © The American Astronomical Society.)

ing the radioactive decay of transition-metal elements created during the thermonuclear disruption of the white dwarf. To avoid problems introduced by local motions, they applied their findings to six supernovae in galaxies with Hubble velocities in excess of 3,000 km/s (d greater than 50 Mpc) and obtained values for H_0 ranging from 39 to 73 km/s/Mpc with a best estimate of 59 km/s/Mpc, corresponding to a Hubble age of 17 Ga.

Another class of distance indicators is the *standard ruler*, which is based on objects of presumably constant dimensions whose apparent size decreases as a function of distance. An example of a standard ruler is the diameter of the giant H_{11} regions surrounding certain very hot newborn stars. Such stars emit enormous quantities of ultraviolet radiation that ionizes the atoms in a spherical region (the H_{11} region) of the surrounding hydrogen gas cloud. The radius of this sphere of ionization is stabilized by the distance over which the dynamic balance between ionization and recombination of the hydrogen atoms is maintained, and for a given type of star the radius is thought to be constant. Because the recombination causes fluorescence, the H_{11} sphere is visible. R. C. Kennicutt (1979), refining the earlier work of A. Sandage and G. A. Tammann (1974a, b) has measured the diameters of H_{11} regions of stars in distant spiral galaxies, calculated their distances, and found a value for H_0 of 65 km/s/Mpc, for which the Hubble age is 15 Ga.

More recently, R. Florentine-Nielsen (1984) of the Copenhagen University Observatory used a distance indicator that depends on neither luminosity nor size but instead on the time delay between double images of a gravitationally lensed quasar with the somewhat unglamorous designation of QSO0957+561. As viewed from the Earth, this quasar has two images because the gravity of a cluster of galaxies in the line of sight is acting as a lens. Because the light that produces each of the images travels a separate path and hence a different distance through this gravitational lens, there is a time delay between identical events displayed by the two images. Like all quasars this "double" quasar shows occasional increases in brightness. In 1982, the apparent brightness of the northern image increased by 30% and then returned to its former value the following year. In 1984 the southern image began to repeat the same event. Florentine-Nielsen estimated the time delay between the two images' change in brightness to be 1.55 ± 0.1 years and, from a theoretical model of the gravitational lens, derived a value of H_0 of 77 km/s/Mpc, which corresponds to a Hubble age of 13 Ga.

As is apparent from these few examples, different methods for estimating H_0 and the Hubble age yield quite different results. In fact, recent estimates of H_0 range from about 50 to 110 km/s/Mpc (Table 8.1). Errors in velocity account for some of this variation, but most of the variation is due to uncertainties in the various ways of estimating distance. Calibration of the standard candles and standard rulers, so that they yield absolute rather than relative distances, is an especially knotty problem for several reasons.[5] First, the calibrations depend, to a large extent, on each other. For example, some indicators, like Cepheids and novae, can be calibrated geometrically within the Milky

5. Rowan-Robinson (1985) and Hodge (1981) have given particularly good and easy-to-follow discussions of the distance calibration problems. De Vaucouleurs (1986) also discussed the subject at length but on a more technical level.

TABLE 8.1

Some Recent Representative Estimates of the Hubble Constant (H$_0$) and the Inferred Hubble Ages of the Universe (1,000/H$_0$) Based on Linear Expansion

Distance method[a]	H$_0$ (km/s/Mpc)	1,000/H$_0$ (Ga)	Source
T–F, distant spiral galaxies	50.3 ± 4.3	20 ± 2	Sandage and Tammann, 1976
Super-luminal expansion of four radio galaxies	110 ± 10	9	Lynden-Bell, 1977
Luminosity of globular clusters in Virgo cluster galaxies	80 ± 11	12 ± 2	Hanes, 1979
Luminosity of 159 galaxies in Local Group	50.8	20	Visvanathan, 1979
Diameters of H$_{II}$ regions in 36 galaxies of Virgo cluster	65	15	Kennicutt, 1979
T–F for 34 distant spiral galaxies	95 ± 4	10 ± 0.4	Aaronson et al., 1980
Luminosity of H$_{II}$ regions in 21 galaxies of Virgo cluster	55	18	Kennicutt, 1981
T–F for 23 galaxies in Virgo cluster	65 ± 4	15 ± 1	Mould, Aaronson, and Huchra, 1980
Luminosity of 150 near and distant galaxies	75 ± 15	13 ± 3	Stenning and Hartwick, 1980
Luminosity of 200 spiral galaxies	96 ± 3	10 ± 0.3	de Vaucouleurs and Peters, 1981
R luminosities of galaxies in Andromeda cluster	83	12	van den Bergh, 1981
R luminosities of galaxies in M33 cluster	89	11	van den Bergh, 1981
Rotational velocity–luminosity relationship for three nearby Sc galaxies	60–65 (±12)	15–17 (±3)	V. C. Rubin et al., 1983
Luminosities of 16 type I SN	50 ± 7	20 ± 3	Tammann and Sandage, 1983
Time delay in double image of gravitationally lensed quasar	77 ± 5	13 ± 1	Florentine-Nielsen, 1984
T–F, blue light, 66 galaxies in Virgo cluster	68 ± 17	15 ± 4	van den Bergh, 1984
T–F, IR, 21 galaxies in Virgo cluster	82 ± 18	12 ± 3	van den Bergh, 1984
Luminosities of six Type I SN	59	17	Arnett, Branch, and Wheeler, 1985
Multiplicity of methods	95 ± 10	10 ± 1	de Vaucouleurs, 1986
Luminosities of Cepheids, bright stars, type I SN in 18 near galaxies plus Virgo and Fornax clusters	55	18	Sandage, 1986b
Luminosities of type I SN	50–70	14–20	Woosley, Taam, and Weaver, 1986

[a] T–F, Tully–Fisher relationship; IR, infrared; SN, supernovae.

Way and then used to find distances to bodies within the Local Group of galaxies where other indicators, such as the brightest globular clusters and the diameters of H$_{II}$ regions, can then be calibrated. These indicators are then used to calibrate other indicators with still greater range, and so forth. Thus, errors in the process may be easily magni-

range, and so forth. Thus, errors in the process may be easily magnified as the calibration progresses through successive steps. Second, the calibrations are exceedingly difficult and require numerous assumptions about the physics of stars and associated phenomena that are sometimes difficult to verify. Even the use of the "double quasar" is not without uncertainty, for this method depends on the model used for the gravitational lens. Measuring extragalactic distances is not an easy task and does not yet yield precise and consistent results.

Despite the uncertainties, most astronomers agree that the correct value for the Hubble constant is in the range 50–100 km/s/Mpc, implying a Hubble time of between 10 and 20 Ga. What does this Hubble time mean? If the inflation has been linear over time, then the Hubble time is equal to the age of the universe, i.e. the time elapsed since the "Big Bang" or the start of inflation. But this interpretation is naive because it does not take into account the gravitational effects caused by the mass of the universe. Gravity would tend to cause a decrease in the Hubble flow over time with the result that the Hubble time must be an upper limit to the age of the universe.

The amount of deceleration of the inflation depends on the mass of the universe, a quantity that is the subject of much current debate. If the mass of the universe is sufficient, then the universe is *bounded*, the inflation will continue to decelerate, and the universe will eventually begin to collapse or deflate. If the mass is insufficient, then the universe is *unbounded* and will continue to inflate forever. On the basis of its visible mass, the universe appears to be unbounded, but this conclusion could change if the *neutrino*, vast numbers of which populate the universe, is ever found to have significant rest mass or if new mass is discovered. For a marginally bound universe the age is given by $2/3H_0$, which is also the theoretical lower limit for the age of an unbounded universe.[6] Thus, if the universe is unbounded, as most astronomers and cosmologists now think, then its age lies within the range $2/3H_0$ to $1/H_0$ or, from the range of 50–100 km/s/Mpc for H_0, 7–20 Ga.

Despite the uncertainties in the determination of a precise Hubble constant and in the knowledge of the rate of inflation over time, one thing is clear—the universe is old, somewhere between 7 and 20 Ga in age. This conclusion is in complete accord with the age of 4.5 Ga measured for the bodies of the Solar System.

Globular Clusters and the Oldest Stars

The Milky Way, within which the Solar System resides, is a spiral galaxy composed of billions of stars. Like a giant pinwheel it rotates about its brilliant central bulge once every few hundred million

6. The calculations, which are not difficult, are explained by Shu (1982: 360–62).

years. But not all of the stars are confined to the nucleus and the flattened disk. Scattered around a spherical region surrounding the Milky Way are billions of additional faint stars. Within this "halo" are the *globular clusters*, of which there are some 200, each containing from 10^5 to 10^6 stars. Compared to stars in the clusters of the galactic disk, the stars in the globular clusters have a very low content of metals (0.01–0.1%). Apparently this deficiency is due to the globular clusters' having formed early in the history of the Milky Way, before the Galaxy had collapsed from a larger volume into a flattened disk, and before recycling of the products of nucleosynthesis in high-mass stars had substantially increased the metal content of the galactic material (Eggen, Lynden-Bell, and Sandage, 1962, Sandage, 1986a). Thus, the stars in the globular clusters are thought to be the oldest stars in the Milky Way. A combination of theoretical and observational considerations indicates that the stars in these globular clusters are 14–18 Ga in age, which provides a younger limit for the age of the Milky Way galaxy. Let us see where this age comes from.

Stars are born when gravitational forces within a cloud of gas cause the cloud to collapse (Shu, 1982: 147–57, Iben, 1970). The gravitational energy released by the collapse generates tremendous internal temperatures that eventually cause the hydrogen in the core of the newborn star to "ignite" and "burn." The "burning" that occurs in the interior of a new star is not combustion but rather a sustained series of nuclear reactions that convert the hydrogen into helium.

Within limits, stars come in a variety of sizes, and as stars go the Sun is rather average, being neither very large nor very small. Below about 0.08 $M_\odot$ ($M_\odot$ = mass of Sun) the gravitational energy and resultant internal temperature are insufficient to start the nuclear furnace, and the result is a gaseous planet, of which Jupiter is an example. Evidently, stars with masses above about $60M_\odot$ either do not form or are very unstable, because none have been found.

Stars do not vary greatly in composition. Most are composed of about 70–80% H, 20–30% He, and 0.01–3.0% heavier elements. Stars that are rich in heavy elements, like the Sun, are classified as *Population I* stars, whereas those low in heavy elements, such as the stars in globular clusters, are called *Population II* stars. Since the heavier elements are created within stars and within supernovae, Population II stars are thought to have formed before Population I stars, i.e. before the H–He gas clouds from which the stars formed were enriched in the heavier elements.

To a first approximation, most of the major properties of stars are related in a reasonably straightforward way. Two of these properties are luminosity (L), usually measured relative to the luminosity of the Sun ($L_\odot$), and effective temperature (T_e), which is the surface temperature that the star would have if it were a perfect *blackbody* radiating its luminosity. Thus, L and T_e are related by the blackbody formula

$$L = \sigma\, T_e^4 4\pi R^2 \tag{8.3}$$

where σ is the radiation constant, a measure of the star's effectiveness at radiating energy, and R is the radius of the star. As discussed in the previous section, the luminosity of a star can be determined from its apparent brightness and distance. The effective temperature can be determined from the color of the star or from its emission spectrum.

When L is plotted as a function of T_e, it is found that most stars fall within a very narrow band called the *main sequence* (Fig. 8.2). This

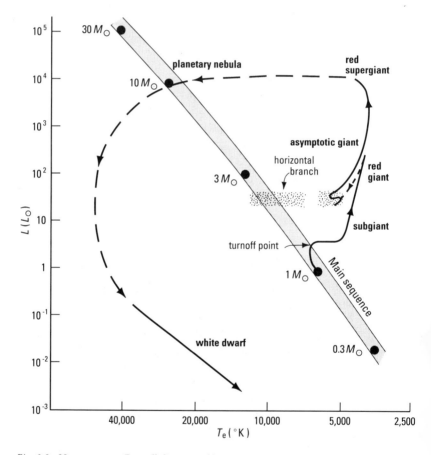

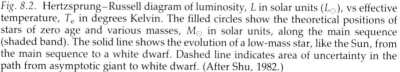

Fig. 8.2. Hertzsprung–Russell diagram of luminosity, L in solar units ($L_\odot$), vs effective temperature, T_e in degrees Kelvin. The filled circles show the theoretical positions of stars of zero age and various masses, $M_\odot$ in solar units, along the main sequence (shaded band). The solid line shows the evolution of a low-mass star, like the Sun, from the main sequence to a white dwarf. Dashed line indicates area of uncertainty in the path from asymptotic giant to white dwarf. (After Shu, 1982.)

relationship was first discovered in 1911 by E. Hertzsprung, a Danish astronomer, and independently in 1913 by H. N. Russell, of Princeton University, and so this type of figure is called the *Hertzsprung–Russell*, or *H–R*, *diagram*. Stars are born, and spend most of their lifetimes, on the main sequence.

Although composition has an effect, the single most important property that determines where any newborn star will fall on the main sequence and how long it will remain there is its mass. Because the fuel of stars on the main sequence is hydrogen and because the proportion of hydrogen among the stars is relatively constant, then it follows that the energy (E) available is proportional to mass:

$$E \propto M \qquad (8.4)$$

The lifetime of a star on the main sequence (t) is a direct function of the energy available and an inverse function of the rate at which that energy is being used. Since the luminosity is a direct expression of energy consumption, then

$$t \propto \frac{E}{L} \propto \frac{M}{L} \qquad (8.5)$$

From the speed and distance at which binary stars orbit each other, it is possible to calculate their mass and from such observations it is known that, on the average

$$L \propto M^4 \qquad (8.6)$$

Then from Equations 8.5 and 8.6

$$t \propto \frac{1}{M^2} \qquad (8.7)$$

From these last two equations, it is easy to see that the greater the mass of a main-sequence star the brighter and shorter lived it will be. The radius of a star also varies with mass, with

$$R \propto M \qquad \text{for lower-mass stars}$$

and

$$R \propto M^{0.6} \qquad \text{for higher-mass stars}$$

Since L, R, and T_e are related by Equation 8.3, then

$$T_e \propto M^{1/2}$$

These (here considerably oversimplified) relationships between M, R, E, L, T_e, and t, which are based primarily on theoretical considerations, are very important because most of the properties of stars cannot be measured directly. Most stars are too distant for their sizes to be measured, and, except for binary stars whose masses can be calculated from their orbital characteristics, star masses (and so also E) are not directly observable. Although the births and deaths of stars have been observed, their lifetimes, t, are much too long for direct measurement. This leaves, as the only commonly measurable quantities, luminosity, which can be found from apparent brightness and distance, and temperature, which can be determined from color or spectral properties. Once L and T_e have been determined, then E, M, R, and t can be estimated by calculation.

Once a main-sequence star has exhausted its hydrogen fuel it leaves the main sequence and rapidly evolves through a series of changes that ultimately lead to its death. These changes are a result of the exhaustion of successive rounds of nuclear fuel, and the delicate balance between the enormous pressure of self-gravity and the flow of heat from the nuclear furnace to the outside. The precise evolutionary path that a star follows depends on its mass and, to a lesser degree, the details of its composition. For purposes of illustration, let us follow the probable evolution of a low-mass star like the Sun, which is still on the main sequence.

As the star exhausts the hydrogen fuel in its core the core begins to collapse. This is caused by the decrease in pressure due to the reduction of the number of particles in the core by the conversion of hydrogen to helium. The collapse of the core releases gravitational energy, which is converted into heat, and so the core temperature rises. The increase in temperature is insufficient to ignite the helium, but it does permit hydrogen in the shell immediately surrounding the core to burn. As the helium core continues to contract because of the addition of material from the burning hydrogen shell, the hydrogen burns ever more rapidly, the flow of energy to the surface is increased, and the radius of the star must increase. At this stage, the increase in R does not keep pace with the increasing energy flow to the surface, so L must also increase. Thus, the star begins to move slightly upward from its initial position on the main sequence (Fig. 8.2).

Not all of the energy generated in the burning hydrogen shell can find its way to the surface because the rate of transfer is governed by the rate at which photons can diffuse outward. The difference between the energy generated within the star and that leaving the surface as luminosity, therefore, goes into heating up the layers above the burning hydrogen shell and the star expands, increasing R. But since L is nearly fixed by the rate at which photons can diffuse outward to the surface, T_e must decrease (Equation 8.3) and the star, having reached the *turnoff point*, leaves the main sequence by moving more or less horizontally to the right, becoming a subgiant and turning redder. T_e, however, does not drop indefinitely because as temperature decreases the ability of photons to stream unimpeded, rather than diffuse, to the surface increases rapidly until a lower limit for T_e is reached. But because the hydrogen shell continues to burn more vigorously the luminosity and the radius must increase and the star rapidly ascends along a vertical track, becoming a *red giant*. Even though the star grows in size, the ascent along the red-giant track is accompanied by significant mass loss.

When the star reaches the tip of the red-giant branch, temperatures in the core are sufficient to start the nuclear reactions that convert helium to carbon and some of the carbon to oxygen. Helium burning, however, occurs under conditions that do not allow the sympathetic regulation of temperature, pressure, and radius that occur in hydrogen-burning stars, so the increase in core temperature results in the runaway production of nuclear energy, otherwise known as the *helium flash*. By now, hydrogen burning in the shell is greatly diminished, the luminosity decreases rapidly, the star shrinks, and the surface temperature rises somewhat, leading the star on a downward track to the *horizontal branch*. The exact evolutionary path from the red-giant branch to the horizontal branch depends on the composition of the star and on the fraction of mass lost during the star's growth into a red giant. There is a gap in the horizontal branch caused by the *instability strip*. Whenever in its lifetime a star crosses this strip it becomes unstable and pulsates. Thus, the gap accounts for the variable stars, whose fluctuating luminosity precludes their being plotted on the H–R diagram.

When the helium fuel in the core is depleted the core must contract, and both the temperature and pressure in the region surrounding the shrinking core then increase. This permits helium to burn in the shell surrounding the carbon–oxygen core and hydrogen to burn in a second shell slightly farther out. As the helium-burning shell contributes more carbon and oxygen to the core, the core continues to contract, the temperature, pressure, and energy production in the regions of the shells continue to increase, and, just as with the single hydrogen-burning shell, the luminosity and size of the star must increase. Accordingly, the star ascends the asymptotic giant branch to become a *red supergiant*. At this point in its evolution, the star is using fuel so rapidly that it cannot survive for long.

Precisely what happens to stars in the red-supergiant stage of their evolution is theoretically quite uncertain, but observations suggest that the loss of mass from a red supergiant may be so prodigious that it becomes a *planetary nebula* surrounding an incipient *white dwarf*. The star soon exhausts the fuel in its hydrogen and helium shells and rapidly descends along an uncertain path to join the senescent white dwarfs.

Both theory and observation indicate that stars with masses less than about $6M_\odot$ follow this evolutionary sequence. High-mass stars, on the other hand, follow a much different sequence of events. They evolve more rapidly along quite different tracks in the H–R diagram, becoming supernovae in less than 10 Ma. The evolution of high-mass stars will not be discussed further because it is the low-mass stars, which evolve much more slowly than the high-mass stars, that are used to determine the ages of globular clusters.

On the basis of the physics of the various nuclear reactions, energy transfer, and gravitational effects that occur in stars of a given mass and composition, astrophysicists can calculate in some detail the evolutionary track of a star and the rate at which it progresses along that track. These calculations are very important because they provide the basis for estimating the ages of globular clusters.

The distribution on a H–R diagram of stars in a typical globular cluster, in this case the globular cluster M3, is shown in Figure 8.3. A striking feature of this distribution of stars is that it looks very similar to the evolutionary track of a single star. This is so because the evolutionary tracks of low-mass stars are very nearly coincident. Thus, the distribution of stars in Figure 8.3 is due primarily to their differences in mass, not age. Because the rate of evolution is a function of mass (Equation 8.7), the more massive stars have progressed farther along their tracks than have the less massive stars.

The one area where there are substantial differences between the evolutionary tracks of different low-mass stars is in the region of the turnoff point. Theoretical tracks for three stars of the same age but different initial mass are shown in Figure 8.4. The three tracks are nearly coincident near the main sequence, distinct at the turnoff point, and converge at the base of the red-giant branch. Because of their mass difference, the rates at which these three hypothetical stars evolve along their respective tracks are greatly different. Fifteen billion years after formation, for example, the star with $M = 0.7M_\odot$ is still on the main sequence, the one with $M = 0.8M_\odot$ is approaching the turnoff point, and the one with $M = 0.9M_\odot$ has already become a subgiant. If the points representing the positions of the three stars after 15 Ga are connected they form an *isochrone*[7] representing the distribution of stars

7. Astronomers use the German *isochrone*, whereas geologists have chosen to adopt the English equivalent, *isochron*. The meaning, i.e. a line of equal time, is the same.

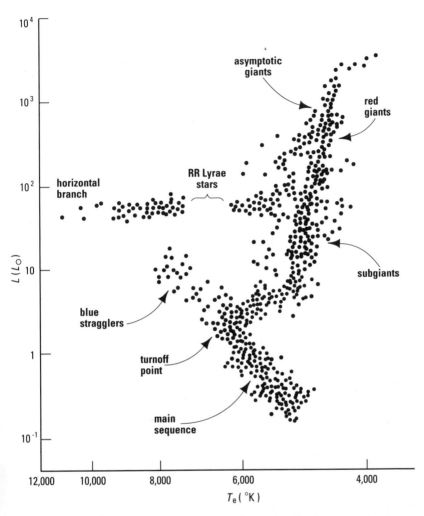

Fig. 8.3. H–R diagram for stars in the globular cluster M3. The distance to the cluster was derived on the assumption that the luminosity of the RR Lyrae stars is $50L_{\odot}$. (After Shu, 1982, and based on data from Johnson and Sandage, 1956.)

of different mass but equal age. Although the isochrone cuts across the evolutionary tracks, it is similar in shape to the tracks, thus explaining the similarity in appearance between the evolutionary tracks of low-mass stars (Fig. 8.2) and the observed locus of stars in a cluster (Fig. 8.3). The isochrone even has a "turnoff point," although its turnoff point does not have the same physical meaning as the turnoff point of an evolutionary track. Inspection of Figure 8.4 indicates that

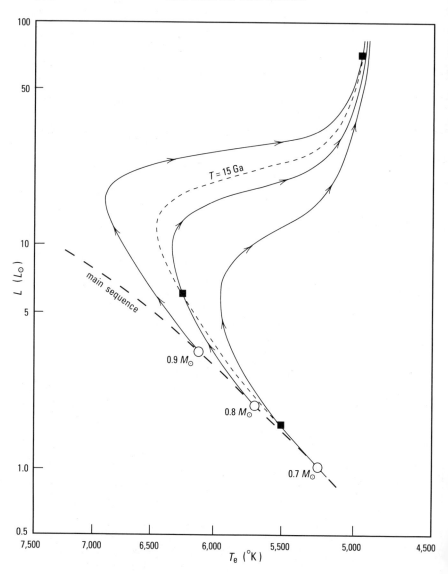

Fig. 8.4. Theoretical evolutionary tracks for stars of three different initial masses (circles) with an initial helium content of 20% and a metal content of 0.01%. An isochrone (short dashed line) is derived by connecting the points of equal time (squares). (After VandenBerg, 1983, © The American Astronomical Society.)

older isochrones are similar in shape but lie to the right of the one shown, while younger isochrones lie to the left. Thus, stars within a cluster fall along an isochrone, and with time the locus of stars will "peel off" of the main sequence to the right much like the peel is pulled from a banana.

We can now see that the ages of star clusters can be estimated by determining their locus on the H–R diagram and comparing the turn-off point of the distribution to theoretical isochrones. One recent comparison by A. Sandage of the Mt. Wilson Observatory for the cluster M92 is shown in Figure 8.5. Note that the turnoff point for the locus of stars coincides with the 18-Ga theoretical isochrone, so 18 Ga is, therefore, the estimate of the age of M92.

While this procedure is conceptually simple, it is complicated by numerous difficulties. The apparent brightness or apparent *magnitude*

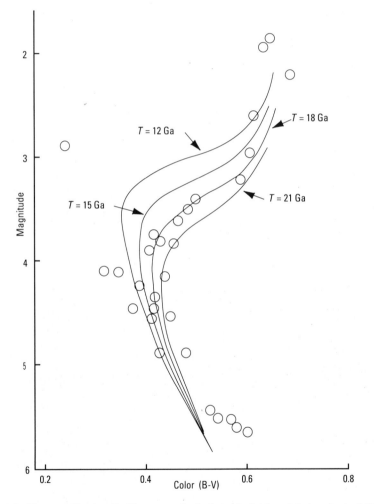

Fig. 8.5. Diagram of color (B–V) vs magnitude for stars in the globular cluster M92 compared with the theoretical isochrones for an initial helium content of 20% and a metal content of 0.01%. The turnoff point matches the 18-Ga isochrone. (After Sandage, 1983.)

TABLE 8.2

Some Recent and Representative Ages for Globular Clusters within the Milky Way Galaxy

(Ga)

M3	M4	M5	M13	M15	M71	M92	47 Tuc	NGC6752	Source
16.2 ± 1.8		12.6 ± 1.6	15.1 ± 1.1	17.2 ± 1.3	8.9 ± 1.7	19.7 ± 2.9	10.5 ± 0.6	15.6 ± 0.6	Carney, 1980
17.2 ± 3.4		19.0 ± 3.8	20.2 ± 4.0	15.1 ± 3.0	14.3 ± 2.9	15.6 ± 3.1	17.2 ± 3.4	14.8 ± 3.0	Sandage, 1982
							15–18		Harris, Hesser, and Atwood, 1983
		15+	15	18	15		16		VandenBerg, 1983
18		18		17–20		18 ± 2	18	15+	Sandage, 1983
				13 ± 3					Caputo, Castellani, and Quarta, 1984
	15 ± 1								Richer and Fahlman, 1984
				16 ± 3					Fahlman, Richer, and VandenBerg, 1985
				18			16		VandenBerg and Bell, 1985
			16 ± 2						Richer and Fahlman, 1986

of a star must be corrected for both distance and the effects of dust and gas along the light path for the luminosity or absolute magnitude of the star to be found. The effective temperature of the star must be found from either the star's spectrum, a difficult measurement at the distances of globular clusters, or from its color as determined from the relative amounts of light emitted by the star within bands of different wavelengths (Iben, 1970, Fahlman, Richer, and VandenBerg, 1985, Sandage, 1983, 1986a). The composition of the star, in particular the initial helium and metal contents, must be estimated. Finally, numerous assumptions about stellar evolution enter into the calculation of the theoretical isochrones (VandenBerg, 1983, VandenBerg and Bell, 1985). The result of these difficulties is that there is still considerable uncertainty in the ages of globular clusters.

Some recent and representative results for the few well-studied globular clusters in the Galactic halo are shown in Table 8.2. The ages of 8.9 Ga for M71 and 10.5 Ga for 47 Tuc are apparently the result of calculations that did not adequately take into account the effects of the high metal contents of the stars in these two globular clusters (Sandage, 1982). The remaining values in the table are either within or not significantly different from the range 14–18 Ga. The variation in the ages is probably due primarily to difficulties with the method rather than to real age differences between the clusters, for there is good evidence to suggest that the globular clusters in the Milky Way do not differ appreciably in age (Sandage, 1986a: 452).

Despite the uncertainties in the method and the fact that no single, precise age has yet been determined for the globular clusters, there is still a remarkable consistency in the results. It is clear, for example, that the globular clusters are about 16 ± 2 Ga in age. Since the globular clusters provide a younger limit for the age of the Milky Way and are thought to have formed early in its history, then the evidence also indicates that our Galaxy is slightly more than 16 ± 2 Ga old, a result that is consistent with the measured age of 4.5 Ga for the Solar System.

Messages from the Elements

In the previous sections the evidence for the antiquity of the cosmos provided by the largest objects in nature—stars, star clusters, galaxies, and the universe—was examined. Now additional evidence from some of the smallest—atoms of the elements—will be summarized. Of particular interest are those observations about the abundances of radioactive nuclides in the Solar System that indicate that

the age of the very material of which the Sun and planets are made is
some two to three times the age of the Solar System.

There is both semi-quantitative and quantitative evidence concerning the age of the elements. The former indicates that the elements in the Earth, Moon, and meteorites must be at least of the order
of billions of years old rather than thousands or millions. The latter
indicates that the most probable age of the elements is between 9 and
16 Ga. Let us examine the semi-quantitative evidence first.

The Case of the Missing Nuclides

Inspection of any chart or table of the nuclides reveals a most
interesting fact: of the radioactive nuclides not currently being produced in the natural environment, only those with half-lives greater
than about 80 Ma occur in nature.

There are 34 unstable nuclides with half-lives greater than 1 Ma
(Table 8.3). Of these, 23 are found in nature. Five of the 23, however,
are continually being produced by natural nuclear reactions. ^{53}Mn
(manganese) and ^{10}Be are produced primarily in dust particles in the
upper atmosphere and in space by cosmic rays. Some of these particles eventually fall to Earth, where they are incorporated into sediment. ^{236}U is rare but is produced by nuclear reactions in some uranium ores where sufficient slow neutrons are available. ^{129}I has been
found in tellurium ores, where it is produced from ^{130}Te (tellurium) by
cosmic-ray muons. ^{237}Np (neptunium) is not found on Earth but is
found in rocks from the Moon, where it is produced by cosmic rays.
The 17 remaining radioactive nuclides all have half-lives of 82 Ma or
more. The absence of short-lived unstable nuclides in nature holds
true even if the list of Table 8.3 is extended to include nuclides with
half-lives less than 1 Ma; only those whose existence is due to continual production by natural processes (or in some cases by man) are
found. How can this most logically be explained?

There are three hypotheses that might explain this curious absence of short-lived radioactive nuclides: (1) the distribution is due to
chance, (2) the processes responsible for formation of the elements
were incapable of producing the short-lived nuclides, or (3) a great
length of time has passed since the elements were created and nuclides
with half-lives less than about 80 Ma have simply decayed away.

The first hypothesis is easily dispatched by considering the probability that the observed distribution could occur by chance as opposed to some random distribution of both long- and short-lived
nuclides. To estimate this probability, one must make an assumption
about the age of the elements. Then one can calculate the probability
that the absent nuclides with half-lives long enough to have survived

TABLE 8.3

Known Radioactive Nuclides with Half-Lives
of 1 Ma or More

Nuclide	Half-life (years)	Found in nature?
^{50}V	6 $\times 10^{15}$	yes
^{144}Nd	2.4 $\times 10^{15}$	yes
^{174}Hf	2.0 $\times 10^{15}$	yes
^{192}Pt	~1 $\times 10^{15}$	yes
^{115}In	6 $\times 10^{14}$	yes
^{152}Gd	1.1 $\times 10^{14}$	yes
^{123}Te	1.2 $\times 10^{13}$	yes
^{190}Pt	6.9 $\times 10^{11}$	yes
^{138}La	1.12 $\times 10^{11}$	yes
^{147}Sm	1.06 $\times 10^{11}$	yes
^{87}Rb	4.88 $\times 10^{10}$	yes
^{187}Re	4.3 $\times 10^{10}$	yes
^{176}Lu	3.5 $\times 10^{10}$	yes
^{232}Th	1.40 $\times 10^{10}$	yes
^{238}U	4.47 $\times 10^{9}$	yes
^{40}K	1.25 $\times 10^{9}$	yes
^{235}U	7.04 $\times 10^{8}$	yes
^{244}Pu	8.2 $\times 10^{7}$	yes
^{146}Sm	7 $\times 10^{7}$	no
^{205}Pb	3.0 $\times 10^{7}$	no
^{236}U	2.39 $\times 10^{7}$	yes-P[a]
^{129}I	1.7 $\times 10^{7}$	yes-P
^{247}Cm	1.6 $\times 10^{7}$	no
^{182}Hf	9 $\times 10^{6}$	no
^{107}Pd	~7 $\times 10^{6}$	no
^{53}Mn	3.7 $\times 10^{6}$	yes-P
^{135}Cs	3.0 $\times 10^{6}$	no
^{97}Tc	2.6 $\times 10^{6}$	no
^{237}Np	2.14 $\times 10^{6}$	yes-P
^{150}Gd	2.1 $\times 10^{6}$	no
^{10}Be	1.6 $\times 10^{6}$	yes-P
^{93}Zr	1.5 $\times 10^{6}$	no
^{98}Tc	1.5 $\times 10^{6}$	no
^{154}Dy	~1 $\times 10^{6}$	no

NOTE: With the exception of those currently produced by natural processes, only nuclides with half-lives longer than 82 Ma are found in nature.
[a] P, currently produced by natural processes.

for that length of time were not created and the long-lived nuclides were created. This assumption of an age for the elements is necessary because it dictates which of the short-lived nuclides must be included in the calculations. Implicit in the calculation also are the additional assumptions that the processes that created the elements were random and that each nuclide has an equal chance of being created in measurable amounts. Neither of these two assumptions is true, but the first is, in fact, a condition of the hypothesis being tested, whereas the second is close enough to being true that the calculated probability should be a reasonable approximation.

Assume that the elements are 10 Ma in age. As a rule of thumb, a radioactive nuclide becomes undetectable by usual counting methods, i.e. effectively "disappears," after about 10 to 20 half-lives have passed. To be conservative, we will use 10. We can presume, therefore, that those with half-lives less than 1 Ma might be absent because of radioactive decay over time and we are left with only those nuclides listed in Table 8.3 to use in the calculations. To be on the safe side, however, we will exclude from consideration the four nuclides we know are produced in nature on a continuing basis. The probability of all 18 of the long-lived nuclides being present and the remaining 12 short-lived ones being absent is then

$$p = \frac{18! \times 12!}{30!} = 1.1 \times 10^{-8}$$

or about one chance in 9×10^{7}. If we take 10 ka as the assumed age of the elements, we then use the nuclides with half-lives of 1 ka or more. As before, we eliminate those that are produced continually by natural processes, leaving 39, so the probability is

$$p = \frac{18! \times 39!}{57!} = 3.2 \times 10^{-15}$$

or only one chance in 3×10^{14}. But even these slim odds are too generous. To be fair, we really must add the 269 stable nuclides to the long-lived ones, since they are all found. The probabilities then become on the order of 10^{-21} for a 10-Ma Earth and 10^{-53} for a 1-ka Earth.

At this point it should be obvious that the hypothesis of pure chance to explain the absence of the short-lived unstable nuclides is very unlikely to be true. Probability, of course, is incapable of completely disproving the hypothesis—there is, after all, that one chance in 3×10^{14}—but it is clearly so unlikely that it can justifiably be excluded from further consideration. Besides, there is a more convincing answer.

The second hypothesis, i.e. that the missing nuclides were never created, can be approached from several directions. The first is not especially rigorous but is still revealing. Nuclei with certain combinations of neutrons and protons seem to be favored in existence and abundance over others with less desirable combinations. The majority of all stable nuclides contain even numbers of both protons and neutrons (Table 8.4). Nuclides with an even number of protons and an odd number of neutrons, or vice versa, are only about one-third as

TABLE 8.4

Relative Abundance of Stable and Naturally Occurring Long-lived Radioactive Isotopes as a Function of Whether A, Z, and N Are Even or Odd

Atomic number (A)	Number of protons (Z)	Number of neutrons (N = A − Z)	Number of stable nuclides	Number of long-lived radioactive nuclides
even	even	even	160	8
odd	even	odd	57	3
odd	odd	even	47	3
even	odd	odd	5	4
TOTAL			269	18

abundant, while those with odd numbers of both protons and neutrons are exceedingly rare. These same observations apply, in a general way, to the 18 long-lived radioactive nuclides as well; those with even numbers of both neutrons and protons prevail. If these same general rules are applied to some of the truant nuclides in Table 8.3, what is the result? Take iodine as an example. Both ^{127}I and ^{129}I have an odd number of protons ($Z = 53$) and even numbers of neutrons ($N = 74$ and 76, respectively). From this we would expect these two isotopes to be of roughly the same abundance, but, except for the rare production in Te ores by cosmic rays, ^{129}I is missing entirely. ^{174}Hf has a half-life of 2.0×10^{15} years and is found in nature, whereas ^{182}Hf, with a half-life of only 9 Ma, is not. Yet both of these isotopes of Hf have even numbers of both protons and neutrons. ^{150}Gd/^{152}Gd (gadolinium), ^{235}U/^{236}U, and ^{146}Sm/^{147}Sm are also anomalous.

While the odd–even argument alone is fairly convincing, a more fruitful approach to assessing the validity of the second hypothesis is to consider the processes by which elements are created: *nucleosynthesis*.

As discussed in Chapter 1 and in the first section of this chapter, the universe in its present form is thought to have originated in a "Big Bang" about 7 to 20 Ga. Some of the very light nuclides, primarily H and He, were formed during the first few minutes of the universe. But the absence of nuclei with masses of 5 and 8 precludes the possibility that a significant proportion of the nuclides heavier than Li (lithium) were formed in this violent beginning when the temperature and density of matter were sufficient to sustain the nuclear reactions necessary for nucleosynthesis. This means that all but the very light nuclides must have been synthesized primarily in stars and supernovae after the universe began, for stars and supernovae provide the only environment in which nucleosynthesis can occur.

There is good evidence that nucleosynthesis occurs in stars today and did so in the past. The spectra of some old stars, for example, reveal the presence of technetium, an element that has no stable nuclide and does not occur either in the Sun or on Earth (Merrill, 1952). The longest-lived isotope of technetium is ^{97}Tc, which has a half-life of 2.6 Ma. From the approximation that a radioactive nuclide "disappears" in about 10 half-lives, the technetium must have been formed within the past 25–30 Ma or so. Because some of the stars in which Tc is observed are much older than that, the logical conclusion is that Tc is formed within these stars. Alternatively, the Tc may have been formed in some older star and incorporated into the newer star when it formed from matter expelled in a supernova explosion. In either case, the Tc must have been formed relatively recently rather than during the Big Bang. Promethium has also been found in stars (Aller, 1971), and yet the longest-lived isotope of Pm has a half-life of only 18 years. A third example of the creation of nuclides in stars is carbon. Certain stars have a ratio of ^{12}C to ^{13}C near 4, which is very much lower than the ratio of 89 observed in the Earth and Sun. The most likely explanation of this difference is that it is due to the synthesis of ^{13}C in these particular stars. Thus, it is reasonable to accept the proposition that the birthplace of elements, at least those heavier than Li, is the stars. How does this happen?

There are nine, perhaps ten, basic processes by which stars synthesize nuclides. Five of these processes involve "burning," in reality nuclear reactions, of specific light elements to form other and slightly heavier elements. H burning, He burning, C burning, O burning, and Si burning occur more or less successively and at progressively higher temperatures. Each burning process proceeds to the next after the star has exhausted its supply of that element, provided that the mass of the star is sufficient to generate the higher temperatures necessary for the nuclear reactions required for the next step in the sequence. These burning processes lead to nuclides with masses up to about 60, i.e. up to and including iron, but are incapable of generating heavier nuclides.

Most of the heavier nuclides are synthesized by the capture of neutrons by lighter nuclei. Neutrons are more effective in nucleosynthesis than are charged particles because they do not have to overcome the Coulomb barrier of the nucleus. The two processes of most importance are the *s-process* and the *r-process* (Burbidge et al., 1957, D. D. Clayton et al., 1961, Seeger, Fowler, and Clayton, 1965, Schramm, 1974b, 1982, Trimble, 1975, R. K. Ulrich, 1982). Both of these processes result in the formation of increasingly neutron-rich nuclei of an element followed by β^- decay to form elements of higher Z. The principal difference between the two processes is the difference between rates of neutron addition and β^- decay.

The s-process (s for slow neutron capture) occurs when the rate of addition of neutrons to the available nuclei is very slow compared to the lifetimes of the unstable nuclides formed by the neutron additions. As successive neutrons are added, isotopes of an element of progressively higher N are formed until an unstable isotope is reached. At this point, the unstable nuclide will decay as quickly as it is formed, so that the isotope of next higher N is preempted, or "blocked," from formation. To demonstrate how this works, let us follow a segment of the s-process path starting at ^{124}Te (tellurium) (Fig. 8.6), which is produced by the addition of a neutron to ^{123}Te. The successive addition of neutrons results in ^{124}Te, ^{125}Te, ^{126}Te, and ^{127}Te. The first three are stable but ^{127}Te is radioactive with a half-life of only 9.4 hours. The ^{127}Te, therefore, decays by β^- emission to ^{127}I before another neutron can be added to form ^{128}Te. Thus, ^{128}Te is "blocked" from the s-process by unstable ^{127}Te. The addition of a neutron to ^{127}I forms ^{128}I, which decays to ^{128}Xe before ^{129}I can be formed. The s-process proceeds from ^{128}Xe through ^{133}Xe, to ^{133}Cs (cesium), and so forth, as shown in Figure 8.6. In this way, a large number of nuclides can be formed from the relatively small variety of "seed" nuclei produced by the "burning" processes discussed above, but because of the blocking by short-lived isotopes, not all elements can be formed in this way. The s-process is thought to occur in red giant stars, where there are sufficient free neutrons to drive the s-process, but insufficient neutrons for r-process production.

The r-process (r for rapid neutron capture) occurs when neutrons are added more rapidly than the unstable isotopes so formed can decay, so that blocking by unstable nuclides cannot occur. The rapid addition of neutrons to existing nuclides forms a wide variety of neutron-rich and unstable isotopes that lie in a band above and to the left of the isotopes in Figure 8.6. The formation of these nuclides continues to progressively higher N until the lifetimes of the isotopes are exceedingly small and an equilibrium with decay is established; only at this point is the r-process blocked by decay. Each of the unstable nuclides formed by this rapid addition of neutrons decays by β^- emission to form nuclides of progressively higher Z and lower N until the process is blocked by a stable nuclide. For example, decay of r-process nuclides leads through ^{128}Sn and ^{128}Sb to ^{128}Te (Fig. 8.6). But because ^{128}Te is stable it blocks the formation of ^{128}I. It is evident that the r-process is capable of forming nuclides (e.g., ^{128}Te) that cannot be formed by the s-process and vice versa (e.g., ^{128}Xe), and also that there are some that are formed by both processes (e.g., ^{127}I). The r-process can occur only where there are exceedingly high neutron fluxes and thus seems to require some sort of astrophysical catastrophe. The most likely sites are supernovae, where the necessary and exceedingly high neutron fluxes exist, but the r-process may also

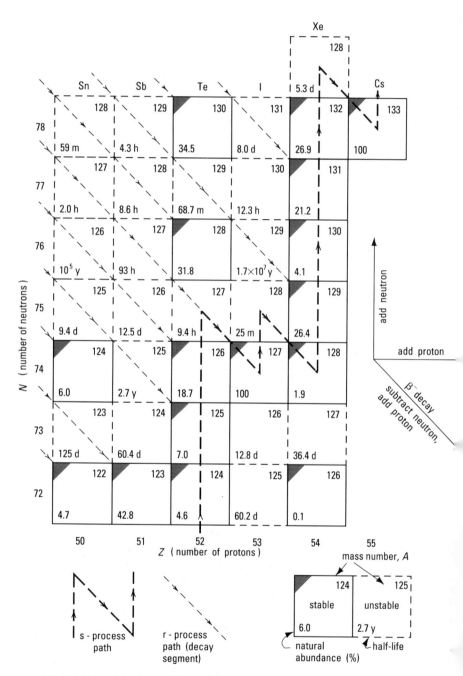

Fig. 8.6. Part of the chart of the nuclides showing the two principal processes (r and s) of nucleosynthesis responsible for the production of the heavy elements. The half-lives of the unstable nuclides are given in years (y), days (d), hours (h), or minutes (m).

occur during other astrophysical catastrophes as well (Norman and Schramm, 1979, Schramm, 1982).

Although the r- and s-processes are the primary synthesizers of the heavy elements, other processes may be responsible for some nuclide production. The n-process (n for neutron) is a mixture of the r- and s-processes and may be capable of producing a variety of heavy elements outside of supernovae. The p-process (p for proton), which occurs in red giant stars, modifies a small proportion of the nuclides produced by the r- and s-processes through the addition of protons to nuclei. The l-process (l for light) forms some light nuclei, like isotopes of Li, B, and Be, by *spallation* of heavier nuclei.

In nature, nucleosynthesis proceeds through the complex interaction of several of the processes operating simultaneously. Which processes occur at any given time depends on the mass and maturity of the star, and on the material from which the star is originally formed. Stars that become supernovae provide material from which other stars may eventually form. Thus, succeeding generations of stars are progressively enriched in the heavier elements (which explains the difference in metal content between the ancient globular clusters and the more recent open clusters that reside in the galactic plane). Alpha decay also plays an important part in r-process nucleosynthesis. Many of the heavier nuclides, such as the U and Th isotopes, decay in this way, forming nuclides of $Z - 2$ and $N - 2$. Thus, it appears that nucleosynthesis is capable of producing a wide variety of nuclides in varying abundances and that virtually every known nuclide can be produced in greater or lesser amounts.

With this elementary understanding of nucleosynthesis, let us now return to hypothesis number two: do the processes by which nuclides form discriminate against the synthesis of nuclides with short half-lives, specifically the missing nuclides in Table 8.3? The answer is no, but let us explore the reasons in more detail by using two isotopes of iodine. ^{129}I has a half-life of 17 Ma and, with the rare exception noted above, is not found today in nature. ^{127}I is stable and constitutes 100% of natural iodine. A glance at Figure 8.6 shows that ^{127}I can be formed by the s-process through neutron enrichment of Te isotopes followed by β^- decay of ^{127}Te.[8] In contrast, ^{129}I is blocked from formation by the s-process because of the instability and short lifetime of ^{128}I. ^{129}I can form, however, by the r-process through beta decay of ^{129}Sb and ^{129}Te. ^{127}I can also be synthesized by the r-process, there being no stable nuclide to block formation. Thus, there are ready mechanisms for the formation of both ^{127}I and ^{129}I. Theoretical considerations show that the r-process is the primary mechanism for

8. About 5% of the ^{128}I decays to ^{128}Te by electron capture.

formation of ^{127}I as well as ^{129}I, and that the production ratio of ^{129}I to ^{127}I is about 1.5 but may be as high as 3 (Seeger, Fowler, and Clayton, 1965, Fowler, 1972, Schramm, 1974a: 391–92). Thus, trace amounts of ^{129}I should be common in nature if the elements from which the Earth and Solar System formed were created as recently as 170 Ma ago (10 half-lives). Except for the rare production of ^{129}I from Te ores, none has ever been found.

Similar arguments can be made for the other missing nuclides listed in Table 8.3. Most occupy advantageous positions in the chart of the nuclides so that ready synthesis by the r- and s-processes is expected. A few are less exposed and are produced in lesser but not negligible amounts by other nucleosynthetic processes. For example, the missing nuclides ^{247}Cm (curium) and ^{182}Hf are formed primarily by the r-process, ^{205}Pb by the s-process, and ^{107}Pd (palladium) and ^{93}Zr (zirconium) by both processes; ^{97}Tc and ^{146}Sm are formed mainly by the p-process. The main point is that there is nothing unusual about the 12 missing nuclides in Table 8.3. All are ready products of nucleosynthesis. This is also true of an additional 27 missing nuclides whose half-lives are between 10^3 and 10^6 years.

It is all well and good to speculate on which nuclides should exist in nature, but is there any evidence that the missing nuclides actually did exist? For some, yes. In the Earth's atmosphere, ^{129}Xe constitutes about one-fourth of total xenon (Fig. 8.6). Yet in many meteorites ^{129}Xe is as much as 30 times more abundant, relative to the other xenon isotopes, than expected (Reynolds, 1967: 294, 1977: 217). As it is very probable that isotopes of the same element were thoroughly mixed when the Solar System formed, where did the excess ^{129}Xe come from? It seems likely that the most obvious answer is the correct one, i.e. the ^{129}Xe was formed by decay of ^{129}I incorporated into the meteorites when they formed. (Recall that the decay of ^{129}I to ^{129}Xe is the basis for measuring formation-interval ages for meteorites, discussed in Chapter 6.)

One final argument for the original existence of the missing nuclides is that all of them are easily made in nuclear reactors. In general they are no more difficult to make by artificial means than are many of the stable and long-lived radioactive nuclides. Because of the abundance and variety of natural nuclear reactors, i.e. the stars and supernovae, it would be surprising indeed if the absent short-lived unstable nuclides were not present when the Solar System first formed.

Thus, the second hypothesis of nonproducton fails, on close inspection, to provide an explanation for the absence of nuclides with half-lives less than 80 Ma. When the missing short-lived nuclides are examined in the light of nucleosynthetic processes and the general rules that describe the existence and abundance of the stable and long-lived nuclides, there is no reason why they should not exist and many reasons why they should. In addition, there is good evidence that ^{129}I was originally present in many meteorites when they formed. It seems that the only distinguishing feature that sets the missing nuclides apart from the rest is the fact that their half-lives are less than 80 Ma.

We are left, then, with the third hypothesis, namely, that the elements were created long ago and the missing nuclides have simply decayed away over time. We are not required, however, to accept this explanation merely by default. We can examine the data in Table 8.3 to see if the hypothesis is reasonable.

Figure 8.7 shows the percentage of an unstable nuclide that remains as a function of elapsed time for various arbitrarily chosen half-lives. Let us use this figure to consider the natural abundance of ^{235}U. This isotope of U constitutes only 0.72% of all U, the rest being ^{238}U. If the Earth's age is 4.5 Ga, then we would expect this 0.72% to represent approximately 1% of the amount present when the Earth was formed. This means that in the original Earth, ^{235}U would have been about 100 times more abundant than it is now, or only slightly less abundant than ^{238}U. Although there is no way to know exactly how

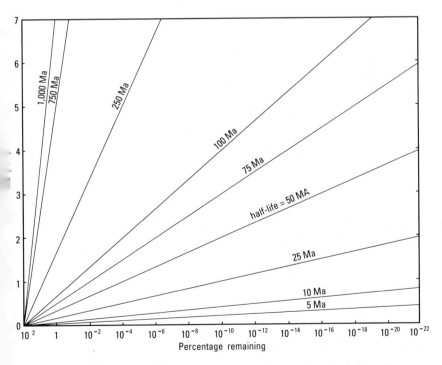

Fig. 8.7. Percentage of a radioactive nuclide remaining as a function of time for nuclides with various half-lives.

much ^{235}U was originally present when the Earth formed without knowing the age of the Earth (which, of course, we do know from other evidence), the predicted sub-equal abundance of the two principal U isotopes is consistent with their calculated r-process production ratio ^{235}U/^{238}U = 1.5 ± 0.5 (Symbalisty and Schramm, 1981: 313, Schramm, 1974a: 391) and so it does not seem unreasonable.

Let us now apply this same logic to ^{146}Sm, the longest-lived of the missing nuclides. Again, we have no way of knowing exactly how much ^{146}Sm might have been present when the Earth formed. Figure 8.7 shows that only about 10^{-19} of the original amount would survive after 4.5 Ga had elapsed. This is an exceedingly small remainder and it is not surprising that none has yet been found in nature. If only 1 Ga had elapsed, however, the remainder becomes approximately 0.01% of the original amount. If the original amount was similar to the present abundance of ^{147}Sm, or even only 0.1% as abundant, then ^{146}Sm would be easily detectable.

What about ^{244}Pu? With a half-life comparable to that of ^{146}Sm, why does it exist in nature? The answer is that it very nearly doesn't! Hoffman et al. (1971), at the Los Alamos Scientific Laboratory and the Knolls Atomic Power Laboratory, were able to separate chemically and identify by mass spectrometry a small amount of ^{244}Pu from a rare-earth mineral found in a California molybdenum mine. The amount of ^{244}Pu was only about 8×10^{-15} g, and this was concentrated from 85 kg of the ore. ^{244}Pu decays by α decay, but the amount of ^{244}Pu was so small that it would emit only one α particle every 6 years and was thus not detectable by ordinary counting methods. If the Earth were only 1 Ga in age, then we would expect this Pu isotope to be more frequently and easily found. Thus, the existence of ^{244}Pu in such an exceedingly small amount is consistent with an age for the Earth of 4.5 Ga.

As an aside, the identification of ^{244}Pu in the Earth's crust was an exciting discovery, for it suggests that this isotope was synthesized at, or at least close to, the time of formation of the Solar System rather than much earlier in the history of the universe. The reasoning is as follows. If the ^{244}Pu was synthesized about the time the Solar System was formed, then the 8×10^{-15} g represents the remainder of an original amount of about 1 kg. This does not seem like an unreasonable amount. If the ^{244}Pu was formed much earlier, however, then this original amount would have to be unreasonably large. If it originated when the Galaxy formed at 16 ± 2 Ga, for example, then the small amount found in California would have to represent the undecayed remainder of 10^{44} g!

Finally, let us return to the probability argument and apply it to a 4.5-Ga Earth. Using the "ten half-lives" rule as before, we would not expect to have any survivors with half-lives shorter than about 450 Ma. The probability of the observed abundances in Table 8.3 is then

$$p = \frac{17!}{17!} = 1$$

which also fits the hypothesis that the absence of the short-lived nuclides is a function of time.

Thus, we find that the abundances of the radioactive nuclides in nature are perfectly consistent with an Earth whose age is 4.5 Ga, but inconsistent with an age much younger. Even an age of 1 Ga is insufficient to explain the absence of ^{146}Sm and the other four nuclides with half-lives between 10 and 80 Ma. Although it is not possible to calculate any precise age for the Earth from these data, they show rather convincingly that the Earth's age must be more than a few but less than about 10 Ga. Although this answer is not very precise, there is a more quantitative way to estimate the age of the elements by using the abundances of certain radioactive nuclides, and that is the subject of the next and final section.

The Age of the Elements

Nucleocosmochronology is the use of certain radioactive nuclides to determine the history of nucleosynthesis, or creation of the elements. The general approach, first attempted by Rutherford (1929) (see also Table 2.1), is to calculate the length of time required to produce the present abundance ratio of a pair of nuclides, called a *chronometer pair*, from its "original," or theoretical, production ratio.[9] To illustrate, consider the ratio ^{232}Th/^{238}U, whose present value is 4.0 (Table 8.5). Because the half-lives of these two radioactive nuclides are known (Table 3.1) the value of the ratio ^{232}Th/^{238}U can be calculated for any time, t', in the past by means of a modification of Equation 3.1:

$$\left(\frac{^{232}\text{Th}}{^{238}\text{U}}\right)_{t'} = R\frac{e^{\lambda_1 t'}}{e^{\lambda_2 t'}} \tag{8.8}$$

where $R = 4.0$, λ_1 and λ_2 are the decay constants of ^{232}Th and ^{238}U, respectively, and t' is positive and measured backward from the present. Conversely, if the ratio at the time of some event in the past, like creation of the elements, is known, then the time elapsed since that event occurred can be calculated. In this example, both of the nuclides are radioactive and change with the passage of time, but the same formula applies if only one of the nuclides of the chronometer pair is ra-

9. There are some excellent reviews of nucleocosmochronology. Among them are Fowler (1972), Schramm (1973, 1974a), and Symbalisty and Schramm (1981). For lighter reading try Schramm (1974b, 1983).

TABLE 8.5

Model-Independent Mean Age for Solidification of the Solar System (Δ^{max}),
Abundance Ratios (R, R_t), and Production Ratios (P)
for Some Chronometer Pairs Used in Nucleocosmochronology

Chronome-ter pair	R (ratio now)	R_t (ratio at 4.54 Ga)	P (production ratio)	Δ^{max} (Ga)
^{187}Re/^{187}Os	(special case; see text)			6.1 ± 2.0
^{232}Th/^{238}U	4.0	2.5	1.9	2.6
^{235}U/^{238}U	137.88	0.313	1.5	1.9
^{244}Pu/^{232}Th	0	0.0062	0.47	0.51
^{244}Pu/^{238}U	0	0.025	0.9	0.43
^{129}I/^{127}I	0	0.00010	1.5	0.24

SOURCE: Symbalisty and Schramm, 1981: 313.

dioactive. In the case of a chronometer pair with one stable nuclide, the decay constant for the stable nuclide is zero and the exponential term for that nuclide is one; the ratio, as expressed by Equation 8.8, will still change over time because of the change in abundance of the radioactive nuclide.

The information required to calculate the time of nucleosynthesis, t', using Equation 8.8 is the present-day abundance ratio (R), the ratio at which the nuclides were produced at nucleosynthesis, $P = (^{232}\text{Th}/^{238}\text{U})_{t'}$, and, of course, the decay constants. Because the ratios of chronometer pairs that involve extinct or nearly extinct radioactive isotopes like ^{129}I and ^{244}Pu can be determined only for the time of so-lidification of the Solar System (they are zero now), it is more conve-nient to use the ratio at the time of solidification of the Solar System, $R_{t'}$, and then add 4.54 Ga to the result. For ^{232}Th/^{238}U, the resulting time since nucleosynthesis using Equation 8.8 and the data of Table 8.5 is 7.05 + 4.54 = 11.6 Ga.

But this approach is oversimplified, and the result is incorrect because implied in Equation 8.8 is a particular model of nucleosynthe-sis. That model calls for creation of the chronometer pair during a single event whose duration was either instantaneous or negligibly short. But is it reasonable to presume that the elements in the Solar System were created in a single event? It is not. As discussed earlier, the elements from which the Solar System was made were undoubt-edly the result of a prolonged process involving the births and deaths of many stars. There is little doubt that there have been many millions of supernovae, each an element factory, throughout the history of the Galaxy. Indeed, the presence of the daughter products of ^{129}I and ^{244}Pu in meteorites and the discovery of a small amount of ^{244}Pu on Earth show that there was some production of these nuclides within at least a few million years of the formation of the Solar System. Thus, the result of 11.6 Ga is only an "average" age whose precise relation to the beginning of nucleosynthesis and the age of the Galaxy depends on knowing something that is not well known, i.e. the detailed history of nucleosynthesis.

To illustrate the relationship between the history of nucleosynthesis, the mean age of a chronometer pair, and the age of the Galaxy, consider Figure 8.8, in which three simple models are shown graphically. In these models,

t = age of Solar system = 4.54 Ga
T_G = beginning of nucleosynthesis = age of Galaxy
T_M = mean age of the chronometer pair
T = Total duration of nucleosynthesis
Δ = time between last nucleosynthetic event and solidification of the Solar System
$\bar{t} = T_G - T_M$

Model (a) is the "one-event" model described by Equation 8.8. Note that T_G and T_M are nearly the same, and for an infinitesimally short event $T = 0$ and so $T_G = T_M$. Model (c) is for continuous nucleosynthesis at a uniform rate starting at the time of formation of the Galaxy and "ending" just before the time of solidification of the Solar System. Note that nucleosynthesis within the Galaxy does not actually stop, but after solidification the products of later nucleosynthetic events cannot be incorporated into the Solar System. Thus, Δ is very short, T is very long, and T_G is much earlier than T_M. Model (b) is one of a number of models that are intermediate between models (a) and (c). In (b), nucleosynthesis decreases exponentially and ends with a final "spike" to account for the quantities of ^{129}I and ^{244}Pu present when the Solar System formed (Fowler, 1972).

In 1970, D. N. Schramm, now of the University of Chicago, and G. J. Wasserburg, of the California Institute of Technology, devised a method of calculating the mean age of a chronometer pair that is essentially independent of the model assumed for nucleosynthesis (Schramm and Wasserburg, 1970; see also Schramm, 1974a, and Symbalisty and Schramm, 1981, for summaries and reviews of the mathematics). They showed that

$$\Delta_{max} = \frac{\log_e(P/R_t)}{\lambda_1 - \lambda_2} \tag{8.9}$$

where P is the production ratio of the two nuclides during nucleosynthesis, R_t is the ratio of the nuclides at the time, t, of solidification of the Solar System 4.54 Ga ago, and λ_1 and λ_2 are the decay constants of the two nuclides. Δ_{max} is so named because it is equal to Δ for the

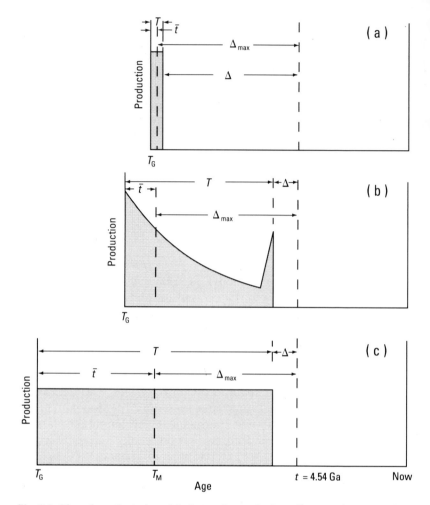

Fig. 8.8. Three hypothetical models for nucleosynthesis to illustrate the terminology and notation of nucleocosmochronology: (a) single-event model, (b) exponential model with terminal spike, and (c) continuous model. The duration and production rate of nucleosynthesis as a function of time are indicated by the shaded areas under the production curves. T_G, age of Galaxy; T_M, mean age of elements; T, duration of nucleosynthesis; Δ, interval between last nucleosynthetic event and time of solidification of Solar System (t); Δ_{max}, mean age of elements measured from solidification of Solar System; $\bar{t}$, $T_G - T_M$. (After Schramm, 1973, 1983.)

limiting case of the one-event model (Fig. 8.8a) and thus is the maximum value possible for Δ. For chronometer pairs whose half-lives are much longer than T, Δ_{max} can be added to the age of the Solar System to find the mean age of the pair (Fig. 8.8)

$$T_M = \Delta_{max} + t$$

which is also model-independent because such long-lived nuclides essentially see nucleosynthesis as a single event. Because virtually all elements heavier than iron are created in either the later phases of giant stars or supernovae, the value of T_M found from a long-lived chronometer pair is also an estimate of the mean age of all heavy elements. For short-lived chronometer pairs, Δ_{max} provides a model-independent estimate of the time of the last nucleosynthetic event. Intermediate-lived chronometer pairs theoretically can "fill in" details about the production history of the elements, although models for the potential intermediate chronometers are still poorly developed. Both the short- and intermediate-lived chronometers provide important information about nucleosynthesis, but they have little to contribute to the subject of this book and so will not be discussed further.

The use of nuclides as chronometer pairs requires relatively precise information about their production ratios, their abundances either now or at the time of solidification of the Solar System, and their decay constants. So far, adequate data are available for only a handful of chronometer pairs (Table 8.5), and of these only two, $^{187}Re/^{187}Os$ and $^{232}Th/^{238}U$, are sufficiently long-lived to provide useful estimates of the mean age of the elements. For ^{232}Th and ^{238}U, both r-process nuclides, the data of Table 8.5 and Equation 8.9 yield $\Delta_{max} = 2.6 (+3.0, -1.7)$ Ga. The relatively large uncertainties in this result are primarily due to two factors (Symbalisty and Schramm, 1981: 315–18). The first factor is the abundance ratio, which is based on analyses of meteorites, lunar rocks, and old terrestrial samples. This quantity is difficult to estimate accurately because it involves nuclides of two different elements that are known to fractionate chemically in nature. The second factor is the half-life of ^{238}U, and to a lesser extent that of ^{232}Th, being not quite long enough to satisfy the condition of model-independent, long-lived chronometer nuclides. To compensate, a model correction is required, but for most models of nucleosynthesis the correction is less than 20% and so Δ_{max} is still, to a first approximation, model-independent.

The $^{187}Re/^{187}Os$ pair is a special case because it involves only one radioactive r-process nuclide and, as for normal radiometric dating, the ratio must be corrected for "initial" osmium, which is a product of the s-process. For this chronometer pair, Schramm (1974a: 393, Symbalisty and Schramm, 1981: 313–14) has shown that

$$\Delta_{max} = \frac{\log_e(1 + {}^{187}Os^*/{}^{187}Re)}{\lambda}$$

where $^{187}Os^*$ is corrected for initial osmium and hence is the contribution only from the decay of ^{187}Re from the beginning of nucleosynthesis to the time of solidification of the Solar System, t; λ is the decay constant of ^{187}Re. Data for the calculation of ^{187}Os$^*/^{187}$Re have been determined from analyses of meteorites (Luck, Birck, and Allègre, 1980) and yield values of 0.07–0.14, and $\Delta_{max} = 6.1 \pm 2.0$ Ga. The uncertainty is primarily in the s-process production ratios of the Os isotopes, the abundances of the relevant nuclides of the two elements, and the decay constant of ^{187}Re, which, because of its long half-life and low-energy decay, is difficult to measure precisely in the laboratory.

Within the uncertainties, the values of Δ_{max} calculated from the ^{232}Th/^{238}U and ^{187}Re/^{187}Os chronometer pairs are concordant and, according to Symbalisty and Schramm (1981: 317), yield combined upper and lower limits for Δ_{max} of

$$\Delta_{max} \geqslant 6.1 - 2.0 = 4.1 \text{ Ga}$$

$$\Delta_{max} \leqslant 2.6 + 3.0 = 5.6 \text{ Ga}$$

Since

$$T_M = \Delta_{max} + t = \Delta_{max} + 4.54 \text{ Ga}$$

the mean age of the r-process elements is within the range

$$T_M = 8.6 \text{ to } 10.1 \text{ Ga}$$

The lower limit of 8.6 Ga for T_M constitutes a model-independent younger limit for the beginning of r-process nucleosynthesis, the age of the Galaxy, T_G (Fig. 8.8a), and the age of the universe, T_U. For an estimate of an older limit for T_G, some estimate of the maximum value of T is required, but T is model-dependent.[10] Evaluations of several models of galactic evolution have shown that T falls within the range $2\bar{t}$ to $6\bar{t}$ and is $2\bar{t}$ for most models (Tinsley, 1975, Hainebach and Schramm, 1977). Now

$$T = \Delta_{max} + \bar{t} - \Delta \simeq \Delta_{max} + t$$

because Δ is very small compared to Δ_{max} and can be neglected. So if $T = 2\bar{t}$, then

$$T \simeq 2\Delta_{max}$$

and the older, model-dependent limit for T_G is

$$T_G \leqslant 2 \times 5.6 + 4.54 = 15.7 \text{ Ga}$$

Thus, the best estimate for the age of the Galaxy from nucleocosmochronology is 8.6–15.7 Ga.

    10. Since the s-process requires heavy "seed" nuclei, and because massive stars form rapidly and have short lifetimes, it is probable that the earliest heavy elements

Summary

Of the radioactive nuclides not continually produced in the natural environment, only those with half-lives greater than 82 Ma presently occur in nature. Theory predicts that the missing short-lived nuclides should have been produced in significant amounts during nucleosynthesis, and the discovery of certain of their decay products in meteorites proves that they did once exist. The reason that the short-lived nuclides are missing is simply that the Solar System is old and they have decayed over time, so that their current abundances have fallen below the limits of detectability. Indeed, the missing nuclides demonstrate that the Solar System is some billions, rather than thousands or millions, of years old.

There are three astrophysical methods by which the age of the Galaxy or the age of the visible universe may be estimated.

1. From the rates at which galaxies at different distances are receding from the Milky Way Galaxy, the time at which the expansion or inflation of the universe began can be calculated. Current data yield values of the Hubble constant, H_0, that fall in the range 50–100 km/s/Mpc. The corresponding Hubble times are 10–20 Ga for an unbounded universe that will expand forever and 7–20 Ga for a bounded universe that is decelerating and will eventually begin to contract.

2. Their very low metal contents indicate that the oldest stars in the Galaxy are those in globular clusters, which reside in the spherical "halo" that surrounds the galactic disk. The ages of stars in a globular cluster can be estimated by comparing their turnoff point from the main sequence on the Hertzsprung–Russell diagram to theoretical isochrones calculated for stars of various masses and compositions. Results from well-studied globular clusters indicate that the Galaxy formed about 14–18 Ga.

3. The mean age of the r-process elements can be calculated from the abundances, production ratios, and decay constants of long-lived nuclides. Results for two long-lived chronometer pairs suggest a model-independent younger limit for the beginning of nucleosynthesis and the age of the Galaxy of 8.6 Ga, and a model-dependent older limit of 15.7 Ga.

Despite the uncertainties, it is quite remarkable that these three astrophysical methods give results that are so similar, i.e. in the range 9–20 Ga (Fig. 8.9). But this range can be further narrowed by requiring that the results from all three methods be consistent. With this constraint, the younger limit, set by the younger limit for the age of

were produced by the r-process very shortly after formation of the Galaxy (Symbalisty and Schramm, 1981).

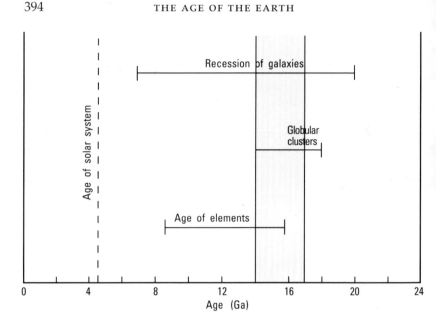

Fig. 8.9. Results from three astrophysical methods of estimating the ages of the Galaxy and the universe. The three methods are consistent with an age of the universe of 12.6–16.7 Ga (shaded band), allowing a maximum of 1 Ga from the Big Bang to the start of r-process nucleosynthesis in the Galaxy.

the globular-cluster stars, is 14 Ga, and the older limit, suggested by the age of the r-process elements, is 15.7 Ga. Because the universe cannot be younger than the globular-cluster stars, the younger limit is acceptable. It must, however, be older than the start of nucleosynthesis in the Milky Way galaxy, but how much older? Probably not much. Calculations based on theory show that star and galaxy formation began about 0.1 Ma or so after the Big Bang, when electrons and protons combined to form neutral atoms, and before the expansion and accompanying decrease in density precluded the concentration of matter into dense regions like galaxies. This means that the ages of the galaxies, including the Milky Way, are not much younger than the age of the universe. Moreover, massive stars, which are the birthplaces of the r-process elements, take only about 10 Ma or so to form and a similar time or less to evolve. Nucleosynthesis, therefore, probably began in the Galaxy within a few tens of millions of years after the Big Bang (Symbalisty and Schramm, 1981: 298). If we allow 1 Ga, just to be on the conservative side, then the upper limit for the age of the universe from r-process nucleosynthesis is no more than 16.7 Ga.

Several workers have applied this consistency argument in a more sophisticated and quantitative way by considering the most reasonable values for the density of nuclear matter and the amount of helium relative to hydrogen created during the Big Bang (Gott et al., 1974, Kazanas, Schramm, and Hainebach, 1978, Symbalisty and Schramm, 1981: 318, Schramm, 1983: 249–52). Such estimates yield a consistent age for the universe of 15 ± 1.5 Ga (Kazanas, Schramm, and Hainebach, 1978, Schramm, 1983: 251–52).

CHAPTER NINE

What We Know and Do Not Know

When the heavens were a little blue arch, stack with stars, methought the universe was too straight and close: I was almost stifled for want of air: but now it is enlarged in height and breadth, and a thousand vortices taken in. I begin to breathe with more freedom, and I think the universe to be incomparably more magnificent than it was before.
BERNARD LE BOVIER DE FONTENELLE (1686)

After more than two centuries of scientific endeavor, we have concluded that the age of the planet on which we live is 4.54 billion years. This value, which is based on the relationships between lead isotopes in meteorites and in the Earth, has an uncertainty of less than 1 percent and is consistent with numerous radiometric age measurements on ancient rocks found on the Earth and Moon as well as on meteorites. In addition, the antiquity of the Earth is consistent with evidence indicating that the Milky Way Galaxy and the universe are of the order of 14–16 billion years in age.

This knowledge did not come easily or swiftly, but required centuries of inquiry. Prior to the mid-eighteenth century, the age of the Earth was based either in whole or in part on religious theory and the interpretation of sacred scripture. The first attempt to find the age of the Earth solely from scientific measurements and principles, in 1748, was by de Maillet, who used the decline of sea level to calculate an age of more than 2 Ga since the presumed decline began.

During the nineteenth and early part of the twentieth centuries four general types of methods were used to estimate the age of the Earth: cooling of the Earth and sun, orbital physics, change in ocean chemistry, and erosion and sedimentation. These methods relied on assumptions of initial conditions now known to be incorrect, the constancy of physical processes now known to be variable, or both. As a result, the methods produced a wide variety of results. For example, the results of cooling calculations ranged from 1.28 Ga, the minimum age of the Earth found by Haughton in 1865, to 5 Ma, calculated for the Sun by Ritter in 1899. Calculations based on erosion and sedimentation yielded values from a low of 3 Ma published by Winchell in 1883 to a high of 15 Ga found by McGee in 1892.

Among the more important of these early age-of-Earth calculations were those of George Darwin, who in 1898 found an age of at least 56 Ma based on the increase in the orbital period of the Moon, of John Joly, who in 1899 calculated an age of 89 Ma based on the increase in the sodium content of the oceans, and of Charles Walcott, who in 1893 figured 55 Ma from the rate of accumulation of Paleozoic sedimentary rocks in the Cordilleran basin of western North America. Probably the most influential calculations were those based on the time required for Earth to cool from a molten state and published by Lord Kelvin and Clarence King. Kelvin's first calculations, published in 1862, yielded a range of 20 to 400 Ma with a probable value of 98 Ma. King refined Kelvin's method in 1893 and found 24 Ma as the most probable age of the Earth, a value that Kelvin endorsed four years later. As a result of the findings of prominent scientists like Kelvin and King, many turn-of-the-century geologists were convinced that all of Earth history had to be accommodated within a period of less than 100 Ma.

There were, however, dissenters. Scientists like Charles Darwin and T. H. Huxley felt that the geologic and fossil record clearly indicated that much more than 100 Ma had passed since the first rocks were deposited in ancient seas. Others, like John Perry and T. C. Chamberlain, challenged Kelvin's and King's basic assumptions and mathematics and showed that their results were anything but exact. In the end, the dissenters proved to be correct. Methods based on cooling, orbital physics, and the accumulations of chemicals in the oceans and sediments in the basins can never be made to yield a valid age for the Earth, because the initial conditions are uncertain and the rates of the processes involved have varied significantly over time.

The discovery of radioactivity by Henri Becquerel in 1896 was the first in a series of developments that would lead to quantitative methods based on the decay of long-lived radioactive nuclides that occur in rocks and minerals, from which a realistic age of the Earth could be found. A few of the other milestones in this quest:

1. The suggestion by E. Rutherford in 1905 that the accumulation of helium and lead in uranium minerals might be used to find the age of the minerals;

2. The demonstrations by B. Boltwood, R. J. Strutt, and A. Holmes over the period 1905–1911 that uranium-bearing minerals contain radiogenic helium and lead and, therefore, that Rutherford's suggestion just might work;

3. The idea advanced by H. N. Russell in 1921 that a large portion of a planetary body, specifically Earth's crust, could be treated as a single geochemical reservoir and dated by the change in the ratio of a radioactive parent isotope to that of its stable daughter isotope;

4. The development of the mass spectrometer and the discovery of isotopes, beginning in 1914 with the parabola mass analyzer of J. J. Thomson, who used it to discover that neon had two isotopes, and culminating in the late 1930's with the development of the first modern mass spectrometer by A. O. Nier, who used his precision instrument to measure the variations of uranium and lead isotopes in lead ores;

5. The idea advanced by E. K. Gerling in 1942 that the age of Earth could be calculated from the composition of lead in lead ores, if the composition of primordial lead is known and the ores represent a fossilized sample of the lead composition of a single-stage reservoir within the Earth;

6. The suggestion by F. G. Houtermans in 1947 that the isotopic composition of primordial lead might be found in iron meteorites, implying that meteorites and the Earth are genetically linked;

7. The demonstration by C. C. Patterson in 1956 that the isotopic composition of lead in two iron meteorites, three stone meteorites, and modern Earth sediment all fall on a single lead–lead isochron indicating an age for the Earth–meteorite system of 4.55 ± 0.07 Ga; and finally,

8. The development and refinement of modern instrumentation and radiometric dating methods based on the decays of isotopes of uranium, potassium, rubidium, samarium, rhenium, and lutetium to their respective daughter products lead, argon, strontium, neodymium, osmium, and hafnium beginning in the mid-1950's and continuing to the present.

These developments finally produced the answer to the question of the antiquity of the Earth.

But scientific conclusions are always tentative, and the age of the Earth is subject to revision should new evidence require it. Therefore it is worthwhile to summarize what we know and what we can infer from that knowledge, and what we would like to know in order to test and refine our conclusions thus far.

What We Know and What We Infer

Radiometric Dating

We know that certain long-lived radioactive nuclides found in rocks and minerals serve as natural clocks and that these clocks can be used in many circumstances to measure the ages of events in the history of the Earth and the other solid bodies in the Solar System. The rates of decay of these radioactive parent elements have all been accurately measured in the laboratory and are known, from both experimental evidence and theoretical considerations, to be constant over the range of temperatures and pressures to which the Earth, Moon, and the meteorites have been subjected since their formation. These radiometric dating systems can be partially or entirely reset, usually through loss of the daughter isotope, by thermal and chemical changes induced by metamorphism. This complicates but does not invalidate radiometric dating because (1) postcrystallization metamorphism usually results in changes that can be recognized from the rocks' mineralogy, texture, chemical composition, or geological setting, and this information is used to guide sample selection, (2) most radiometric dating methods are, to a great extent, self-checking and yield information about the degree to which the isotopic systems within the sample may or may not have been disturbed after crystallization, and (3) several of the radiometric methods can be used to obtain valid crystallization ages from disturbed, as well as undisturbed, samples. When properly applied, radiometric dating is an accurate and reliable method of determining the ages of rocks and sometimes of the complex events that have affected them since formation.

The Earth

We can be confident that the minimum age for the Earth exceeds 4 Ga—the evidence is abundant and compelling. Rocks exceeding 3.5 Ga in age are found on all the continents, but there are four especially well-studied areas on Earth: the Superior region of North America, the Isua–Godthaab area of western Greenland, the Pilbara block in the northern part of Western Australia, and Swaziland in southern Africa, where rocks 3.5 Ga or more in age have been found, carefully mapped, thoroughly studied, and dated by more than one radiometric method.

In the Minnesota River Valley and northern Michigan ancient granitoid gneisses have been dated by U–Pb and Rb–Sr methods at 3.5–3.7 Ga. The progenitors of these gneisses are thought to have been sedimentary rocks, lava flows, and igneous intrusive rocks.

In Swaziland ancient lava flows and sediments in the lowest part of the Barberton Greenstone belt have Sm–Nd, Rb–Sr, Pb–Pb, and ^{40}Ar/^{39}Ar ages of 3.4–3.5 Ga. These supracrustal rocks are enveloped by gneisses whose radiometric ages are also 3.4–3.5 Ga.

In the Pilbara block of Western Australia volcanic rocks of the Warrawoona Group, a thick accumulation of metamorphosed supracrustal rocks, have been dated at 3.4–3.56 Ga by Rb–Sr, Sm–Nd, and

U–Pb methods. The Warrawoona Group is intruded by granitoid gneiss that has a U–Pb age of 3.5 Ga.

In western Greenland supracrustal rocks at Isua, including a mafic intrusive body and sedimentary rocks deposited in shallow Archean seas, have been dated at 3.7–3.8 Ga by U–Pb, Pb–Pb, Rb–Sr, and Sm–Nd methods. The Isua supracrustal rocks are enveloped by younger granitoid gneisses that have Rb–Sr, Lu–Hf, U–Pb, and Pb–Pb ages of 3.5–3.7 Ga.

There are other areas in the world where rocks exceeding 3.5 Ga in age have been reported, including Venezuela, China, Zimbabwe, Antarctica, and Canada. The most important of these are at Enderby Land in Antarctica and the Slave Province in northwestern Canada, where gneisses have been dated by U–Pb methods at 3.93 ± 0.01 and 3.962 ± 0.003 Ga, respectively. The progenitors of the gneisses at Enderby Land are uncertain, but those of the Slave Province gneisses are thought to have been granitoids.

The gneisses of the Slave Province are currently the oldest rocks known on Earth, but single zircon crystals found in sedimentary rocks within the Yilgarn block of Western Australia have been dated by the U–Pb method and found to have ages of 4.0–4.3 Ga. The source rocks from which these crystals were eroded are as yet unknown, but the discovery of these ancient mineral grains offers the hope that rocks whose ages are greater than 4.0 Ga may yet be found.

It is intriguing that most of the oldest rocks found to date are metamorphosed sedimentary rocks and lava flows rather than igneous crust. Certainly, Earth's "primordial" crust, if such ever existed, has not yet been (and may never be) found, but the nature of the oldest rocks discovered so far clearly indicates that the Earth's age must exceed 4.0 Ga.

The Moon

In contrast to the Earth, a planet that has been highly active since its formation, the Moon experienced most of the significant events that shaped it early in its history. As a result nearly all Moon rocks are old by Earth standards, and radiometric ages in the range 3.5–4.0 Ga are common. Shortly after formation of the lunar crust, the Moon was subjected to bombardment by asteroid-sized objects that brecciated and fractured the crust to depths of as much as 20 km and formed the large ringed basins that dominate the topography of the Moon's near side. Current evidence indicates that this violent episode in lunar history occurred within the first 500–600 Ma. Because of the early bombardment and the resulting disturbance of the isotopic systems in the lunar rocks, only about a dozen of the returned lunar samples, all from the highlands, have radiometric ages exceeding 4.2 Ga. The oldest of these are three cumulates that are thought to have crystallized during formation of the lunar crust and have Rb–Sr and Sm–Nd ages in the range 4.42–4.51 Ga. Thus, we can infer that the minimum age of the Moon is 4.4–4.5 Ga.

The currently favored and only viable hypothesis for the formation of the Moon is that the Moon aggregated from the debris formed when a planetoid collided with the Earth. If this hypothesis is correct, then the minimum age for the Moon is also a minimum age for the Earth, and the Earth–Moon system must be at least 4.5 Ga in age.

The Meteorites

Meteorites are the oldest and most primitive rocks available for Earthbound scientists to study. Their compositions and orbits indicate that nearly all meteorites are fragments of asteroids (the rare exceptions are from Mars and the Moon) freed from their parent bodies by collisions with other asteroids and thrown into Earth-crossing orbits by gravitational resonance with Jupiter.

Meteorites vary widely in composition. Some, the chondrites, were formed by accretion of silicates from the Solar Nebula and represent primitive asteroids. Others, like the achondrites, the stony-irons, and the iron meteorites represent, respectively, superficial lava flows, the mantles, and the cores of asteroids that were large enough to undergo internal chemical segregation.

A number of the meteorites were metamorphosed and their isotopic systems were disturbed by the collisions that fragmented their parent bodies. The majority of the 70 or so well-dated meteorites, however, have individual Rb–Sr, Sm–Nd, Pb–Pb and $^{40}Ar/^{39}Ar$ ages of 4.4–4.6 Ga. Internal and whole-rock isochron ages determined by Rb–Sr, Lu–Hf, Sm–Nd, Pb–Pb, and Re–Os methods, as well as other isotopic evidence, show that the major meteorite types were formed within a few tens of millions of years between 4.5 and 4.6 Ga.

All hypotheses for formation of the Solar System call for the planets, including the Earth and the asteroids, to be formed within a very short interval of time after condensation of silicates from the Solar Nebula. Thus, the ages of meteorites are relevant to the age of the Earth and suggest that the Earth and the other solid bodies of the Solar System formed about 4.5–4.6 Ga.

Lead Isotopes in the Earth and Meteorites

The isotopic composition of lead when the matter of the Solar System first segregated into discrete bodies is thought to be preserved in the uranium-free phase (troilite) of iron meteorites. Internal Pb–Pb isochrons for individual meteorites and for several meteorite groups all fall within the range of 4.53 to 4.58 Ga and pass through the value of primordial lead composition found in the iron meteorites.

The isochron formed by meteorites and the primordial lead composition, known as the meteoritic geochron, represents the present-day lead-isotopic composition of all bodies that began with the primordial lead composition and evolved as separate U–Pb isotopic systems since the Solar System formed. The isotopic composition of lead in modern sediments and in young single-stage or near-single-stage lead ores falls very close to the meteoritic geochron. In addition, the growth curve of radiogenic lead isotopes in conformable lead ores passes close to the lead isotopic composition of troilite in iron meteorites. These observations indicate that, to a first approximation, meteorites and the Earth are part of the same lead-isotopic system and are of the same age.

The best value of the age of the Earth is based on the time required for the isotopic composition of lead in the oldest (2.6–3.5 Ga) terrestrial lead ores, of which there are only four, to evolve from the primordial composition to their compositions at the time the ores separated from their mantle reservoirs. These calculations result in ages of the Earth of from 4.52 to 4.56 Ga with a best value of 4.54 Ga.

The Galaxies and the Universe

The age of the Milky Way Galaxy has been calculated in two ways. One method compares the observed stage of evolution of globular cluster stars with theoretical star-evolution calculations. This method yields an age of 14–18 Ga. The second method is based on the present abundances, the production ratios in supernovae, and the decay constants of a few long-lived radioactive nuclides. This method yields ages of 9 to 16 Ga depending on the assumptions used for the frequency and timing of production of elements in supernovae.

The age of the universe can be estimated from the observed velocities at which distant galaxies are receding from the Milky Way and by calculating the time required for the matter of the universe to have expanded from a point of infinite density. The results of this method range from 7 to 20 Ga, depending on whether the universe is assumed to be unbounded, expanding forever, or bounded, eventually to collapse.

A "best age" for the universe can be found by requiring that the three methods for calculating the ages of the universe and the Milky Way Galaxy produce a consistent result. This best age is 15 ± 1.5 Ga, with a conservative 1 Ga allowed for the elapsed time between the "Big Bang" (start of expansion) and the formation of the Milky Way.

The ages found for the Milky Way and the universe are consistent with the age of the Earth and Solar System as determined from the ages of Earth rocks, Moon rocks, meteorites, and the isotopic composition of lead in meteorites and ancient lead ores.

What We Would Like to Know

An important endeavor, and one that is being actively pursued, is the search for the Earth's oldest rocks. No doubt this search will result in evidence of older rocks and minerals than have yet been found, but how much older cannot be predicted. The evidence for the existence of a lunar crust as early as 4.4–4.5 Ga suggests that the Earth's crust also may have formed as early as 4.5 Ga, but the Earth has been a tectonically active planet since birth and it is also possible, if not probable, that the Earth's earliest crust has long since been destroyed by recycling.

There is little doubt that the Earth's age is about 4.5–4.6 Ga, but currently the "best age" for the Earth (4.54 Ga) rests on lead-isotopic data from only four ancient lead ores. It would be highly desirable to increase this data set by a factor of two or three, especially by the addition of lead ores in excess of 3 Ga in age. Whether a significant number of additional ancient lead ores will be found, however, is highly questionable.

There is much to be learned about lunar history that is relevant to the age of the Earth. In particular, we know far too little about the nature and age of the early lunar crust simply because we have too few samples and too little ground-based geologic information from the lunar highlands. It is quite likely that the age of the earliest lunar crust could be determined rather precisely from the proper samples, but such samples are not in the existing lunar collections. The Moon is our closest neighbor, however, and additional visits are neither impossible, impractical, nor prohibitively expensive. Remote landers capable of returning samples would help fill this gap some, but better yet would be to send a few geologists. One of the severe shortcomings of the Apollo missions (Apollo 17 excepted) was the lack of geological expertise among the crews. We can hope that NASA will not make the same mistake with future lunar missions.

There is still much to be learned from the study of meteorites. In particular, there are thousands of unstudied meteorites in the new collections from Antarctica. Eventually, these specimens will provide new and more precise information about the times of formation of the various classes of meteorites. Especially useful will be isotopic analyses using the new generation of microprobes, such as the ion microprobe and the continuous laser probe, which are capable of yielding precise age data on individual mineral grains.

Samples from the other planets and planetoids would help to clarify the early history of the Solar System immensely. Samples from Mars, Venus, and Mercury would all be of interest and, because of the expense and danger, could best be obtained by remote landers (perhaps except on Venus, where, because of excessive temperatures and pressures, even machines have brief lifetimes). The single most productive mission relevant to the age and earliest history of the Solar System, however, would return samples from one or more asteroids of chondritic composition, especially those similar to the primitive carbonaceous chondrites. Spectral studies indicate that asteroids with surface compositions resembling that of carbonaceous chondrites are common. Since carbonaceous chondrites and their parent bodies are thought to be among the most primitive bodies in the Solar System, samples undisturbed by a stressful passage through the Earth's atmosphere are highly desirable. Analyses of such samples would test the conclusion that asteroids are the sources of meteorites and would provide age and petrologic information for the Solar System's most primitive remaining objects.

To supplement our knowledge of the age of the Earth and the Solar System, refined estimates of the ages of the Milky Way Galaxy and the universe are desirable. A more accurate age for the Milky Way will require refined models of star evolution, additional observations of the distributions of globular-cluster stars on the H–R diagram, and more precise knowledge about the rate of production of the long-lived radioactive nuclides in supernovae and other "element-creating" events in the Galaxy. Still narrower limits for the age of the universe require (1) better ways to measure distances to faraway galaxies, so that the Hubble constant and the resulting Hubble age can be more precisely calculated, and (2) more knowledge about the mass of the universe, so that the deceleration of the inflation can be more accurately estimated.

Despite the information we are lacking and the various uncertainties in the precise ages of the Earth, the Moon, the meteorites, the Galaxy, and the universe, the past three decades have produced a rich, consistent, and convincing body of evidence showing that the solid bodies of the Solar System formed 4.5–4.6 Ga ago and that the universe has been in existence for approximately 15 Ga. Our current understanding of the chronology of the universe, Galaxy, and Solar System represents the fulfillment of a quest that required more than two centuries of endeavor and surely is one of the most notable and spectacular achievements of modern science.

References Cited

References Cited

Aaronson, M., J. Mould, J. Huchra, W. T. Sullivan, III, R. A. Schommer, and G. D. Bothun. 1980. A distance scale from the infrared magnitude/HI velocity–width relation. III. The expansion rate outside the local supercluster. *Astrophysical Journal*, vol. 239, pp. 12–37.

Albritton, C. C., Jr. 1980. *The Abyss of Time.* San Francisco: Freeman, Cooper.

Alexander, E. C., Jr., and P. K. Davis. 1974. ^{40}Ar–^{39}Ar ages and trace element contents of Apollo 14 breccias: An interlaboratory cross-calibration of ^{40}Ar–^{39}Ar standards. *Geochimica et Cosmochimica Acta*, vol. 38, pp. 911–28.

Alexander, E. C., Jr., P. K. Davis, and R. S. Lewis. 1972. Argon-40–Argon-39 dating of Apollo sample 15555. *Science*, vol. 175, pp. 417–19.

Alexander, E. C., Jr., P. K. Davis, and P. H. Reynolds. 1972. Rare-gas analyses of neutron irradiated Apollo 12 samples. *Proceedings of the Third Lunar Science Conference*, pp. 1787–95.

Alexander, E. C., Jr., N. M. Evensen, and V. R. Murthy. 1973. ^{40}Ar–^{39}Ar and Rb–Sr studies of the Fiskenaesset complex, West Greenland. [abstract] *Eos*, vol. 54, p. 1227.

Allaart, J. H. 1976. The pre-3760 m. y. old supracrustal rocks of the Isua area, central West Greenland, and the associated occurrence of quartz-banded ironstone. pp. 177–89 in B. F. Windley, editor. *The Early History of the Earth.* London, New York, Sydney, and Toronto: Wiley.

Allègre, C. J., J. L. Birck, S. Fourcade, and M. P. Semet. 1975. Rubidium-87/strontium-87 age of Juvinas basaltic achondrite and early igneous activity in the solar system. *Science*, vol. 187, pp. 436–38.

Aller, M. F. 1971. Promethium in the star HR465. *Sky and Telescope*, vol. 41, pp. 220–22.

Allsopp, H. L., H. R. Roberts, G. D. I. Schreiner, and D. R. Hunter. 1962. Rb–Sr age measurements on various Swaziland granites. *Journal of Geophysical Research*, vol. 62, pp. 5307–13.

Allsopp, H. L., T. J. Ulrych, and L. O. Nicolaysen. 1968. Dating some significant events in the history of the Swaziland System by the Rb–Sr isochron method. *Canadian Journal of Earth Sciences*, vol. 5, pp. 605–19.

Alpher, R. A., and R. C. Herman. 1951. The primeval lead isotopic abundances and the age of the earth's crust. *Physical Review*, vol. 84, pp. 1111–14.

Alvarez, L. W. 1987. Mass extinctions caused by large bolide impacts. *Physics Today*, vol. 40, no. 7 (July), pp. 24–33.

Anderson, J. L. 1972. Non-Poisson distributions observed during counting of certain carbon-14-labeled organic (sub)monolayers. *The Journal of Physical Chemistry*, vol. 76, pp. 3603–12.

Anderson, J. L., and G. W. Spangler. 1973. Serial statistics: Is radioactive decay random? *The Journal of Physical Chemistry*, vol. 77, pp. 3114–21.

Anhaeusser, C. R. 1973. The evolution of the early Precambrian crust of southern Africa. *Philosophical Transactions of the Royal Society of London,* series A, vol. 273, pp. 359–88.

———. 1978. The geologic evolution of the primitive Earth—evidence from the Barberton Mountain Land. pp. 71–106 in D. H. Tarling. *Evolution of the Earth's Crust.* London, New York, and San Francisco: Academic Press.

———, editor. 1983. *Contributions to the Geology of the Barberton Mountain Land.* The Geological Society of South Africa Special Publ. No. 9, 223 pp.

Anhaeusser, C. R., L. J. Robb, and M. J. Viljöen. 1983. Notes on the provisional geologic map of the Barberton Greenstone Belt and surrounding granitic terrane, eastern Transvaal and Swaziland (1:250,000 colour map). pp. 221–23 in C. R. Anhaeusser, editor. *Contributions to the Geology of the Barberton Mountain Land.* The Geological Society of South Africa Special Publ. No. 9, 223 pp.

Anhaeusser, C. R., and J. F. Wilson. 1981. The granitic–gneiss greenstone shield. pp. 423–99 in D. R. Hunter, editor. *Precambrian of the Southern Hemisphere.* Amsterdam, Oxford, and New York: Elsevier Scientific.

Anonymous. 1984a. New solar system. *Eos*, vol. 65, p. 1177.

Anonymous. 1984b. Planet discovered outside Solar System. *Eos*, vol. 65, p. 1241.

Apollo Soil Survey. 1971. Apollo 14: Nature and origin of rock types in soil from the Fra Mauro Formation. *Earth and Planetary Science Letters*, vol. 12, pp. 49–54.

Arnett, W. D., D. Branch, and J. C. Wheeler. 1985. Hubble's constant and exploding carbon–oxygen white dwarf models for Type I supernovae. *Nature*, vol. 314, pp. 337–38.

Arriens, P. A. 1971. The Archaean geochronology of Australia. *Geological Society of Australia Special Publ.* 3, pp. 11–23.

Aston, F. W. 1919. A positive ray spectrograph. *Philosophical Magazine*, series 6, vol. 38, pp. 707–14.

———. 1927. The constitution of ordinary lead. *Nature*, vol. 120, p. 224.

———. 1929. The mass-spectrum of uranium lead and the atomic weight of protactinium. *Nature*, vol. 123, p. 313.

———. 1933. The isotopic composition and atomic weight of lead from different sources. *Proceedings of the Royal Society of London*, series A, vol. 140, pp. 535–43.

Austin, S. A. 1979. Uniformitarianism—a doctrine that needs rethinking. *The Compass*, vol. 56, no. 2, pp. 29–45.

Baadsgaard, H. 1973. U–Th–Pb dates on zircons from the early Precambrian Amîtsoq gneisses, Godthaab district, West Greenland. *Earth and Planetary Science Letters*, vol. 19, pp. 22–28.

———. 1976. Further U–Pb dates on zircons from the early Precambrian rocks of the Godthaabsfjord area, West Greenland. *Earth and Planetary Science Letters*, vol. 33, pp. 261–67.

———. 1983. U–Pb isotope systematics on minerals from the gneiss complex at Isukasia, West Greenland. *The Geological Survey of Greenland*, Rept. No. 112, pp. 35–42.

Baadsgaard, H., K. D. Collerson, and D. Bridgwater. 1979. The Archean gneiss complex of northern Labrador. 1. Preliminary U–Th–Pb geochronology. *Canadian Journal of Earth Sciences*, vol. 16, pp. 951–61.

Baadsgaard, H., R. St. J. Lambert, and J. Krupicka. 1976. Mineral isotopic age relationships in the polymetamorphic Amîtsoq gneisses, Godthaab district, West Greenland. *Geochimica et Cosmochimica Acta*, vol. 40, pp. 513–27.

Baadsgaard, H., and V. R. McGregor. 1981. The U–Th–Pb systematics of zircons from the type Nûk gneisses, Godthaabsfjord, West Greenland. *Geochimica et Cosmochimica Acta*, vol. 45, pp. 1099–109.

Baadsgaard, H., A. P. Nutman, D. Bridgwater, M. Rosing, V. R. McGregor, and J. H. Allaart. 1984. The zircon geochronology of the Akilia association and Isua supracrustal belt, West Greenland. *Earth and Planetary Science Letters*, vol. 68, pp. 221–28.

Baksi, A. K., D. A. Archibald, S. N. Sarkar, and A. K. Saha. 1984. ^{40}Ar–^{39}Ar incremental heating studies on mineral separates from the older metamorphic group of rocks, Singbhum, India. *Eos*, vol. 65, pp. 1132–33.

Barker, J. L., Jr., and E. Anders. 1968. Accretion rate of cosmic matter from iridium and osmium contents of deep-sea sediments. *Geochimica et Cosmochimica Acta*, vol. 32, pp. 627–45.

Barrell, J. 1917. Rhythms and the measurement of geologic time. *Bulletin of the Geological Society of America*, vol. 28, pp. 745–904.

Barton, J. M., Jr. 1975. Rb–Sr isotopic characteristics and chemistry of the 3.6-B.Y. Hebron gneiss, Labrador. *Earth and Planetary Science Letters*, vol. 27, pp. 427–35.

———. 1981. The pattern of Archaean crustal evolution in southern Africa as deduced from the evolution of the Limpopo mobile belt and the Barberton granite–greenstone terrain. pp. 21–31 in J. E. Glover and D. I. Groves, editors. *Archaean Geology*. Geological Society of Australia Special Publ. No. 7.

———. 1982. "A reappraisal of the Rb–Sr systematics of early Archaean gneisses from Hebron, Labrador" by K. D. Collerson et al.—a reply. *Earth and Planetary Science Letters*, vol. 60, pp. 337–38.

———. 1983. Isotopic constraints on possible tectonic models for crustal evolution in the Barberton granite–greenstone terrane, southern Africa. pp. 73–79 in C. R. Anhaeusser, editor. *Contributions to the Geology of the Barberton Mountain Land*. The Geological Society of South Africa Special Publ. No. 9.

Barton, J. M., Jr., D. R. Hunter, M. P. A. Jackson, and A. C. Wilson. 1980. Rb–Sr age and source of the Bimodal Suite of the Ancient Gneiss Complex, Swaziland. *Nature*, vol. 283, pp. 756–58.

———. 1983a. Geochronologic and Sr-isotopic studies of certain units in the Barberton granite–greenstone terrane, Swaziland. *Transactions of the Geological Society of South Africa*, vol. 86, pp. 71–80.

Barton, J. M., Jr., L. J. Robb, C. R. Anhaeusser, and D. A. Van Nierop. 1983b. Geochronologic and Sr-isotopic studies of certain units in the Barberton granite–greenstone terrane, South Africa. pp. 63–72 in C. R. Anhaeusser, editor. *Contributions to the Geology of the Barberton Mountain Land*. The Geological Society of South Africa Special Publ. No. 9.

Barton, J. M., Jr., B. Ryan, and R. E. P. Fripp. 1978. The relationship between Rb–Sr and U–Th–Pb whole-rock and zircon systems in the >3790 M. Y. old Sand River Gneisses, Limpopo mobile belt, southern Africa.

pp. 27–28 in R. E. Zartman, editor. *Short Papers of the Fourth International Conference: Geochronology, Cosmochronology, Isotope Geology, 1978*. U. S. Geological Survey Open-File Report 78-701.

———. 1983. Rb–Sr and U–Th–Pb isotopic studies of the Sand River Gneisses, central zone, Limpopo mobile belt. pp. 9–18 in W. J. Van Biljon and J. E. Legg, editors. *The Limpopo Belt*. The Geological Society of South Africa Special Publ. 8.

Barus, C. 1891. The contraction of molten rock. *American Journal of Science*, 3rd series, vol. 42, pp. 498–99.

———. 1892. The relation of melting point to pressure in case of igneous rock fusion. *American Journal of Science*, 3rd series, vol. 43, pp. 56–57.

Basaltic Volcanism Study Project. 1981. *Basaltic Volcanism on the Terrestrial Planets*. New York: Pergamon Press.

Basu, A. R., S. L. Ray, A. K. Saha, and S. N. Sarkar. 1981. Eastern Indian 3800-million-year-old crust and early mantle differentiation. *Science*, vol. 212, pp. 1502–06.

Baxter, J., and T. Atkins. 1976. *The Fire Came By: The Riddle of the Great Siberian Explosion*. Garden City: Doubleday.

Beatty, J. K. 1984. Lunar meteorites: Three and counting. *Sky and Telescope*, vol. 68, p. 224.

Becker, G. F. 1908. Relations of radioactivity to cosmogony and geology. *Bulletin of the Geological Society of America*, vol. 19, pp. 113–46.

———. 1910a. Halley on the age of the ocean. *Science*, vol. 31, pp. 459–61.

———. 1910b. Reflections on Joly's method of determining the ocean's age. *Science*, vol. 31, pp. 509–12.

———. 1910c. The age of the Earth. *Smithsonian Miscellaneous Collections*, vol. 56, no. 6, pp. 1–28.

Becker, R. H., and R. O. Pepin. 1984. The case for a Martian origin of the shergottites: Nitrogen and noble gases in EETA79001. *Earth and Planetary Science Letters*, vol. 69, pp. 225–42.

Berlage, H. P. 1968. *The Origin of the Solar System*. Oxford: Pergamon Press.

Bibikova, E. B. 1984. The most ancient rocks in the USSR territory by U–Pb data on accessory zircons. pp. 233–50 in A. Kroner, G. N. Hanson, and A. M. Goodwin, editors. *Archaean Geochemistry*. Berlin, Heidelberg, New York, and Tokyo: Springer-Verlag.

Bickle, M. J., L. F. Bettenay, C. A. Boulter, D. I. Groves, and P. Morant. 1980. Horizontal tectonic interaction of an Archean gneiss belt and greenstones, Pilbara Block, Western Australia. *Geology*, vol. 8, pp. 525–29.

Bickle, M. J., P. Morant, L. F. Bettenay, C. A. Boulter, T. S. Blake, and D. I. Groves. 1985. Archean tectonics of the Shaw Batholith, Pilbara Block, Western Australia. Structural and metamorphic tests of the batholith concept. Geological Association of Canada, Special Paper 28, pp. 325–41.

Binder, A. B. 1982. The mare basalt magma source region and mare basalt magma genesis. *Proceedings of the 13th Lunar Science Conference*, Part 1 (*Journal of Geophysical Research*, vol. 87 Supplement), pp. A37–53.

Bingley, W. 1796. Stones fallen from the air a natural phenomenon. Gentleman's Magazine, vol. 66, pp. 726–28.

Birck, J. L., and C. J. Allègre. 1978. Chronology and chemical history of the parent body of basaltic achondrites studied by the ^{87}Rb–^{87}Sr method. *Earth and Planetary Science Letters*, vol. 39, pp. 37–51.

———. 1981. ^{87}Rb/^{87}Sr study of diogenites. *Earth and Planetary Science Letters*, vol. 55, pp. 116–22.

Birck, J. L., S. Fourcade, and C. J. Allègre. 1975. ^{87}Rb–^{86}Sr age of rocks from the Apollo 15 landing site and significance of internal isochrons. *Earth and Planetary Science Letters*, vol. 26, pp. 29–35.

Black, L. P., N. H. Gale, S. Moorbath, R. J. Pankhurst, and V. R. McGregor. 1971. Isotopic dating of very early Precambrian amphibolite facies gneisses from the Godthaab district, West Greenland. *Earth and Planetary Science Letters*, vol. 12, pp. 245–59.

Black, L. P., I. S. Williams, and W. Compston. 1986. Four zircon ages from one rock: The history of a 3930 Ma-old granulite from Mount Sones, Enderby Land, Antarctica. *Contributions to Mineralogy and Petrology*, vol. 94, pp. 427–37.

Blockley, J. G. 1975. Pilbara Block. pp. 33–54 in *Geology of Western Australia*. Western Australia Geological Survey Memoir 2.

Bogard, D. D., D. Burnett, P. Eberhardt, and G. J. Wasserburg. 1968. ^{40}Ar–^{40}K ages of silicate inclusions in iron meteorites. *Earth and Planetary Science Letters*, vol. 3, pp. 275–83.

Bogard, D. D., and W. C. Hirsch. 1980. ^{40}Ar/^{39}Ar dating, Ar diffusion properties, and cooling rate determination of severely shocked chondrites. *Geochimica et Cosmochimica Acta*, vol. 44, pp. 1667–82.

Bogard, D. D., and L. Husain. 1977. A new 1.3 aeon-young achondrite. *Geophysical Research Letters*, vol. 4, pp. 69–71.

Bogard, D. D., L. Husain, and L. E. Nyquist. 1979. ^{40}Ar–^{39}Ar age of the Shergotty achondrite and implications for its post-shock thermal history. *Geochimica et Cosmochimica Acta*, vol. 43, pp. 1047–55.

Bogard, D. D., L. Husain, and R. J. Wright. 1976. ^{40}Ar–^{39}Ar dating of collisional events in chondrite parent bodies. *Journal of Geophysical Research*, vol. 81, pp. 5664–78.

Bogard, D. D., L. E. Nyquist, B. M. Bansal, H. Wiesmann, and C.-Y. Shih. 1975. 76535: An old lunar rock? *Earth and Planetary Science Letters*, vol. 26, pp. 69–80.

Bogard, D. D., D. M. Unruh, and M. Tatsumoto. 1983. ^{40}Ar/^{39}Ar and U–Th–Pb dating of separated clasts from the Abee E4 chondrite. *Earth and Planetary Science Letters*, vol. 62, pp. 132–46.

Bok, B. J. 1982. Galaxy. *McGraw-Hill Encyclopedia of Science and Technology*, vol. 6, pp. 6–9.

Boltwood, B. B. 1905. On the ultimate disintegration products of the radioactive elements. *American Journal of Science*, vol. 20, pp. 253–67.

———. 1907. On the ultimate disintegration products of the radio-active elements. Part II. The disintegration products of uranium. *American Journal of Science*, series 4, vol. 23, pp. 77–88.

Bowell, E., C. R. Chapman, J. C. Gradie, D. Morrison, and B. Zellner. 1978. Taxonomy of asteroids. *Icarus*, vol. 35, pp. 313–35.

Bowring, S. A., I. S. Williams, and W. Compston. 1989. 3.96 Ga gneisses from the Slave Province, Northwest Territories, Canada. *Geology*, vol. 17, pp. 971–75.

Brereton, N. R. 1970. Corrections for interfering isotopes in the ^{40}Ar/^{39}Ar dating method. *Earth and Planetary Science Letters*, vol. 8, pp. 427–33.

Brévart, O., B. Dupré, and C. J. Allègre. 1986. Lead–lead age of komatiitic lavas and limitations on the structure and evolution of the Precambrian mantle. *Earth and Planetary Science Letters*, vol. 77, pp. 293–302.

Brewer, A. K. 1937. Radioactivity of potassium and geological time. *Science,* vol. 86, pp. 198–99.

———. 1938. Age of matter as determined by the radioactivity of potassium and rubidium. *Journal of the Washington Academy of Sciences,* vol. 28, p. 416.

Brewer, P. G. 1975. Minor elements in sea water. pp. 415–96 in J. P. Riley and G. Skirrow, editors. *Chemical Oceanography,* vol. 1, 2nd edition. London, New York, and San Francisco: Academic Press.

Brice, W. R. 1982. Bishop Ussher, John Lightfoot and the age of creation. *Journal of Geological Education,* vol. 30, pp. 18–24.

Bridgwater, D., and K. D. Collerson. 1976. The major petrological and geochemical character of the 3600 m.y. Uivak gneisses from Labrador. *Contributions to Mineralogy and Petrology,* vol. 54, pp. 43–59.

Bridgwater, D., K. D. Collerson, R. W. Hurst, and C. W. Jesseau. 1975. Field characters of the Early Precambrian rocks from Saglek, coast of Labrador. *Geological Survey of Canada,* Paper 75-1a, pp. 287–96.

Bridgwater, D., K. D. Collerson, and J. S. Myers. 1978. The development of the Archaean gneiss complex of the North Atlantic Region. pp. 19–69 in D. H. Tarling, editor. *Evolution of the Earth's Crust.* London, New York, and San Francisco: Academic Press.

Bridgwater, D., V. R. McGregor, and J. S. Myers. 1974. A horizontal tectonic regime in the Archean of Greenland and its implications for early crustal thickening. *Precambrian Research,* vol. 1, pp. 179–97.

Bridgwater, D., J. Watson, and B. F. Windley. 1973. The Archean craton of the North Atlantic region. *Philosophical Transactions of the Royal Society of London,* series A, vol. 273, pp. 493–512.

Brown, H. 1947. An experimental method for the estimation of the age of the elements. *Physical Review,* vol. 72, p. 348.

Brush, S. G. 1982. Nickel for your thoughts: Urey and the origin of the Moon. *Science,* vol. 217, pp. 891–98.

Buchwald, V. F. 1975. *Handbook of Iron Meteorites* (Vols. I–III). Berkeley and Los Angeles: University of California Press.

Bugnon, M.-F., F. Tera, and L. Brown. 1979. Are ancient lead deposits chronometers of the early history of the Earth? *Carnegie Institution of Washington Year Book for 1978,* pp. 346–52.

Bullard, E. C., and J. P. Stanley. 1949. The age of the Earth. *Verröffentlichungen des Finnischen Geodätischen Institutes,* vol. 36, p. 33.

Burbidge, E. M., G. R. Burbidge, W. A. Fowler, and F. Hoyle. 1957. Synthesis of the elements in stars. *Reviews of Modern Physics,* vol. 29, pp. 547–650.

Burchfield, B. C. 1983. The continental crust. *Scientific American,* vol. 249, no. 3, pp. 130–42.

Burchfield, J. D. 1975. *Lord Kelvin and the Age of the Earth.* New York: Science History Publications.

Burnett, D. S., and G. J. Wasserburg. 1967a. ^{87}Rb–^{87}Sr ages of silicate inclusions in iron meteorites. *Earth and Planetary Science Letters,* vol. 2, pp. 397–408.

———. 1967b. Evidence for the formation of an iron meteorite at 3.8×10^9 years. *Earth and Planetary Science Letters,* vol. 2, pp. 137–47.

Cadogan, P. H. 1981. *The Moon—Our Sister Planet.* Cambridge, London, and New York: Cambridge University Press.

Cadogan, P. H., and G. Turner. 1977. ^{40}Ar–^{39}Ar dating of Luna 16 and Luna

20 samples. *Philosophical Transactions of the Royal Society of London*, vol. 284, pp. 167–77.

Cameron, A. G. W. 1975. The origin and evolution of the Solar System. *Scientific American*, vol. 233, pp. 33–41.

———. 1982. Solar System. *McGraw-Hill Encyclopedia of Science and Technology*, vol. 12, pp. 603–06.

Cameron, M., K. D. Collerson, W. Compston, and R. Morton. 1981. The statistical analysis and interpretation of imperfectly fitted Rb–Sr isochrons from polymetamorphic terrains. *Geochimica et Cosmochimica Acta*, vol. 45, pp. 1087–97.

Caputo, F., V. Castellani, and M. L. Quarta. 1984. A self-consistent approach to the age of globular cluster M15. *Astronomy and Astrophysics*, vol. 138, pp. 457–63.

Carlson, R. W. 1983. Sm–Nd age of the Ancient Gneiss Complex, Swaziland, southern Africa. *Annual Report of the Director, Department of Terrestrial Magnetism, Carnegie Institution of Washington, 1982–1983*, pp. 550–55.

Carlson, R. W., D. R. Hunter, and F. Barker. 1983. Sm–Nd age and isotopic systematics of the bimodal suite, ancient gneiss complex, Swaziland. *Nature*, vol. 305, pp. 701–04.

Carlson, R. W., and G. W. Lugmair. 1981. Time and duration of lunar highlands crust formation. *Earth and Planetary Science Letters*, vol. 52, pp. 227–38.

Carney, B. W. 1980. The ages and distances of eight globular clusters. *Astrophysical Journal Supplement*, vol. 42, pp. 481–500.

Carozzi, M. 1983. *Voltaire's Attitude toward Geology*. Geneva: Société de Physique et d'Histoire naturelle.

Catanzero, E. J. 1963. Zircon ages in southwestern Minnesota. *Journal of Geophysical Research*, vol. 68, pp. 2045–48.

Chaikin, A. 1983. A stone's throw from the planets. *Sky and Telescope*, vol. 65, pp. 122–23.

Chamberlain, T. C. 1899a. Editorial [on Ritter's age of Earth]. *Journal of Geology*, vol. 7, pp. 92–94.

———. 1899b. On Lord Kelvin's address on the age of the Earth as an abode fitted for life. *Annual Report of the Smithsonian Institution, 1899*, pp. 223–46.

Chapman, C. R. 1975. The nature of asteroids. *Scientific American*, vol. 232, no. 1, pp. 24–33.

———. 1976. Asteroids as meteorite parent-bodies: The astronomical perspective. *Geochimica et Cosmochimica Acta*, vol. 40, pp. 701–19.

Chapman, C. R., D. Morrison, and B. Zellner. 1975. Surface properties of asteroids: A synthesis of polarimetry, radiometry and spectrophotometry. *Icarus*, vol. 25, pp. 104–30.

Chen, J. H., and G. R. Tilton. 1976. Isotopic lead investigations on the Allende carbonaceous chondrite. *Geochimica et Cosmochimica Acta*, vol. 40, pp. 635–43.

Chen, J. H., and G. J. Wasserburg. 1981. The isotopic composition of uranium and lead in Allende inclusions and meteoritic phosphates. *Earth and Planetary Science Letters*, vol. 52, pp. 1–15.

Chladni, E. E. F. 1794. *Über den Ursprung der von Pallas gefunden und anderer ihr anlicher Eisenmassen, und über einige damit in Verbindung stehende Naturerscheinungen* [Observations on a Mass of Iron Found in Siberia by Pallas and on Other Masses of Like Kind, with Some Conjectures Respecting Their Connection with Certain Natural Phenomena]. Riga: J. F. Hartknoch.

Chow, T. J., and C. C. Patterson. 1961. On the primordial lead of the Canyon Diablo meteorite. *Geokhimiya*, no. 12, pp. 1124–25 (in Russian).

Claoue-Long, J. C., M. F. Thirlwall, and R. W. Nesbitt. 1984. Revised Sm–Nd systematics of Kambalda greenstones, Western Australia. *Nature*, vol. 307, pp. 697–701.

Clarke, F. W. 1924. Data of geochemistry. *U. S. Geological Survey Bulletin 770*, pp. 1–841.

Clayton, D. D., W. A. Fowler, T. E. Hall, and B. A. Zimmerman. 1961. Neutron capture chains in heavy element synthesis. *Annals of Physics*, vol. 12, pp. 331–408.

Clayton, R. N. 1978. Isotopic anomalies in the early Solar System. *Annual Reviews of Nuclear and Particle Science*, vol. 28, pp. 501–22.

Clayton, R. N., and T. K. Mayeda. 1975. Genetic relations between the Moon and meteorites. *Proceedings of the Sixth Lunar Science Conference*, pp. 1761–69.

Cloud, P. E. 1968. Atmospheric and hydrospheric evolution on the primitive earth. *Science*, vol. 160, pp. 729–36.

―――. 1973. Paleoecological significance of the banded iron-formation. *Economic Geology*, vol. 68, pp. 1135–43.

Collerson, K. D., C. Brooks, A. B. Ryan, and W. Compston. 1982. A reappraisal of the Rb–Sr systematics of early Archaean gneisses from Hebron, Labrador. *Earth and Planetary Science Letters*, vol. 60, pp. 325–36.

Collerson, K. D., C. W. Jesseau, and D. Bridgwater. 1976. Crustal development of the Archaean gneiss complex, eastern Labrador. pp. 237–53 in B. F. Windley, editor. *The Early History of the Earth*. London, New York, Sydney, and Toronto: Wiley.

Collerson, K. D., A. Kerr, and W. Compston. 1981. Geochronology and evolution of Late Archaean gneisses in Northern Labrador: An example of reworked sialic crust. pp. 205–22 in J. E. Glover and D. I. Groves, editors. *Archaean Geology*. Geological Society of Australia Special Publ. 7.

Collerson, K. D., and M. T. McCullouch. 1982. Field and isotopic relationships in the Archaean Pilbara Block, Western Australia. *The Australian National University Research School of Earth Sciences Annual Report for 1981*, pp. 172–73.

Collerson, K. D., and M. T. McCulloch. 1982. Field and isotopic relationships in the Archaean Pilbara Block, Western Australia. *The Australian National University Research School of Earth Sciences Annual Report for 1981*, pp. 172–73.

Collins, C. B., R. D. Russell, and R. M. Farquhar. 1953. The maximum age of the elements and the age of the earth's crust. *Canadian Journal of Physics*, vol. 31, pp. 420–28.

Compston, W., and P. A. Arriens. 1968. The Precambrian geochronology of Australia. *Canadian Journal of Earth Science*, vol. 5, pp. 561–83.

Compston, W., J. J. Foster, and C. M. Gray. 1977. Rb–Sr systematics in clasts and aphanites from consortium breccia 73215. *Proceedings of the Eighth Lunar Science Conference*, pp. 2525–49.

Compston, W., D. O. Froude, T. R. Ireland, P. D. Kinney, I. S. Williams, I. R. Williams, and J. S. Myers. 1985. The age of (a tiny part of) the Australian continent. *Nature*, vol. 317, pp. 559–60.

Compston, W., and R. T. Pidgeon. 1986. Jack Hills, evidence of more very old detrital zircons in Western Australia. *Nature*, vol. 321, pp. 766–69.

Compston, W., I. S. Williams, D. Froude, and J. J. Foster. 1983. The age and metamorphic history of the Mt. Narryer banded gneiss. *The Australian National University Research School of Earth Sciences Annual Report for 1982*, pp. 196–99.

Compston, W., I. S. Williams, and C. Meyer. 1984. U–Pb geochronology of zircons from lunar breccia 73217 using a sensitive high mass-resolution ion microprobe. Proceedings of the 14th Lunar and Planetary Science Conference, *Journal of Geophysical Research*, vol. 89 Supplement, pp. B525–34.

Condie, K. C. 1976. *Plate Tectonics and Crustal Evolution*. New York: Pergamon Press.

———. 1981. *Archean Greenstone Belts*. Amsterdam, Oxford, and New York: Elsevier Scientific.

Cooper, J. A., P. R. James, and R. W. R. Rutland. 1982. Isotopic dating and structural relationships of granitoids and greenstones in the east Pilbara, Western Australia. *Precambrian Research*, vol. 18, pp. 199–236.

Cooper, J. A., P. H. Reynolds, and J. R. Richards. 1969. Double-spike calibration of the Broken Hill standard lead. *Earth and Planetary Science Letters*, vol. 6, pp. 467–78.

Cumming, G. L. 1969. A recalculation of the age of the Solar System. *Canadian Journal of Earth Sciences*, vol. 6, pp. 719–35.

Cumming, G. L., and J. R. Richards. 1975. Ore lead isotope ratios in a continuously changing Earth. *Earth and Planetary Science Letters*, vol. 28, pp. 155–71.

Dalrymple, G. B. 1969. Potassium–argon ages of Recent rhyolites of the Mono and Inyo Craters, California. *Earth and Planetary Science Letters*, vol. 3, pp. 289–98.

———. 1983a. Can the Earth be dated from decay of its magnetic field? *Journal of Geological Education*, vol. 31, pp. 124–33.

———. 1983b. Radiometric dating and the age of the Earth: A reply to scientific creationism. *Federation Proceedings*, vol. 42, pp. 3033–38.

———. 1984. How old is the Earth? A reply to "scientific creationism." *Proceedings of the Annual Meeting, Pacific Division, American Association for the Advancement of Science* (63rd, April, 1984), vol. 1, pp. 66–131.

———. 1986. Radiometric dating, geologic time, and the age of the Earth: A reply to scientific creationism. U. S. Geological Survey Open-File Report 86-110.

Dalrymple, G. B., E. C. Alexander, Jr., M. A. Lanphere, and G. P. Kraker. 1981. Irradiation of samples for $^{40}Ar/^{39}Ar$ dating using the Geological Survey TRIGA reactor. *U. S. Geological Survey Professional Paper* 1176.

Dalrymple, G. B., and M. A. Lanphere. 1969. *Potassium–Argon Dating*. San Francisco: Freeman.

———. 1971. $^{40}Ar/^{39}Ar$ technique of K–Ar dating: A comparison with the conventional technique. *Earth and Planetary Science Letters*, vol. 12, pp. 300–08.

Dana, J. D. 1846. On the volcanoes of the Moon. *American Journal of Science*, 2nd series, vol. 2, pp. 335–53.

———. *Manual of Geology* (4th edition). New York: American Book Co.

Darwin, G. H. 1879. On the precession of a viscous spheroid, and on the remote history of the earth. *Philosophical Transactions of the Royal Society of London*, vol. 170, pp. 447–538.

———. 1880. On the secular changes in the elements of the orbit of a satellite revolving about a tidally distorted planet. *Philosophical Transactions of the Royal Society of London*, vol. 171, pp. 713–891.

———. *The Tides and Kindred Phenomena in the Solar System*. San Francisco: Freeman (reprinting, 1962).

Davies, R. D., H. L. Allsopp, A. J. Erlank, and W. I. Manton. 1970. Sr-isotopic studies on various layered mafic intrusions in southern Africa. The Geological Society of South Africa Special Publ. No. 1, pp. 576–93.

Dean, D. R. 1981. The age of the Earth controversy: Beginnings to Hutton. *Annals of Science*, vol. 38, pp. 435–56.

de Laeter, J. R., and J. G. Blockley. 1972. Granite ages within the Archaean Pilbara Block, Western Australia. *Journal of the Geological Society of Australia*, vol. 19, pp. 363–70.

de Laeter, J. R., I. R. Fletcher, K. J. R. Rosman, I. R. Williams, R. D. Gee, and W. G. Libby. 1981. Early Archaean gneisses from the Yilgarn block, Western Australia. *Nature*, vol. 292, pp. 322–24.

de Laeter, J. R., J. D. Lewis, and J. G. Blockley. 1975. Granite ages within the Shaw Batholith of the Pilbara Block. *The Geological Survey of Western Australia Annual Report for 1974*, pp. 73–79.

de Laeter, J. R., W. G. Libby, and A. F. Trendall. 1981. The older Precambrian geochronology of Western Australia. pp. 145–57 in J. F. Glover and D. J. Groves, editors. *Archaean Geology*. Geological Society of Australia Special Publ. 7.

de Maillet, B. 1748. Telliamed, or Conversations between an Indian Philosopher and a French Missionary on the Diminution of the Sea (A. V. Carozzi, editor). Urbana: University of Illinois Press (reprinting, 1968).

DePaolo, D. J. 1981. Nd isotopic studies: Some new perspectives on Earth structure and evolution. *Eos*, vol. 62, pp. 137–40.

———. 1988. *Neodymium Isotope Geochemistry*. Berlin, New York: Springer-Verlag.

DePaolo, D. J., and G. J. Wasserburg. 1976. Nd isotopic variations and petrogenetic models. *Geophysical Research Letters*, vol. 3, pp. 249–52.

Dermott, S. F., editor. 1978. *The Origin of the Solar System*. New York: Wiley.

de Vaucouleurs, G. 1986. The cosmic distance scale and the Hubble constant. pp. 1–61 in F. Melchiorri and R. Ruffini, editors. *Gamow Cosmology* (Proceedings of the International School of Physics, "Enrico Fermi," Course 86). Amsterdam, Oxford, New York, and Tokyo: North-Holland.

de Vaucouleurs, G., and W. L. Peters. 1981. Hubble ratio and solar motion from 200 spiral galaxies having distances derived from the luminosity index. *Astrophysical Journal*, vol. 248, pp. 395–407.

Dodd, R. T. 1981. *Meteorites: A Petrological–Chemical Synthesis*. Cambridge, London, and New York: Cambridge University Press.

Doe, B. R. 1970. *Lead Isotopes*. Berlin, Heidelberg, and New York: Springer-Verlag.

Doe, B. R., and M. H. Delevaux. 1980. Lead-isotope investigations in the Minnesota River Valley—late-tectonic and posttectonic granites. pp. 105–12 in G. B. Morey and G. N. Hanson, editors. *Selected Studies of Archean Gneisses and Lower Proterozoic Rocks, Southern Canadian Shield*. The Geological Society of America Special Paper No. 182.

Doe, B. R., and J. S. Stacey. 1974. The application of lead isotopes to the problems of ore genesis and ore prospect evaluation: A review. *Economic Geology*, vol. 69, pp. 757–76.

Dohnanyi, J. S. 1972. Interplanetary objects in review: Statistics of their masses and dynamics. *Icarus*, vol. 17, pp. 1–48.

Dominik, B., and E. K. Jessberger. 1978. Early lunar differentiation: 4.42-AE-old plagioclase clasts in Apollo 16 breccia 67435. *Earth and Planetary Science Letters*, vol. 38, pp. 407–15.

Dominik, B., E. K. Jessberger, Th. Staudacher, K. Nagel, and A. E. L. Goresy. 1978. A new type of white inclusion in Allende: Petrography, mineral chemistry, $^{40}Ar–^{39}Ar$ ages and genetic implications. *Proceedings of the Ninth Lunar and Planetary Conference*, pp. 1249–66.

Drozd, R. J., and F. A. Podosek. 1977. Systematics of iodine–xenon dating. *Geochemical Journal*, vol. 11, pp. 231–37.

Dymek, R. F., A. L. Albee, and A. A. Chodos. 1975. Comparative petrology of lunar cumulate rocks of possible primary origin: Dunite 72415, troctolite 76535, norite 78235 and anorthosite 62237. *Proceedings of the Sixth Lunar Science Conference*, pp. 301–41.

Eberhardt, P., J. Geiss, N. Grogler, and A. Stettler. 1973. How old is the crater Copernicus? *The Moon*, vol. 8, pp. 104–14.

Eggen, O. J., D. Lynden-Bell, and A. Sandage. 1962. Evidence from the motions of old stars that the Galaxy collapsed. *Astrophysical Journal*, vol. 136, pp. 748–66.

El-Baz, F. 1978. A stratigraphic approach to the evolution of the lunar crust. pp. 103–39 in C. Ponnamperuma, editor. *Comparative Planetology*. New York, San Francisco, and London: Academic Press.

Emery, G. T. 1972. Perturbation of nuclear decay rates. *Annual Reviews of Nuclear Science*, vol. 22, pp. 165–202.

Enright, M. C., and G. Turner. 1977. History and size of chondrite parent bodies from $^{40}Ar–^{39}Ar$ ages. *Meteoritics*, vol. 12, p. 217.

Evensen, N. M., P. J. Hamilton, G. E. Harlow, R. Klimentidis, R. K. O'Nions, and M. Prinz. 1979. Silicate inclusions in Weekeroo Station: Planetary differentiates in an iron meteorite. *Lunar and Planetary Science*, vol. X, pp. 376–78.

Evensen, N. M., V. R. Murthy, and M. R. Coscio, Jr. 1973. Rb–Sr ages of some mare basalts and the isotopic and trace element systematics in lunar fines. *Proceedings of the Fourth Lunar Science Conference*, pp. 1707–24.

Fahlman, G. G., H. B. Richer, and D. A. VandenBerg. 1985. Deep CCD photometry in globular clusters. III. M15. *Astrophysical Journal Supplement*, vol. 58, pp. 225–54.

Farhat, J. S., and G. W. Wetherill. 1975. Interpretation of apparent ages in Minnesota. *Nature*, vol. 257, pp. 721–22.

Faul, H. 1966. *Ages of Rocks, Planets, and Stars*. New York: McGraw-Hill.

Faure, G. 1986. *Principles of Isotope Geology* (2nd ed.). New York: Wiley.

Faure, G., and J. L. Powell. 1972. *Strontium Isotope Geology*. Berlin, Heidelberg, and New York: Springer-Verlag.

Fenner, C. N., and C. S. Piggot. 1929. The mass-spectrum of lead from Broggerite. *Nature*, vol. 123, pp. 793–94.

Fisher, D. E. 1965. Anomalous Ar^{40} contents of iron meteorites. *Journal of Geophysical Research*, vol. 70, pp. 2445–52.

Florentine-Nielsen, R. 1984. Determination of differences in light travel time for QSO0957+561A,B. *Astronomy and Astrophysics*, vol. 138, pp. L19–20.

Fontenelle, B. B. 1686. *Entretiens sur la pluralité des mondes* (Conversations on the Plurality of Worlds), translation by J. Glanville, 1688. London: Bentley and Magues.

Fowler, W. A. 1972. New observations and old nucleocosmochronologies. pp. 67–123 in F. Reines, editor. *Cosmology, Fission and Other Matters—A Memorial to George Gamow.* Boulder: Colorado Associated University Press.

French, B. M. 1977. *The Moon Book.* New York: Penguin.

Froude, D. O., T. R. Ireland, P. D. Kinney, I. S. Williams, and W. Compston. 1983. Ion microprobe identification of 4,100–4,200 Myr-old terrestrial zircons. *Nature*, vol. 304, pp. 616–18.

Fyfe, W. S. 1980. Crust formation and destruction. pp. 77–88 in D. W. Strangway, editor. *The Continental Crust and Its Mineral Deposits.* Geological Association of Canada Special Paper 20.

Gaffey, M. J., and T. B. McCord. 1978. Asteroid surface materials: Mineralogic characterization from reflectance spectra. *Space Science Reviews*, vol. 21, pp. 555–628.

Gale, N. H. 1972. Uranium–lead systematics in lunar basalts. *Earth and Planetary Science Letters*, vol. 17, pp. 65–78.

Gale, N. H., J. W. Arden, and M. C. B. Abranches. 1980. Uranium–lead age of the Bruderheim L6 chondrite and the 500-Ma shock event in the L-group parent body. *Earth and Planetary Science Letters*, vol. 48, pp. 311–24.

Gale, N. H., J. W. Arden, and R. Hutchison. 1975. The chronology of the Nakhla achondritic meteorite. *Earth and Planetary Science Letters*, vol. 26, pp. 195–206.

———. 1979. U–Pb studies of the Appley Bridge meteorite. *Naturwissenschaften*, vol. 66, pp. 419–20.

Gale, N. H., and A. E. Mussett. 1973. Episodic uranium–lead models and the interpretation of variations in the isotopic composition of lead in rocks. *Reviews of Geophysics and Space Physics*, vol. 11, pp. 37–86.

Gancarz, A. J., and G. J. Wasserburg. 1977. Initial Pb of the Amîtsoq gneiss, west Greenland, and implications for the age of the Earth. *Geochimica et Cosmochimica Acta*, vol. 41, pp. 1283–1301.

Gee, R. D. 1979. Structure and tectonic style of the Western Australia shield. *Tectonophysics*, vol. 58, pp. 327–69.

Geikie, A. 1892. Address of the President. *Report of the 62nd Meeting of the British Association for the Advancement of Science, Edinburgh*, pp. 3–26.

———. 1899. Address of the President, geology section. *Report of the 69th Meeting of the British Association for the Advancement of Science, Dover*, pp. 718–30.

———. 1903. *Text-Book of Geology*, vol. 1 (4th edition). London: Macmillan.

———. 1905. *The Founders of Geology* (2nd edition). London: Macmillan.

Geiss, J., P. Eberhardt, N. Grögler, S. Guggisberg, P. Maurer, and A. Stettler. 1977. The filling of the mare basins. *Philosophical Transactions of the Royal Society of London*, series A, vol. 285, pp. 151–58.

Gerling, E. K. 1942. Age of the Earth according to radioactivity data. *Comptes Rendus (Doklady) de l'Académie des Sciences de l'URSS*, vol. 34, pp. 259–61.

Gilbert, G. K. 1893. The Moon's face: A study of the origin of its features. *Philosophical Society of Washington Bulletin*, vol. 12, pp. 241–92.

Gilkey, L. 1985. *Creationism On Trial—Evolution and God at Little Rock.* Minneapolis: Winston Press.

Gill, R. C. O., D. Bridgwater, and J. H. Allaart. 1981. The geochemistry of the earliest known basic metavolcanic rocks at Isua, West Greenland: A preliminary investigation. pp. 313–25 in J. E. Glover and D. I. Groves, editors. *Archaean Geology.* Geological Society of Australia Special Publ. No. 7.

Glikson, A. Y. 1979. Early Precambrian tonalite–trondhjemite sialic nuclei. *Earth-Science Reviews*, vol. 15, pp. 1–73.

Goldberg, E. D. 1965. Minor elements in sea water. pp. 163–96 in J. P. Riley and G. Skirrow, editors. *Chemical Oceanography*, vol. 1 (1st edition). London and New York: Academic Press.

Goldich, S. S., and L. B. Fischer. 1986. Air-abrasion experiments in U–Pb dating of zircon. *Chemical Geology* (Isotope Geoscience Section), vol. 58, pp. 195–215.

Goldich, S. S., and C. E. Hedge. 1974. 3,800-Myr granite gneiss in southwestern Minnesota. *Nature*, vol. 252, pp. 467–68.

———. 1975. Interpretation of apparent ages in Minnesota (reply to Farhat and Wetherill, 1975). *Nature*, vol. 257, p. 722.

Goldich, S. S., C. E. Hedge, and T. W. Stern. 1970. Age of the Morton and Montevideo gneisses and related rocks, southwestern Minnesota. *Geological Society of America Bulletin*, vol. 81, pp. 3671–96.

Goldich, S. S., C. E. Hedge, T. W. Stern, J. L. Wooden, J. B. Bedkin, and R. M. North. 1980. Archean rocks of the Granite Falls area, southwestern Minnesota. pp. 19–43 in G. B. Morey and G. N. Hanson, editors. *Selected Studies of Archean Gneisses and Lower Proterozoic Rocks, Southern Canadian Shield.* The Geological Society of America Special Paper No. 182.

Goldich, S. S., and J. L. Wooden. 1980. Origin of the Morton Gneiss, southwestern Minnesota: Part 3. Geochronology. pp. 77–94 in G. B. Morey and G. N. Hanson, editors. *Selected Studies of Archean Gneisses and Lower Proterozoic Rocks, Southern Canadian Shield.* The Geological Society of America Special Paper No. 182.

Goodchild, J. G. 1897. Some geological evidence regarding the age of the Earth. *Proceedings of the Royal Society of Edinburgh*, vol. 13, pp. 259–308.

Gooley, R., R. Brett, and J. Warner. 1974. A lunar rock of deep crustal origin: Sample 76535. *Geochimica et Cosmochimica Acta*, vol. 38, pp. 1329–39.

Gopolan, K., and G. W. Wetherill. 1968. Rubidium–strontium age of hypersthene (L) chondrites. *Journal of Geophysical Research*, vol. 73, pp. 7133–36.

———. 1969. Rb–Sr age of amphoterite (LL) chondrites. *Journal of Geophysical Research*, vol. 74, pp. 4349–58.

———. 1970. Rubidium–strontium studies on enstatite chondrites: Whole-rock and mineral isochrons. *Journal of Geophysical Research*, vol. 75, pp. 3457–67.

———. 1971. Rubidium–strontium studies on black hypersthene chondrites: Effects of shock and reheating. *Journal of Geophysical Research*, vol. 76, pp. 8484–92.

Gott, J. R., III, J. E. Gunn, D. N. Schramm, and B. M. Tinsley. 1974. An unbound universe. *Astrophysical Journal*, vol. 194, pp. 543–53.

Gould, S. J. 1967. Is uniformitarianism necessary? *American Journal of Science*, vol. 263, pp. 223–28.

Graham, A. L., A. W. R. Bevan, and R. Hutchison. 1985. *Catalog of Meteorites* (4th edition). Tucson: University of Arizona Press.

Gray, C. M., D. A. Papanastassiou, and G. J. Wasserburg. 1973. The identification of early condensates from the solar nebula. *Icarus*, vol. 20, pp. 213–39.

Green, D. H. 1972. Archean greenstone belts may include terrestrial equivalents of lunar maria? *Earth and Planetary Science Letters*, vol. 15, pp. 263–70.

Grieve, R. A. F. 1982. The record of impact on Earth: Implications for a major Cretaceous/Tertiary impact event. pp. 25–37 in L. T. Silver and P. H. Schultz, editors. *Geological Implications of Impacts of Large Asteroids and Comets on the Earth*. Geological Society of America Special Paper 190.

———. 1987. Terrestrial impact structures. *Annual Reviews of Earth and Planetary Science*, vol. 15, pp. 245–70.

Griffin, W. L., V. R. McGregor, A. Nutman, P. N. Taylor, and D. Bridgwater. 1980. Early Archaean granulite-facies metamorphism south of Ameralik, West Greenland. *Earth and Planetary Science Letters*, vol. 50, pp. 59–74.

Guest, J. E., and R. Greeley. 1977. *Geology on the Moon*. London: Wykeham.

Guggisberg, S., P. Eberhardt, J. Geiss, N. Grögler, A. Stettler, G. M. Brown, and A. Peckett. 1979. Classification of the Apollo-11 mare basalts according to $Ar^{39}-Ar^{40}$ ages and petrological properties. *Proceedings of the Tenth Lunar and Planetary Science Conference*, pp. 1–39.

Haber, F. C. 1959. *The Age of the Earth: Moses to Darwin*. Baltimore: Johns Hopkins Press.

Hainebach, K. L., and D. N. Schramm. 1977. Comments on galactic evolution and nucleo-cosmochronology. *Astrophysical Journal*, vol. 212, pp. 347–59.

Hallam, A. 1983. *Great Geological Controversies*. New York: Oxford University Press.

Hallberg, J. A., and A. Y. Glikson. 1981. Archaean granite–greenstone terranes of Western Australia. pp. 33–103 in D. R. Hunter, editor. *Precambrian of The Southern Hemisphere*. Amsterdam, Oxford, and New York: Elsevier Scientific.

Halley, E. 1715. On the cause of the saltness of the ocean, and of the several lakes that emit no rivers; with a proposal, by means thereof, to discover the age of the world. *Philosophical Transactions of the Royal Society of London*, vol. 29.

Hamet, J., N. Nakamura, D. M. Unruh, and M. Tatsumoto. 1978. Origin and history of the adcumulate eucrite Moama, as inferred from REE abundances, Sm–Nd and U–Pb systematics. *Proceedings of the Ninth Lunar and Planetary Science Conference*, pp. 1115–36.

Hamilton, P. J., N. M. Evenson, and R. K. O'Nions. 1979. Chronology and chemistry of Parnallee (LL-3) chondrules. *Lunar and Planetary Science*, vol. X, pp. 494–96.

Hamilton, P. J., N. M. Evensen, R. K. O'Nions, A. Y. Glikson, and A. H. Hickman. 1981. Sm–Nd dating of the North Star Basalt, Warrawoona Group, Pilbara Block, Western Australia. pp. 187–92 in J. E. Glover and D. J. Groves, editors. *Archaean Geology*. Geological Society of Australia Special Publ. 7.

Hamilton, P. J., N. M. Evensen, R. K. O'Nions, H. S. Smith, and A. J. Erlank. 1979. Sm–Nd dating of Onverwacht Group volcanics, southern Africa. *Nature*, vol. 279, pp. 298–300.

Hamilton, P. J., R. K. O'Nions, D. Bridgwater, and A. Nutman. 1983. Sm–Nd studies of Archean metasediments and metavolcanics from West Greenland and their implications for the Earth's early history. *Earth and Planetary Science Letters*, vol. 62, pp. 263–72.

Hamilton, P. J., R. K. O'Nions, N. M. Evensen, D. Bridgwater, and J. H. Allaart. 1978. Sm–Nd isotopic investigations of Isua supracrustals and implications for mantle evolution. *Nature*, vol. 272, pp. 41–43.

Hanan, B. B., and G. R. Tilton. 1985. Early planetary metamorphism in chondritic meteorites. *Earth and Planetary Science Letters*, vol. 74, pp. 209–19.

Hanes, D. A. 1979. A new determination of the Hubble constant. *Monthly Notices of the Royal Astronomical Society*, vol. 188, pp. 901–09.

Hargraves, R. B. 1978. Punctuated evolution of tectonic style. *Nature*, vol. 276, pp. 459–61.

Harland, W. B., A. V. Cox, P. G. Llewellyn, C. G. A. Pickton, A. G. Smith, and R. Walters. 1982. *A Geologic Time Scale.* Cambridge: Cambridge University Press.

Harris, W. E., J. E. Hesser, and B. Atwood. 1983. Color-magnitude photometry of 47 Tucanae to Mv = +9. *Astrophysical Journal*, vol. 268, pp. L111–15.

Hartmann, W. K. 1970. Lunar cratering chronology. *Icarus*, vol. 13, pp. 299–301.

———. 1972. Paleocratering of the Moon: A review of post-Apollo data. *Astrophysics and Space Sciences*, vol. 17, pp. 48–64.

———. 1983. *Moons and Planets* (2nd edition). Belmont: Wadsworth.

Hartmann, W. K., and D. R. Davis. 1975. Satellite-sized planetesimals. *Icarus*, vol. 24, pp. 504–15.

Hartmann, W. K., and C. A. Wood. 1971. Moon: Origin and evolution of multi-ring basins. *The Moon*, vol. 3, pp. 3–78.

Haughton, S. 1865. *Manual of Geology.* London: Longman, Green, Longman, Roberts, and Green.

———. 1878. A geological proof that the changes in climate in past times were not due to changes in position of the pole; with an attempt to assign a minor limit to the duration of geological time. *Nature*, vol. 18, pp. 266–68.

Hawkesworth, C. J., S. Moorbath, R. K. O'Nions, and J. F. Wilson. 1975. Age relationships between greenstone belt and "granites" in the Rhodesian Archaean craton. *Earth and Planetary Science Letters*, vol. 25, pp. 251–62.

Hawking, S. W. 1988. *A Brief History of Time.* Toronto and New York: Bantam.

Head, J. W., III. 1976. Lunar volcanism in space and time. *Reviews of Geophysics and Space Physics*, vol. 14, pp. 265–300.

Heiken, G. 1975. Petrology of lunar soils. *Reviews of Geophysics and Space Physics*, vol. 13, pp. 567–87.

Helin, E. F., and E. M. Shoemaker. 1979. The Palomar planet-crossing asteroid survey, 1973–1978. *Icarus*, vol. 40, pp. 321–28.

Hensley, W. K., W. A. Bassett, and J. R. Huizenga. 1973. Pressure dependence of the radioactive decay constant of beryllium-7. *Science*, vol. 181, pp. 1164–65.

Hickman, A. H. 1975. Precambrian structural geology of part of the Pilbara region. *Geological Survey of Western Australia Annual Report for 1974*, pp. 68–72.

———. 1981. Crustal evolution of the Pilbara block, Western Australia. pp. 57–69 in J. E. Glover and D. J. Groves, editors. *Archaean Geology.* Geological Society of Australia Special Publ. 7.

————. 1983. Geology of the Pilbara Block and its environs. *Geological Survey of Western Australia*, Bulletin 127.

Hickman, M. H. 1974. 3500 Myr old granite in southern Africa. *Nature*, vol. 251, pp. 295–96.

Hodge, P. W. 1981. The extragalactic distance scale. *Annual Reviews of Astronomy and Astrophysics*, vol. 19, pp. 357–72.

Hoffman, D. C., F. O. Lawrence, J. L. Mewherter, and F. M. Rourke. 1971. Detection of plutonium-244 in nature. *Nature*, vol. 234, pp. 132–34.

Hohenberg, C. M., B. Hudson, B. M. Kennedy, and F. A. Podosek. 1981. Noble gas retention chronologies for the St. Severin meteorite. *Geochimica et Cosmochimica Acta*, vol. 45, pp. 535–46.

Holmes, A. 1911a. The association of lead with uranium in rock-minerals, and its application to the measurement of geological time. *Proceedings of the Royal Society of London*, series A, vol. 85, pp. 248–56.

————. 1911b. The duration of geological time. *Nature*, vol. 87, pp. 9–10.

————. 1913. *The Age of the Earth*. London and New York: Harper and Brothers.

————. 1927. *The Age of the Earth: An Introduction to Geological Ideas*. London: Ernest Benn (Benn's Sixpenny Library, No. 102).

————. 1931. Radioactivity and geological time. pp. 124–466 in *Physics of the Earth*, Part 4, The Age of the Earth. Washington, D.C.: National Research Council of the National Academy of Sciences, Bulletin 80.

————. 1946. An estimate of the age of the earth. *Nature*, vol. 157, pp. 680–84.

————. 1947a. A revised estimate of the age of the earth. *Nature*, vol. 159, pp. 127–28.

————. 1947b. The age of the earth. *Endeavor*, vol. 6, pp. 99–108.

————. 1948. The age of the Earth. *Annual Report of the Board of Regents of the Smithsonian Institution, 1948*, pp. 227–39.

Hood, L. L. 1986. Impact model featured on film. *Geotimes*, vol. 31, pp. 15–17.

Hopke, P. K. 1974. Extranuclear effects on nuclear decay rates. *Journal of Chemical Education*, vol. 51, pp. 517–19.

Horn, P., E. K. Jessberger, T. Kirsten, and H. Richter. 1975. ^{39}Ar–^{40}Ar dating of lunar rocks: Effects of grain size and neutron irradiation. *Proceedings of the Sixth Lunar Science Conference*, pp. 1563–91.

Houtermans, F. G. 1946. Die Isotopenhaufigkeiten im Naturlichen Bloi und das Alter des Urans [The isotopic abundances in natural lead and the age of uranium]. *Naturwissenschaften*, vol. 33, pp. 185–86, 219 [translation in C. T. Harper, editor. *Geochronology* (Benchmark Papers In Geology). Stroudsburg, Pa.: Dowden, Hutchinson, and Ross, 1973].

————. 1947. Das Alter des Urans. *Zeitschrift für Naturforschung*, vol. 2a, pp. 322–28.

————. 1953. Determination of the age of the earth from the isotopic composition of meteoritic lead. *Nuovo Cimento*, series 9, vol. 10, no. 12, pp. 1623–33.

Howard, K. A., D. E. Wilhelms, and D. H. Scott. 1974. Lunar basin formation and highland stratigraphy. *Reviews of Geophysics and Space Physics*, vol. 12, pp. 309–27.

Hubbert, M. K. 1967. Critique of the principle of uniformity. pp. 3–33 in C. C. Albritton, Jr., editor. *Uniformity and Simplicity*. Geological Society of America Special Paper 89.

Hubble, E. P. 1929. A relation between distance and radial velocity among extra-galactic nebulae. *Proceedings of the National Academy of Sciences*, vol. 15, pp. 168–73.

Huey, J. M., and T. P. Kohmar. 1973. [207]Pb–[206]Pb isochron and age of chondrites. *Journal of Geophysical Research*, vol. 78, pp. 3227–44.

Hughes, C. J. 1982. *Igneous Petrology*. Amsterdam, Oxford, and New York: Elsevier Scientific.

Huneke, J. C., E. K. Jessberger, F. A. Podosek, and G. J. Wasserburg. 1973. [40]Ar/[39]Ar measurements in Apollo 16 and 17 samples and the chronology of metamorphic and volcanic activity in the Taurus–Littrow region. *Proceedings of the Fourth Lunar Science Conference*, pp. 1725–56.

Huneke, J. C., F. A. Podosek, and G. J. Wasserburg. 1972. Gas retention and cosmic ray exposure ages of a basalt fragment from Mare Fecunditatis. *Earth and Planetary Science Letters*, vol. 13, pp. 375–83.

Huneke, J. C., and G. J. Wasserburg. 1975. Trapped [40]Ar in troctolite 76535 and evidence for enhanced [40]Ar–[39]Ar age plateaux. *Lunar Science*, vol. VI, pp. 417–19.

———. 1979. K/Ar evidence from Luna 20 rocks for luna differentiation prior to 4.51 AE ago. *Lunar and Planetary Science*, vol. X, pp. 598–600.

Hunter, D. R. 1970. The ancient gneiss complex in Swaziland. *Transactions of the Geological Society of South Africa*, vol. 73, pp. 107–50.

———. 1974. Crustal development in the Kaapvaal craton, I. The Archaean. *Precambrian Research*, vol. 1, pp. 259–94.

Hunter, D. R., F. Barker, and H. T. Millard. 1978. The geochemical nature of the Archean ancient gneiss complex and granodiorite suite, Swaziland: A preliminary study. *Precambrian Research*, vol. 7, pp. 105–27.

Hurley, P. M., H. W. Fairbairn, and H. E. Gaudette. 1976. Progress report on early Archean rocks in Liberia, Sierra Leone, and Guayana, and their general stratigraphic setting. pp. 511–21 in B. F. Windley, editor. *The Early History of the Earth*. London, New York, Sydney, and Toronto: Wiley.

Hurley, P. M., W. H. Pinson, Jr., B. Nagy, and T. M. Teska. 1972. Ancient age of the Middle Marker horizon, Onverwacht Group, Swaziland sequence, South Africa. *Earth and Planetary Science Letters*, vol. 14, pp. 360–66.

Hurst, R. W., D. Bridgwater, K. D. Collerson, and G. W. Wetherill. 1975. 3600-m.y. Rb–Sr ages from very early Archaean gneisses from Saglek Bay, Labrador. *Earth and Planetary Science Letters*, vol. 27, pp. 393–403.

Husain, L., and O. A. Schaeffer. 1975. Lunar evolution: The first 600 million years. *Geophysical Research Letters*, vol. 2, pp. 29–32.

Husain, L., O. A. Schaeffer, J. Funkhouser, and J. Sutter. 1972. The ages of lunar material from Fra Mauro, Hadley Rille, and Spur Crater. *Proceedings of the Third Lunar Science Conference*, pp. 1557–68.

Hutchison, R. 1983. *The Search for Our Beginning*. Oxford: Oxford University Press.

Huxley, T. H. 1869. The anniversary address of the President. *Quarterly Journal of the Geological Society of London*, vol. 25, pp. xxviii–liii.

Iben, I., Jr. 1970. Globular cluster stars. *Scientific American*, vol. 223, pp. 27–39.

Jackson E. D., R. L. Sutton, and H. G. Wilshire. 1975. Structure and petrology of a cumulus norite boulder sampled by Apollo 17 in Taurus–Littrow Valley, the Moon. *Geological Society of America Bulletin*, vol. 86, pp. 433–42.

Jacobsen, S. B., and R. F. Dymek. 1988. Nd and Sr isotopic systematics of clastic metasediments from Isua, West Greenland: Identification of pre-3.8Ga differentiated crustal components. *Journal of Geophysical Research*, vol. 93, pp. 338–54.

Jacobsen, S. B., and G. J. Wasserburg. 1984. Sm–Nd isotopic evolution of chondrites and achondrites, II. *Earth and Planetary Science Letters*, vol. 67, pp. 137–50.

Jäger, E., and J. C. Hunziker, editors. 1979. *Lectures in Isotope Geology*. Berlin, Heidelberg, and New York: Springer-Verlag.

Jahn, B. M., A. Y. Glikson, J. J. Peucat, and A. H. Hickman. 1981. REE geochemistry and isotopic data of Archean silicic volcanics and granitoids from the Pilbara Block, Western Australia: Implications for the early crustal evolution. *Geochimica et Cosmochimica Acta*, vol. 45, pp. 1633–52.

Jahn, B. M., and C. Y. Shih. 1974. On the age of the Onverwacht Group, Swaziland sequence, South Africa. *Geochimica et Cosmochimica Acta*, vol. 38, pp. 873–85.

Jahn, B. M., and Z.-Q. Zhang. 1984. Radiometric ages (Rb–Sr, Sm–Nd, U–Pb) and REE geochemistry of Archaean granulite gneisses from eastern Hebei Province, China. pp. 204–34 in A. Kroner, G. N. Hanson, and A. M. Goodwin, editors. *Archaean Geochemistry*. Berlin, Heidelberg, New York, and Tokyo: Springer-Verlag.

James, O. B. 1973. Crystallization history of lunar feldspathic basalt 14310. *U.S. Geological Survey Professional Paper* 841.

———. 1981. Petrologic and age relations of the Apollo 16 rocks: Implications for subsurface geology and the age of the Nectaris Basin. *Proceedings of the 12th Lunar and Planetary Science Conference*, pp. 209–33.

Jeffreys, H. 1918. On the early history of the Solar System. *Monthly Notices of the Royal Astronomical Society*, vol. 78, pp. 424–41.

———. 1948. Lead isotopes and the age of the earth. *Nature*, vol. 162, pp. 822–23.

Jessberger, E. K., B. Dominik, T. Staudacher, and G. F. Herzog. 1980. [40]Ar–[39]Ar ages of Allende. *Icarus*, vol. 42, pp. 380–405.

Jessberger, E. K., P. Horn, and T. Kirsten. 1975. [39]Ar–[40]Ar dating of lunar rocks: A methodical investigation of mare basalt 75075. *Lunar Science*, vol. VI, pp. 441–43.

Jessberger, E. K., T. Kirsten, and T. Staudacher. 1977. One rock and many ages—Further K–Ar data on consortium breccia 73215. *Proceedings of the Eighth Lunar Science Conference*, pp. 2567–80.

Jessberger, E. K., T. Staudacher, B. Dominik, and T. Kirsten. 1978. Argon-argon ages of aphanite samples from consortium breccia 73255. *Proceedings of the Ninth Lunar and Planetary Science Conference*, pp. 841–54.

Johnson, H. L., and A. R. Sandage. 1956. Three-color photometry in the globular cluster M3. *Astrophysical Journal Supplement*, vol. 124, pp. 379–89.

Joly, J. 1899. An estimate of the geological age of the Earth. *Annual Report of the Smithsonian Institution for 1899*, pp. 247–88.

———. 1900. On the geological age of the Earth. *British Association for the Advancement of Science, Report of the Seventeenth Meeting*, pp. 369–79.

Jordan, J., T. Kirsten, and H. Richter. 1980. [129]I/[127]I: A puzzling early solar system chronometer. *Zeitschrift für Naturforschung*, vol. 35a, pp. 145–70.

Kanasewich, E. R. 1968. The interpretation of lead isotopes and their geological significance. pp. 147–223 in E. I. Hamilton and R. M. Farquhar, editors. *Radiometric Dating for Geologists*. New York: Interscience.

Kaneoka, I. 1980. [40]Ar–[39]Ar ages of L and LL chondrites from Allan Hills,

Antarctica: ALHA 77015, 77214 and 77304. *Memoirs of the National Institute of Polar Research*, Special Issue No. 17, pp. 177–88.

———. 1981. ^{40}Ar–^{39}Ar ages of Antarctic meteorites: Y-74191, Y-75258, Y-7308, Y-74450 and ALH-765. *Memoirs of the National Institute of Polar Research*, Special Issue No. 20, pp. 250–63.

———. 1983. Investigation of the weathering effect on the ^{40}Ar–^{39}Ar ages of Antarctic meteorites. *Memoirs of the National Institute of Polar Research*, Special Issue no. 30, pp. 259–74.

Kaushal, S. K., and G. W. Wetherill. 1969. Rb^{87}–Sr^{87} age of bronzite (H group) chondrites. *Journal of Geophysical Research*, vol. 74, pp. 2717–26.

———. 1970. Rubidium-87–strontium-87 age of carbonaceous chondrites. *Journal of Geophysical Research*, vol. 75, pp. 463–68.

Kazanas, D., D. N. Schramm, and K. L. Hainebach. 1978. A consistent age for the universe. *Nature*, vol. 274, pp. 672–73.

Kazansky, V. I., and V. M. Moralev. 1981. Archaean geology and metallogeny of the Aldan Shield, USSR. pp. 111–20 in J. E. Glover and D. I. Groves, editors. *Archaean Geology*. Geological Society of Australia Special Publ. No. 7.

Kelvin, Lord (W. Thomson). 1897. The age of the Earth as an abode fitted for life. *Annual Report of the Smithsonian Institution*, pp. 337–57.

Kempe, W., and O. Muller. 1969. The stony meteorite Krahenberg. Its chemical composition and the Rb–Sr age of the light and dark portions. pp. 418–28 in P. M. Millman, editor. *Meteorite Research*. Dordrecht: D. Reidel.

Kennicutt, R. C. 1979. H$_{\text{II}}$ regions as extragalactic distance indicators. II. Application of isophotal diameters. *Astrophysical Journal*, vol. 228, pp. 696–703.

———. 1981. H$_{\text{II}}$ regions as extragalactic distance indicators. IV. The Virgo cluster. *Astrophysical Journal*, vol. 247, pp. 9–16.

Kesson, S. E., and D. H. Lindsley. 1976. Mare basalt petrogenesis—A review of experimental petrogenesis. *Reviews of Geophysics and Space Physics*, vol. 14, pp. 361–73.

King, C. 1893. The age of the Earth. *American Journal of Science*, 3rd series, vol. 45, pp. 1–20.

Kinney, P. D., I. S. Williams, and W. Compston. 1984. Insight into the geological history of ancient gneisses in U.S.A. from microanalysis of zircon. *Annual Report of the Australian National University for 1984*, pp. 98–100.

Kirkaldy, J. F. 1971. *Geological Time*. Edinburgh: Oliver and Boyd.

Kirsten, T. 1973. Isotope studies in the Mundrabilla iron meteorite. *Meteoritics*, vol. 8, pp. 400–03.

Kirsten, T., and P. Horn. 1974. Chronology of the Taurus–Littrow region III: Ages of mare basalts and highland breccias and some remarks about the interpretation of lunar highland rock ages. *Proceedings of the Fifth Lunar Science Conference*, pp. 1451–76.

Knopf, A. 1931a. Age of the Ocean. pp. 65–72 in *Physics of the Earth*, Part 4, The Age of the Earth. Washington, D.C.: National Research Council of the National Academy of Sciences, Bulletin 80.

———. 1931b. The age of the Earth—summary of principal results. pp. 3–9 in *Physics of the Earth*, Part 4, The Age of the Earth. Washington, D.C.: National Research Council of the National Academy of Sciences, Bulletin 80.

————. 1949. The geologic records of time. pp. 1–27 in *Time and its Mysteries,* series III. New York: New York University Press.

Koczy, F. F. 1943. The "age" of terrestrial matter and the geochemical uranium/lead ratio. *Nature,* vol. 151, p. 24.

Kopal, Z. 1974. *The Moon in the Post-Apollo Era.* Dordrecht and Boston: D. Reidel.

Kratz, K., and F. Mitrofanov. 1980. Main type reference sequences of the early Precambrian in the USSR. *Earth-Science Reviews,* vol. 16, pp. 295–301.

Langevin, Y., and J. R. Arnold. 1977. The evolution of the lunar regolith. *Annual Reviews of Earth and Planetary Sciences,* vol. 5, pp. 449–89.

LeConte, J. 1884. *A Compend of Geology.* New York, Cincinnati, and Chicago: American Book Co.

Lederer, C. M., J. M. Hollander and I. Perlman. 1967. *Table of Isotopes* (6th edition). New York: Wiley.

Lightfoot, J. 1647. *Harmony, Chronicle, and Order of the Old Testament.* London: Printed by R. Cotes for J. Clark.

Lipple, S. L. 1975. Definitions of new and revised stratigraphic units of the eastern Pilbara region. *Geological Survey of Western Australia Annual Report for 1974,* pp. 58–63.

Lipschutz, M. E., and W. A. Cassidy. 1986. Antarctic meteorites: A progress report. *Eos,* vol. 67, pp. 1339–41.

Lopez Martinez, M., D. York, C. M. Hall, and J. A. James. 1984. Oldest reliable $^{40}Ar/^{39}Ar$ ages for terrestrial rocks: Barberton Mountain komatiites. *Nature,* vol. 307, pp. 352–54.

Louderback, G. D. 1936. The age of the earth from sedimentation. *Scientific Monthly,* vol. 42, pp. 240–46.

Lubimova, E. A., and O. Parphenuk. 1981. Terrestrial heat flow history and temperature profiles. pp. 217–28 in R. J. O'Connell and W. S. Fyfe, editors. *Evolution of the Earth.* Washington, D.C.: American Geophysical Union, Geodynamics Series, vol. 5.

Luck, J. M., J. L. Birck, and C. J. Allègre. 1980. $^{187}Re-^{187}Os$ systematics in meteorites: Early chronology of the solar system and age of the galaxy. *Nature,* vol. 283, pp. 256–59.

Lugmair, G. W. 1974. Sm–Nd ages: A new dating method [abstract]. *Meteoritics,* vol. 9, p. 369.

Lugmair, G. W., and R. W. Carlson. 1978. The Sm–Nd history of KREEP. *Proceedings of the Ninth Lunar and Planetary Science Conference,* pp. 689–704.

Lugmair, G. W., and K. Marti. 1977. Sm–Nd–Pu timepieces in the Angra dos Reis meteorite. *Earth and Planetary Science Letters,* vol. 35, pp. 273–84.

————. 1978. Lunar initial $^{143}Nd/^{144}Nd$: Differential evolution of the lunar crust and mantle. *Earth and Planetary Science Letters,* vol. 39, pp. 349–57.

Lugmair, G. W., K. Marti, J. P. Kurtz, and N. B. Scheinin. 1976. History and genesis of lunar troctolite 76535 or: How old is old? *Proceedings of the Seventh Lunar Science Conference,* pp. 2009–33.

Lugmair, G. W., and N. B. Scheinin. 1975. Sm–Nd systematics of the Stannern meteorite [abstract]. *Meteoritics,* vol. 10, pp. 447–48.

Lugmair, G. W., N. B. Scheinin, and R. W. Carlson. 1977. Sm–Nd systematics of the Serra de Mage eucrite [abstract]. *Meteoritics,* vol. 12, pp. 300–301.

Lugmair, G. W., N. B. Scheinin, and K. Marti. 1975. Sm–Nd age and history of Apollo 17 basalt 75075: Evidence for early differentiation of the lunar exterior. *Proceedings of the Sixth Lunar Science Conference,* pp. 1419–29.

Lunatic Asylum. 1970. Ages, irradiation history, and chemical composition of lunar rocks from the Sea of Tranquility. *Science*, vol. 167, pp. 463–66.

Lynden-Bell, D. 1977. Hubble's constant determined from superluminal radio sources. *Nature*, vol. 270, pp. 396–99.

Macdonald, G. A. 1968. Composition and origin of Hawaiian lavas. pp. 477–522 in R. R. Coats, R. L. Hay, and C. A. Anderson, editors. *Studies in Volcanology*. Geological Society of America Memoir 116.

Manhes, G., and C. J. Allègre. 1978. Time differences as determined from the ratio of lead 207 to lead 206 in concordant meteorites. *Meteoritics*, vol. 13, pp. 543–48.

Manhes, G., C. J. Allègre, B. Dupré, and B. Hamelin. 1979. Lead–lead systematics, the "age of the Earth" and the chemical evolution of our planet in a new representation space. *Earth and Planetary Science Letters*, vol. 44, pp. 91–104.

Manhes, G., C. J. Allègre, and A. Provost. 1984. U–Th–Pb systematics of the eucrite "Juvinas": Precise age determination and evidence for exotic lead. *Geochimica et Cosmochimica Acta*, vol. 48, pp. 2247–64.

Manhes, G., J.-F. Minster, and C. J. Allègre. 1978. Comparative uranium–thorium–lead and rubidium–strontium study of the Saint Severin amphoterite: Consequences for early Solar System chronology. *Earth and Planetary Science Letters*, vol. 39, pp. 14–24.

Marble, J. P. 1940. Annotated bibliography of selected articles dealing with the measurement of geologic time. *Report of the National Research Council Committee on the Measurement of Geologic Time, 1939–40*, pp. 4–54.

———. 1950. Annotated bibliography of papers related to the measurement of geologic time. *Report of the National Research Council Committee on the Measurement of Geologic Time, 1949–1950*, pp. 52–112.

Marvin, U. B. 1983. The discovery and characterization of Allan Hills 81005: The first lunar meteorite. *Geophysical Research Letters*, vol. 10, pp. 775–78.

———. 1984. A meteorite from the Moon. *Smithsonian Contributions to Earth Science*, No. 26, pp. 95–103.

———. 1986. Meteorites, the Moon and the history of geology. *Journal of Geological Education*, vol. 34, pp. 140–65.

Mason, B. 1962. *Meteorites*. New York: Wiley.

———. 1967. Meteorites. *American Scientist*, vol. 55, pp. 429–55.

Mason, R. 1973. The Limpopo mobile belt—southern Africa. *Philosophical Transactions of the Royal Society of London*, series A, vol. 273, pp. 463–85.

Masursky, H., G. W. Colton, and F. El-Baz, editors. 1978. *Apollo over the Moon*. NASA Special Publ. SP-362.

McCall, G. J. H. 1973. *Meteorites and Their Origins*. New York: Wiley.

McCord, T. B., J. B. Adams, and T. V. Johnson. 1970. Asteroid Vesta: Spectral reflectivity and compositional implications. *Science*, vol. 168, pp. 1445–47.

McCulloch, M. T., K. D. Collerson, and W. Compston. 1983. Growth of Archaean crust within the western gneiss terrain, Yilgarn block, Western Australia. *Journal of the Geological Society of Australia*, vol. 30, pp. 155–60.

McCulloch, M. T., and G. J. Wasserburg. 1978. Sm–Nd and Rb–Sr chronology of continental crust formation. *Science*, vol. 200, pp. 1003–11.

———. 1980. Early Archean Sm–Nd model ages from a tonalitic gneiss, northern Michigan. pp. 135–38 in G. B. Morey and G. N. Hanson, editors. *Selected Studies of Archean Gneisses and Lower Proterozoic Rocks, Southern Canadian Shield*. Geological Society of America Special Paper 182.

McDougall, I., and T. M. Harrison. 1988. *Geochronology and Thermochronology by the $^{40}Ar/^{39}Ar$ Method.* Oxford: Oxford University Press.

McGee, W. J. 1893. Note on the "Age of the Earth." *Science,* vol. 21, pp. 309–10.

McGetchin, T. R., M. Settle, and J. W. Head. 1973. Radial thickness variation in impact crater ejecta: Implications for lunar basin deposits. *Earth and Planetary Science Letters,* vol. 2, pp. 226–36.

McGregor, V. R. 1973. The early Precambrian gneisses of the Godthaab district, West Greenland. *Philosophical Transactions of the Royal Society of London,* series A, vol. 273, pp. 343–58.

McGregor, V. R., and B. Mason. 1977. Petrogenesis and geochemistry of metabasaltic and metasedimentary enclaves in the Amîtsoq gneisses, West Greenland. *American Mineralogist,* vol. 62, pp. 887–904.

McSween, H. Y., Jr. 1979. Are carbonaceous chondrites primitive or processed? A review. *Reviews of Geophysics and Space Physics,* vol. 17, pp. 1059–78.

———. 1984. SNC meteorites: Are they Martian rocks? *Geology,* vol. 12, pp. 3–6.

———. 1987. *Meteorites and Their Parent Planets.* Cambridge and New York: Cambridge University Press.

Melosh, H. J. 1985. Ejection of rock fragments from planetary bodies. *Geology,* vol. 13, pp. 144–48.

Merrill, P. W. 1952. Technetium in stars. *Science,* vol. 115, p. 484.

Michard-Vitrac, A., J. Lancelot, J. C. Allègre, and S. Moorbath. 1977. U–Pb ages on single zircons from the early Precambrian rocks of West Greenland and the Minnesota River Valley. *Earth and Planetary Science Letters,* vol. 35, pp. 449–53.

Minster, J.-F., and C. J. Allègre. 1979a. $^{87}Rb-^{87}Sr$ chronology of H chondrites: constraint and speculations on the early evolution of their parent body. *Earth and Planetary Science Letters,* vol. 42, pp. 333–47.

———. 1979b. $^{87}Rb-^{87}Sr$ dating of L chondrites: Effects of shock and brecciation. *Meteoritics,* vol. 14, pp. 235–48.

———. 1981. $^{87}Rb-^{87}Sr$ dating of LL chondrites. *Earth and Planetary Science Letters,* vol. 56, pp. 89–106.

Minster, J.-F., J.-L. Birck, and C. J. Allègre. 1982. Absolute age of formation of chondrites studied by the $^{87}Rb-^{87}Sr$ method. *Nature,* vol. 300, pp. 414–19.

Minster, J.-F., L.-P. Ricard, and C. J. Allègre. 1979. $^{87}Rb-^{87}Sr$ chronology of enstatite meteorites. *Earth and Planetary Science Letters,* vol. 44, pp. 420–40.

Mittlefehldt, D. W., and G. W. Wetherill. 1979. Rb–Sr studies of CI and CM chondrites. *Geochimica et Cosmochimica Acta,* vol. 43, pp. 201–06.

Montgomery, C. W. 1979. Uranium–lead geochronology of the Archean Imataca Series, Venezuelan Guayana shield. *Contributions to Mineralogy and Petrology,* vol. 69, pp. 167–76.

Moorbath, S. 1975a. Evolution of Precambrian crust from strontium isotopic evidence. *Nature,* vol. 254, pp. 395–98.

———. 1975b. The geological significance of early Precambrian rocks. *Proceedings of the Geologists Association,* vol. 86, pp. 259–79.

———. 1975c. Geological interpretation of whole-rock isochron dates from high grade gneiss terrain. *Nature,* vol. 255, p. 391.

———. 1977a. Ages, isotopes and evolution of Precambrian continental crust. *Chemical Geology,* vol. 20, pp. 151–87.

————. 1977b. The oldest rocks and the growth of continents. *Scientific American*, vol. 236, no. 3, pp. 92–104.

————. 1980. Aspects of the chronology of ancient rocks related to continental evolution. pp. 89–115 in D. W. Strangway, editor. *The Continental Crust and Its Mineral Deposits*. Geological Association of Canada Special Paper 20.

Moorbath, S., J. H. Allaart, D. Bridgwater, and V. R. McGregor. 1977a. Rb–Sr ages of early Archean supercrustal rocks and Amîtsoq gneisses at Isua. *Nature*, vol. 270, pp. 43–45.

Moorbath, S., R. K. O'Nions, and R. J. Pankhurst. 1973. Early Archean age for the Isua iron formation, West Greenland. *Nature*, vol. 245, pp. 138–39.

————. 1975. The evolution of early Precambrian crustal rocks at Isua, West Greenland—geochemical and isotopic evidence. *Earth and Planetary Science Letters*, vol. 27, pp. 229–39.

Moorbath, S., R. K. O'Nions, R. J. Pankhurst, N. H. Gale, and V. R. McGregor. 1972. Further rubidium–strontium age determinations on the very early Precambrian rocks of the Godthaab district, West Greenland. *Nature Physical Science*, vol. 240, pp. 78–82.

Moorbath, S., and R. J. Pankhurst. 1976. Further rubidium–strontium age and isotope evidence for the nature of the late Archaean plutonic event in West Greenland. *Nature*, vol. 262, pp. 124–26.

Moorbath, S., P. N. Taylor, and R. Goodwin. 1981. Origin of granitic magma by crustal remobilization: Rb–Sr and Pb/Pb geochronology and isotope geochemistry of the late Archean Qorqut granite complex of southern West Greenland. *Geochimica et Cosmochimica Acta*, vol. 45, pp. 1051–60.

Moorbath, S., J. F. Wilson, and P. Cotterill. 1976. Early Archaean age for the Sebakwian Group at Selukwe, Rhodesia. *Nature*, vol. 264, pp. 536–38.

Moorbath, S., J. F. Wilson, R. Goodwin, and M. Humm. 1977b. Further Rb–Sr age and isotope data on Early and Late Archaean rocks from the Rhodesian craton. *Precambrian Research*, vol. 5, pp. 229–39.

Morey, G. B., and P. K. Sims. 1976. Boundary between two Precambrian W terranes in Minnesota and its geologic significance. *Bulletin of the Geological Society of America*, vol. 87, pp. 141–52.

Mould, J., M. Aaronson, and J. Huchra. 1980. A distance scale from the infrared magnitude/HI velocity–width relation. II. The Virgo Cluster. *Astrophysical Journal*, vol. 238, pp. 458–70.

Müller, O., and J. Zähringer. 1966. K–Ar Altersbestimmungen an Eisenmeteoriten III. *Geochimica et Cosmochimica Acta*, vol. 30, pp. 1075–92.

Murthy, V. R., and W. Compston. 1965. Rb–Sr ages of chondrites and carbonaceous chondrites. *Journal of Geophysical Research*, vol. 70, pp. 5297–307.

Murthy, V. R., N. M. Evensen, B.-M. Jahn, and M. R. Coscio, Jr. 1972. Apollo 15 and 16 samples: Rb–Sr ages, trace elements, and lunar evolution. *Proceedings of the Third Lunar Science Conference*, pp. 1503–14.

Murthy, V. R., and C. C. Patterson. 1962. Primary isochron of zero age for meteorites and the Earth. *Journal of Geophysical Research*, vol. 67, pp. 1161–67.

Mussett, A. E. 1970. The age of the Earth and meteorites. *Comments on Earth Sciences—Geophysics*, vol. 1, pp. 65–73.

Myers, J. S. 1976. The early Precambrian gneiss complex of Greenland. pp. 165–76 in B. F. Windley, editor. *The Early History of the Earth*. London, New York, Sydney, and Toronto: Wiley.

Nagasawa, M., and B.-M. Jahn. 1976. REE distribution and Rb–Sr systematics of Allende inclusions. *Lunar Science*, vol. VII, pp. 591–93.

Nagy, B. 1975. *Carbonaceous Meteorites*. New York: Elsevier.

Nakamura, N., M. Tatsumoto, and D. Coffraut. 1983. Sm–Nd isotopic systematics and REE abundance studies of the ALH-765 eucrite. *Memoirs of the National Institute of Polar Research*, Special Issue No. 30, pp. 323–31.

Nakamura, N., M. Tatsumoto, P. D. Nunes, D. M. Unruh, A. P. Schwab, and T. R. Wildeman. 1976. 4.4 b.y.-old clast in boulder 7, Apollo 17: A comprehensive chronological study by U–Pb, Rb–Sr and Sm–Nd methods. *Proceedings of the Seventh Lunar Science Conference*, pp. 2309–33.

Nakamura, N., D. M. Unruh, M. Tatsumoto, and R. Hutchison. 1982. Origin and evolution of the Nakhla meteorite inferred from the Sm–Nd and U–Pb systematics and REE, Ba, Sr, Rb, and K abundances. *Geochimica et Cosmochimica Acta*, vol. 46, pp. 1555–73.

Newson, H. E. 1984. The lunar core and the origin of the Moon. *Eos*, vol. 65, pp. 369–70.

Newton, I. 1687. *Philosophiae Naturalis Principia Mathematica*, vols. I–III. [A. Motte, translator, 1729, *Sir Isaac Newton's Mathematical Principles of Natural Philosophy and His System of the World*. London.]

Nicolaysen, L. O. 1961. Graphic interpretation of discordant age measurements on metamorphic rocks. *Annals of the New York Academy of Sciences*, vol. 91, pp. 198–206.

Niemeyer, S. 1979. ^{40}Ar–^{39}Ar dating of inclusions from IAB iron meteorites. *Geochimica et Cosmochimica Acta*, vol. 43, pp. 1829–40.

———. 1980. I–Xe and ^{40}Ar–^{39}Ar dating of silicate from Weekeroo Station and Netschaevo IIE iron meteorites. *Geochimica et Cosmochimica Acta*, vol. 44, pp. 33–44.

———. 1983. I–Xe and ^{40}Ar–^{39}Ar analyses of silicate from the Eagle station pallasite and the anomalous iron meteorite Enon. *Geochimica et Cosmochimica Acta*, vol. 47, pp. 1007–12.

Nier, A. O. 1938. Variations in the relative abundances of the isotopes of common lead from various sources. *Journal of the American Chemical Society*, vol. 60, pp. 1571–76.

———. 1939. The isotopic constitution of radiogenic leads and the measurement of geological time. II. *Physical Review*, vol. 55, pp. 153–63.

Nier, A. O., R. W. Thompson, and B. F. Murphey. 1941. The isotopic constitution of lead and the measurement of geological time. III. *Physical Review*, vol. 60, pp. 112–16.

Nininger, H. H. 1952. *Out of the Sky*. New York: Dover.

Norman, E. B., and D. N. Schramm. 1979. On the conditions required for the r-process: *Astrophysical Journal*, vol. 228, pp. 881–92.

Nunes, P. O., M. Tatsumoto, and D. M. Unruh. 1974. U–Th–Pb systematics of some Apollo 17 lunar samples and implications for lunar basin chronology. *Proceedings of the Fifth Lunar Science Conference*, pp. 1487–1514.

Nutman, A. P., J. H. Allaart, D. Bridgwater, E. Dimroth, and M. Rosing. 1984. Stratigraphic and geochemical evidence for the depositional environment of the early Archean Isua supracrustal belt, West Greenland. *Precambrian Research*, vol. 25, pp. 365–96.

Nutman, A. P., D. Bridgwater, E. Dimroth, R. C. O. Gill, and M. Rosing. 1983. Early (3700 Ma) Archaean rocks of the Isua supracrustal belt and adjacent gneisses. *The Geological Survey of Greenland*, Report No. 112, pp. 5–22.

Nyquist, L. E., B. M. Bansal, and H. Wiesmann. 1975. Rb–Sr ages and initial $^{87}Sr/^{86}Sr$ for Apollo 17 basalts and KREEP basalt 15386. *Proceedings of the Sixth Lunar Science Conference*, pp. 1445–65.

Nyquist, L. E., B. M. Bansal, J. L. Wooden, and H. Wiesmann. 1977. Sr-isotopic constraints on the petrogenesis of Apollo 12 mare basalts. *Proceedings of the Eighth Lunar Science Conference*, pp. 1383–415.

Nyquist, L. E., N. J. Hubbard, P. W. Gast, S. E. Church, B. M. Bansal, and H. Wiesmann. 1972. Rb–Sr systematics for chemically defined Apollo 14 breccias. *Proceedings of the Third Lunar Science Conference*, pp. 1515–30.

Nyquist, L. E., W. U. Reimold, D. D. Bogard, J. L. Wooden, B. M. Bansal, H. Wiesmann, and C.-Y. Shih. 1981. A comparative Rb–Sr, Sm–Nd, and K–Ar study of shocked norite 78236: Evidence of slow cooling in the lunar crust? *Proceedings of the 12th Lunar and Planetary Science Conference*, pp. 67–97.

Nyquist, L. E., C.-Y. Shih, J. L. Wooden, B. M. Bansal, and H. Wiesmann. 1979a. The Sr and Nd isotopic record of Apollo 12 basalts: Implications for lunar geochemical evolution. *Proceedings of the Tenth Lunar and Planetary Science Conference*, pp. 77–114.

Nyquist, L. E., H. Takeda, B. M. Bansal, C.-Y. Shih, H. Wiesmann, and J. L. Wooden. 1986. Rb–Sr and Sm–Nd internal isochron ages of a subophitic basalt clast and a matrix sample from the Y75011 eucrite. *Journal of Geophysical Research*, vol. 91, pp. 8137–50.

Nyquist, L. E., J. Wooden, B. Bansal, H. Wiesmann, G. McKay, and D. Bogard. 1979b. Rb–Sr age of the Shergotty achondrite and implications for metamorphic resetting of isochron ages. *Geochimica et Cosmochimica Acta*, vol. 43, pp. 1057–74.

Oberli, F., J. C. Huneke, and G. J. Wasserburg. 1979. U–Pb and K–Ar systematics of cataclysm and precataclysm lunar impactites. *Lunar and Planetary Science*, vol. X, pp. 940–42.

Oberli, F., M. T. McCulloch, F. Tera, D. A. Papanastassiou, and G. J. Wasserburg. 1978. Early lunar differentiation constraints from U–Th–Pb, Sm–Nd, and Rb–Sr model ages. *Lunar Science*, vol. IX, pp. 832–34.

O'Nions, R. K., S. R. Carter, N. M. Evensen, and P. J. Hamilton. 1979. Geochemical and cosmochemical applications of Nd isotope analysis. *Annual Reviews of Earth and Planetary Science*, vol. 7, pp. 11–38.

Ostic, R. G., R. D. Russell, and P. H. Reynolds. 1963. A new calculation for the age of the earth from abundances of lead isotopes. *Nature*, vol. 199, pp. 1150–52.

Ostic, R. G., R. D. Russell, and R. L. Stanton. 1967. Additional measurements of the isotopic composition of lead from stratiform deposits. *Canadian Journal of Earth Sciences*, vol. 4, pp. 245–69.

Oversby, V. M. 1970. The isotopic composition of lead in iron meteorites. *Geochimica et Cosmochimica Acta*, vol. 34, pp. 65–75.

———. 1974. A new look at the lead isotope growth curve. *Nature*, vol. 248, pp. 132–33.

———. 1976. Isotopic ages and geochemistry of Archaean acid igneous rocks from the Pilbara, Western Australia. *Geochimica et Cosmochimica Acta*, vol. 40, pp. 817–29.

Oversby, V. M., and P. W. Gast. 1968. Oceanic basalt leads and the age of the Earth. *Science*, vol. 162, pp. 925–27.

Palmer, A. R. 1983. The decade of North American geology 1983 geologic time scale. *Geology*, vol. 11, pp. 503–04.

Pankhurst, R. J., S. Moorbath, and V. R. McGregor. 1973. Late event in the geological evolution of the Godthaab district, West Greenland. *Nature Physical Science*, vol. 243, pp. 24–26.

Pankhurst, R. J., S. Moorbath, D. C. Rex, and G. Turner. 1973. Mineral age patterns in ca. 3700 m.y. old rocks from West Greenland. *Earth and Planetary Science Letters*, vol. 20, pp. 157–70.

Papanastassiou, D. A., D. J. DePaolo, and G. J. Wasserburg. 1977. Rb–Sr and Sm–Nd chronology and genealogy of mare basalts from the Sea of Tranquility. *Proceedings of the Eighth Lunar Science Conference*, pp. 1639–72.

Papanastassiou, D. A., and G. J. Wasserburg. 1969. Initial strontium isotopic abundances and the resolution of small time differences in the formation of planetary objects. *Earth and Planetary Science Letters*, vol. 5, pp. 361–76.

———. 1971a. Lunar chronology and evolution from Rb–Sr studies of Apollo 11 and 12 samples. *Earth and Planetary Science Letters*, vol. 11, pp. 37–62.

———. 1971b. Rb–Sr ages of igneous rocks from the Apollo 14 mission and the age of the Fra Mauro Formation. *Earth and Planetary Science Letters*, vol. 12, pp. 36–48.

———. 1972a. Rb–Sr systematics of Luna 20 and Apollo 16 samples. *Earth and Planetary Science Letters*, vol. 17, pp. 52–63.

———. 1972b. Rb–Sr age of a Luna 16 basalt and model age of lunar soils. *Earth and Planetary Science Letters*, vol. 13, pp. 368–74.

———. 1973. Rb–Sr ages and initial strontium in basalts from Apollo 15. *Earth and Planetary Science Letters*, vol. 17, pp. 324–37.

———. 1974. Evidence for late formation and young metamorphism in the achondrite Nakhla. *Geophysical Research Letters*, vol. 1, pp. 23–26.

———. 1975. Rb–Sr study of a lunar dunite and evidence for early lunar differentiation. *Proceedings of the Sixth Lunar Science Conference*, pp. 1467–89.

———. 1976. Rb–Sr age of troctolite 76535. *Proceedings of the Seventh Lunar Science Conference*, pp. 2035–54.

Papike, J. J., F. N. Hodges, A. E. Bence, M. Cameron, and J. M. Rhodes. 1976. Mare basalts: Crystal chemistry, mineralogy, and petrology. *Reviews of Geophysics and Space Physics*, vol. 14, pp. 475–540.

Parkin, D. W., R. A. L. Sullivan, and J. N. Andrews. 1980. Further studies on cosmic spherites from deep-sea sediments. *Philosophical Transactions of the Royal Society of London*, series A, vol. 297, pp. 495–518.

Patchett, P. J., and M. Tatsumoto. 1980. Lu–Hf total-rock isochron for the eucrite meteorites. *Nature*, vol. 288, pp. 571–74.

Patterson, C. C. 1953. The isotopic composition of meteoritic, basaltic and oceanic leads, and the age of the Earth. Proceedings of the Conference on Nuclear Processes in Geologic Settings, Williams Bay, Wisconsin, Sept. 21–23, 1953, pp. 36–40.

———. 1955. The Pb[207]/Pb[206] ages of some stone meteorites. *Geochimica et Cosmochimica Acta*, vol. 7, pp. 151–53.

———. 1956. Age of meteorites and the earth. *Geochimica et Cosmochimica Acta*, vol. 10, pp. 230–37.

Patterson, C. C., H. Brown, G. R. Tilton, and M. G. Inghram. 1953. Concentration of uranium and lead and the isotopic composition of lead in meteoritic material. *Physical Review*, vol. 92, pp. 1234–35.

Patterson, C. C., E. D. Goldberg, and M. G. Inghram. 1953. Isotopic compositions of Quaternary leads from the Pacific Ocean. *Geological Society of America Bulletin*, vol. 64, pp. 1387–88.

Patterson, C. C., G. R. Tilton, and M. G. Inghram. 1955. Age of the Earth. *Science*, vol. 121, pp. 69–75.

Perry, J. 1895a. On the age of the Earth. *Nature*, vol. 51, pp. 224–27.

———. 1895b. The age of the Earth. *Nature*, vol. 51, pp. 582–85.

Peterman, Z. E., R. E. Zartman, and P. K. Sims. 1980. Tonalitic gneiss of early Archean age from northern Michigan. pp. 125–34 in G. B. Morey and G. N. Hanson, editors. *Selected Studies of Archean Gneisses and Lower Proterozoic Rocks, Southern Canadian Shield*. Geological Society of America Special Paper No. 182.

Pettingill, H. S., and P. J. Patchett. 1981. Lu–Hf total-rock age for the Amîtsoq gneisses, West Greenland. *Earth and Planetary Science Letters*, vol. 55, pp. 150–56.

Pidgeon, R. T. 1978a. 3450-m.y.-old volcanics in the Archaean layered greenstone succession of the Pilbara Block, Western Australia. *Earth and Planetary Science Letters*, vol. 37, pp. 421–28.

———. 1978b. Geochronological investigation of granite batholiths of the Archaean granite–greenstone terrain of the Pilbara Block, Western Australia. pp. 360–62 in I. E. M. Smith and J. G. Williams, editors. *Proceedings of the 1978 Archaean Geochemistry Conference, University of Toronto*.

———. 1978c. Big Stubby and the early history of the Earth. pp. 334–35 in R. E. Zartman, editor. *Short Papers of the Fourth International Conference, Geochronology, Cosmochronology, Isotope Geology*. U.S. Geological Survey Open-File Report 78-701.

Pidgeon, R. T., M. Aftalion, and F. Kalsbeek. 1976. The age of the Ilivertalik granite in the Fiskenaesset area. *The Geological Survey of Greenland*, Report no. 73, pp. 31–33.

Pidgeon, R. T., and A. M. Hopgood. 1975. Geochronology of Archean gneisses and tonalites from north of the Frederikshaabs isblink, S. W. Greenland. *Geochimica et Cosmochimica Acta*, vol. 39, pp. 1333–46.

Podosek, F. A. 1970. Dating of meteorites by the high-temperature release of iodine-correlated Xe^{129}. *Geochimica et Cosmochimica Acta*, vol. 34, pp. 341–61.

———. 1973. Thermal history of Nakhlites by the $^{40}Ar-^{39}Ar$ method. *Earth and Planetary Science Letters*, vol. 19, pp. 135–44.

Podosek, F. A., J. C. Huneke, and G. J. Wasserburg. 1972. Gas-retention and cosmic-ray exposure ages of lunar rock 15555. *Science*, vol. 175, pp. 423–25.

Ramsay, W., and F. Soddy. 1903. Experiments in radioactivity, and the production of helium from radium. *Proceedings of the Royal Society of London*, vol. 72, pp. 204–07.

Rancitelli, L., D. E. Fisher, J. Funkhouser, and O. A. Schaeffer. 1967. Potassium–argon dating of iron meteorites. *Science*, vol. 155, pp. 999–1000.

Reade, T. M. 1876. President's address. *Proceedings of the Liverpool Geological Society*, vol. 3, pp. 211–35.

———. 1879. *Chemical Denudation in Relation to Geological Time*. London: David Bogue.

———. 1893. Measurement of geological time. *The Geological Magazine*, vol. 30, pp. 97–100.

Reese, R. L., S. M. Everett, and E. D. Craun. 1981. The chronology of Archbishop James Ussher. *Sky and Telescope*, vol. 62, pp. 404–05.

Reichhardt, T. 1984. New Solar Systems? *Eos*, vol. 65, p. 97.

Reid, A. M. 1971. Rock types present in lunar highland soil. *The Moon*, vol. 9, pp. 141–46.

Reynolds, J. H. 1967. Isotopic abundance anomalies in the Solar System. *Annual Reviews of Nuclear Science*, vol. 17, pp. 253–316.

———. 1977. Isotope cosmochemistry: The rare gas story and related matters. *Proceedings of the Robert A. Welch Foundation on Chemical Research*. Cosmochemistry (Houston, Texas, Nov. 7–9, 1977), vol. 21, pp. 201–44.

Richards, J. R., I. R. Fletcher, and J. G. Blockley. 1981. Pilbara galenas: Precise isotopic assay of the oldest Australia leads; model ages and growth-curve implications. *Mineralium Deposita*, vol. 16, pp. 7–30.

Richer, H. B., and G. G. Fahlman. 1984. Deep CCD photometry in globular clusters. I. The main sequence of M4. *Astrophysical Journal*, vol. 277, pp. 227–34.

———. 1986. Deep CCD photometry in globular clusters. IV. M13. *Astrophysical Journal*, vol. 304, pp. 273–82.

Ridley, W. I. 1976. Petrology of lunar rocks and implications to lunar evolution. *Annual Review of Earth and Planetary Sciences*, vol. 4, pp. 15–48.

Ringwood, A. E. 1979. *Origin of the Earth and Moon*. New York: Springer-Verlag.

Robertson, I. D. M., M. C. duToit, P. Joubert, P. E. Matthews, N. H. Lockett, F. Mendelsohn, J. J. Broderick, K. Bloomfield, and R. Mason. 1981. Mobile belts. pp. 641–802 in D. R. Hunter, editor. *Precambrian of the Southern Hemisphere*. Amsterdam, Oxford, and New York: Elsevier Scientific.

Rose, J. L., and R. K. Stranathan. 1936. Geologic time and isotopic constitution of radiogenic lead. *Physical Review*, vol. 50, pp. 792–96.

Rowan-Robinson, M. 1985. *The Cosmological Distance Ladder—Distance and Time in the Universe*. New York: Freeman.

Rubin, A. E. 1984. Whence came the Moon? *Sky and Telescope*, vol. 68, pp. 389–93.

Rubin, V. C., D. Burstein, N. Thonnard, and W. K. Ford, Jr. 1983. A new evaluation of the Hubble constant. *Annual Report of the Director, Department of Terrestrial Magnetism, Carnegie Institution of Washington, for 1982–83*, pp. 563–66.

Russell, H. N. 1921. A superior limit to the age of the Earth's crust. *Proceedings of the Royal Society of London*, series A, vol. 99, pp. 84–86.

Russell, R. D. 1956. Interpretation of lead isotope abundances. pp. 68–78 in *Nuclear Processes in Geological Settings*. Washington, D.C.: National Research Council of the National Academy of Sciences, Publ. 400.

———. 1972. Evolutionary model for lead isotopes in conformable ores and in ocean volcanics. *Reviews of Geophysics and Space Physics*, vol. 10, pp. 529–49.

Russell, R. D., and R. M. Farquhar. 1960. *Lead Isotopes in Geology*. New York: Interscience.

Russell, R. D., and P. H. Reynolds. 1965. The age of the earth. pp. 35–48 in N. I. Khitarov, editor. *Problems of Geochemistry* (Vinogradov Volume). Moscow: Academy of Sciences of the USSR (English translation, 1969).

Rutherford, E. 1906. *Radioactive Transformations*. New York: Charles Scribner's Sons.

———. 1929. Origin of actinium and age of the Earth. *Nature*, vol. 123, pp. 313–14.

Rutherford, E., and F. Soddy. 1902a. The cause and nature of radioactivity. Part I. *Philosophical Magazine*, series 6, vol. 4, pp. 370–96.

————. 1902b. The cause and nature of radioactivity. Part II. *Philosophical Magazine,* series 6, vol. 4, pp. 569–85.

————. 1902c. The radioactivity of thorium compounds. I. An investigation of the radioactive emanation. *Journal of the Chemical Society of London,* vol. 81, pp. 321–50.

————. 1902d. The radioactivity of thorium compounds. II. The cause and nature of radioactivity. *Journal of the Chemical Society of London,* vol. 81, pp. 837–60.

————. 1903. Radioactive change. *Philosophical Magazine,* series 6, vol. 5, pp. 576–91.

Rutland, R. W. R. 1981. Structural framework of the Australian Precambrian. pp. 1–32 in D. R. Hunter, *Precambrian of the Southern Hemisphere.* Amsterdam, Oxford, and New York: Elsevier Scientific.

Saager, R., and V. Koppel. 1976. Lead isotopes and trace elements from sulfides of Archean greenstone belts in South Africa—A contribution to the knowledge of the oldest known mineralization. *Economic Geology,* vol. 71, pp. 44–57.

Salop, L. J. 1983. *Geological Evolution of the Earth during the Precambrian.* Berlin, Heidelberg, and New York: Springer-Verlag.

Sandage, A. 1982. The Oosterhoff period groups and the age of globular clusters. III. The age of the globular cluster system. *Astrophysical Journal,* vol. 252, pp. 553–73.

————. 1983. On the age of M92 and M15. *Astronomical Journal,* vol. 88, pp. 1159–65.

————. 1986a. The population concept, globular clusters, subdwarfs, ages, and the collapse of the Galaxy. *Annual Reviews of Astronomy and Astrophysics,* vol. 24, pp. 421–58.

————. 1986b. The redshift–distance relation IX. Perturbation of the very nearby velocity field by the mass of the Local Group. *Astrophysical Journal,* vol. 307, pp. 1–19.

Sandage, A., and G. A. Tammann. 1974a. Steps toward the Hubble constant. I. Calibration of the linear sizes of extragalactic H_{u} regions. *Astrophysical Journal,* vol. 190, pp. 525–38.

————. 1974b. Steps toward the Hubble constant. IV. Distances to 39 galaxies in the general field leading to a calibration of the galaxy luminosity classes and a first hint of the value of H_0. *Astrophysical Journal,* vol. 194, pp. 559–68.

————. 1976. Steps toward the Hubble constant. VII. Distances to NGC2403, M101, and the Virgo cluster. *Astrophysical Journal,* vol. 210, pp. 7–24.

Sangster, D. F., and W. A. Brook. 1977. Primitive lead in an Australian Zn–Pb–Ba deposit. *Nature,* vol. 270, p. 423.

Sanz, H. G., D. S. Burnett, and G. J. Wasserburg. 1970. A precise [87]Rb/[87]Sr age and initial [87]Sr/[86]Sr for the Colomera iron meteorite. *Geochimica et Cosmochimica Acta,* vol. 34, pp. 1227–39.

Sanz, H. G., and G. J. Wasserburg. 1969. Determination of an internal [87]Rb–[87]Sr isochron for the Olivenza chondrite. *Earth and Planetary Science Letters,* vol. 6, pp. 335–45.

Schaeffer, O. A., and L. Husain. 1974a. Chronology of lunar basin formation. *Proceedings of the Fifth Lunar Science Conference,* pp. 1541–55.

————. 1974b. Chronology of lunar basin formation and ages of lunar anorthositic rocks. *Lunar Science,* vol. V, pp. 663–65.

Schaeffer, O. A., L. Husain, and G. A. Schaeffer. 1976. Ages of highland rocks: The chronology of lunar basin formation revisited. *Proceedings of the Seventh Lunar Science Conference*, pp. 2067–92.

Schaeffer, G. A., and O. A. Schaeffer. 1977. ^{39}Ar–^{40}Ar ages of lunar rocks. *Proceedings of the Eighth Lunar Science Conference*, pp. 2253–300.

Schärer, U., and C.-J. Allègre. 1985. Determination of the age of the Australian continent by single-grain zircon analysis of Mt. Narryer metaquartzite. *Nature*, vol. 315, pp. 52–55.

Schramm, D. N. 1973. Nucleo-cosmochronology. *Space Science Reviews*, vol. 15, pp. 51–67.

———. 1974a. Nucleo-cosmochronology. *Annual Reviews of Astronomy and Astrophysics*, vol. 12, pp. 383–406.

———. 1974b. The age of the elements. *Scientific American*, vol. 230, no. 1, pp. 69–77.

———. 1982. The r-process and nucleocosmochronology. pp. 325–53 in C. A. Barnes, D. D. Clayton, and D. N. Schramm, editors. *Essays in Nuclear Astrophysics*. Cambridge: Cambridge University Press.

———. 1983. Nuclear constraints on the age of the universe. pp. 241–53 in R. M. West, editor. *Highlights of Astronomy*, vol. 6. Dordrecht, Boston, and London: Reidel.

Schramm, D. N., and G. J. Wasserburg. 1970. Nucleochronologies and the mean age of the elements. *Astrophysical Journal*, vol. 162, pp. 57–69.

Schuchert, C. 1931. Geochronology or age of the Earth on the basis of sediments and life. pp. 10–64 in *Physics of the Earth*, Part 4. The Age of the Earth. Washington, D.C.: National Research Council of the National Academy of Sciences, Bulletin 80.

Schultz, P. H., and P. D. Spudis. 1983. Beginning and end of lunar mare volcanism. *Nature*, vol. 302, pp. 233–36.

Scoville, N., and J. S. Young. 1984. Molecular clouds, star formation, and galactic structure. *Scientific American*, vol. 250, no. 4, pp. 42–53.

Sears, D. W. 1978. *The Nature and Origin of Meteorites*. New York: Oxford University Press.

Seeger, P. A., W. A. Fowler, and D. D. Clayton. 1965. Nucleosynthesis of heavy elements by neutron capture. *Astrophysical Journal Supplement*, vol. 11, pp. 121–66.

Shea, J. H. 1982. Twelve fallacies of uniformitarianism. *Geology*, vol. 10, pp. 455–60.

Shcherbak, N. P., E. N. Bartnitsky, E. V. Bibikova, and V. L. Boiko. 1984. Age and evolution of the early Precambrian continental crust of the Ukrainian shield. pp. 251–61 in A. Kroner, G. N. Hanson, and A. M. Goodwin, editors. *Archaean Geochemistry*. Berlin, Heidelberg, New York, and Tokyo: Springer-Verlag.

Shirley, D. N. 1983. A partially molten magma ocean model. *Proceedings of the 13th Lunar and Planetary Science Conference*, Part 2 (*Journal of Geophysical Research*, vol. 88 Supplement), pp. A519–27.

Shoemaker, E. M., and R. J. Hackman. 1962. Stratigraphic basis for a lunar time scale. pp. 289–300 in Z. Kopal and Z. K. Mikhailov, editors. *The Moon*. Symposium 14 of the International Astronomical Union. New York: Academic Press.

Shu, F. S. 1982. *The Physical Universe: An Introduction to Astronomy*. Mill Valley, Calif.: University Science Books.

Silver, L. T., and M. B. Duke. 1971. U–Th–Pb isotope relations in some basaltic achondrites [abstract]. *Eos,* vol. 52, p. 269.

Silver, L. T., and P. H. Schultz, editors. 1982. *Geological Implications of Impacts of Large Asteroids and Comets on the Earth.* Geological Society of America Special Paper No. 190.

Sims, P. K. 1980. Boundary between Archean greenstone and gneiss terranes in northern Wisconsin and Michigan. pp. 113–24 in G. B. Morey and G. N. Hanson, editors. *Selected Studies of Archean Gneisses and Lower Proterozoic Rocks, Southern Canadian Shield.* Geological Society of America Special Paper No. 182.

Sims, P. K., and Z. E. Peterman. 1976. Geology and Rb–Sr ages of reactivated gneiss and granite in the Marenisco–Watersmeet area, northern Michigan. *U.S. Geological Survey Journal of Research,* vol. 4, pp. 405–14.

———. 1981. Archaean rocks in the southern part of the Canadian shield—a review. pp. 85–98 in J. E. Glover and D. I. Groves, editors. *Archaean Geology.* Geological Society of Australia Special Publ. No. 7.

Sims, P. K., Z. E. Peterman, and W. C. Prinz. 1977. Geology and Rb–Sr age of Precambrian W Puritan Quartz Monzonite, northern Michigan. *U.S. Geological Survey Journal of Research,* vol. 5, pp. 185–92.

Sinha, A. K., and G. R. Tilton. 1973. Isotopic evolution of common lead. *Geochimica et Cosmochimica Acta,* vol. 37, pp. 1823–49.

Sleep, N. H., and R. T. Langan. 1981. Thermal evolution of the Earth: Some recent developments. *Advances in Geophysics,* vol. 23, pp. 1–23.

Soderblom, L. A., C. D. Condit, R. A. West, B. M. Herman, and T. J. Kreidler. 1974. Martian planetwide crater distributions: Implications for geologic history and surface processes. *Icarus,* vol. 22, pp. 239–63.

Sollas, W. J. 1900. Address of the President, geology section. *Report of the 70th Meeting of the British Association for the Advancement of Science, Bradford,* pp. 711–30.

———. 1909. The anniversary address of the President. *Quarterly Journal of the Geological Society of London,* vol. 65, pp. l–cxxii.

Spicer, H. C. 1937. Tables of temperature, geothermal gradient, and age of a non-radioactive earth. *Bulletin of the Geological Society of America,* vol. 48, pp. 75–92.

Stacey, F. D. 1977a. *Physics of the Earth* (2nd edition). New York: Wiley.

———. 1977b. A thermal model of the Earth. *Physics of the Earth and Planetary Interiors,* vol. 15, pp. 341–48.

———. 1981. Cooling of the Earth—a constraint on paleotectonic hypotheses. pp. 272–76 in R. J. O'Connell and W. S. Fyfe, editors. *Evolution of the Earth.* Washington, D.C.: American Geophysical Union, Geodynamics Series, vol. 5.

Stacey, J. S., M. E. Delevaux, and T. J. Ulrych. 1969. Some triple-filament lead isotope ratio measurements and an absolute growth curve for single-stage leads. *Earth and Planetary Science Letters,* vol. 6, pp. 15–25.

Stacey, J. S., and J. D. Kramers. 1975. Approximation of terrestrial lead isotope evolution by a two-stage model. *Earth and Planetary Science Letters,* vol. 26, pp. 207–21.

Stanton, R. L., and R. D. Russell. 1959. Anomalous leads and the emplacement of lead sulfide ores. *Economic Geology,* vol. 54, pp. 588–607.

Starik, I. 1937. Earth's age by radio-active data. *Seventeenth International Geological Congress, Moscow,* Abstracts of Papers, p. 192.

Staudacher, T., E. K. Jessberger, I. Flohs, and T. Kirsten. 1979. ^{40}Ar–^{39}Ar age systematics of consortium breccia 73255. *Proceedings of the Tenth Lunar and Planetary Science Conference*, pp. 745–62.

Steiger, R. H., and E. Jäger. 1977. Subcommission on geochronology: Convention on the use of decay constants in geo- and cosmochronology. *Earth and Planetary Science Letters*, vol. 36, pp. 359–62.

Stenning, M., and F. D. A. Hartwick. 1980. The local value of the Hubble constant from luminosity classification of Sb galaxies. *The Astronomical Journal*, vol. 85, pp. 101–16.

Stephenson, F. R. 1981. Tidal recession of the Moon from ancient and modern data. *Journal of the British Astronomical Association*, vol. 91, pp. 136–47.

Stettler, A., P. Eberhardt, J. Geiss, and N. Grögler. 1974. ^{39}Ar–^{40}Ar ages of samples from the Apollo 17 station 7 boulder and implications for its formation. *Earth and Planetary Science Letters*, vol. 23, pp. 453–61.

Stettler, A., P. Eberhardt, J. Geiss, N. Grögler, and S. Guggisberg. 1978. Chronology of the Apollo 17 station 7 boulder and the south Serenitatis impact. *Lunar Science*, vol. IX, pp. 1113–15.

Stettler, A., P. Eberhardt, J. Geiss, N. Grögler, and P. Maurer. 1973. Ar^{39}–Ar^{40} ages and Ar^{37}–Ar^{38} exposure ages of lunar rocks. *Proceedings of the Fourth Lunar Science Conference*, pp. 1865–88.

Stevenson, D. J. 1987. Origin of the Moon—the collision hypothesis. *Annual Reviews of Earth and Planetary Science*, vol. 15, pp. 271–315.

Stoenner, R. W., and J. Zähringer. 1958. Potassium–argon age of iron meteorites. *Geochimica et Cosmochimica Acta*, vol. 15, pp. 40–50.

Strutt, R. J. 1905. On the radio-active minerals. *Proceedings of the Royal Society of London*, series A, vol. 76, pp. 88–101.

———. 1908. On the accumulation of the helium in geological time. *Proceedings of the Royal Society of London*, series A, vol. 81, pp. 272–77.

Subsidiary Committee on the Age of the Earth. 1931. Physics of the Earth—IV. The age of the Earth. *Bulletin of the National Research Council*, No. 80. Washington, D.C.: National Research Council of the National Academy of Sciences.

Sun, D., and C. Wu. 1981. The principal geological and geochemical characteristics of the Archaean greenstone—gneiss sequences in north China. pp. 121–32 in J. E. Glover and D. I. Groves, editors. *Archaean Geology*. Geological Society of Australia Special Publ. No. 7.

Sutton, J., and B. F. Windley. 1974. The Precambrian. *Science Progress*, vol. 61, pp. 401–20.

Symbalisty, E. M. D., and D. N. Schramm. 1981. Nucleocosmochronology. *Reports in Progress in Physics*, vol. 44, pp. 293–328.

Tammann, G. A., and A. Sandage. 1983. The value of H_0. pp. 301–13 in R. West, editor. *Highlights of Astronomy*, vol. 6. Dordrecht, Boston, and London: D. Reidel.

Tankard, A. J., M. P. A. Jackson, K. A. Eriksson, D. K. Hobday, D. R. Hunter, and W. E. L. Minter. 1982. *Crustal Evolution of Southern Africa, 3.8 Billion Years of Earth History*. New York, Heidelberg, and Berlin: Springer-Verlag.

Tatsumoto, M. 1966a. Isotopic composition of lead in volcanic rocks from Hawaii, Iwo Jima, and Japan. *Journal of Geophysical Research*, vol. 71, pp. 1721–33.

———. 1966b. Genetic relations of oceanic basalts as indicated by lead isotopes. *Science*, vol. 153, pp. 1094–101.

Tatsumoto, M., R. J. Knight, and C. J. Allègre. 1973. Time differences in the formation of meteorites as determined from the ratio of lead-207 to lead-206. *Science,* vol. 180, pp. 1279–83.

Tatsumoto, M., P. D. Nunes, R. J. Knight, C. E. Hedge, and D. M. Unruh. 1973. U–Th–Pb, Rb–Sr, and K measurements of two Apollo 17 samples. *Eos,* vol. 54, p. 614.

Tatsumoto, M., and D. M. Unruh. 1975. Formation and early brecciation of the Juvinas achondrite inferred from U–Th–Pb systematics [abstract]. *Meteoritics,* vol. 10, pp. 500–01.

Tatsumoto, M., D. M. Unruh, and G. A. Desborough. 1976. U–Th–Pb and Rb–Sr systematics of Allende and U–Th–Pb systematics of Orgueil. *Geochimica et Cosmochimica Acta,* vol. 40, pp. 617–34.

Tatsumoto, M., D. M. Unruh, and P. J. Patchett. 1981. U–Pb and Lu–Hf systematics of Antarctic meteorites. Proceedings of the Sixth Symposium on Antarctic Meteorites. *Memoirs of the National Institute of Polar Research,* Special Issue 20, pp. 237–49.

Taylor, G. J. 1985. Lunar origin meeting favors impact theory. *Geotimes,* vol. 30, no. 4, pp. 16–17.

Taylor, L. A., J. W. Shervais, R. H. Hunter, C.-Y. Shih, B. M. Bansal, J. Wooden, L. E. Nyquist, and L. C. Laul. 1983. Pre-4.2 AE mare-basalt volcanism in the lunar highlands. *Earth and Planetary Science Letters,* vol. 66, pp. 33–47.

Taylor, P. N., S. Moorbath, R. Goodwin, and A. C. Petrykowski. 1980. Crustal contamination as an indicator of the extent of early Archaean continental crust: Pb isotopic evidence from the late Archaean gneisses of West Greenland. *Geochimica et Cosmochimica Acta,* vol. 44, pp. 1437–53.

Taylor, S. R. 1975. *Lunar Science: A Post-Apollo View.* New York, Toronto, Oxford, Sydney, and Braunschweig: Pergamon Press.

———. 1979. Structure and evolution of the Moon. *Nature,* vol. 281, pp. 105–10.

Tera, F. 1980. Reassessment of the "Age of the Earth." *Carnegie Institution of Washington Year Book,* vol. 79, pp. 524–31.

———. 1981. Aspects of isochronism in Pb isotope systematics—application to planetary evolution. *Geochimica et Cosmochimica Acta,* vol. 45, pp. 1439–48.

Tera, F., D. A. Papanastassiou, and G. J. Wasserburg. 1974. Isotopic evidence for a terminal lunar cataclysm. *Earth and Planetary Science Letters,* vol. 22, pp. 1–21.

Tera, F., and G. J. Wasserburg. 1972. U–Th–Pb systematics in three Apollo 14 basalts and the problem of initial Pb in lunar rocks. *Earth and Planetary Science Letters,* vol. 14, pp. 281–304.

———. 1974. U–Th–Pb systematics on lunar rocks and inferences about lunar evolution and the age of the Moon. *Proceedings of the Fifth Lunar Science Conference,* pp. 1571–99.

———. 1975. The evolution and history of mare basalts as inferred from U–Th–Pb systematics. *Lunar Science,* vol. VI, pp. 807–09.

ter Haar, D. 1967. On the origin of the Solar System. *Annual Reviews of Astronomy and Astrophysics,* vol. 5, pp. 267–78.

Thomson, J. J. 1914. Rays of positive electricity. *Proceedings of the Royal Society of London,* series A, vol. 89, pp. 1–20.

———. 1936. *Recollections and Reflections.* London: Bell.

Thomson, W. 1852. On a universal tendency in nature to the dissipation of mechanical energy. *Philosophical Magazine*, series 4, vol. 4, pp. 304–06.

———. 1862a. On the age of the sun's heat. *Macmillan's Magazine*, March, pp. 388–93 (reprinted in W. Thomson and P. G. Tait. 1890. *Treatise on Natural Philosophy*, Part II. Cambridge: Cambridge University Press, pp. 485–94).

———. 1862b. On the secular cooling of the Earth. *Royal Society of Edinburgh Transactions*, vol. 23, pp. 157–59 (reprinted in W. Thomson and P. G. Tait. 1890. *Treatise on Natural Philosophy*. Part II. Cambridge: Cambridge University Press, pp. 468–85).

———. 1871. On geological time. *Transactions of the Geological Society of Glasgow*, vol. 3, part 1, pp. 1–28.

Tilton, G. R. 1973. Isotopic lead ages of chondritic meteorites. *Earth and Planetary Science Letters*, vol. 19, pp. 321–29.

———. 1988. Age of the Solar System. pp. 259–75 in J. F. Kerridge and M. S. Matthews, editors. *Meteorites and the Early Solar System*. Tucson: University of Arizona Press.

Tilton, G. R., and R. H. Steiger. 1965. Lead isotopes and the age of the Earth. *Science*, vol. 150, pp. 1805–08.

———. 1969. Mineral ages and isotopic composition of primary lead at Manitouwadge, Ontario. *Journal of Geophysical Research*, vol. 74, pp. 2118–32.

Tinsley, B. M. 1975. Nucleochronology and chemical evolution. *Astrophysical Journal*, vol. 198, pp. 145–50.

Toksöz, M. N., A. M. Dainty, S. C. Solomon, and K. R. Anderson. 1974. Structure of the Moon. *Reviews of Geophysics and Space Physics*, vol. 12, pp. 539–67.

Trimble, V. 1975. The origin and abundance of the chemical elements. *Reviews of Modern Physics*, vol. 47, pp. 877–976.

Tsunakawa, H., and M. Yanagisawa. 1981. Rb–Sr and ^{40}Ar–^{39}Ar geochronological studies on the Precambrian rocks of the Minnesota River Valley. *Geochemical Journal*, vol. 15, pp. 17–23.

Turner, G. 1968. The distribution of potassium and argon in chondrites. pp. 387–98 in L. N. Ahrens, editor. *Origin and Distribution of the Elements*. Oxford and New York: Pergamon Press.

———. 1969. Thermal histories of meteorites by the Ar^{39}–Ar^{40} method. pp. 407–17 in P. M. Millman, editor. *Meteorite Research*. Dordrecht: D. Reidel.

———. 1970. Argon-40/argon-39 dating of lunar rock samples. *Proceedings of the First Lunar Science Conference*, pp. 1665–84.

———. 1971. ^{40}Ar–^{39}Ar ages from the lunar maria. *Earth and Planetary Science Letters*, vol. 11, pp. 169–91.

Turner, G., and P. H. Cadogan. 1975. The history of lunar bombardment inferred from ^{40}Ar–^{39}Ar dating of highland rocks. *Proceedings of the Sixth Lunar Science Conference*, pp. 1509–38.

Turner, G., P. H. Cadogan, and C. J. Yonge. 1973. Argon selenochronology. *Proceedings of the Fourth Lunar Science Conference*, pp. 1889–914.

Turner, G., M. C. Enright, and P. H. Cadogan. 1978. The early history of chondrite parent bodies inferred from ^{40}Ar–^{39}Ar ages. *Proceedings of the Ninth Lunar and Planetary Science Conference*, pp. 989–1025.

Ulrich, R. K. 1982. The s-process. pp. 301–23 in C. A. Barnes, D. D. Clayton, and D. N. Schramm, editors. *Essays in Nuclear Astrophysics*. Cambridge: Cambridge University Press.

Ulrych, T. J. 1967. Oceanic basalt leads: A new interpretation and an independent age for the Earth. *Science*, vol. 158, pp. 252–56.

Unruh, D. M. 1982. The U–Th–Pb age of equilibrated L chondrites and a solution to the excess radiogenic Pb problem in chondrites. *Earth and Planetary Science Letters*, vol. 58, pp. 75–94.

Unruh, D. M., R. Hutchison, and M. Tatsumoto. 1979. U–Th–Pb age of the Barwell chondrite: Anatomy of a "discordant" meteorite. *Proceedings of the Tenth Lunar and Planetary Science Conference*, pp. 1011–30.

Unruh, D. M., N. Nakamura, and M. Tatsumoto. 1977. History of the Pasamonte achondrite: Relative susceptibility of the Sm–Nd, Rb–Sr, and U–Pb systems to metamorphic events. *Earth and Planetary Science Letters*, vol. 37, pp. 1–12.

Upham, W. 1893. Estimates of geologic time. *American Journal of Science*, 3rd series, vol. 45, pp. 209–20.

Ussher, J. 1658. *The Annals of the World. Deduced from the origin of time, and continued to the beginning of the Emperour Vespasian's reign, and the totall destruction and abolition of the temple and commonwealth of the Jews. Containing the historie of the Old and New Testament, with that of the Macchabees. Also all the most memorable affairs of Asia and Egypt, and the rise of the empire of the Roman caesars, under C. Julius, and Octavianus. Collected from all history, as well sacred, as prophane and methodically digested by the Most Reverend James Ussher, Archbishop of Armagh, and Primate of Ireland.* London: Printed by E. Tyler for J. Crook and G. Bedell.

———. 1984. The age and size of the universe. *Quarterly Journal of the Royal Astronomical Society*, vol. 25, pp. 137–46.

VandenBerg, D. A. 1983. Star clusters and stellar evolution. I. Improved synthetic color-magnitude diagram for the oldest clusters. *Astrophysical Journal Supplement*, vol. 51, pp. 29–66.

VandenBerg, D. A., and R. A. Bell. 1985. Theoretical isochrons for globular clusters with predicted BVRI and stromgren photometry. *Astrophysical Journal Supplement*, vol. 58, pp. 561–621.

van den Bergh, S. 1981. The size and age of the universe. *Science*, vol. 213, pp. 825–30.

van Niekerk, C. B., and A. J. Burger. 1969. A note on the minimum age of the acid lava of the Onverwacht series of the Swaziland system. *Transactions of the Geological Society of South Africa*, vol. 72, part 1, pp. 9–21.

Van Schmus, W. R., and J. A. Wood. 1967. A chemical–petrologic classification for the chondritic meteorites. *Geochimica et Cosmochimica Acta*, vol. 31, pp. 747–65.

Viljöen, M. J., and R. P. Viljöen. 1969a. An introduction to the geology of the Barberton granite–greenstone terrain. *Geological Society of South Africa Special Publ.* No. 2, pp. 9–28.

———. 1969b. The geology and geochemistry of the lower ultramafic unit of the Onverwacht Group and a proposed new class of igneous rock. *Geological Society of South Africa Special Publ.* No. 2, pp. 55–85.

Viljöen, M. J., R. P. Viljöen, N. S. Smith, and A. J. Erlank. 1983. Geological, textural and geochemical features of komatiitic flows from the Komati Formation. pp. 1–20 in C. R. Anhaeusser, editor. *Contributions to the Geology of the Barberton Mountain Land.* The Geological Society of South Africa Special Publ. No. 9.

Visvanathan, N. 1979. The Hubble diagram for E and S0 galaxies in the local region. *Astrophysical Journal*, vol. 228, pp. 81–94.

Walcott, C. D. 1893. Geologic time, as indicated by the sedimentary rocks of North America. *Journal of Geology*, vol. 1, pp. 639–76.

Walgate, R. 1983. Emerging solar systems in view. *Nature*, vol. 304, p. 681.

Wanless, R. K., D. Bridgwater, and K. D. Collerson. 1979. Zircon age measurements for Uivak II gneisses from the Saglek area, Labrador. *Canadian Journal of Earth Sciences*, vol. 16, pp. 962–65.

Wasserburg, G. J., A. L. Albee, and M. A. Lanphere. 1964. Migration of radiogenic strontium during metamorphism. *Journal of Geophysical Research*, vol. 69, pp. 4395–401.

Wasserburg, G. J., D. S. Burnett, and C. Frondel. 1965. Strontium–rubidium age of an iron meteorite. *Science*, vol. 150, pp. 1814–18.

Wasserburg, G. J., D. A. Papanastassiou, and H. G. Sanz. 1969. Initial strontium for a chondrite and the determination of a metamorphism or formation age. *Earth and Planetary Science Letters*, vol. 7, pp. 33–43.

Wasserburg, G. J., F. Tera, D. A. Papanastassiou, and J. C. Huneke. 1977. Isotopic and chemical investigations on Angra dos Reis. *Earth and Planetary Science Letters*, vol. 35, pp. 294–316.

Wasson, J. T. 1974. *Meteorites, Classification and Properties*. New York, Heidelberg, and Berlin: Springer-Verlag.

Weisskopf, V. F. 1983. The origin of the universe. *American Scientist*, vol. 71, pp. 473–80.

Wetherill, G. W. 1956. Discordant uranium–lead ages I. *Transactions of the American Geophysical Union*, vol. 37, pp. 320–26.

———. 1975. Radiometric chronology of the early Solar System. *Annual Reviews of Nuclear Science*, vol. 25, pp. 283–328.

———. 1980. Formation of the terrestrial planets. *Annual Reviews of Astronomy and Astrophysics*, vol. 18, pp. 77–113.

———. 1984. Orbital evolution of impact ejecta from Mars. *Meteoritics*, vol. 19, pp. 1–13.

———. 1985a. Asteroidal source of ordinary chondrites. *Meteoritics*, vol. 20, pp. 1–20.

———. 1985b. Occurrence of giant impacts during the growth of the terrestrial planets. *Science*, vol. 228, pp. 877–79.

Wetherill, G. W., and E. M. Shoemaker. 1982. Collision of astronomically observable bodies with Earth. pp. 1–13 in L. T. Silver and P. H. Schultz, editors. *Geological Implications of Impacts of Large Asteroids and Comets on the Earth*. Geological Society of America Special Paper 190.

Wilhelms, D. E. 1984. Moon. pp. 107–205 in M. H. Carr, editor. *The Geology of the Terrestrial Planets*. National Aeronautics and Space Administration Special Publ. SP-469.

Wilhelms, D. E., and J. F. McCauley. 1971. Geologic map of the near side of the Moon. *U. S. Geological Survey Miscellaneous Investigations Map* I-703.

Wilkins, T. 1958. *Clarence King*. New York: Macmillan.

Williams, I. S., R. W. Page, J. J. Foster, W. Compston, K. D. Collerson, and M. T. McCulloch. 1983. Zircon U–Pb ages from the Shaw Batholith, Pilbara block, determined by ion microprobe. *The Australian National University Research School of Earth Sciences Annual Report for 1982*, pp. 199–203.

Windley, B. F. 1984. *The Evolving Continents* (2nd edition). Chichester, New York, Brisbane, Toronto, and Singapore: Wiley.

Wood, C. A., and L. D. Ashwal. 1981. SNC meteorites: Igneous rocks from

Mars. *Proceedings of the 12th Lunar and Planetary Science Conference,* pp. 1359–75.

Wood, J. A. 1968. *Meteorites and the Origin of Planets.* New York, St. Louis, and San Francisco: McGraw-Hill.

———. 1975. The Moon. *Scientific American,* vol. 233, no. 3, pp. 92–102.

Wooden, J. L., H. Takeda, L. E. Nyquist, H. Wiesmann, and B. Bansal. 1983. A Sr and Nd isotopic study of five Yamato polymict eucrites and a comparison to other Antarctic and ordinary eucrites. *Memoirs of the National Institute of Polar Research,* Special Issue no. 30, pp. 315–22.

Woolsey, S. E., R. E. Taam, and T. A. Weaver. 1986. Models for type I supernova. I. Detonations in white dwarfs. *Astrophysical Journal,* vol. 301, pp. 601–23.

Xuan, H., B. Ziwei, and D. J. DePaolo. 1986. Sm–Nd isotope study of early Archean rocks, Qianan, Hebei Province, China. *Geochimica et Cosmochimica Acta,* vol. 50, pp. 625–31.

York, D., and R. M. Farquhar. 1972. *The Earth's Age and Geochronology.* Oxford and New York: Pergamon Press.

York, D., W. J. Kenyon, and R. J. Doyle. 1972. ^{40}Ar–^{39}Ar ages of Apollo 14 and 15 samples. *Proceedings of the Third Lunar Science Conference,* pp. 1613–22.

Young, D. A. 1982. *Christianity and the Age of the Earth.* Grand Rapids, Mich.: Zondervan.

Zartman, R. E. 1965. The isotopic composition of lead in microclines from the Llano uplift, Texas. *Journal of Geophysical Research,* vol. 70, pp. 965–75.

Glossary and Index

Glossary

Words defined in the Glossary are italicized in the text at first use. Those that
are italicized in the Glossary are defined elsewhere in the Glossary.

achondrite   One of the two main types of *stony meteorites*. Achondrites do
not contain *chondrules*.

age-diagnostic diagram   A graphic device for analyzing the *isotopic* composition of a sample with the goal of determining its *radiometric* age.
Age-diagnostic diagrams include *isochron, concordia–discordia*, and lead-
evolution diagrams and *age spectra*.

age spectrum   Diagram showing a sample's apparent $^{40}Ar/^{39}Ar$ age as a
function of the percentage of ^{39}Ar in increments of gas released from
the sample as it is heated to progressively higher temperatures. Also
called an Ar-release diagram.

alpha decay   *Radioactive decay* in which an alpha particle, consisting of two
neutrons and two *protons*, is ejected from a nucleus.

aluminous basalt   Lunar *basalt* that closely resembles terrestrial basalt except
for being higher in titanium. Aluminous basalt is higher in aluminum
than other types of lunar basalt because it contains more *plagioclase feldspar*, less of iron–titanium oxides, and no *olivine*.

amphibole   A group of silicate *minerals* of which *hornblende* is the most common member.

amphibolite   A *metamorphic rock* containing primarily *amphibole* and *plagioclase feldspar*.

andesite   A fine-grained volcanic *rock* that contains *plagioclase feldspar* and
one or more of the *mafic minerals pyroxene, hornblende*, and *biotite*. Andesite is higher in silica and less mafic than is *basalt*. It is commonly
erupted from the volcanos along active continental margins, such as the
Cascades of North America and the Andes of South America.

angrite   A type of *achondrite meteorite* of which Angra dos Reis, a magmatic
cumulate composed primarily of *pyroxene* and *olivine*, is the only
example.

angular momentum   A vector quantity that is the measure of the intensity
of rotational motion. It is the sum of the product of the position vectors
and the linear momentum of all bodies in the system under consideration.

angular velocity   The rate of change of angular displacement, and a common means of describing the rate at which a point or body moves
around another.

anion   A negatively charged *ion*.

anorthosite   A *plutonic rock* composed almost entirely of *plagioclase feldspar*.

anorthositic gabbro   A *plutonic rock* intermediate in composition between *anorthosite* and *gabbro*.

anticlinal   Pertaining to an anticline, or upfold (convex up), in *rock* strata. Compare *synclinal*.

antineutrino   The antimatter counterpart of the *neutrino*.

ANT suite   A collective term for the lunar *cumulate rocks anorthosite, norite,* and *troctolite*.

aphanitic   A term that describes the texture of a rock in which the crystals are too small to see with the unaided eye.

aphelion (pl. aphelia)   The point on an orbit farthest from the sun. Compare *perihelion*.

Apollo asteroids   *Asteroids* whose orbits intersect the Earth's orbit.

Ar-release diagram   See *age spectrum*.

asteroid   A small planet. Most of the asteroids occur in the asteroid belt between Mars and Jupiter and have diameters ranging from a fraction of a kilometer to about 1,000 km (Ceres).

astronomical unit   A unit of distance equivalent to the mean distance from the Earth to the Sun, about 1.50×10^8 km.

aubrite   A type of *achondrite meteorite* composed primarily of the *pyroxene mineral* enstatite.

banded-iron formation   A chemically precipitated *sedimentary rock* consisting of alternating layers of iron-rich and iron-poor *silica*. Banded-iron formation was deposited only in the Precambrian. Some of these formations are mined for iron.

basalt   A fine-grained volcanic *rock* composed primarily of *plagioclase feldspar* and *pyroxene*. Basalt is a common rock on the Earth and the other solid bodies of the Solar System. The floors of the oceans, Hawaiian volcanoes, the lunar *maria*, and the surfaces of some *asteroids* are composed primarily of basalt. See also *gabbro*.

basaltic achondrite   A group of *achondrite meteorites* that originated in *basalt* lava flows, probably on the larger *asteroids*.

basin (lunar)   A very large lunar impact crater, typically hundreds of kilometers in diameter, characterized by having one or more concentric rings of uplifted material.

batholith   A body of *intrusive rock* at least 150 km^2 in area.

beta decay   Radioactive decay in which a nucleus ejects a beta particle (an *electron*) and a *neutron* is converted into a *proton*.

beta emission   The emission of a beta particle (an electron) from the nucleus of a radioactive *nuclide* during the process of *beta decay*.

biogenic   Resulting from the activity of living organisms, said of sediment or *sedimentary rock*.

biotite   A dark mica *mineral* rich in potassium, iron, and magnesium. It is a common constituent in *felsic* volcanic, *plutonic*, and *metamorphic rocks* and is also found in *sedimentary rocks*.

blackbody   A body that absorbs all radiation striking it, reflecting none. Nothing in nature behaves as a perfect blackbody, although some objects come close.

block   See *craton*.

branching ratio   Some radioactive *isotopes*, like ^{40}K, decay by two different modes, such as *beta decay* and *electron capture*, and so produce two different stable *daughter nuclides*. The branching ratio is the fraction of the decay by one mode divided by the fraction of the decay by the other.

breccia A coarse-grained *rock* composed primarily of angular, broken rock fragments. Breccia may originate by either volcanic, *sedimentary*, or *tectonic* processes. A rock is said to be brecciated if it has endured the processes that form a breccia.

bronzite A mineral of the *pyroxene* group.

capture hypothesis The hypothesis that the Moon originated elsewhere and was captured by the Earth's gravity field.

carbonaceous chondrite The most primitive class of *meteorite*. Carbonaceous chondrites are high in volatile elements and may contain several percent of carbon, including organic molecules.

cation A positively charged *ion*.

Cepheid A class of variable, or pulsating, stars whose *luminosities* are a function of their periods.

chassignite A type of *achondrite meteorite* composed primarily of *olivine*. Chassignites resemble terrestrial *dunite*, a magmatic *cumulate*.

chemical age A radiometric age based on the ratios of two elements, without the *isotopic* composition of the elements being taken into consideration. Chemical ages were used before isotopes were discovered and are invalid.

chert An extremely fine-grained *sedimentary rock* consisting primarily of *silica*. Chert is formed either from the remains of siliceous microorganisms or by direct precipitation of *silica* from sea water.

chloride A chemical compound in which chlorine is an anion. Common table salt (NaCl) is a chloride.

chondrite One of the two main types of *stony meteorites*. Chondrites are so named because they contain *chondrules*.

chondritic Resembling a *chondrite*.

chondrules Small spherules of *silicate minerals* that distinguish the *chondrite meteorites*. They are thought to have condensed from the *Solar Nebula* as liquid droplets.

chronometer pair A pair of *isotopes*, at least one of which is radioactive, used to calculate the mean age of the elements by the method of *nucleocosmochronology*.

Clapeyron–Clausius equation A *thermodynamic* equation that describes the relationship between pressure, temperature, *latent heat*, and volume in a one-component system.

clast An individual *rock* or *mineral* constituent in a *sedimentary* rock or volcanic *breccia*.

closed system A system in which matter neither enters nor leaves. Compare *isolated system* and *open system*.

cogenetic Two or more *rocks* originating at the same time from the same source are said to be cogenetic.

collision hypothesis The hypothesis that the Moon originated as the result of a collision between the Earth and a large asteroid.

comet A small celestial body consisting primarily of ices and dust. Most comets are thought to reside in the Oort cloud beyond the orbits of Neptune and Pluto, but some have orbits that take them into the inner Solar System, where radiation from the Sun causes them to develop a "tail" of ionized gases.

common lead Lead in *minerals* that are very low in uranium and thorium, with the consequence that the *isotopic* composition of the lead is frozen and does not change significantly over time.

compound A substance consisting of two or more elements in chemical combination.

concordia The locus of all concordant U–Pb ages on a graph of the ratios $^{206}Pb/^{238}U$ vs $^{207}Pb/^{235}U$.

concordia diagram A graph of the ratios $^{206}Pb/^{238}U$ vs $^{207}Pb/^{235}U$ on which *concordia* is drawn.

concordia–discordia method The method of U–Pb *radiometric dating* using the *concordia diagram*.

conduction The transmission of energy, such as heat, through a substance without movement of the substance itself. See also *convection*.

conformable ore A sedimentary lead ore, formed by precipitation from sea water, that conforms to the geometry of the *sedimentary rocks* with which it is interbedded.

conglomerate A coarse-grained *sedimentary rock* consisting of rounded rocks or pebbles cemented together by finer-grained material.

conservation of angular momentum A law of physics stating that the *angular momentum* of a system of particles (or bodies) will not change as long as no external forces are applied to the system. Thus, a rotating cloud of dust and gas must rotate faster as it contracts.

convection Mass movement in a fluid in a gravitational field because of differences in temperature. Convection moves both the fluid and the energy (heat) it contains.

Cordilleran Sea The inland sea that covered much of western North America during the Paleozoic Era.

core The central part of the Earth, beginning about 2900 km below the surface. The core is probably composed of nickel and iron and consists of an outer liquid part and an inner solid part.

correlation diagram A plot of the ratio of *parent* and *daughter isotopes*, normalized to a stable isotope of the daughter element, from which a radiometric age can be found. For example, a plot of $^{87}Rb/^{86}Sr$ vs $^{87}Sr/^{86}Sr$. Also known as an *isochron* or isotope-evolution diagram.

cosmic ray Natural radiation in space consisting of high-energy subatomic particles.

cosmic-ray exposure age The time that a *rock* has been exposed to *cosmic rays* as calculated from *isotopic* changes induced by the *cosmic rays* over the time of the exposure.

craton A relatively immobile part of a continent, usually of large size. Cratons are composed primarily of Precambrian rocks. Also called blocks or shields.

crust The outermost part of the Earth. The continental crust is about 30–50 km thick, whereas the oceanic crust is only about 5–10 km thick.

crustal plate A discrete unit of the *lithosphere* that is in constant motion relative to the other plates. There are about a half-dozen major plates, each about 100 km thick, that make up the Earth's lithosphere.

cumulate An *igneous rock* consisting of crystals that solidified in *magma* and then were concentrated by either sinking or floating in the magma.

dacite A volcanic *rock* consisting primarily of *plagioclase, quartz, hornblende,* or *pyroxene*. Dacite is higher in *silica* and more *felsic* than *andesite*.

daughter isotope An *isotope* produced by *radioactive decay* of a *parent* isotope.

decay constant A quantity that expresses the probability per unit time that a radioactive *nuclide* will decay.

decay series The chain from a radioactive *parent* nuclide through one or more intermediate radioactive *daughter* nuclides to a stable daughter nuclide.

deformation A change in the form or volume of a mass of *rock* due to *tectonic* forces. Rocks are deformed most commonly by faulting, folding, and plastic flow.

diabase A shallow intrusive rock with a *basaltic* composition and a distinctive texture, with discrete grains of *pyroxene* between *feldspar* laths. Diabase commonly occurs as *dikes* and *sills*.

diagenesis The processes that cause physical and chemical changes in sediment and thereby convert the sediment to *sedimentary rock*. Compaction, cementation, recrystallization, and replacement are common processes of diagenesis.

diapiric A form of *intrusion* by plastic flow in which the intruding body is squeezed up through the overlying rocks. The term is also applied to sedimentary rocks in which the plastic core of a fold ruptures the overlying strata.

differentiation The process by which *igneous rocks* of different compositions are formed from a single parent magma. Common mechanisms of differentiation include fractional crystallization and the concentration of crystals by sinking or floating. See also *cumulate*.

dike A tabular body of *igneous rock* that cuts across the bedding or structure of the rocks it *intrudes*. See also *sill*.

diogenite An *achondrite meteorite* composed primarily of the *pyroxene bronzite*. Diogenite is a *cumulate* and related to the *eucrites*.

diorite A *granitoid* consisting primarily of *plagioclase feldspar, hornblende, biotite*, or *pyroxene*.

discordia A straight line on a *concordia diagram* formed by connecting data points from *cogenetic minerals* that have lost lead. The upper intersection of a discordia with *concordia* indicates the crystallization age of the samples.

disturbed system An *open system* of rocks or minerals.

double-planet hypothesis The hypothesis that the Moon and Earth formed (aggregated) as a double planet.

dunite An *igneous cumulate rock* formed almost entirely of *olivine*.

ecliptic plane The imaginary plane that contains the Earth's orbit. The ecliptic is the intersection of the plane of the Earth's orbit with the celestial sphere.

electron A particle with negative charge and mass of 1/1,836 that of a *proton*. Electrons orbit the *nucleus* of an atom and largely control its chemical behavior.

electron capture A form of radioactive decay in which an *electron* from the innermost shell of electrons falls into the nucleus and a *proton* is converted into a *neutron*.

enstatite A mineral of the *pyroxene* group.

enstatite chondrite A type of *chondrite* that is *brecciated* and composed primarily of the *pyroxene enstatite*.

eucrite A type of *basaltic achondrite* that resembles terrestrial and lunar basalts.

fast neutron A *neutron* having kinetic energy greater than some arbitrary value, usually about 100 electron volts. A neutron with energy below this value is called a slow, or thermal, neutron.

feldspar A group of common *rock*-forming aluminum-*silicate minerals* that contain potassium, sodium, or calcium.

felsic rock A qualitative term for light-colored *rocks*. Felsic rocks contain an abundance of light-colored *minerals* like *feldspar* and *quartz* and a low proportion of dark minerals like *pyroxene, hornblende,* and *olivine*. The opposite of *mafic*.

fergusonite An yttrium oxide *mineral*. Fergusonite may also contain niobium, tantalum, iron, titanium, or uranium.

fission hypothesis The hypothesis that the Moon was "thrown off" from the Earth.

formation intervals The difference between the formation times of individual *meteorites* as determined by certain types of *radiometric dating*.

Fra Mauro basalts *Lunar highland basalts,* so named because they are common in the Fra Mauro Formation, which is the ejecta from the formation of the Imbrium Basin.

Ga See *giga-annum*.

gabbro A *plutonic rock* consisting primarily of *plagioclase feldspar* and *pyroxene*. The *intrusive* equivalent of *basalt*.

galaxy An organized system of stars rotating about a common center.

Galaxy (The) The *galaxy* in which the Solar System resides. Also known as the Milky Way Galaxy.

gamma ray Very-high-frequency electromagnetic radiation emitted by the nucleus of a radioactive *nuclide* to dispose of excess energy. See also *X ray*.

garbenschiefer A term describing the texture of a platy rock with distinctive spots resembling caraway seeds.

geodesy The study of the shape and dimensions of the Earth.

geothermal gradient The change in temperature with depth in the Earth.

giga-annum (Ga) One billion (10^9) years.

globular cluster A group of stars that resides in a spherical halo surrounding a galactic disk rather than in the disk itself. The stars in globular clusters are thought to be the oldest in the *Galaxy*.

gneiss (pron. "nice") A coarse-grained *metamorphic rock* in which layers rich in granular *minerals* alternate with layers rich in platy minerals.

granite A granitoid consisting primarily of potassium *feldspar* and *quartz*.

granitoid A general term for *felsic* and *intermediate plutonic rocks*. *Granite, diorite,* and *quartz monzonite,* to name a few specific rock types, are granitoids.

granodiorite A specific type of *granitoid* consisting primarily of calcic *plagioclase,* potassium *feldspar,* and *quartz* with *hornblende, biotite,* or *pyroxene*.

greenstone belt A stratified accumulation of interbedded lava flows and sediments that typically occurs in a symmetrical basin-like structure. All known greenstone belts are Precambrian in age and most are thought to have been deposited in shallow seas.

groundmass The fine-grained material, *minerals* and glass, between the phenocrysts (larger crystals) of an *igneous rock*.

half-life The time it takes one-half of a quantity of a particular radioactive *nuclide* to decay.

helium flash The runaway production of nuclear energy from the conversion of helium to carbon in a *red giant* star.

Hertzsprung–Russell diagram A plot of stellar *luminosity* vs effective temperature on which stars' evolution can be traced.

high-grade metamorphism *Metamorphism* at high temperature and pressure.

high-titanium basalt A type of lunar *mare basalt* that is high in titanium.

Holmes–Houtermans model A graphical and mathematical model describing the change over time in $^{206}Pb/^{204}Pb$ vs $^{207}Pb/^{204}Pb$ in a closed uranium-bearing system.

horizontal branch A stage of evolution of a star on the *Hertzsprung–Russell diagram* between the *red giant* and *red supergiant* phases.

hornblende A ferromagnesian *silicate mineral* of the *amphibole* group. Hornblende typically occurs as dark needle-like crystals and is common in *granitoids, gneisses,* and *andesites.*

howardite A *brecciated basaltic achondrite meteorite.*

Hubble age An age for the universe calculated from the rate at which the *galaxies* are receding from each other due to expansion of the universe. Also called Hubble time.

Hubble constant (H_0) The constant of proportionality specifying the relationship between velocity and distance in the expanding universe, in units of kilometers per second per *megaparsec.*

Hubble flow The velocity of expansion of the universe at a specified distance.

Hubble's law The mathematical relationship that states that the velocity of expansion is proportional to distance in the expanding universe.

Hubble time See *Hubble age.*

igneous rock A rock that forms by cooling and solidification of *magma.*

ilmenite An iron titanium oxide *mineral* that is a minor but common constituent of *igneous rocks.*

immature sandstone A *sandstone* that contains a high proportion of easily decomposed *minerals,* such as *feldspar* and *mica.* A mature sandstone consists primarily of *quartz.*

impact glass Glass formed by the rapid cooling of *rock* melted by the impact of a *meteorite* or *asteroid.*

inclusion A fragment of an older *rock* or *mineral* enclosed in a younger *igneous rock.*

instability strip A gap in the horizontal branch of the *Hertzsprung–Russell diagram* left by unstable, pulsating stars.

intermediate rock An *igneous rock* containing about 52–66% silica. Intermediate rocks are neither extremely *felsic* nor extremely *mafic.*

internal conversion A mechanism by which a *nuclide* containing excess energy goes from a higher to a lower energy state by the emission of a *gamma ray.* Also called isomeric transition.

intrinsic brightness The quantity of light actually emitted by a star, irrespective of its apparent brightness as seen from Earth. Also called luminosity.

intrusion A body of *igneous rock* that invades older rock. The younger rock is said to intrude the older.

intrusive rock An *igneous rock* that *intrudes* older rock.

ion An atom or molecule with a positive or negative charge as a result of having lost or gained one or more *electrons.* See *anion, cation.*

ion microprobe A type of *mass spectrometer* that uses a narrow beam of high-energy primary *ions* to extract and ionize *nuclides* from a small spot on the surface of a sample. The secondary ions emitted from the sample surface are accelerated into the mass spectrometer's analyzer, where their *isotopic* compositions are precisely measured.

iron meteorite   A *meteorite* composed primarily or entirely of nickel–iron alloy. Also called siderite. See also *stony meteorite*.

isochron   A line of equal time on an isotope *correlation diagram*. Used in *radiometric dating* to determine the ages of *rocks* and *minerals*.

isochron diagram   See *correlation diagram*.

isochrone   A line of equal time on a diagram of stellar color vs magnitude, on which stars' evolution can be traced. Used to determine the ages of stars. See also *Hertzsprung–Russell diagram*.

isolated system   A system in which neither matter nor energy enters or leaves. Compare *open system* and *closed system*.

isomeric transition   See *internal conversion*.

isotope   A *nuclide* of a specific element. Isotopes of an element differ only in the number of *neutrons* in their nuclei.

isotope-evolution diagram   See *correlation diagram*.

ka   See *kilo-annum*.

Kelvin scale   One Kelvin equals one degree Celsius, but the Kelvin scale has its zero-point at absolute zero, which is $-273.16°C$. Absolute zero is the temperature at which all molecular motion ceases; a body at that temperature would have no heat energy.

kilo-annum (ka)   One thousand (10^3) years.

KREEP   One category of *lunar highland* (*Fra Mauro*) *basalt* higher in potassium (K), the rare-earth elements (REE), and phosphorus (P) than the *aluminous* highland *basalts*.

latent heat of crystallization   The heat (energy) released when a substance crystallizes, as when a *mineral* crystallizes from a *magma*.

limestone   A *sedimentary rock* consisting primarily of calcium carbonate. Most limestones are formed by the accumulation of *biogenic* remains (shells) or by direct precipitation from seawater.

lithosphere   The rigid part of the *crust* and upper *mantle* consisting of the *crustal plates* that are in constant motion relative to one another. The lithosphere is about 100 km thick.

lodranite   A *stony-iron meteorite* composed of nickel–iron alloy and the silicates *olivine* and *pyroxene*.

low-K Fra Mauro basalt   A type of *lunar highland basalt* that is lower in potassium and higher in aluminum than the *KREEP basalts*. Also called *aluminous basalt*.

luminosity   See *intrinsic brightness*.

lunar basin   See *basin*.

lunar highlands   The older and more heavily cratered upland areas of the Moon, as opposed to the younger, lower, and smoother *maria*. Also called terrae.

Ma   See *mega-annum*.

mafic rock   A qualitative term for dark-colored *rocks*. Mafic rocks contain an abundance of dark-colored *minerals* like *hornblende*, *pyroxene*, and *olivine* and a low proportion of light-colored minerals like *quartz* and *feldspar*. The opposite of *felsic*.

magma   Molten *rock*.

magnetite   An iron oxide *mineral* that is a minor but common constituent of *igneous* and *metamorphic rocks*.

magnitude (stellar)   A measure of either the apparent or the absolute brightness of a star.

main sequence A band on the *Hertzsprung–Russell* star-evolution *diagram* within which most stars lie and in which all stars start their lives.

mantle That part of Earth between the *crust* and *core*.

mare (pl. maria) The lunar lowlands or "seas," smoother and younger than the *lunar highlands* (terrae).

mass extinction The extinction of a large number of plant or animal species, such as occurred at the boundary between the Cretaceous and Tertiary periods.

mass spectrometer An instrument for precisely measuring according to their mass the relative amounts in a sample of various *nuclides*.

mega-annum (Ma) One million (10^6) years.

megaparsec (Mpc) One million (10^6) *parsecs*.

megaregolith The layer of *basin* ejecta that underlies the *regolith* or "soil" on the Moon's surface.

mesosiderite A *brecciated stony-iron meteorite* consisting of fragments of *diogenite, eucrite,* and nickel–iron alloy.

metaclastic A *metamorphic rock* composed of *clastic* (broken) fragments of *minerals* and rocks. A metamorphosed *breccia* is a metaclastic rock.

metamorphic Pertaining to *metamorphism*, e.g. metamorphic rock, metamorphic process.

metamorphism The process by which the composition, texture, or structure of *rocks* is altered in the solid state by heat, pressure, and the introduction of new chemical constituents. Rocks experiencing metamorphism become metamorphic rocks.

meteor The streak of light created by the passage of a *meteoroid* through the Earth's upper atmosphere.

meteorite A *meteoroid* that has fallen to the surface of the Earth or another planet.

meteoritic dust Very small particles or fragments of meteoroids that fall to Earth.

meteoritic geochron The *isochron* on a plot of $^{206}Pb/^{204}Pb$ vs $^{207}Pb/^{204}Pb$ on which the lead-isotopic compositions of undisturbed *meteorites* fall.

meteoroid A small object orbiting the Sun. A meteoroid that falls to Earth or the surface of another planet becomes a *meteorite*.

metric ton 1,000 kilograms or approximately 2,204 pounds.

mica Any of a group of minerals characterized by having good platy cleavage.

microprobe See *ion microprobe*.

Milky Way See *Galaxy*.

mineral A naturally occurring element or *compound*, usually with a characteristic crystal structure. The crystalline constituents of *rocks*.

mobile belt An elongate portion of Earth's *crust* that is more mobile, as evidenced by abundant folds and faults, than the stable continental *block* it borders.

model age A radiometric age calculated when the value of one of the unknown quantities in the age equation is assumed.

monzonite A *granitoid* containing, primarily, equal amounts of potassium *feldspar* and *plagioclase*.

multi-stage growth The change (growth) in lead-isotopic composition of a system in which the ratio of uranium to lead has changed one or more times by means other than *radioactive decay*.

multi-stage lead   Lead whose isotopic composition is the result of multi-stage growth. Compare *single-stage lead*.

muscovite   A light-colored *mica* that is rich in potassium. Muscovite is a common *mineral* in *plutonic* and *metamorphic rocks*.

nakhlite   A type of *achondrite meteorite* composed primarily of *pyroxene*. Nakhlites are *cumulates* and probably originated on Mars.

near-infrared   Infrared radiation with wavelengths nearest the visible spectrum's, i.e. between 0.75 and about 2.5 micrometers.

nebula   A cloud of interstellar dust and gas. See *Solar Nebula*.

neutrino   An elementary particle with no charge and little or no rest mass.

neutron   An uncharged particle with mass about equal to a *proton's*. Present in the *nucleus* of all nuclides except the commonest isotope of hydrogen.

norite   A type of *gabbro*. The lunar norites are *cumulates*.

norm   A system of classifying *rocks* based on an ideal mineralogy that is calculated from measured chemical compositions.

nucleocosmochronology   A method of calculating the average age of the elements based on the decay of long-lived radioactive nuclides. See *chronometer pair*.

nucleosynthesis   The various means by which elements are created during astrophysical events.

nucleus   The central, positively charged, dense portion of an atom.

nuclide   A species of atom characterized by the number of *protons* and *neutrons* in its *nucleus*.

olivine   An iron–magnesium *silicate* that is common in *mafic* and *ultramafic igneous* and *metamorphic rocks*.

olivine basalt   A *basalt* that either contains *olivine* or has *olivine* in its *norm*.

open system   A system that is neither *closed* nor *isolated*.

ordinary chondrite   The most common type of *chondrite meteorite*. Ordinary chondrites are similar to the *carbonaceous chondrites* but are lower in the volatile elements.

pallasite   A class of *stony-iron meteorite* composed primarily of *olivine* and nickel–iron alloy.

parent isotope   A radioactive *isotope* that decays to a *daughter* isotope.

parsec (pc)   The distance from which an object has a parallax of one second of an arc when viewed from opposite sides of the Earth's orbit about the Sun. It is equivalent to 3.09×10^{13} km.

partial melting   The process of forming a *magma* from a parent *rock* by selective melting of some of the constituents. The selective melting is controlled by temperature, pressure, and thermodynamic properties.

perihelion   The point on an orbit nearest the Sun. Compare *aphelion*.

petrography (adj. petrographic)   The description and classification of *rocks*, primarily by their composition and texture.

phosphate nodule   A rounded mass of calcium phosphate. Phosphate nodules range in diameter from a few millimeters to as much as 30 cm and occur in *sedimentary rocks*. They are formed largely by precipitation from sea water and commonly contain shell fragments or other debris.

phosphatized bone   Fossil bone that has been preserved by replacement of the original bone material by calcium phosphate.

pillow lava   A lava that exhibits pillow structure. The pillows are rounded masses of chilled *rock* that form only from lava that flows and cools under water.

plagioclase   A calcium–sodium *feldspar* and one of the most common *rock*-forming *minerals*.

plane of the ecliptic   See *ecliptic plane*.

planetary nebula   A gaseous envelope surrounding a star.

planetesimal   A small solid body orbiting the Sun from which the planets aggregated.

pluton   A body of *igneous rock* that formed beneath the surface by cooling and crystallization of *magma*.

plutonic rock   A *rock* formed beneath the surface by cooling and crystallization of *magma*. Compare *volcanic rock*.

population I star   A star, such as the Sun, relatively rich in the heavy elements.

population II star   A star relatively poor in the heavy elements.

positron   The antimatter equivalent of an *electron*.

Poynting–Robertson effect   The process that causes small dust particles to spiral inward to the Sun as they absorb and reemit radiation from the Sun.

primary terrestrial growth curve   The curve on a $^{206}Pb/^{204}Pb$ vs $^{207}Pb/^{204}Pb$ diagram followed by the Earth's lead-isotopic composition as it evolved from the primeval to its present-day ratio.

primeval lead   Lead whose isotopic composition is the same as that in the Solar System when solid matter first aggregated. Also called primordial lead.

primordial lead   See *primeval lead*.

proton   A particle with positive charge and mass about 1,836 times that of an *electron*. Present in the nucleus of all atoms.

pyroxene   A group of minerals whose principal constituents are usually iron, magnesium, calcium, alumina, and *silica*. Pyroxenes are common in *igneous* and *metamorphic mafic* and *ultramafic rocks*.

quartz   A *mineral* consisting entirely of silicon and oxygen.

quartz basalt   A *basalt* that either contains *quartz* or has quartz in its *norm*.

quartz monzonite   A *monzonite* that also contains *quartz* in significant proportions.

quartzite   A *metamorphic rock* consisting primarily or entirely of *quartz*. The progenitor of quartzite is most commonly *sandstone*.

radioactive decay   The spontaneous process by which *nuclides* reach stability by ridding themselves of excess *neutrons, protons,* or energy. Radioactive decay commonly results in the formation of an element different from that of the *parent* nuclide.

radiometric dating   The family of methods for dating *rocks* and *minerals* using the decay of naturally occurring radioactive *nuclides*.

raft   A small body of rock "floating" in a larger and surrounding rock body.

rare earth element   The group of elements with atomic numbers from 57 (lanthanum) to 71 (lutetium).

red giant   A star in a vigorous phase of hydrogen "burning" in which it expands significantly and increases its luminosity. See also *red supergiant*.

red supergiant   A star in a vigorous phase of helium "burning" in which it expands significantly and increases its luminosity.

regolith   The lunar "soil," composed primarily of comminuted (pulverized) *rock* and glass.

relativistic speed   A velocity approaching the speed of light.

remobilization   The formation and emplacement of a new *igneous rock* by re-melting, re*intruding*, and recrystallizing of a preexisting *igneous* or *meta-morphic* rock.

residence time   The average time that a specific element remains in solution in the oceans before it is deposited in sediments or otherwise removed.

Roche limit   The orbital distance inside of which a satellite will be disrupted by tidal forces.

rock   An aggregate of *minerals*.

r-process   One of the several processes of nucleosynthesis (r for rapid). See also *s-process*.

sandstone   A *sedimentary rock* consisting of sand-sized (0.062–2.0 mm) particles.

schist   A *metamorphic rock* with a laminate structure caused by the subparallel alignment of *mica*, a dominant constituent.

secular equilibrium   The state in a *radioactive decay series* when the produc-tion rate of the stable *daughter* is exactly equal to the decay rate of the radioactive *parent* and the quantity of each of the intermediate radioac-tive nuclides is inversely proportional to its decay constant.

sedimentary rock   A *rock* formed by the accumulation of any type of sedi-ment in water or air.

shale   A *sedimentary rock* consisting primarily of clay-sized particles.

shergottite   A type of *basaltic achondrite meteorite* and a *cumulate* that re-sembles *diabase* in composition.

shield   See *craton*.

siderite   See *iron meteorite*.

siderophile   The elements that associate with iron in *rock*-forming processes, including nickel, osmium, iridium, rhenium, molybdenum, palladium, and cobalt.

silica   The *compound* silicon dioxide.

silicate   Any mineral containing *silica*.

sill   A tabular body of *igneous rock* that is parallel to the bedding or structure of the rocks it intrudes. See also *dike*.

single-stage lead   Lead whose isotopic composition differs from that of pri-mordial lead only because of the radioactive decay of uranium and thor-ium in a single, closed system. Compare *multi-stage lead*.

single-stage system   A system closed to uranium, thorium, and lead from the time of origin of the Solar System to the time of interest.

Solar Nebula   The cloud of gas and dust that condensed to form the Solar System.

spallation   A nuclear reaction in which light particles are ejected from a nu-cleus bombarded by energetic particles, such as *cosmic rays*.

specific heat   The ratio of the amount of heat required to raise the tempera-ture of a given mass of a substance by 1°C to that required to raise the temperature of the same mass of water by 1°C.

spontaneous fission   A reaction in which a nucleus spontaneously splits into nuclei of two or more lighter elements of similar atomic mass.

s-process   One of several processes of *nucleosynthesis* (s for slow). See also *r-process*.

standard candle   An astronomical distance indicator based on the concept that objects of certain types give off the same amount of radiant energy, and thus have the same luminosity, regardless of their distance from the Earth.

standard ruler   An astronomical distance indicator based on the concept that objects of certain types are of constant dimensions and so their apparent diameters decrease proportionately with their distance from the Earth.

stellar magnitude   See *magnitude*.

stony–iron meteorite   A *meteorite* composed of a mixture of *silicate minerals* and nickel-iron alloy. Intermediate between a *stony meteorite* and an *iron meteorite*.

stony meteorite (stone)   A meteorite composed primarily or entirely of *silicate minerals*. See also *iron meteorite*.

stratiform anorthosite   A layered *anorthosite*.

stratiform ore   See *conformable ore*.

stratigraphy   The study of the composition, formation, sequence, and correlation of Earth's *sedimentary rocks*.

sulfate   Any compound containing the sulfur tetraoxide radical, $(SO_4)^{-2}$.

sulfide   Any compound containing sulfur and one other element as a *cation*. The *mineral* pyrite (FeS) is a sulfide.

supergiant   See *red supergiant*.

supernova   An exploding star of large mass. The intense nuclear processes in supernova are responsible for creation of most of the heavier elements.

supracrustal   Rocks deposited on continental or oceanic crust.

supracrustal sequence   A sequence or accumulation of rocks, usually lava flows and sediments, deposited on continental or oceanic crust.

synclinal   Pertaining to a syncline, or downfold (concave up), in *rock* strata. Compare *anticlinal*.

tectonic   Pertaining to deformation of the Earth's *crust*. Folding and faulting are tectonic processes.

terra (pl. terrae).   See *lunar highlands*.

terrane   The area over which a particular *rock* or group of rocks is prevalent.

thermal conductivity   A measure of the ability of a material to conduct heat.

thermodynamics   The mathematical treatment of heat and its relation to chemical and physical phenomena.

tonalite   A *granitoid* consisting primarily of *plagioclase feldspar, quartz*, and *hornblende* or *biotite*. Synonymous with quartz diorite.

trap rock   An archaic term applied to *mafic dikes* and lava flows, principally of *basalt* and *diabase*.

troctolite   A *rock* composed primarily of *pyroxene* and *olivine*.

Trojan asteroids   Asteroids that orbit the sun at the Lagrangian points 60° ahead of and behind Jupiter.

tuff   A *volcanic rock* composed of small *mineral* and glass fragments. Formed by ash falls or ash flows.

tuffaceous   A rock having all or most of the properties of a tuff.

turnoff point   The point on a *Hertzsprung–Russell diagram* where a star of a particular mass leaves the *main sequence*.

ultramafic   An *igneous* or *metamorphic rock* consisting almost entirely of dark ferromagnesian *minerals*. See also *mafic*.

uraninite   A uranium oxide mineral. Also called pitchblende.

ureilite   An *achondrite meteorite* consisting primarily of *cumulate* fragments of *pyroxene* in a carbonaceous matrix. Some ureilites contain diamonds.

viscosity   A fluid's resistance to flow, caused by internal friction.

volcanic rock   A *rock* formed above the surface (in air or water or vacuum) by cooling and crystallization of magma. Compare *plutonic*.

volcanism   The processes that result in the formation of volcanos and *volcanic rocks*.

white dwarf   A small whitish star of low *luminosity* and very high density. The final phase in the evolution of most stars.

X ray   High-frequency electromagnetic radiation that arises from within an atom but outside of the atomic nucleus. See also *gamma ray*.

zircon   A zirconium *silicate mineral*.

Index

Index

Library of Congress Cataloging-in-Publication Data

Dalrymple, G. Brent.
   The age of the earth / G. Brent Dalrymple.
      p.   cm.
   Includes bibliographical references and index.
  ISBN 0-8047-1569-6 (cl.): ISBN 0-8047-2331-1 (pb.)
    1. Geological time—Measurement. 2. Earth—Age. 1. Title.
QE508.D28   1991
551.1—dc20   90-47051
                                               CIP

∞This book is printed on acid-free paper.